Introduction to
**Modern
Inorganic
Chemistry**

Introduction to
Modern
Inorganic
Chemistry
Second Edition

K. M. Mackay
and
R. Ann Mackay

INTERTEXT BOOKS—LONDON

Published by
International Textbook Company Limited
24, Market Square, Aylesbury, Bucks, HP20 1TL

© K. M. Mackay and R. A. Mackay 1973

First published 1968
Reprinted with corrections 1969
Reprinted 1970
This edition published 1972
Reprinted with corrections 1973
ISBN 0 7002 0181 5

Printed by photolithography and bound in Great
Britain at The Pitman Press, Bath

Contents

Foreword to the First Edition

Modern Inorganic Chemistry is a rapidly growing and rapidly evolving subject and the last few years have seen the production of a number of excellent textbooks which present the subject at a high level to the Special Honours and postgraduate reader. It has seemed to us, however, that a gap is growing between this level of presentation and the standard of understanding reached in the schools. Moreover, the advanced texts are less suitable for those who are studying chemistry as part of a more general course or who wish to proceed to a less sophisticated level. It is to meet the needs of such readers that we present this book.

The growth of inorganic chemistry has a two-fold basis. The theoretical understanding of the subject has greatly improved and the experimental basis of the subject has been greatly extended. The theories of inorganic chemistry provide an important framework but the subject is too complex to evolve from improvements in theory alone. General experience of the reactions of the elements and of classes of compounds, together with a knowledge of the patterns of behaviour found in different areas of the Periodic Table, are vitally important to understanding the progress of the subject. A number of chemists, including commentators on Royal Institute of Chemistry examinations, have recently found it necessary to warn against tendencies to neglect systematic chemistry.

We have dealt with the theoretical side in the earlier chapters and in Chapter 12. Our aim has been to explain the basis of the more important theories and to illustrate the application, usually in a qualitative way. The treatment of systematic chemistry is perhaps the most difficult problem in studying modern inorganic chemistry for a great deal of information is available and the total is rapidly increasing. One approach is to grasp the overall patterns of behaviour of the elements in relation to the Periodic Table and to supplement this by a close study of a few limited regions. Once the student feels himself expert in a small area of chemistry—even of a single element or class or compound—it becomes comparatively easy to extend these 'islands' of knowledge as

the chemistry of related species will be found to correlate with that which is already known. We have attempted to describe the patterns of behaviour and to provide sufficient material to allow the reader to build up a fuller acquaintance with the properties of the elements. We hope the reader will be able to study selected areas in depth with the help of the reading lists.

In common with most authors of text books we owe a considerable debt to the work of others. All the sources listed in Appendix A have proved invaluable. We have also been greatly helped by discussions with many inorganic chemists, particularly members of staff of the Department of Chemistry of the University of Nottingham. We are particularly grateful to Dr D. B. Sowerby who read through Chapter 15 and made many helpful comments. We also thank Mr Fred Whetsone for his sterling work on the diagrams.

K. M. MACKAY
R. ANN MACKAY

A Note to the Reader

Three sections of this book outline the general plan of the treatment of certain topics. The position of these key sections is as follows:

(2) In Chapter 7, on the general properties of the elements, Sections 7.5 and 7.6 (pages 107–110) indicate the basis of the detailed treatment of systematic chemistry adopted in the succeeding chapters.
(3) Sections 12.1 (page 152) and 15.1 (page 229) outline in more detail the treatment of the transition elements and the p block elements respectively.

Certain topics are rather more advanced than the general level of the text, and will be better understood when the basic ideas are well assimilated. Some sections are also intended to bridge on to a more advanced treatment and may be omitted at first reading. Included under these heads are sections 3.9, parts of 6.4 and 6.5, 8.6, 12.6, 12.7, 12.9, 12.11, 12.12, the last part of 15.9 and Appendix C.

Finally, we would suggest that systematic chemistry, i.e. chapters 8, 9, 10, 11, 13, 14, 15, is not presented to be 'just learnt' in all its detail. This material is the basis of all Inorganic Chemistry and may be best mastered if the reader makes the effort to pick out the patterns for himself and works at cross-relating the different groups and periods of the Periodic Table, and in relating the theoretical and systematic sections.

Foreword to the Second Edition

In this edition, our basic aim has been to preserve the standard and level of presentation of the first. The ever-continuing advance of the subject has led to many changes of detail, and we have also tried to reflect some of the recent changes of emphasis in teaching inorganic chemistry. These are reflected in an expansion and rearrangement of Chapter 12, on the general properties of transition elements, and in additions to Chapter 15 of a section on reaction mechanisms at silicon and a fuller discussion of d orbital participation in Main Group chemistry. We have also expanded the account of spectroscopic methods in Chapter 6.

The Honours student now expects to use Group Theory and symmetry ideas as an integral part of his course and a necessary preliminary to this is the ability to assign molecules to their point groups. As this follows on closely from the discussion of molecular shapes, given in Chapter 3, we have introduced a new Appendix C which covers the determination of point groups and gives a sketch of the applications of symmetry theory. For the convenience of readers who will follow this topic further, we include character tables for the chemically more important point groups.

Since the first edition, the adoption of SI units has become general in scientific publications. Although all the details of conventional usage have not been finally settled, we have converted entirely to the SI system in this edition and have tried to follow the most generally-adopted view on points of controversy.

The wide acceptance of the first edition owes much to the appealing layout produced under the auspices of John Brooks, and we are equally grateful to Ann Drybrough-Smith who has greatly aided us with this edition. We also thank Dr. Keith Stead of the University of Exeter, for his advice on units and thermodynamics, especially in Chapter 5.

We also thank all those who have commented on the text and pointed out errors and infelicities.

K. M. MACKAY
R. ANN MACKAY

School of Science,
University of Waikato,
Hamilton, New Zealand.

SI Units

Many international scientific and engineering bodies have recommended a unified system of units and the SI (Système International d'Unités) has been proposed for general adoption. This recommendation is under consideration at national level and has been adopted in Britain by the Royal Society Conference of Editors for use in scientific journals. Unfortunately, acceptance is not universal, and modifications to the scheme may be made.

The system uses six basic units:

Physical quantity	Name of unit	Symbol for unit
length	metre	m
mass	kilogramme	kg
time	second	s
electric current	ampere	A
temperature	kelvin	K
luminous intensity	candela	cd

So-called supplementary units to be used with these are:

plane angle	radian	rad
solid angle	steradian	sr
amount of substance	mole	mol

This system differs from the metric system, as currently used, in replacing the centimetre and the gramme by the metre and the kilogramme.

Multiples of these units are normally restricted to steps of a thousand, and fractions by steps of a thousandth: i.e. multiples of $10^{\pm 3}$. The multiples 10, 100, 1/10 and 1/100 are retained at present though they may ultimately be dropped. Thus the standard fractions and multiples are

Fraction	Prefix	Symbol	Multiple	Prefix	Symbol
10^{-1}	deci	d	10	deka	da
10^{-2}	centi	c	10^2	hecto	h
10^{-3}	milli	m	10^3	kilo	k
10^{-6}	micro	μ	10^6	mega	M
10^{-9}	nano	n	10^9	giga	G
10^{-12}	pico	p	10^{12}	tera	T
10^{-15}	femto	f			
10^{-18}	atto	a			

It is important to note that 1 km^2 implies 1 (km)^2, that is 10^6 m^2 and *not* 10^3 m^2.

A range of units derive from these basic units, or are supplementary or allowed in conjunction with them, and those commonly used by the chemist are shown in Table 1. It is to be noted that the unit of force, the newton, is independent of the Earth's gravitation and avoids the introduction of g (the gravitational acceleration) into equations. The unit of energy is the joule (newton × metre) and of power, the joule per second (watt).

The gramme will be used until a new name is adopted for the kilogramme as the basic unit of mass, both as an elementary unit to avoid 'millikilogramme', and with prefixes, e.g. mg.

It is recommended that all units not compatible with SI should be abandoned progressively. For most units, this will rapidly happen and the majority of the incompatible units will probably disappear from the literature in the next few years. There are, however, a few units with particular advantages in certain contexts whose use is likely to persist. In this respect, it should be recognized that a system of units should combine logical construction and consistency with a reasonable degree of convenience. Some particularly difficult cases are indicated:

Litre. If the proposal to restrict multiples to $10^{\pm 3}$ is applied, then lengths would be confined to the metre and millimetre, giving volume units of m^3 and mm^3. It seems unlikely that a system restricted to values differing by 10^9 will find favour and the cm^3 and dm^3 are likely to be long with us. It is probable that the litre will survive as a convenient name for the cubic decimetre.

Ångstrom. The ångstrom unit ($10^{-8} \text{ cm} = 10^{-10} \text{ m} = 10^{-1} \text{ nm} = 100 \text{ pm}$) was originally introduced as a unit of length for use on the interatomic scale. It is being superceded by the picometre, pm.

Atomic mass unit. Atomic weights expressed relative to $^{12}\text{C} = 12 \cdot 000\ 0$ present a further case of a non-SI unit which will be retained for its convenience.

Energy. As the joule is of the same order of magnitude as the calorie, there are no reasons for retaining the latter. Energies of chemical processes commonly fall into the convenient range of 10 to 10^3 kJ mol^{-1} and the older unit of kcal mol^{-1} will rapidly disappear. Chemists commonly use two other units in energy measurement: the *wave number*, *reciprocal centimetre* or *Kayser*, written as cm^{-1}, and the *electron volt*, eV. The latter is strictly the energy acquired by one electron falling through a potential of one volt, but eV is commonly used to signify the molar quantity found by multiplying by Avogadro's constant. The cm^{-1} is not a unit of energy, but

TABLE 1. Units derived from the basic SI units, or supplementary to them

Physical quantity	Name of unit	Symbol and definition
force	newton	$N = kg\ m\ s^{-2}$ or $J\ m^{-1}$
energy	joule	$J = kg\ m^2\ s^{-2}$ or $N\ m$
power	watt	$W = kg\ m^2\ s^{-3}$ or $J\ s^{-1}$
electric charge	coulomb	$C = A\ s$
potential difference	volt	$V = kg\ m^2\ s^{-3}\ A^{-1}$ or $J\ A^{-1}\ s^{-1}$
resistance	ohm	$\Omega = kg\ m^2\ s^{-3}\ A^{-2}$ or $V\ A^{-1}$
capacitance	farad	$F = A^2\ s^4\ kg^{-1}\ m^{-2}$ or $A\ s\ V^{-1}$
frequency	hertz	Hz = cycle per second
temperature, t	degree Celsius (centigrade)	$^{\circ}C$ where $t/^{\circ}C = T/K - 273.15$
area	square metre	m^2
volume	cubic metre	m^3
density	kilogramme per cubic metre	$kg\ m^{-3}$
velocity	metre per second	$m\ s^{-1}$
angular velocity	radian per second	$rad\ s^{-1}$
acceleration	metre per (second)^2	$m\ s^{-2}$
pressure	newton per square metre	$N\ m^{-2}$
magnetic flux density	tesla	$T = kg\ s^{-2}\ A^{-1}$ or $V\ s\ m^{-2}$
time	hour, year, etc will continue to be used	

is used by spectroscopists. The relations underlying this application are described in Chapter 6, page 93.

The electron volt has changed status in the recent discussions of SI. It was initially listed as compatible but is now to be progressively abandoned. Its main advantage is its close relation to the methods for measuring certain parameters, such as ionization potentials, and its larger size. At about 10^2 kJ mol^{-1}, it is convenient for the larger chemical energies and will probably continue to be used until its status is completely clarified.

The SI system does not, of course, require that other units should never be used. It is available for reporting precise measurements in a fundamentally self-consistent way. It is not intended to preclude 'working units' and units used in a nonrigorous context. Thus we shall continue to fractionally distil at pressures measured in millimetres of mercury, autoclave at hundreds of atmospheres and read temperatures in Celsius degrees.

In Table 1 are listed the units which are derived from the basic SI units or their supplements.

In Table 2, we list those units which are contrary to SI and whose use is discouraged. The first part includes those which differ only by powers of 10 while the remainder have exactly defined conversion factors except the electron volt and atomic mass unit which are given in terms of the best experimentally-determined conversion factors. Some of these units will disappear quickly but must be converted in the pre-1970 literature while others, for various reasons as in the examples discussed above, will probably persist indefinitely. The values of physical constants in SI units are listed in Table 3.

Units used in this edition

A textbook written wholly in SI units would have a number of disadvantages at the present time. All the details are not completely settled and the system has not yet been formally recognized by all scientific bodies in a form that is agreed to be final. Most of the literature sources available to the student are not in SI units and any scientist must be able to convert data into the system. Finally, it is possible that the vigorous representations of various bodies will lead to the incorporation of certain current units which do not fall into the strict scheme.

With these considerations in mind, and having regard to the discussion above of certain units, the following convention has been adopted.

Length. Basically the SI system of m, mm, and smaller fractions is used. Notice that interatomic distances (given originally in ångstroms = 10^{-10} m) are usually known to an accuracy expressed by two decimal places in ångstroms, e.g. 1·07 Å. It follows that the most convenient SI-allowed multiple is the picometre, pm, so that most values become whole numbers of picometres, e.g. 107 pm, and unusually accurate values are distinguished by having figures after the decimal point as in 107·6 pm (= 1·076 Å). This usage is more convenient than using fractions of the nanometre, as in 0·107 nm. To avoid sudden changes in unit, we have retained the picometre for all interatomic distances so that the very few values for larger distances which are listed will appear as thousands of picometres, rather than as nanometres.

Energy. Most values are given in kJ mol^{-1} and kcal are not used. Where appropriate, cm^{-1} and eV are also used

TABLE 2. Commonly occurring units which are contrary to SI

Unit	Quantity	Equivalent
(A) Units differing from SI units by powers of 10		
ångstrom (Å or A)	length	10^{-10} m $= 10^{-1}$ nm $= 10^2$ pm
litre (l)	volume	10^{-3} m^3 = dm^3
dyne (dyn)	force	10^{-5} N
erg	energy	10^{-7} J
mho (siemens or reciprocal ohm)	conductance	Ω^{-1}
bar	pressure	10^5 N m^{-2}
poise	viscosity	10^{-1} kg m^{-1} s^{-1}
(B) Other units		
calorie (cal)	energy	I.T. cal $= 4\cdot186\,8$ J: $15°$ cal $= 4\cdot185\,5$ J: thermochemical cal $= 4\cdot184$ J
electron volt (eV)	energy	$1\cdot602\,1 \times 10^{-19}$ J
(electron volt per mole, also symbolized eV $= 96\cdot484$ kJ mol^{-1})		
atmosphere (atm)	pressure	$101\cdot325$ kN m^{-2}
millimetre of mercury (mmHg) or torr (Torr)	pressure	$133\cdot322$ N m^{-2}
atomic mass number (amu or u $= 1/12$ mass of ^{12}C)	mass	$1\cdot660\,41 \times 10^{-27}$ kg

TABLE 3. Values of physical constants

Physical constant	Symbol	Recommended value
Speed of light in a vacuum	c	$2\cdot997\,9 \times 10^8$ m s^{-1}
Mass of hydrogen atom	m_H	$1\cdot673\,4 \times 10^{-27}$ kg
Mass of electron	m_e	$9\cdot109\,1 \times 10^{-31}$ kg
Charge on proton	e	$1\cdot602\,1 \times 10^{-19}$ C
Boltzmann constant	k	$1\cdot380\,5 \times 10^{-23}$ J K^{-1}
Planck constant	h	$6\cdot625\,6 \times 10^{-34}$ J s
Permeability of a vacuum	μ_0	$4\pi \times 10^{-7}$ J s^2 C^{-2} m^{-1}
Rydberg constant	$R_\infty = \dfrac{\mu_0^2 m_e e^4 c^3}{8h^3}$	$1\cdot097\,4 \times 10^7$ m^{-1}
Bohr magneton	$\mu_B = \dfrac{eh}{4\pi m_e}$	$9\cdot273\,2 \times 10^{-24}$ A m^2
Avogadro constant	N_A	$6\cdot022\,5 \times 10^{23}$ mol^{-1}
Gas constant	R	$8\cdot314$ J K^{-1} mol^{-1}
'Ice-point' temperature	T_{ice}	$273\cdot150$ K
Permittivity of a vacuum	ϵ_0	$8\cdot854\,185 \times 10^{-12}$ J^{-1} C^2 m^{-1}
Faraday constant	F	$9\cdot648\,7 \times 10^4$ C mol^{-1}
$\dfrac{RT \ln 10}{F}$ at 298 K		$5\cdot916 \times 10^{-2}$ V
Molar volume of ideal gas (273$\cdot$15 K, 1 atm)	V_m	$2\cdot241\,4 \times 10^{-2}$ m^3 mol^{-1}

TABLE 4. Conversion factors

	kJ mol^{-1}	cm^{-1}	eV	MHz	kcal mol^{-1}
kJ mol^{-1}	1	83·626	$1·036\ 4 \times 10^{-2}$	$2·506\ 2 \times 10^{6}$	0·239 4
cm^{-1}	$1·195\ 7 \times 10^{-2}$	1	$1·239\ 4 \times 10^{-4}$	$2·997\ 9 \times 10^{4}$	$2·858 \times 10^{-3}$
eV	96·484	8068·3	1	$2·418\ 8 \times 10^{8}$	23·063
MHz	$3·990\ 3 \times 10^{-7}$	$3·335\ 6 \times 10^{-5}$	$4·134\ 4 \times 10^{-9}$	1	$9·534\ 5 \times 10^{-8}$
kcal mol^{-1}	4·184	349·83	$4·335\ 9 \times 10^{-2}$	$1·048\ 7 \times 10^{7}$	1

and conversion factors are included in tables in these units. Occasionally, particularly when dealing with magnetic resonance, frequencies in Hz are found in place of wave numbers in cm^{-1}. As 1 Hz is one wave per second while 1 cm^{-1} is one wave per cm, the two are connected by the speed of light. Table 4 shows the interconversion factors for all these units.

Temperatures. Temperatures are mainly quoted in degrees Celsius.

Electrical units. These do not occur widely in this text. Note that the interaction between charges is modified by the permittivity of a vacuum given with the symbol ϵ_0. Thus the non-SI factors e^2/r, which occur in the wave equation for example, now become $e^2/4\pi\epsilon_0 r$.

Other parameters. SI units are used where exact values are stated. Note that the litre is used instead of the strict SI unit of the dm^3 as we are not quoting any highly accurate values. There is some discussion that the gramme-molecule might be replaced by the kilogramme-molecule but most chemists would find this unacceptable.

In the text, where numerical values for physical properties, such as atomic radii or oxidation-reduction potentials, are quoted, they are chosen from consistent sets and are quoted for purposes of comparison only. Much more accurate values are often available for particular data, and any calculation of physical significance should use such values, which can be found in a number of critical compilations of data or in the original literature. Formulae are commonly used in place of chemical names where the former are clearer and less clumsy. Equations which are written unbalanced are used either to show the major product of a reaction, or to indicate the variety of products without being definite about their relative proportions.

In addition to the symbols for fundamental constants given in Table 3, a number of other symbols and abbreviations are to be found in the text. These are:

S	Entropy
H	Enthalpy (heat content)
K	Equilibrium constant
ccp	Cubic close packed
hcp	Hexagonal close packed
bcc	Body centred cube
m or m.p.	Melts or melting point
b or b.p.	Boils or boiling point
d	Decomposes
subl	Sublimes

1 Introduction

1.1 Inorganic chemistry

Inorganic chemistry is that branch of chemistry which is concerned with the properties and reactions of elements and their compounds. Excluded from this branch are the compounds of carbon which form organic chemistry, with the exception of a few simple compounds such as the oxides and carbonates, which are commonly allotted to inorganic chemistry. Also excluded from the general definition above is the detailed study of energy changes and mechanisms involved in reactions, which comes into the field of physical chemistry. It is, of course, impossible to draw strict demarcation lines between these three major parts of chemistry, nor is it wise to attempt to do so. Many fields where advance is most rapid lie on the borderline between two of the major divisions, or between chemistry and other sciences. Some obvious examples are organo-metallic chemistry, radio-chemistry, geo-chemistry, biochemistry and metallurgy all of which overlap with inorganic chemistry.

The origins of inorganic chemistry are ancient and it was the first of the chemical sciences to flower in the course of the Scientific Revolution. Most of the classical chemistry of the eighteenth and early nineteenth centuries, which led to the formulation of the atomic theory, was carried out on inorganic systems, especially on the simple gases and ionic solids such as carbonates and sulphates. By 1820, some fifty of the elements were known and many of their simple compounds had been studied, including some, such as hydrogen fluoride and nitrogen trichloride, whose study requires considerable resource. By contrast, only a few simple organic compounds were known and progress in organic chemistry was slow; largely because very complex materials, such as blood or milk, were studied. In the middle of the nineteenth century came a period of spectacular advance in organic chemistry, followed at the turn of the century by a great upsurge of interest in physical chemistry. All this meant a period of comparative neglect of inorganic chemistry which came to an end only in the last two or three decades. Of course, very important advances were made, including the formulation of the Periodic Table, the discovery and exploitation of radio-chemistry, and the classical work on non-aqueous solvents and on the complex chemistry of the transition elements, but it was not until the nineteen-thirties that the modern upsurge of interest in inorganic chemistry got under way. Among the seeds of this renaissance were the work of Stock and his school on volatile hydrides of boron and silicon, of Werner and others on the chemistry of transition metal complexes, of Kraus and Walden on non-aqueous solvents, and the work of a number of groups on radioactive decay processes. At the same time, the theories which play an important part in modern inorganic chemistry were being formulated and applied to chemical problems. The discovery of the fundamental particles and the structure of the atom culminated in the development of wave mechanics, which is the basis of all modern approaches to valency and bonding. This theory is outlined in Chapter 2, and its application to molecular structure is given in Chapter 3. A little later, the effect of an atom's environment on the energy of its d electrons was brought into the treatment of transition metal compounds in the crystal field theory which is discussed in Chapter 12.

All these developments prepared the ground for the post-war expansion in inorganic chemistry. This was stimulated both by new discoveries in academic chemistry and by demands for new materials. An important factor was the advent of atomic energy, which created a vital interest in the properties of many elements hitherto scarcely studied. Other industrial developments had a similar effect, including the exploitation of semi-conducting devices which has led to the growth of germanium, gallium, and indium chemistry, and the whole aero-space development which has prompted studies of possible rocket fuels (e.g., boron hydrides), inorganic polymers, and many other fields.

This expansion in inorganic chemistry is, of course, paralleled by a continuing expansion in all fields of science. The effect has been particularly marked in inorganic chemistry, however, as it follows a period of comparative neglect and therefore builds on a slighter basis. This rapid growth of inorganic chemistry has made it a very lively and exciting subject in which to work and to teach, but it does lead to problems from the student's point of view. Textbooks tend to be out of date by the time they are published, and the treatment of each subject changes as new discoveries are made. One striking example is provided by the compounds of the rare gases. In all textbooks published up to the beginning of 1963, the treatment of the rare gases was confined to a discussion of a few clathrates and an emphasis on their inertness. However, after the first report of a rare gas compound in late 1962, interest in this topic grew explosively and getting on for a hundred papers were published on rare gas compounds in 1963 and 1964. The modern inorganic textbook thus has a section on rare gas chemistry comparable to its discussion of iodine chemistry, for example. More recently, topics like metal-metal bonding and metal atom clusters, or activation of small molecules by complex-formation such as the N_2 complexes, have shown similar rapid development. These are striking examples, but a significant advance in any field may provoke a similar response. The total body of experimental data is growing at a rapid rate, and part, at least, of this new knowledge has to be reflected in the textbooks.

Changes in experimental data are easier to accommodate than changes in theory, and inorganic chemistry has been absorbing two major variations in its theoretical apparatus. One has been a marked shift in emphasis in the ways of applying wave mechanics to chemistry. The two approaches which have been popular are the method of *molecular orbitals* and the *valence bond* theory. In many simple cases, these approaches may be refined to give almost identical descriptions—as they must when applied accurately. However, in polyatomic species, only approximate theories can be applied and the valence bond theory was the preferred one, as it was closer to the classical picture of a molecule as linked by discrete electron-pair bonds. In particular, the partial double-bonding in a species such as the nitrate ion was described in terms of 'resonance' between contributing forms, each of which was described in terms of single and double bonds:

This approach is now favoured rather less than the molecular orbital approach, which discusses such species in terms of *delocalized* bonds extending over all the molecule (see Section 3.9). In other areas, such as the properties of excited states and of species with delocalized bonds, the molecular orbital theory is more satisfactory. In this text, the structures of molecules and ions are described largely in terms of the molecular orbital theory.

A second major change has been the introduction of *ligand field theory* to deal with compounds which have valency electrons in *d* orbitals. This includes crystal field theory, mentioned above, and wave-mechanical aspects which lead again to molecular orbital descriptions of transition metal complexes with multi-centred bonds. The development of this theory accompanied the huge expansion of transition metal chemistry in the last two decades. From an almost negligible position about 1950, transition element chemistry

has grown to become the major field of inorganic chemistry, largely as a result of the strong mutual stimuli of new experimental and theoretical advances.

There are, however, signs that the rate of expansion in the transition metal area is slowing down, perhaps because the theories are becoming more elaborate and sophisticated, and the last two or three years has seen advances in the theoretical background to Main Group chemistry, particularly in the discussion of the participation of *d* orbitals in this area of chemistry. All these changes are reflected in the succeeding pages, but the reader is asked to realize that the rapid rate of growth means that any text is somewhat out of date by the time it is published and the review literature should be consulted for recent advances.

1.2 Inorganic nomenclature

The nomenclature of inorganic chemistry was put on a definitive basis by the publication of the IUPAC (International Union of Pure and Applied Chemistry) Rules in 1957. These rules define a systematic method for naming all inorganic compounds, but they also allow the retention of a number of trivial names which are well established.

The detailed application of the systematic nomenclature need not concern the reader too much as it is usually easy to translate the systematic name into the more familiar forms, and the reader will be less likely to have to form the systematic name of a compound. The principles of systematic naming are straightforward and are outlined below:

(i) The cation, or electropositive component, of a compound has its name unmodified.

(ii) If the anion, or electronegative constituent, is monatomic, its name is modified to end in -ide.

(iii) If the anion is polyatomic, its name is modified to end in -ate.

(iv) Where oxidation states are to be indicated, they are shown by means of Roman numerals following the name of the element.

(v) The stoichiometric proportions of constituents are denoted by Greek prefixes (mono, di, tri, tetra, penta, hexa, hepta, octa, nona, deca, undeca, and dodeca). Alternatively, numerals may be used, as in $B_{10}H_{14}$—decaborane-14. In addition, the multiplicative prefixes (bis, tris, tetrakis, etc.) may be used to indicate a multiplicity of complex groups, especially when these already contain a numeral, as in $Ni(PPh_3)_4$—tetrakis (triphenylphosphine) nickel.

(vi) In extended structures, a bridging group is indicated by the prefix μ, for example, $[(NH_3)_5Cr-OH-Cr(NH_3)_5]Cl_5$ which is μ-hydroxo-bis [penta-amminechromium(III)] chloride.

Notice that, according to rule (iii), all polyatomic ions have names ending with -ate. This must not be confused with the trivial naming of oxygen anions where the endings -ate, -ite, etc., are used to indicate the oxidation state (see below). In systematic naming, all such anions end with -ate and the oxidation state is shown by the stoichiometry or by Roman numerals. Thus, $SnCl_6^{2-}$ is hexachlorostannate(IV), $SnCl_3^-$ is trichlorostannate(II), SO_4^{2-} is tetraoxosulphate

TABLE 1.1 Examples of systematic inorganic nomenclature. Of the detailed rules for the order of citation of ligands in complexes we need only note that anionic ligands come before neutral and cationic ones and 'oxo' is often dropped out of the names of familiar oxyions.

$NaCl$	Sodium chloride	
SiC	Silicon carbide	Rules i, ii and v: note that mono-
As_4S_4	Tetra-arsenic tetrasulphide	is usually omitted as a prefix
Cl_2O	Dichlorine oxide	
OF_2	Oxygen difluoride	
$KICl_4$	Potassium tetrachloroiodate	
$FeCl_2$	Iron dichloride (or iron(II) chloride)	Rules i, iii, iv and v. Notice that
$Pb_2^{II}Pb^{IV}O_4$	Trilead tetroxide (or dilead(II) lead(IV) oxide)	the use of roman numerals in rule
$K_4[Fe(CN)_6]$	Potassium hexacyanoferrate(II)	iv, and the prefixes of rule v, often
$K_3[Fe(CN)_6]$	Potassium hexacyanoferrate(III)	provide alternative names; super-
$Na(SO_3F)$	Sodium trioxofluorosulphate	fluous information is avoided
NaH_2PO_4	Sodium dihydrogen tetraoxophosphate	
$Na(NH_4)HPO_4,4H_2O$	Sodium ammonium hydrogen phosphate tetrahydrate	Note that multiple groups are
$BiOCl$	Bismuth oxide chloride	written separately (often all run
$VOSO_4$	Vanadium(IV) oxide sulphate	together) in an order which is
$ZrOCl_2,8H_2O$	Zirconium oxide dichloride octahydrate	defined by the rules. Notice also
$Li(AlH_4)$	Lithium tetrahydroaluminate	that oxo- is often dropped out of
$NH_4[Cr(SCN)_4(NH_3)_2]$	Ammonium tetrathiocyanatodiamminechromate(III)	the names of familiar oxyanions
$[Co(CO_3)(NH_3)_4]Cl$	Carbonatotetra-amminecobalt(III) chloride	
$[Be_4O(CH_3COO)_6]$	(see Figure 9.8) μ_4-oxo-hexa-μ-acetatotetraberyllium	

(or tetraoxosulphate(VI)) and SO_3^{2-} is trioxosulphate. Some examples are given in Table 1.1.

The system outlined above gives a means of providing an unambiguous systematic name for any inorganic compound. However, many of the systematic names of familiar compounds are clumsy, and a considerable number of common names are retained for general use. The following points may be particularly noted.

(i) The terminations -ous and -ic may be retained for cations of elements with only two oxidation states, as in ferrous and ferric or stannous and stannic.

(ii) In anions, the termination -ite to distinguish a lower oxidation state is retained in such cases as nitrite, sulphite, phosphite and chlorite. Similarly the hypo- . . . -ite method of showing an even lower oxidation state is retained for hyponitrite, hypophosphite and hypochlorite. Corresponding acid names end in -ous and hypo- . . . -ous.

(iii) The term thio- is used to denote the replacement of an oxygen atom by a sulphur one, as in $PSCl_3$—thiophosphoryl chloride. Similar use is made of seleno- and telluro-.

(iv) The terms ortho- and meta- are retained to indicate different 'water contents' of acids, as in orthophosphoric acid, H_3PO_4, and metaphosphoric acid, $(HPO_3)_n$, or H_5IO_6 orthoperiodic acid and HIO_4 periodic acid.

(v) As the last example shows, the prefix per- is used to indicate an oxidation state above the one indicated by the normal -ate or -ic termination of an anion or acid. (Per- should not be used for metals and cations.) This usage should be confined to the cases of perchlorate, perbromate, periodate, permanganate and per-rhenate. The prefix peroxo-

should be used to denote the presence of the $-O-O-$ group derived from hydrogen peroxide, as in peroxodisulphuric acid (HO_3SOOSO_3H), although per- is often used instead.

On the whole, the tendency is to use the simplest name available for well-known compounds, as long as this is accurate. New classes of compounds tend to be given their systematic name unless, as in the case of ferrocene for example, the discoverer happens to have hit on a suitable trivial name that has become widely accepted. Inorganic nomenclature need not cause any difficulty to the beginner who only has to translate the systematic name into a more familiar form, as long as the details above, especially about the various uses of the termination -ate, are familiar.

1.3 Approach to inorganic chemistry and further reading

The starting point for an understanding of inorganic chemistry lies in the electronic structure of the atom. The basis of this is provided by quantum theory and wave mechanics and is outlined in Chapter 2. The approach here has been to indicate the mathematical basis of the theory and then to develop the applications in a pictorial manner. A more rigorous approach can be found in the references cited at the end of the book. In Chapter 3, the results for atomic structures are applied to study the formation of molecules, first diatomic species and then polyatomic ones. An outline of a semi-empirical theory whereby the structures of polyatomic molecules and ions may be derived from simple ideas of electron repulsion is also given. Chapter 4 deals with the solid state and the formation of ionic compounds, while Chapter 5 outlines the approach to solution chemistry, both in water and in non-aqueous solvents. Chapter 6 gives a brief account of experimental methods in inorganic chemistry. These five chapters make up the first section of the book, and attempt to give the reader a feeling for the theories and concepts the inorganic chemist uses in his work and to indicate his methods.

The second, and longer, part of the book deals with the chemistry of the elements. It is shown in Chapter 7 how the Periodic Table may be split up into a small number of 'blocks' of elements with general properties in common, and a number of properties which summarize the chemistry of the elements is discussed. A general outline of methods of preparing the elements from their ores is also given. The remaining chapters deal with the elements in blocks, giving a general account of their chemistry set in the context of the properties common to all the elements of the block. The systematic chemistry of most of the elements is now known in considerable detail and it is quite impossible to give a reasonably full account of such chemistry in a book of this type. Instead, the aim has been to give a sufficient account of the general properties of the elements, illustrated by the behaviour of simple compounds like oxides and halides, to allow the reader to build up an overall picture of the variations in chemistry in the Periodic Table. This can then be supplemented by a more detailed study of particular groups with the aid of the standard textbooks. To give some indication of current work in the chemistry of the elements, an account of some compounds of special interest is given for each block of elements. The selection of material to be included here is to a large extent arbitrary and may be supplemented from the reviews cited.

The next step in finding out more about any topic discussed here is to look up one or other of the standard textbooks which treat inorganic chemistry to an advanced level.

These are cited in the lists of references given in Appendices at the end of the book. More detailed information about particular sections of the subject can be derived from the many reviews available. The number of review volumes covering inorganic chemistry has increased rapidly in recent years, and the level at which the articles are written varies widely, even within a single volume. Among the more general presentations of inorganic topics are the Royal Institute of Chemistry monographs, especially the series *Monographs for Teachers*, a number of articles in the journals, *Education in Chemistry* and *Journal of Chemical Education*, and some articles in the *School Science Review*. Some useful general articles appear in *Chemistry in Britain*. On a generally more advanced level are the *Quarterly Reviews* of the Chemical Society, the Academic Press annual series *Advances in Inorganic Chemistry and Radiochemistry* and the similar Interscience series *Progress in Inorganic Chemistry*. Similar, more specialized review volumes such as *Progress in Stereochemistry* (Butterworth), *Advances in Organometallic Chemistry* (Academic Press) and similar series are listed in detail in Appendix A. A full, but brief, account of the year's publications in inorganic chemistry is to be found in the appropriate section of the Chemical Society's *Annual Reports on the Progress of Chemistry*.

There is also a number of compendia of inorganic chemistry which give more extensive treatment than a textbook, though often more limited than a review. These suffer also from the disadvantage that they may be badly out of date: the current ones are *Mellor*, *Gmelin*, and *Pascal* listed in Appendix A. A similar but shorter work now in course of preparation is *Comprehensive Inorganic Chemistry* (Pergamon), probably in five volumes, which will be the most convenient initial source for the modern chemistry of the elements when it is published.

2 The Electronic Structure and the Properties of Atoms

Introduction

2.1 Background

Although some 200 different fundamental particles have been discovered by the physicists, only three, the proton, the neutron, and the electron are of direct interest to the chemist. The masses and charges of these sub-atomic particles are so minute that it is convenient to define much smaller units than the gramme and the coulomb which are used on the macroscopic scale. The unit of mass used on the atomic scale is approximately the mass of the proton or neutron (which are very nearly equal and about 10^{-24} g) and it is called the atomic mass unit, abbreviated a.m.u. The mass of the electron is very much less than that of the other two particles and is of the order of one two-thousandth a.m.u. The charges on the proton and electron are equal in size though opposite in sign, that on the electron being negative. (There is also a short-lived particle of the same mass as the electron, but with unit positive charge, called the positron or positive electron.) This electronic charge, which equals about -1.6×10^{-19} C, is taken as the unit of charge on the atomic scale and is given the symbol $-e$ (but note that e is also used to represent the electron itself). The neutron has no charge. Thus, the proton has unit mass and unit positive charge, the neutron has unit mass and zero charge while the electron has negligible mass and unit negative charge. The exact values are given in Table 2.1, but this approximation suffices for most purposes.

The foundation of the modern theory of atomic structure was laid by the work of Rutherford on the scattering of α-particles by very thin metal targets. He found that, when a

beam of α-particles (mass = 4 a.m.u., charge = $+2e$) was directed at a target of thin metal foil, nearly all the particles passed through the target with scarcely any deflection, a few were deflected through large angles, and an even smaller proportion were reversed along their paths. These observations suggested a model of the atom as a small, dense, positively-charged core surrounded by a much larger and more tenuously-occupied region of electrons. The α-particles are so much more massive than the electrons that they would be relatively unaffected when they passed through the outer regions of the atoms and deflected only if they came close to the core. Since most of the α-particles passed straight through the target, which was about a thousand atoms thick, it followed that the cores must be very small. When the α-particle passed close to a core, it was strongly deflected by the charge repulsion, while the occasional particle which happened to be heading straight at the core (with a mass and charge about thirty times its own) was repelled back along its path. The effect is illustrated in a very diagrammatic form in Figure 2.1.

This work has been fully substantiated by later investiga-

TABLE 2.1 Properties of the fundamental particles

Particle		Charge, coulomb	Mass, a.m.u.
Proton	$^{1}_{1}\mathrm{p}$	$+e = 1.602\ 1 \times 10^{-19}$	1.007 57
Neutron	$^{1}_{0}\mathrm{n}$	zero	1.008 93
Electron	$^{0}_{-1}\mathrm{e}$ or β^{-}	$-e = -1.602\ 1 \times 10^{-19}$	0.000 548 6

*These symbols are explained on page 6. The positron has the symbol $^{0}_{+1}\mathrm{e}$ or β^{+}.

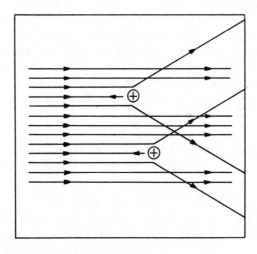

FIGURE 2.1 *Rutherford's experiment*
Diagrammatic representation of the deflections observed when a beam of alpha-particles hits a metal target.

tions. In the modern view, the atom consists of a tiny, dense, positively-charged nucleus containing protons and neutrons, surrounded by a much larger, more tenuous, cloud of electrons. Since the electron mass is so much smaller than the masses of the other particles, the *atomic weight* is approximately equal to the mass number, A (in a.m.u.), which equals the sum of the number of protons, Z, and the number of neutrons $(A-Z)$ or N. Since there are Z protons, the nucleus bears a charge of $+Ze$ and this is balanced by having Z electrons surrounding the nucleus so that the atom is electrically neutral. Z is termed the *atomic number* and is the most important single property of an atom.

The volume of an atom is essentially the space occupied by its electron cloud, the nucleus filling only a minute proportion of the whole (about 10^{-15} of the atomic volume). The radius of an atom is of the order of 100 pm, while the nuclear radius is about 10^{-3} pm. The radii of molecules extend to about 1000 pm, or even more for very complex or polymeric species. For comparison, the present-day electron microscope can resolve down to about 1000 pm and can thus be used to 'see' molecules of moderate size.

2.2 Isotopes

Chemical behaviour is determined by the interaction between the electron clouds of atoms whose character, in turn, depends on Z, the number of electrons present. That is, the atomic number determines the chemical properties of an element, and all the atoms of a particular element have Z protons in their nuclei and Z electrons. It is found, however, that these Z protons may be accompanied in the nucleus by varying numbers of neutrons so that atoms of the same atomic number may have different atomic weights, A. For example, chlorine with $Z = 17$ has two forms, one with eighteen neutrons and atomic weight $A = 35$, and the other with twenty neutrons and atomic weight 37. Such atoms with the same atomic number Z, but with different atomic weights A, are termed *isotopes*.* Since the chemical properties of an element depend only on Z, all the isotopes of an element undergo identical chemical reactions although the rates of reaction (and other effects which depend on mass) may show small differences. These mass effects are negligible except for the lightest elements such as hydrogen. The number of naturally-occurring isotopes of an element varies widely. Some elements such as $^{9}_{4}Be$, $^{31}_{15}P$, or $^{197}_{79}Au$ are found in only one isotopic form while others may form up to ten stable isotopes. For example:

Element	Z	A
Cadmium	48	106, 108, 110, 112, 113, 114, 116
Tin	50	112, 114, 115, 116, 117, 118, 119, 120, 122, <u>124</u>
Tellurium	52	120, 122, 123, <u>124</u>, 125, 126, 128, 130
Xenon	54	<u>124</u>, 126, 128, 130, 131, 132, 134, 136

This illustrates that, as well as isotopes with the same Z value and different A values, there also exist atoms with the same

*These nuclear properties are commonly shown by left sub- and super-scripts on the element symbol: $^{A}_{Z}X$. Thus the α-particle, which is the helium nucleus, is written $^{4}_{2}He^{2+}$. (The alternative form, $_Z X^A$ is also found, especially in American publications.)

mass numbers but differing atomic numbers. Such atoms are called *isobars*. One example is provided by the isotopes of tin, tellurium, and xenon of mass 124 which are underlined in the table. Isobars are of less importance to the chemist than are isotopes.

The atomic weight, as determined chemically, is the mean weight of the naturally-occurring mixture of isotopes. Thus chlorine, which consists of 75·4 per cent ^{35}Cl and 24·6 per cent ^{37}Cl, has a chemical atomic weight of 35·457. Very extensive studies have shown that, where elements are not involved in natural radioactive decay processes, the isotopic composition and chemical atomic weight of samples from widely varied sources is nearly always constant. The exceptions are those elements, e.g. copper and lead, with variable isotopic composition in naturally obtained samples, and also lithium, boron, and uranium in which an isotopic separation may result from methods used in commercial preparation of a sample.

2.3 Radioactive isotopes and tracer studies

Only certain combinations of protons and neutrons form stable isotopes. If the neutron/proton ratio becomes too large or too small, the nucleus is unstable and radioactive, and some nuclear process takes place to restore the balance. For example, the isotope of hydrogen of mass three called tritium, $^{3}_{1}H$, has too many neutrons to be stable. It spontaneously converts a neutron into a proton by emitting a negative electron to form a helium isotope, $^{3}_{2}He$. This type of process is written out as a nuclear equation; in this case

$$^{3}_{1}H = {}_{-1}^{0}e + {}^{3}_{2}He$$

(In nuclear equations, both the charge on the nucleus and the mass are shown and each of these quantities must balance.)

The converse case is illustrated by magnesium-23 which has excess protons and converts a proton to a neutron by the emission of the positive electron to form a sodium isotope:

$$^{23}_{12}Mg = {}_{+1}^{0}e + {}^{23}_{11}Na$$

Nuclei are not stabilized only by electron emission: other ways include the emission of an alpha-particle or the capture of an orbital electron (K capture). Sometimes a nucleus is so far from a stable ratio of protons to neutrons that the daughter nucleus produced by the first radioactive decay step is itself unstable and rearranges, and this process continues until a stable nucleus results.

All the atoms in a sample of an unstable material do not undergo a nuclear reaction simultaneously and instantaneously. It is found that a particular radioactive isotope is characterized by a half-life, given the symbol $t_{\frac{1}{2}}$, which is the time in which half the atoms present in a given sample have undergone transformation. Half-lives range from minute fractions of a second to millions of years. The half-life is a constant, characteristic of the isotope, and is not varied by any change in the environment of the atom. For tritium, $t_{\frac{1}{2}} = 12\cdot4$ years and for magnesium-23 it is 11·6 seconds. An example of an isotope with a very long half-life is provided by uranium-238 where $t_{\frac{1}{2}} = 4 \cdot 5 \times 10^9$ years.

Since all the isotopes of an element have identical chemical properties, it is possible to study the course of many reactions by using an element in an isotopic form other than the natural one. To choose a very simple example, it could be shown that an acid ionizes in water by dissolving it in 'heavy water' formed from the isotope of hydrogen of mass two (heavy hydrogen or deuterium, ^2_1H, often given the special symbol D). When the acid is recovered from solution all the ionizable hydrogen atoms in the molecule will have been replaced by deuterium ions which are present in great excess in the solution. For example, acetic acid—CH_3COOH—is recovered from heavy water as CH_3COOD, showing that only the carboxylic hydrogen atom is acidic.

Radioactive atoms may be detected, by means of their characteristic radiations, in extremely small amounts and are therefore valuable for tracer studies such as these. One illustration is provided by a recent study of hypophosphorous acid, H_3PO_2. When this is dissolved in heavy water, only one of the three hydrogen atoms is replaced rapidly by a deuterium atom and this confirms that the other two hydrogen atoms are directly bonded to the phosphorus atom and do not ionize:

$$\begin{array}{c} H \quad OH \\ \diagdown \diagup \\ P \\ \diagup \diagdown \\ H \quad O \end{array} + D_2O \; \rightleftharpoons \; \begin{array}{c} H \quad OD \\ \diagdown \diagup \\ P \\ \diagup \diagdown \\ H \quad O \end{array} + HDO$$

However, when the radioactive form of hydrogen, tritium, was used to study the exchange, a further, very slow, reaction was discovered in which the hydrogen bonded to the phosphorus atom exchanged with the solvent by means of an isomerization reaction:

$$\begin{array}{c} H \quad O^- \\ \diagdown \diagup \\ P \\ \diagup \diagdown \\ H \quad O \end{array} \underset{slow}{\longleftrightarrow} \; H-P \underset{OH}{\overset{O^-}{\diagdown}} \; \underset{fast}{\overset{T_2O}{\longrightarrow}} \; H-P \underset{OT}{\overset{O^-}{\diagdown}} \underset{slow}{\longleftrightarrow} \begin{array}{c} H \quad O^- \\ \diagdown \diagup \\ P \\ \diagup \diagdown \\ T \quad O \end{array} \; etc.$$

Theory of the Electronic Structure of Hydrogen

2.4 Introduction

As the chemical properties of elements depend on the interaction of the electron clouds of their atoms, any fundamental theory of chemistry must start by examining the electronic structure of atoms. The remainder of this chapter is devoted to this theme while the succeeding ones discuss the ways in which the electron clouds interact in the formation of molecules and ions. The modern theory of atomic structure—based on the work of Heisenberg and Schrödinger—attempts to describe the shape and arrangement of the electron clouds and to calculate the energy of any given configuration of electrons. If this aim could be carried out completely and accurately, the result would be a complete

description of chemical phenomena derived purely from theory—and perhaps the end of chemistry as an experimental science! Thus, the course of a reaction or the shape of a molecule could be derived entirely from theory by calculating the energies of all the alternatives and then choosing the most favourable. The present state of theoretical chemistry is, perhaps fortunately, far from this idealized position. It is not possible to carry out complete *ab initio* calculations for any systems other than the simplest. The difficulties are formidable as the energy of a reaction is a relatively small difference between two large quantities. For example, hydrogen atoms combine to form hydrogen molecules in an extremely vigorous and exothermic reaction:

$$2H = H_2$$

The theory yields a value for the *total electronic energy* of the atoms or molecules. This is the gain in energy when, in this case, two protons and two electrons are brought from infinite separation to form the two hydrogen atoms on the left of the equation or else the hydrogen molecule on the right. The energy of formation of hydrogen molecules from hydrogen atoms is the difference between these total electronic energies: 2652 kJ mol^{-1} for 2H and 3109 kJ mol^{-1} for H_2, equal to $434 \cdot 3$ kJ mol^{-1} after allowing for minor effects. It can be seen that, even in this simple case, an error of one per cent in each of the total electronic energies can produce an error of about thirteen per cent in the resultant energy of formation.

In general, the total electronic energies of the reactants and products are much greater, and the energies of reaction are smaller, than in the case of hydrogen. Consider, for example, the question whether carbon and hydrogen combine as:

(a) $C_{(solid)} + H_{2(gas)} = CH_{2(gas)}$

or (b) $C_{(solid)} + 2H_{2(gas)} = CH_{4(gas)}$

The total electronic energy of methane has been calculated as $104\,850$ kJ mol^{-1} and this differs from the experimental value by about $1\frac{1}{2}$ per cent, 1360 kJ mol^{-1}. On the other hand, the difference in the heats of the reactions (a) and (b) is less than 42 kJ mol^{-1} (in favour of the formation of methane). It can be seen, therefore, that the accuracy of the theoretical calculations must be improved about a thousandfold before they are of value in complete *ab initio* predictions.

The great value of the theory of the electronic structure of atoms and molecules in its present form lies, not in its use for absolute calculations, but in the correlation and rationalization it provides for a great mass of experimental results. In current theory, much use is made of experimental data (such as bond lengths and angles) to help to simplify and solve the equations and these solutions in turn help to suggest further experimental work. This provision of a wide, stimulating, and flexible theoretical framework and the strong interaction between theory and experiment has been one of the most exciting and vital aspects of chemistry in the last few decades. In the next few sections an attempt has been made to outline the basic steps in this theory.

2.5 The dual nature of the electron

Before examining the electronic structure of the atom, two important properties of the electron itself must be discussed. These are:

(i) the dual nature of the electron which partakes of the properties both of a particle and of a wave, and

(ii) the effect of Heisenberg's Uncertainty Principle when applied to the electron.

The classical picture of the electron is of a tiny particle whose position in space can be accurately defined by its co-ordinates x, y, and z. Its motion in an atom is then described by the variation of (x, y, z) with time. This was the way in which Rutherford and Bohr described the atom on a 'planetary' model with the massive central nucleus and the light electrons moving in 'orbits' around it. However, it was shown in the nineteen-twenties that moving particles should behave in some ways as waves and that this effect should be particularly marked for a particle as light as the electron. Experimental support for this prediction was soon found. It was shown, for example, that a beam of electrons could be diffracted by a suitable grating in exactly the same way as a beam of light. These wave properties were introduced into the theory of atomic structure by Schrödinger in his *wave mechanics* in which the electron in an atom is described by a wave equation.

The Uncertainty Principle places an absolute limit on the accuracy with which the position and motion of a particle may be known. A formal statement is that the product of the uncertainty in the position Δx and of the uncertainty in the momentum Δp of a particle cannot be less than the modified Planck's constant, $h/2\pi$. (Planck's constant $h = 6.624 \times 10^{-34}$ J.) That is $\Delta x \Delta p \not< h/2\pi$. This limit is so small that the uncertainty is negligible for normal bodies but it is large for a particle as light as the electron. As a result, the idea of the electron's having a definite position (for example, of its following a definite orbit) must be replaced by the concept of a probability distribution for the electron. In other words, the answer to the question, 'where is the electron in an atom?', becomes a statistical one.

When both these concepts—the wave nature of the electron and the Uncertainty Principle—were taken into account, it was found that a satisfactory description of an atom resulted from Schrödinger's Wave Equation whose solutions, named *wave functions*, were given the symbol ψ. The value of the square of the wave function at any given point $P(x, y, z)$—written $\psi^2_{(x, y, z)}$—gave the probability of finding the electron at P. ψ^2 is sometimes termed the probability density of the electron. Put in rather crude terms, the old particle-in-an-orbit picture of classical atomic theory is replaced in wave mechanics by an electron 'smeared out' into a charge cloud, and the function ψ^2 describes how the density of the charge cloud is distributed in space.

2.6 The hydrogen atom

The theory of wave mechanics may now be examined in more detail for the simplest case, the hydrogen atom, where there is only one electron moving in the field of a singly-charged nucleus. Once the hydrogen atom is understood, the results for more complex systems may be derived similarly.

The time independent form of the Schrödinger Wave Equation for the hydrogen atom looks like this:*

$$\frac{-h^2}{8\pi^2 m}\nabla^2\psi - \frac{e^2}{4\pi\epsilon_0 r}\psi = E\psi \qquad (2.1)*$$

In this expression, e and m are the charge and mass, respectively, of the electron, h is Planck's constant, r is the distance from the nucleus which is taken as the origin of coordinates, ψ is the wave function, ϵ_0 is the permittivity of a vacuum and E is the total energy of the system. This equation looks rather formidable but it is just a statement in wave-mechanical terms of the principle of conservation of energy. The del-squared term corresponds to the kinetic energy, the term $-e^2\psi/r$ is the potential energy (the attraction between the charges on the electron $(-e)$ and on the nucleus $(+e)$ at a separation r), while the right hand side of the equation is the total energy.

A number of wave functions which satisfy equation (2.1) may be found. These are written ψ_1, ψ_2, ψ_3, etc. and are called *orbitals* by analogy with the orbits of the old planetary theory. To each solution or orbital there corresponds a certain value—E_1, E_2, E_3, etc.—of the total energy of the system. That orbital, say ψ_1, which corresponds to the lowest value of the total energy describes the electron distribution in the normal, most stable state of the hydrogen atom (called the *ground state*), and this is the orbital where the electron density is concentrated most closely to the nucleus. The expression for the ground state orbital is:

$$\psi_1 = \frac{1}{\sqrt{(\pi a_0^3)}}\exp(-r/a_0) \dots\dots\dots\dots(2.2)$$

where $a_0 = 0.529$ Å $= 52.9$ pm is the atomic length unit or Bohr radius. The corresponding energy E_1 is:

$$E_1 = -e^2/8\pi\epsilon_0 a_0 = -1310 \text{ kJ mol}^{-1} \qquad (2.3)$$
$$= -13.60 \text{ eV}$$

The other solutions to the wave equation describe states of hydrogen of higher energies than the ground state (called *excited states*) where the electron is in one of the orbitals concentrated further out from the nucleus. The hydrogen atom will be excited from the ground state to one of these higher-energy states if it absorbs energy, equal in amount to the difference between E_1 and the value of E of the higher orbital, which will promote the electron to this higher orbital. This energy absorption may be observed in the electronic spectrum of the hydrogen atom and the frequencies of electromagnetic radiation absorbed give experimental values for the various energy states E_1, E_2, E_3, etc., of the atom, which agree very well with the calculated E values.

---

*∇^2 is an abbreviation for $\frac{\partial^2}{\partial x^2} + \frac{\partial^2}{\partial y^2} + \frac{\partial^2}{\partial z^2}$, where (x, y, z) are the space-coordinates of the electron. It is read 'del-squared'.

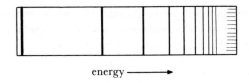

energy ⟶

FIGURE 2.2 *The spectrum of hydrogen*
This diagram shows the Balmer series in the visible spectrum. Transitions to upper states become increasingly close together until the continuum on the right of the diagram is reached. The Lyman, Paschen, Brackett, and Pfund series have similar appearances.

The wave equation (2.1) for hydrogen holds for all other 'hydrogen-like' species, that is for those with only one electron such as He^+ or Li^{2+}. The only modification to equation (2.1) required is to allow for the different nuclear charge Z in the potential energy term. Thus the general form of the wave equation for hydrogen and hydrogen-like atoms is:

$$\frac{-h^2}{8\pi^2 m}\nabla^2\psi - \frac{Ze^2}{4\pi\epsilon_0 r}\psi = E\psi \quad \ldots\ldots\ldots(2.4)$$

and the solutions are exactly the same as for hydrogen except that the value of Z will carry through the working.

Since the value of ψ^2 at a given point P (x, y, z) is the probability of finding the electron at P; it follows that acceptable solutions of the wave equation must have certain properties. Suitable functions must, for example, be single-valued at all points P as there cannot be two or more answers to the question 'what is the probability of finding the electron at P?' Similarly, ψ must be continuous and finite. Further, the total value of ψ^2 summed over all the points in space (i.e. $\int_{-\infty}^{+\infty} \psi^2 \, dx \, dy \, dz$) must equal one, since the probability of finding the electron somewhere in space must be certainty (which is unity by definition). The result of these restrictions is to limit the acceptable solutions of the wave equation to those which can be determined by three quantum numbers n, l and m which may take only those values shown in Table 2.2.

TABLE 2.2 The quantum numbers n, l, and m

Quantum number	Allowed values
n	1, 2, 3, 4, 5,....
l	$(n-1), (n-2), \ldots.2, 1, 0$
m	$+l, +(l-1), +(l-2), \ldots.2, 1, 0, -1, -2,$ $\ldots.-(l-2), -(l-1), -l$

A set of three quantum numbers is required to describe each orbital. Thus the ground state of the hydrogen atom (equation 2.2), has the electron in the orbital where $n = 1$, $l = 0$, $m = 0$. These quantum numbers arise naturally in the course of the mathematics because of the requirement that acceptable solutions are well-behaved functions. This is in contrast to the older theory where the quantum numbers had

to be added, apparently arbitrarily, to the classical description. The detail of these calculations is too complicated to be shown here but the interested reader is referred to the sources cited at the end of the book.

When the complete set of allowed solutions of the wave equation is examined, it becomes apparent that the orbitals fall into families. The first type, of which ψ_1 is an example, is of the form $\psi = f(r)$. In other words, the value of the wave function, and hence of its square, the probability of finding the electron, depends only on the distance from the nucleus and is the same in all directions in space. These orbitals are spherical and they correspond to the cases where the quantum number l (and therefore m also) is zero. Orbitals are usually designated by a number equal to the value of n, and a letter corresponding to the value of l as follows:

$$l = 0, 1, 2, 3, 4, 5, \ldots\ldots\ldots$$
$$s \quad p \quad d \quad f \quad g \quad h \ldots\ldots\ldots$$

where the rather odd selection of letters at the beginning arises for historical reasons. Thus these orbitals which are functions of r only and have $l = 0$, are s orbitals and ψ_1 is the $1s$ orbital. There is an s orbital for each value of n and they increase in energy as n increases (see Table 2.3 and Figure 2.4).

A second type of solution to the wave equation has the form $\psi = f(r) f(x)$. Clearly these orbitals now have directional properties and will have different magnitudes in the $\pm x$ direction than in the rest of space. There are two other exactly similar types of orbitals, $\psi = f(r) f(y)$ and $\psi = f(r) f(z)$ which are concentrated in the y and z directions respectively. The most stable representatives of these orbital types for hydrogen are:

$$\begin{aligned}\psi_x &= k. \, x \exp{(-r/2a_0)} \\ \psi_y &= k. \, y \exp{(-r/2a_0)} \quad \ldots\ldots\ldots(2.5) \\ \psi_z &= k. \, z \exp{(-r/2a_0)}\end{aligned}$$

where k is a constant compounded of a number of fundamental constants. These three orbitals are equal in energy and for each of them:

$$E = -e^2/32\pi\epsilon_0 a_0 \ldots\ldots\ldots\ldots(2.6)$$

This equality of energy for orbitals of this second type, in sets of three, is generally true. The three orbitals are entirely equivalent except for their direction in space.* They have $l = 1$ and are therefore p orbitals. The set of lowest energy in hydrogen, ψ_x, ψ_y and ψ_z, are $2p$ orbitals.† It follows from Table 2.2 that the p orbitals must occur in sets of three for each value of n as, when $l = 1$, m may take the three values $+1, 0, -1$.

The theory requires only that there should be three independent p orbitals and any set of three may be chosen. It is usually most convenient to choose the above set, ψ_x, ψ_y, and ψ_z, which coincide with the co-ordinate axes but different sets of three p orbitals may be useful on occasion. For example, a set making different angles with the axes may be

*Such a set of equal energy levels is termed *degenerate*.
†Since $l = n-1$, there are no p orbitals for $n = 1$.

chosen and these will be formed from appropriate combinations of ψ_x, ψ_y, and ψ_z. In particular, it should be noted that ψ_x, ψ_y, and ψ_z do not correspond directly to the m values, ± 1 and 0. On the normal convention of axial directions, the z-orbital is that for which $m = 0$ but the two orbitals for which $m = \pm 1$ have to be rearranged to give the two orbitals ψ_x and ψ_y.* There are further sets of higher energy p orbitals for the higher n values. These $3p$, $4p$, $5p$, etc., sets of orbitals again contain three independent orbitals of equal energy.

The next type of solution of the wave equation to be considered is the set of orbitals which are functions of two directions as well as of r, of the type $\psi = f(x)\,f(y)\,f(r)$. Such orbitals have l values equal to two and there are five independent orbitals of equal energy in a set, corresponding to the five m values, ± 2, ± 1, 0. The set of $3d$ orbitals has the lowest energy.

Further types are the set of seven f orbitals and the even more complex g, h, etc., orbitals. Fortunately, the chemist usually needs to work only with the s, p, d, and f types and even then he is not much concerned with the directional properties of the f orbitals. Table 2.3 lists all the orbitals with n values up to four.

TABLE 2.3 Atomic orbitals with n values up to four

n	l	m	symbol	Number of levels with this n value
1	0	0	$1s$	1
2	0	0	$2s$	4
	1	± 1, 0	$2p$	
3	0	0	$3s$	9
	1	± 1, 0	$3p$	
	2	± 2, ± 1, 0	$3d$	
4	0	0	$4s$	16
	1	± 1, 0	$4p$	
	2	± 2, ± 1, 0	$4d$	
	3	± 3, ± 2, ± 1, 0	$4f$	

In more detailed treatments of the p, d, and f orbitals (compare references given for this chapter on page 280), it is convenient to change from Cartesian co-ordinates (x, y, z) to polar coordinates (r, θ, ϕ). This does not alter the principles but simplifies the mathematics and makes it possible to separate the wave functions expressed in polar coordinates into a radial part involving only r, the distance from the nucleus, and an angular part $f(\theta, \phi)$ which expresses the directional properties of the orbitals. For example, equations (2.5) then have the form

$$\psi_x = k'.r(\exp r/2a_0).\cos\theta$$
$$\psi_y = k'.r(\exp r/2a_0).\sin\theta\cos\phi$$
$$\psi_z = k'.r(\exp r/2a_0).\sin\theta\sin\phi$$

*When the direction of a p orbital is to be distinguished a subscript is added to the symbol. For example, the orbitals of equation (2.5) are respectively, p_x, p_y and p_z.

In hydrogen and hydrogen-like atoms, the energy of the electron depends only on the n value of the orbital in which it is found. That is, the order of energies is:

$$1s < 2s = 2p < 3s = 3p = 3d < 4s = 4p = 4d = 4f$$
$$< 5s = 5p = 5d = 5f < \ldots\ldots\ldots\ldots$$

The electron in the ground state is in the $1s$ orbital and is found in one of the higher energy orbitals only if the atom has been excited by the absorption of energy. The lowest excited state is where the electron is in an orbital of n value $= 2$, and this corresponds to the absorption of energy equal to $(e^2/8\pi\epsilon_0 a_0 - e^2/32\pi\epsilon_0 a_0)$, (see equations 2.3 and 2.6) which equals 10·20 eV (984·3 kJ mol^{-1}) or absorption of radiation of wave number 82 273 cm^{-1}. The energies required for excitation to higher states can be calculated similarly.

These excitations can also be observed experimentally in the spectrum of hydrogen and the first major test of the theory to be applied was to see if the experimental and theoretical energies matched. The electronic spectrum of hydrogen was observed many years before the development of Schrödinger's theory, or before Bohr's earlier theory, and both theories do give energy levels which agree with the observed ones. Schrödinger's theory also gives close agreement for many-electron atoms where Bohr's theory is less successful.

If electromagnetic radiation consisting of a continuous range of frequencies—'white' radiation—is shone on to an absorbing system, then those frequencies which correspond to the difference, ΔE, in energy between two states of the system will be absorbed. (The relation between energy and frequency, v, is $E = hv$.) Thus the absorption spectrum shows lines at frequencies which correspond to energy level differences in the irradiated species. Lines at the same frequencies may be *emitted* when the species return from the higher to the lower energy level and it is easier, for atoms, to observe this *emission spectrum*. When the hydrogen spectrum is examined several series of lines like that in Figure 2.2 are observed. Each series, named after its discoverer and falling in the ultraviolet, visible, or infrared regions is similar in showing lines which become progressively less intense and closer together towards higher frequencies. They all show an upper frequency limit for line absorption, above which continuum is observed. In each series the wave numbers $\tilde{v}$ of the lines fit the general formula:

$$\tilde{v} = R(1/m^2 - 1/m'^2).$$

In this formula, m has an integral value which is different for each series of lines while $m' = (m+1)$, $(m+2)$, $(m+3)\ldots.\infty$. R is a constant, the Rydberg Constant, which equals 109 740 cm^{-1} approximately. The values of m and examples of the transitions for the various series are shown in Table 2.4. These various series may be portrayed on an energy level diagram, as in Figure 2.3, where the zero value for the energy is taken as the start of the continuum corresponding to $m' = \infty$. The correlation between this diagram and the electronic levels of the atom, as derived from the wave equation, is obvious. The states whose line spectra correspond to $m = 1, 2, 3, \ldots$ are those whose n quantum number

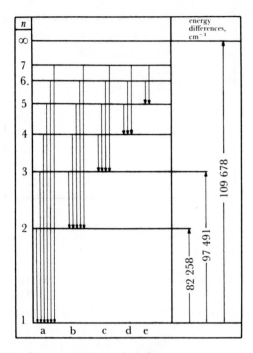

FIGURE 2.3 *Energy level diagram for hydrogen*
Energy level diagram showing the transitions corresponding to the various series of lines in the electronic spectrum of hydrogen. The transitions are lettered corresponding to the different series of lines in the hydrogen spectrum: (a) Lyman, (b) Balmer, (c) Paschen, (d) Brackett, (e) Pfund.

takes the values, 1, 2, 3, ... respectively. The transition in the Lyman series corresponding to

$$\tilde{\nu} = R(1/1^2 - 1/2^2)$$

(i.e. with $m = 1$, $m' = 2$) is the transition from the $1s$ level to the $2s$ (or $2p$) level. The Lyman series corresponds to transitions from the ground state, $1s$, to levels with higher n values; while the Balmer, Paschen, etc., series, for which $m = 2, 3$, etc., respectively, correspond to transitions of electrons which are already excited being further excited to higher levels. Thus the frequencies of the Lyman series correspond to the energy gaps between the $1s$ level and the higher levels and, in particular, the start of the continuum corresponding to $m' = \infty$ corresponds to the complete removal of the $1s$ electron. Similar transitions from the $2s$ level are observed in the Balmer series, from the $3s$ level in the Paschen series, and so forth. As an illustration of the correlation between the calculated and observed energy differences, it is worth calculating the transitions in the Lyman series for $m' = 2$ (equal to the $1s - 2s$ energy gap) and $m' = \infty$, (equal to the energy of the $1s$ orbital with respect to complete dissociation). For $1s$ to $2s$ the value is $\tilde{\nu} = R(1/1^2 - 1/2^2) = 3/4 \times 109\,740$ cm^{-1} $= 82\,305$ cm^{-1} (10·20 eV). The dissociation transition is $\tilde{\nu} = R(1/1^2 - 1/\infty^2) = R = 109\,740$ cm^{-1} (13·61 eV). The values calculated from the wave equation agree exactly.

The Bohr theory gives an equally good correlation with experiment for the hydrogen atom but wave mechanics is to be preferred for dealing with many-electron atoms.

TABLE 2.4 The electronic spectrum of hydrogen

Series	m	m'	$\tilde{\nu}$	Corresponding orbital description
Lyman	1	2	82 303	$n = 1 \rightarrow n = 2$
		3	97 544	$n = 1 \rightarrow n = 3$
		4	102 879	$n = 1 \rightarrow n = 4$
		∞	109 737	Total electronic energy when electron is in the $1s$ orbital: dissociation energy for the ground state
Balmer	2	3	15 241	$n = 2 \rightarrow n = 3$
		4	20 576	$n = 2 \rightarrow n = 4$
		∞	27 434	Total electronic energy when electron is in $2s$ or $2p$ orbitals
Paschen	3	4	5 334	$n = 3 \rightarrow n = 4$
		5	7 803	$n = 3 \rightarrow n = 5$
		∞	12 193	Total electronic energy when electron is in $3s$, $3p$ or $3d$ orbitals
Brackett	4	5	2 469	$n = 4 \rightarrow n = 5$
		∞	6 859	Total electronic energy when electron is in any orbital with $n = 4$
Pfund	5	6	1 341	$n = 5 \rightarrow n = 6$
		∞	4 390	Total electronic energy of H atom when electron is in any orbital with $n = 5$

The Lyman series occurs in the ultra-violet, the Balmer series in the visible, the Paschen series in the near infra-red, and the Brackett and Pfund series in the far infra-red regions of the spectrum.

Many Electron Atoms

2.7 The approach to the wave equation

When the wave equation for helium with two electrons and $Z = 2$ is considered, it is found to be more complicated than the wave equation for the hydrogen atom both in the kinetic energy term and in the potential energy term. As both electrons contribute to the kinetic energy there are now two del-squared terms, one applying to each electron. The potential energy term consists of three parts in place of the single term in equation (2.1). In helium there are two attractions—between each electron and the nucleus—and there is also a repulsion term between the two electrons.

The wave equation for helium is:

$$k(\nabla_1^2 + \nabla_2^2)\psi - \left(\frac{Ze^2}{4\pi\epsilon_0 r_1} + \frac{Ze^2}{4\pi\epsilon_0 r_2} - \frac{e^2}{4\pi\epsilon_0 r_{12}}\right)\psi = E\psi \dots (2.7)$$

where k is a product of universal constants, r_1 and r_2 are the distances of each electron from the nucleus and r_{12} is the interelectronic distance. This increase in complexity adds considerably to the difficulty of the calculation. Particular difficulty arises when dealing with the repulsion between the two electrons. These problems become even more complex as the analysis is extended to atoms with larger numbers of electrons and it is at present impossible to solve the wave equation of a many-electron atom directly. Methods of approximation have to be sought and many of these methods attempt to get round the problem of the interelectron terms by considering only one electron at a time. In other words, the many-electron atom is treated as a series of problems based on hydrogen-like situations and this, in turn, means that many of the results of the last section carry over for many-electron atoms with only minor modifications.

One method of approximation, for example, starts with the wave equation for only one of the electrons moving in a potential field determined by the nuclear charge and the averaged-out field of all the other electrons. This case is then like that of the hydrogen atom with a modified value for the nuclear charge and may be solved to give a 'first approximation' distribution function for this electron. Each electron is considered in turn and a set of first approximation distribution functions is obtained. These functions are then used to refine the value of the potential field of the nucleus plus electrons and then the analysis is repeated giving 'second approximation' functions. The whole process is continued until self-consistent results are obtained and a solution to the many-electron problem is found using the method for the hydrogen-like atom at each step in the calculation. The above method is called the *self-consistent field method*. Other, more accurate, approaches are available, some of which start off from the self-consistent field answers, but these cannot be followed out here. An introduction is given in Coulson's book referred to in the references on page 280.

Since this and other systems of approximation use hydrogen-like fields, most of the results for the exactly-solved hydrogen atom apply in a qualitatively similar form. The electron distributions in many-electron atoms are described by wave functions ψ similar to those for hydrogen, and the square of the wave function again describes the probability distribution of the electron in one of these orbitals. These atomic orbitals are defined by the three quantum numbers n, l and m which are restricted to certain integral values by the same rules as for hydrogen, as in Table 2.2. One difference between many-electron atoms and the hydrogen atom is that the energy of an electron in an orbital depends on l as well as on n: that is, the order of energies now includes $s < p < d < f$ well as $n = 1 < 2 < 3 \dots$. As a result, when there are d and

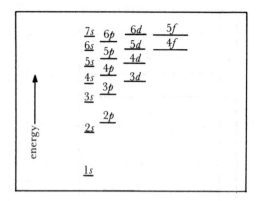

FIGURE 2.4 *Energy levels in many-electron atoms*
This diagram shows the relative energy levels of the orbitals at the Z values where they are about to be filled.

f levels corresponding to an n value, these overlap in energy with the s and p levels of higher n values. The order of energies for a many-electron atom is approximately $1s < 2s < 2p < 3s < 3p < 4s = 3d < 4p < 5s = 4d < 5p < 6s = 5d = 4f < 6p < 7s = 6d = 5f$. This is illustrated in Figure 2.4. Two points are of importance in this figure: first, that for higher n values the ns level is approximately equal in energy to the $(n-1)d$ level and to the $(n-2)f$ level, and second, that the energy gaps between successive levels with the same l value become smaller as n values increase. This has important effects on the chemistry of the heavier elements.

Although the energy of an orbital in a many-electron atom depends on both the n and l values, there is no dependence on the value of m in the free atom, and the three p orbitals or the five d orbitals of a given n value are equal in energy as they are in the hydrogen atom. The total electronic energy of an atom is the sum of the energies of each electron in its orbital, with a correction for the interaction between the electrons. The spectra of many-electron atoms are much more complicated than that of hydrogen, as there are more energy levels and some overlap. However, most atomic spectra have now been successfully analyzed and these give experimental values of the energy levels which agree with the calculated ones.

2.8 The electronic structures of atoms

The electronic structure of an atom may be built up by placing the electrons in the atomic orbitals. Clearly, the orbitals of lowest energy will be the first to be filled and the questions which have to be answered in order to carry out the building-up, or *aufbau*, process are:
(i) how many electrons in each orbital?
(ii) what happens when there are a number of orbitals of equal energy, such as the three $2p$ orbitals?

The answer to (i) depends on one other property of the electron, its *spin*. This property was discovered during a study of the spectra of the alkali metals, when it was found that the absorption lines had a fine structure which could be explained only if the electron was regarded as spinning on its axis and able to take up one of two orientations with respect to a given direction. This spin is included in the description of the state of an electron by introducing a fourth quantum number, m_s, which may take one of the two values $\pm\frac{1}{2}$. The spin is usually indicated in diagrams by using arrows, $\uparrow$ or $\downarrow$. The distribution of electrons in an atom is then determined by Pauli's Principle that no two electrons may have all four quantum numbers the same. Thus the answer to (i) is that each orbital, which is defined by the three quantum numbers n, l and m, can hold only two electrons and these will have $m_s = +\frac{1}{2}$ and $-\frac{1}{2}$. Such a pair of electrons of opposite spin in one orbital is termed 'spin-paired'.

The answer to (ii) is given by Hund's Rules which may be stated:

(a) electrons tend to avoid as far as possible being in the same orbital,

(b) electrons in different orbitals of the same energy have parallel spins.

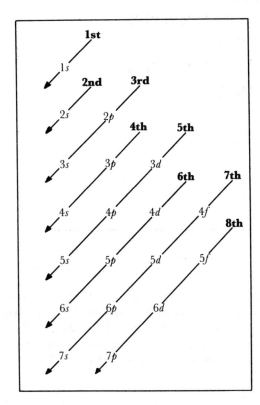

FIGURE 2.5 *The order of filling energy levels in an atom*
The levels are written out in a square array and filled in the order in which they are cut by a series of diagonals as shown.

The electronic structure of any atom may be worked out taking account of these factors as follows:

1. Orbitals are filled in order of increasing energy. This may be remembered from the diagram shown in Figure 2.5 where the orbitals corresponding to a given value of n are written out in horizontal rows and then filled in the order in which they are cut by the series of diagonal lines as shown. This device gives an order which is substantially correct, though there are a few slight anomalies for heavier atoms. In particular, one electron enters the $5d$ level before the $4f$ level is filled and some uncertainty exists about the distribution of electrons between the $6d$ and $5f$ levels in the heaviest atoms.

2. Each orbital (which is defined by specific values of n, l and m) may hold only two electrons with $m_s = \pm\frac{1}{2}$. In other words, each electron is described by the four quantum numbers, n, l, m and m_s, which may not all have the same values.

3. Where a number of orbitals of equal energy is available, the electrons fill each singly, keeping their spins parallel, before spin-pairing starts.

The use of these rules to build up the electronic structures of the elements may be illustrated by a few examples, but first two useful notations for describing these structures must be defined. A convenient way of showing the electronic structures on a diagram is to place cells on each level of the appropriate part of Figure 2.4, corresponding to the number of orbitals in each level. The electrons are indicated by arrows and a cell with a pair of arrows shows a filled orbital. The diagrams for a number of the lighter elements are shown in Figure 2.6. A second notation is to write out the orbitals in order of increasing energy and to indicate the number of electrons by superscripts. To save this latter notation becoming too clumsy, it is convenient to show the configuration of inner, completed levels by the symbol of the appropriate rare gas element. Thus, helium is written, $He = (1s)^2$, lithium is $Li = (1s)^2(2s)^1$ or $Li = [He](2s)^1$ and sodium which is, in full, $(1s)^2(2s)^2(2p)^6(3s)^1$ is shortened to $Na = [Ne](3s)^1$.

The electron in the hydrogen atom is described by the four

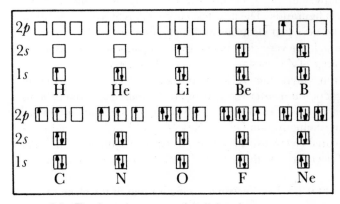

FIGURE 2.6 *The electronic structures of the lighter elements*
The electronic structures of the elements where the $1s$, $2s$, and $2p$ levels are being filled, illustrating the *aufbau* process.

values of the quantum numbers $(1, 0, 0, \frac{1}{2})$. Then, in helium, the second electron enters the lowest energy orbital (rule 1) which is the $1s$ orbital singly-occupied in the hydrogen atom. This second electron must have its spin antiparallel with the first (rule 2) and corresponds to the quantum numbers $(1, 0, 0, -\frac{1}{2})$. In the next element, lithium, the third electron must enter the next lowest orbital, $2s$, as in Figure 2.6.

The operation of the third rule is illustrated by the case of carbon with $Z = 6$. Starting from the three electrons in lithium, the fourth completes the $2s$ orbital and the fifth enters the next lowest level, $2p$. The sixth electron then enters one of the two remaining empty $2p$ orbitals with its spin parallel to that of the previous electron. The third $2p$ level is occupied in the next element, nitrogen, and not until there is a fourth electron to be accommodated in the $2p$ level—at oxygen—does spin-pairing occur in that level. The orbitals with $n = 2$ are completely filled at neon, $Z = 10$. Sets of orbitals with the same value of the n quantum number are known as quantum *shells* and the electrons in that shell which is partly filled in a particular atom are termed the *valency electrons*.

The rules for deriving atomic structures are further illustrated below for some of the heavier elements.

Consider first iron, Fe with $Z = 26$, whose electronic structure is shown in Figure 2.7. The first ten electrons fill up to the neon structure while the next eight fill the $3s$ and $3p$ levels, repeating the pattern of the second shell, to form the argon core. This accounts for eighteen electrons. Figure 2.4 shows that the level which comes next in energy to the $3p$ level is $4s$ which is a little more stable than $3d$. The nineteenth and twentieth electrons fill the $4s$ level, leaving six electrons to be accommodated in the five $3d$ orbitals. The first five electrons enter these orbitals singly with all their spins parallel while the last electron pairs up with one of these. Thus the configuration is Fe $= [\text{Ar}](3d)^6(4s)^2$ with four unpaired spins. Note that $3d$ is now more stable than $4s$.

As a second example, take gadolinium, Gd with $Z = 64$, see Figure 2.7. Continuing from the configuration of iron, the next ten electrons fill the $3d$ and $4p$ levels to give the configuration of krypton, $Z = 36$. The next eighteen repeat this pattern in the $5s$, $4d$ and $5p$ levels giving the xenon ($Z = 54$) configuration. The next level in energy is $6s$ which is filled by the next two electrons and then come the $5d$ and $4f$ levels which are very close in energy, being almost identical in the range of Z values around $Z = 60$. In the event, the first electron enters the $5d$ level and the last seven electrons in gadolinium enter the $4f$ orbitals. As there are seven orbitals in an f level, each one is singly occupied and gadolinium has the configuration Gd $= [\text{Xe}](4f)^7(5d)^1(6s)^2$ with eight unpaired electrons (the single d electron and the seven parallel f electrons). As the $5d$ and $4f$ orbitals are so similar in energy,

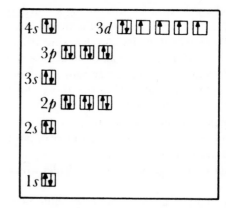

(a)

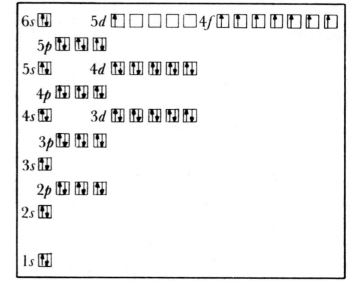

(b)

FIGURE 2.7 (a) *The electronic structure of iron*
(b) *The electronic structure of gadolinium*

the elements in this region of the Periodic Table vary in the distribution of their electrons between the two. There is never more than one electron in the $5d$ level, but often there is none at all. Thus, europium with $Z = 63$, which precedes gadolinium, has the configuration Eu $= [\text{Xe}](4f)^7(5d)^0(6s)^2$.

TABLE 2.5 The electronic configurations of the elements

Element	Symbol	Z	A	Electron configuration Inner shells	Valency shell	
					$1s$	
Hydrogen	H	1	1·008 0		1	
Helium	He	2	4·002 60		2	
					$2s$	$2p$
Lithium	Li	3	6·94	He	1	
Beryllium	Be	4	9·012 18	He	2	
Boron	B	5	10·81	He	2	1
Carbon	C	6	12·011	He	2	2
Nitrogen	N	7	14·006 7	He	2	3
Oxygen	O	8	15·999 4	He	2	4
Fluorine	F	9	18·998 5	He	2	5
Neon	Ne	10	20·179	He	2	6

Element	Symbol	Z	A	Inner shells	Electron configuration — Valency shell			
							3s	3p
Sodium	Na	11	22·989 8	Ne			1	
Magnesium	Mg	12	24·305	Ne			2	
Aluminium	Al	13	26·981 5	Ne			2	1
Silicon	Si	14	28·086	Ne			2	2
Phosphorus	P	15	30·973 8	Ne			2	3
Sulphur	S	16	32·06	Ne			2	4
Chlorine	Cl	17	35·453	Ne			2	5
Argon	Ar	18	39·948	Ne			2	6
						3d	4s	4p
Potassium	K	19	39·102	Ar			1	
Calcium	Ca	20	40·08	Ar			2	
Scandium	Sc	21	44·955 9	Ar		1	2	
Titanium	Ti	22	47·90	Ar		2	2	
Vanadium	V	23	50·941 4	Ar		3	2	
Chromium	Cr	24	51·996	Ar		5	1	
Manganese	Mn	25	54·938 0	Ar		5	2	
Iron	Fe	26	55·847	Ar		6	2	
Cobalt	Co	27	58·933 2	Ar		7	2	
Nickel	Ni	28	58·71	Ar		8	2	
Copper	Cu	29	63·546	Ar		10	1	
Zinc	Zn	30	65·37	Ar		10	2	
Gallium	Ga	31	69·72	Ar		10	2	1
Germanium	Ge	32	72·59	Ar		10	2	2
Arsenic	As	33	74·921 6	Ar		10	2	3
Selenium	Se	34	78·96	Ar		10	2	4
Bromine	Br	35	79·904	Ar		10	2	5
Krypton	Kr	36	83·80	Ar		10	2	6
						4d	5s	5p
Rubidium	Rb	37	85·467 8	Kr			1	
Strontium	Sr	38	87·62	Kr			2	
Yttrium	Y	39	88·905 9	Kr		1	2	
Zirconium	Zr	40	91·22	Kr		2	2	
Niobium	Nb	41	92·906 4	Kr		4	1	
Molybdenum	Mo	42	95·94	Kr		5	1	
Technetium	Tc	43	98·906 2*	Kr		6	1	
Ruthenium	Ru	44	101·07	Kr		7	1	
Rhodium	Rh	45	102·905 5	Kr		8	1	
Palladium	Pd	46	106·4	Kr		10	0	
Silver	Ag	47	107·868	Kr		10	1	
Cadmium	Cd	48	112·40	Kr		10	2	
Indium	In	49	114·82	Kr		10	2	1
Tin	Sn	50	118·69	Kr		10	2	2
Antimony	Sb	51	121·75	Kr		10	2	3
Tellurium	Te	52	127·60	Kr		10	2	4
Iodine	I	53	126·904 5	Kr		10	2	5
Xenon	Xe	54	131·30	Kr		10	2	6
					4f	5d	6s	6p
Cesium	Cs	55	132·905 5	Xe			1	
Barium	Ba	56	137·34	Xe			2	
Lanthanum	La	57	138·905 5	Xe		1	2	
Cerium	Ce	58	140·12	Xe	2		2	
Praseodymium	Pr	59	140·907 7	Xe	3		2	

Element	Symbol	Z	A	Inner shells	Electron configuration Valency shell			
					4f	5d	6s	6p
Neodymium	Nd	60	144·24	Xe	4		2	
Promethium	Pm	61	(147)	Xe	5		2	
Samarium	Sm	62	150·4	Xe	6		2	
Europium	Eu	63	151·96	Xe	7		2	
Gadolinium	Gd	64	157·25	Xe	7	1	2	
Terbium	Tb	65	158·925 4	Xe	9		2	
Dysprosium	Dy	66	162·50	Xe	10		2	
Holmium	Ho	67	164·930 3	Xe	11		2	
Erbium	Er	68	167·26	Xe	12		2	
Thulium	Tm	69	168·934 2	Xe	13		2	
Ytterbium	Yb	70	173·04	Xe	14		2	
Lutetium	Lu	71	174·97	Xe	14	1	2	
Hafnium	Hf	72	178·49	Xe	14	2	2	
Tantalum	Ta	73	180·947 9	Xe	14	3	2	
Tungsten	W	74	183·85	Xe	14	4	2	
Rhenium	Re	75	186·2	Xe	14	5	2	
Osmium	Os	76	190·2	Xe	14	6	2	
Iridium	Ir	77	192·22	Xe	14	7	2	
Platinum	Pt	78	195·09	Xe	14	9	1	
Gold	Au	79	196·966 5	Xe	14	10	1	
Mercury	Hg	80	200·59	Xe	14	10	2	
Thallium	Tl	81	204·37	Xe	14	10	2	1
Lead	Pb	82	207·2	Xe	14	10	2	2
Bismuth	Bi	83	208·980 6	Xe	14	10	2	3
Polonium	Po	84	(209)	Xe	14	10	2	4
Astatine	At	85	(210)	Xe	14	10	2	5
Radon	Rn	86	222	Xe	14	10	2	6
					5f	6d	7s	
Francium	Fr	87	(223)	Rn			1	
Radium	Ra	88	226·025 4*	Rn			2	
Actinium	Ac	89	227	Rn		1	2	
Thorium	Th	90	232·038 1*	Rn		2	2	
Protoactinium	Pa	91	231·035 9*	Rn	2	1	2	
Uranium	U	92	238·03	Rn	3	1	2	
Neptunium	Np	93	237·048 2*	Rn	5		2	
Plutonium	Pu	94	(242)	Rn	6		2	
Americium	Am	95	(243)	Rn	7		2	
Curium	Cm	96	(247)	Rn	7	1	2	
Berkelium	Bk	97	(249)	Rn	8	1	2	
Californium	Cf	98	(251)	Rn	10		2	
Einsteinium	Es	99	(254)	Rn	11		2	
Fermium	Fm	100	(253)	Rn	12		2	
Mendelevium	Md	101	(256)	Rn	13		2	

(a) Values for atomic weights are based on 1969 revision. Recent studies of isotopic composition have shown variations from different sources (H. Li, B, C, O, Si, S, Ar, Cu, Pb) or variations introduced by commercial isolation (Li, B, U) which limits the accuracy of the atomic weight or requires that the isotopic composition of a particular sample should be determined. Elements involved in radioactive decay processes (Ar, Sr, Pb, Ra) may have different isotopic compositions in different geological specimens. The other changes are mainly increases in precision.

(b) Weights marked with an asterisk are those of the commonest long-lived isotope of a radioactive element: those in brackets indicate the most accessible isotope of the heavier elements.

(c) Recent years have seen the discovery of, and controversy about, a number of elements beyond 101. Elements 102, 103, and 105 seem relatively established but there is still doubt about the claimed synthesis of 104 and Russian claims to have made 105 and 106 preceded the American announcement of Hahnium.

(d) Atomic weights based on $^{12}C = 12·000\ 0$: in SI, $^{12}C = 12·000\ 0 \times 1·660\ 41 \times 10^{-27}$ kg.

Element	Symbol	Z	A	Electron configuration			
				Inner shells		Valency shell	
Nobelium	No	102	(254)	Rn	14		2
Lawrencium	Lr	103	(257)	Rn	14	1	2
Element 104	?	104	(260)	Rn	14	2?	2
Hahnium	Ha	105	(260)	Rn	14	3?	2?

The rest of the elements up to $Z = 86$ complete the $4f$, $5d$, and $6p$ levels to give the configuration of the heaviest rare gas, radon. The next two electrons fill the $7s$ level and then there is a close correspondence between the energies of the $5f$ and $6d$ levels, similar to that between the $4f$ and $5d$ levels. Here the energy gap is even smaller and there has been considerable difficulty and confusion about the levels being occupied in the heaviest elements. The present conclusion is that these elements are best regarded as paralleling the $4f$ ones in filling mainly the $5f$ level, but the first members of the set make more use of the d level than their lighter congeners. The full electronic structures of all the elements is given in Table 2.5.

Shapes of Atomic Orbitals

In the last section, the electronic structures of the elements were built up from a knowledge of the energy levels of the orbitals derived from the wave equation. Since the chemist is primarily concerned with the outermost electrons and these are found in s, p, or d orbitals, the shape and extension in space of these orbitals is of primary importance. Indeed, the general aim of understanding and predicting the shape and reactions of ions and molecules may be carried quite a long way by qualitative reasoning which is based largely on diagrams of the relevant atomic orbitals, as the next chapter shows.

2.9 The s orbital

The solution of the wave equation for the $1s$ orbital of hydrogen is the wave function given in equation (2.2):

$$\psi_1 = \sqrt{(1/\pi a_0^3)} \exp(-r/a_0)$$

where r is the distance from the nucleus. The probability density of the electron distribution is the square of this wave function:

$$\psi_1^2 = 1/\pi a_0^3 \exp(-2r/a_0)$$

As these expressions are functions of r, the distance from the nucleus, only, they are spherically symmetrical around the nucleus. As they are exponential functions, it follows that the values of the wave function and of the electron density fall off smoothly and rapidly with increasing distance from the nucleus although they never quite reach zero (see Figure 2.8a). The contour diagram which results from joining points in space with the same value of ψ or ψ^2 has the general appearance of Figure 2.8b (where the values decrease outwards from the nucleus). As exponential functions never quite fall to

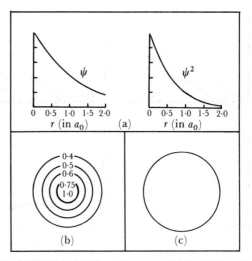

FIGURE 2.8 *The 1s orbital of hydrogen:* (a) *Plots of ψ_{1s} and ψ_{1s}^2 against r;* (b) *Contour representation of the 1s orbital* (c) *Boundary contour representation*

zero, there is always a finite electron density outside any given one of these contour lines but it is possible to include the major part of the electron cloud within a boundary surface close to the nucleus, and it is common to represent the orbital by a single boundary contour (as in Figure 2.8c) enclosing an arbitrary fraction—say 90 per cent—of the electron density. Such a boundary diagram may be used to represent the electron density, ψ^2, or it may be the corresponding diagram of the orbital, ψ. Since both functions are exponential ones, the boundary diagrams for ψ and ψ^2 are similar in appearance, that for ψ being distinguished by showing the variation in the sign of the wave function in different regions of space (see Figures 2.10 and 2.11 for examples, the $1s$ orbital has the same sign throughout). As discussed in the next chapter, a bond is formed by the combination of atomic orbitals, and the way in which they may combine depends on the signs of the various wave functions. For this reason, the chemist usually works with boundary diagrams of the wave functions (ψ) rather than with those of the electron density distributions (ψ^2).

A further important function used in representing an orbital is the *radial density function*, $4\pi r^2 \psi^2(r)$. The volume of a spherical shell of thickness dr at a distance r from the nucleus is $4\pi r^2 dr$. Hence, $4\pi r^2 \psi^2(r) dr$ is the probability of the electron's being found at a distance between r and $(r+dr)$ from the nucleus. The plot of the variation of this function with distance from the nucleus for the hydrogen $1s$ orbital is shown in Figure 2.9a. The radial density in hydrogen is at a

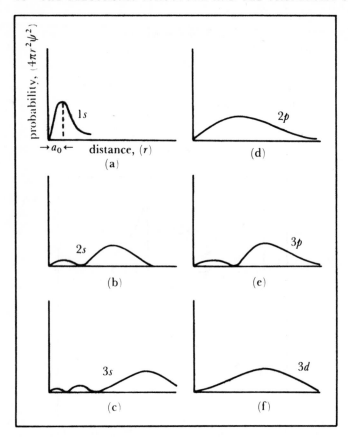

FIGURE 2.9 *Radial density plots of hydrogen orbitals:* (a) $1s$, (b) $2s$, (c) $3s$, (d) $2p$, (e) $3p$, (f) $3d$

maximum at that distance, a_0, from the nucleus that Bohr calculated as the radius of the most stable orbit in the planetary theory.

The s orbitals of higher n value resemble the $1s$ orbital, being spherically symmetrical and having the same sign for the wave function in all directions. They extend further into space and there are changes close in towards the nucleus where spherical nodes appear across which the wave function changes sign. As only the outer regions are of chemical interest, these nodes are of no direct interest. They appear as minima in the radial density plot; compare for example, the hydrogen $2s$ orbital in Figure 2.9b.

2.10 The p orbitals

The p orbitals of lowest energy are the three $2p$ orbitals which in hydrogen are described by the equations (2.5). These orbitals do have directional properties and are typically two-lobed distributions of the electron density along the positive and negative directions of their characteristic axes. The 90 per cent boundary surfaces are shown in Figure 2.10 and the radial density plot along the characteristic axis in Figure 2.9d. The orbitals in three dimensions are the figures of rotation of these diagrams about their respective axes. Apart from their directional properties, these p orbitals differ from the s orbitals in possessing a nodal plane through the nucleus, across which the wave function changes sign. The square of the wave function, the electron density, accordingly drops

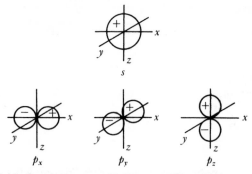

FIGURE 2.10 *Boundary contour of hydrogen s and $2p$ orbitals*

to zero across this plane and the p orbitals differ from the s orbitals in having no electron density at the nucleus.

It must be noted that the two-lobed p orbital is a complete entity and there is no question, for example, of the electrons having to pass through the nucleus to get from one side to the other. The interpretation of ψ^2 in terms of the probability of finding the electron at a given point is helpful in becoming accustomed to this idea.

The higher p orbitals with $n = 3, 4, \ldots$ are qualitatively similar to the $2p$ orbitals in the outer regions of chemical interest. They also have radial nodes as well as the nodal plane through the nucleus, and this may be seen in the radial density plot of the $3p$ orbital as shown in Figure 2.9e.

2.11 The d orbitals

The five $3d$ orbitals are shown in Figure 2.11. Like the p orbitals they have directional properties, and the basic shape of the d orbital is a four-lobed figure with two nodal planes at right angles through the nucleus and therefore zero electron density at the nucleus. The d_{z^2} orbital does not fit this description and it arises in the following way. The three orbitals d_{xy}, d_{yz}, and d_{zx} are identical except for their orientation in space, and the $d_{x^2-y^2}$ orbital is the same as the d_{xy} orbital but rotated through 45 degrees. There are two other possible orbitals corresponding to the $d_{x^2-y^2}$ orbital and directed along the other two pairs of axes (in an obvious notation, $d_{y^2-z^2}$ and $d_{z^2-x^2}$) but these would give six d orbitals in all, although there are only five independent solutions of the d type to the wave equation. It is easy to show that the three solutions of the d_{xy} type are independent while

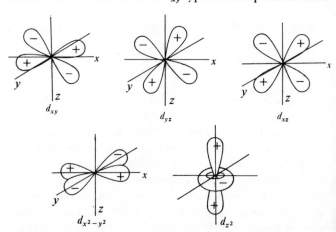

FIGURE 2.11 *Boundary contour diagrams of the hydrogen 3d orbitals*

the three of the $d_{x^2-y^2}$ type are not. (If the orbital diagrams of this latter set of three are superimposed, taking account of the signs, the lobes all cancel out.) Any two of the three would be acceptable along with the three orbitals of the d_{xy} type, but it is preferable for theoretical reasons to use one of them, say $d_{x^2-y^2}$, together with a combination of the other two, d_{z^2}, where:

$$\psi_{d_{z^2}} = 1/\sqrt{3}(\psi_{d_{z^2-x^2}} - \psi_{d_{y^2-z^2}})$$

This gives a satisfactory set of five independent d orbitals and this is the set in common use. Naturally, the z-direction may be chosen to suit the situation. For example, if the atom is in an external magnetic field, it is useful to have the z-axis parallel to the field. The higher d orbitals are of the same basic shape as the $3d$ orbitals.

The basic shape of the f orbital is an eight-lobed figure with three nodal planes at right angles through the nucleus. As these orbitals are little used in bonding they need not be discussed further.

The basic shapes of orbitals are unchanged by changes in the nuclear charge, although their extensions into space depend inversely on the value of Z. Figure 2.12 shows the

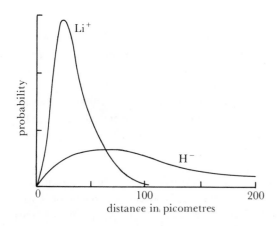

FIGURE 2.12 *Radial density curves for ions containing two electrons and with different nuclear charges*
This plot for isoelectronic ions illustrates the effect of increasing the nuclear charge from $Z = 1$ to $Z = 3$ on the spatial distribution of the electron probability density.

radial density curve for the $1s$ orbital of helium-like species with different values of the nuclear charge. The radius of the electron cloud around an atom is the resultant of this contraction of the charge cloud with increasing Z, and the fact that orbitals further and further out from the nucleus are being occupied as the atomic number rises. There is a general increase in atomic size as Z increases, but this change occurs in a very irregular manner (compare the plot of atomic radius against atomic number in Figure 7.9). The greatest jumps in radius come when the outermost electron starts to fill the s level of a new quantum shell and there are less sharp increases when any new level starts to be occupied and when spin-pairing occurs at p^4 and d^6 configurations.

2.12 The Periodic Table

As chemical behaviour depends on the interaction of the electron clouds of atoms, and especially on the interaction of the outermost parts of these clouds, atoms which have their outer electrons in the same type of orbital should have similar chemical behaviour. For example, a configuration such as s^2p^3 implies the same shape of electron cloud, whatever the n values of the orbitals, although the extension of the electron cloud, and hence the atomic radius, clearly depends on the atomic number. If the elements are arranged so that those with the same outer electron configuration fall into Groups, the result is the Periodic Table of the elements. The electronic configurations of the elements, derived from the theory of atomic structure, thus provide the theoretical explanation of the Periodic Law first proposed by Mendeléef and by Lothar Meyer about sixty years before the electronic theory was evolved.

The Periodic Table reflects the order of energy levels in the atoms as they are derived from the wave equation, and the form of the Periodic Table follows from the allowed values of the quantum numbers as given in Table 2.2. This is illustrated by the block form of the Periodic Table shown in Figure 2.13. The different 'blocks' hold sets of two, six, ten, and

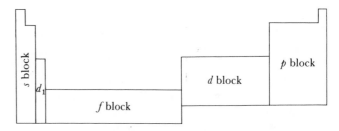

FIGURE 2.13 *Block diagram of the Periodic Table*

fourteen elements; these being, respectively, the elements where the s, p, d, and f levels are filling and these blocks thus follow from the number of orbitals of each type which are allowed by the quantum rules. The order of the blocks across the Table from left to right follows from the general order of energy levels: $ns < (n-1)d \leqslant (n-2)f < np$. The modern 'long' form of the Periodic Table is given as Table 2.6.

It is difficult to realize, nowadays when all this has become accepted, just how important a contribution a knowledge of electronic structure made to the chemist's understanding and use of the Periodic Table. In particular, the Table produced by Mendeléef, based on valency considerations, was in the 'short' form where the transition elements were introduced as sub-groups into the main groups of the s and p blocks. This practice was justifiable at the time as there are some resemblances, particularly when the transition elements are showing their maximum valencies, and there were many blanks in the knowledge of the chemistry of these elements. As such knowledge increased, it became apparent that more anomalies than analogies were introduced by the short form of presentation. The long form removes many of these difficulties, and the case for its adoption became overwhelming when the electronic structures of the elements were worked

TABLE 2.6 Periodic Table of the elements

		1	2						Groups							3	4	5	6	7	0	Rare gas electrons
Period 1	1s	H																		H	He	2
2	2s	Li	Be												2p	B	C	N	O	F	Ne	2, 8
3	3s	Na	Mg												3p	Al	Si	P	S	Cl	Ar	2, 8, 8
4	4s	K	Ca	3d	Sc	Ti	V	Cr	Mn	Fe	Co	Ni	Cu	Zn	4p	Ga	Ge	As	Se	Br	Kr	2, 8, 18, 8
5	5s	Rb	Sr	4d	Y	Zr	Nb	Mo	Tc	Ru	Rh	Pd	Ag	Cd	5p	In	Sn	Sb	Te	I	Xe	2, 8, 18, 18, 8
6	6s	Cs	Ba	5d	La*	Hf	Ta	W	Re	Os	Ir	Pt	Au	Hg	6p	Tl	Pb	Bi	Po	At	Rn	2, 8, 18, 32, 18, 8
7	7s	Fr	Ra	6d	Ac†	104																
	*Lanthanides	4f	Ce	Pr	Nd	Pm	Sm	Eu	Gd	Tb	Dy	Ho	Er	Tm	Yb	Lu						
	†Actinides	5f	Th	Pa	U	Np	Pu	Am	Cm	Bk	Cf	Es	Fm	Md	No	Lw						

out. There are still a number of minor anomalies. In particular, no form of the Table completely reflects the differences in the ground state electron distribution between the s and d levels among the transition elements, nor those between the d and f levels among the inner transition elements (see Table 2.5), but these differences have no effect on the chemistry of these elements as the anomalies disappear in the valency states. In fact, all the problems and objections to the long form of the Periodic Table disappear when it is regarded as a very successful broad generalization about properties, based on the electronic structures of the elements but not reflecting them in every detail.

A number of special names are given to particular sections of the Periodic Table. There is, unfortunately, some confusion in nomenclature and Table 2.7 lists both the special names approved for general use and some of the cases where conflicting usages appear in text-books. In this book, Groups which do not have a trivial name listed in the table will be named by the lightest element of the group—carbon Group, titanium Group, etc.

Further Properties of the Elements

In this section a number of important atomic properties are defined and discussed.

2.13 Ionization potential

If sufficient energy is available, it is possible to detach one or more electrons from an atom, molecule, or ion. The minimum amount of energy required to remove one electron from a gaseous atom, leaving both the electron and the resulting ion without any kinetic energy, is termed the *ionization potential*. Since energy has to be provided to remove the electron against the attraction of the nucleus, ionization potentials are always positive. The energies required to remove the first, second, third, etc., electrons from an atom are its first, second, third, etc., ionization potentials. It is clear that the successive ionization potentials will increase in size as it becomes increasingly difficult to remove further electrons from the positively charged ions. Ionization potentials of molecules and ions are defined in a similar way to those for atoms.

The ionization potentials of atoms reflect the binding energies of their outermost electrons and will be lowest for those elements where the valency electrons have just started to enter a new quantum level. If Table 2.8 of ionization potentials and Figures 7.5 and 7.6 are examined, it will be seen that the lowest first ionization potentials are shown by the alkali metals where the last electron has entered a new quantum shell and has its main probability density markedly further out from the nucleus than the preceding electrons. As a quantum shell fills, on going across a Period from the alkali metal, the outermost electrons become more and more tightly bound and the ionization potentials rise to a maximum at the rare gas element where that quantum shell is completed. The stability of the completed shell is also shown when the successive ionization potentials of any particular element are examined. As electrons are removed from a partly-filled shell the successive ionization potentials rise steadily, but there is a great leap in the energy required to remove an electron when all the valency electrons have been removed and the underlying complete quantum shell has to be broken. For example, it requires 518 kJ mol^{-1} to remove the single 2s electron from lithium but, when the closed $1s^2$ group has to be broken in order to remove a second electron, the second ionization potential shoots up to 7280 kJ mol^{-1}. Similarly, the successive ionization potentials of aluminium are:

1st	Al[Ne]$(3s)^2(3p)^1$	$\rightarrow$	Al$^+$[Ne]$(3s)^2$ needs 577 kJ mol^{-1}
2nd	Al$^+$[Ne]$(3s)^2$	$\rightarrow$	Al^{2+}[Ne]$(3s)^1$ needs 1815 kJ mol^{-1}
3rd	Al^{2+}[Ne]$(3s)^1$	$\rightarrow$	Al^{3+}[Ne] needs 2740 kJ mol^{-1}
but 4th	Al^{3+}[Ne]	$\rightarrow$	Al^{4+}[He]$(2s)^2(2p)^5$ needs 11 590 kJ mol^{-1}

Other examples are shown in the Table of ionization potentials. For all atoms the energies required to remove electrons from filled shells below the valence shell are so great that they cannot be provided in the course of a chemical reaction, and ions such as Li^{2+} or Al^{4+} are found only in high energy discharges. Thus the energy gap between

TABLE 2.7 Nomenclature of the elements

General sections of the Periodic Table

Main Group elements (or Typical elements or Representative elements)	Those elements with the outermost electrons in the s or p levels—the Groups headed by Li, Be, B, C, N, O, F, and He plus H. Sometimes the two lighter elements in each group are excluded.
Transition elements	Those elements where the d or f levels are filling. On the fuller definition of Main Group elements, this class then includes the rest of the elements.
Inner Transition elements	Those elements where an f shell is filling. With this usage, 'transition elements' is then confined to elements of the d block.
A and B subgroups	This terminology is derived from the older short form. The Main and Transition elements were divided into subgroups, A and B. In Groups I and II the Main Group elements (the alkali and alkaline earth metals respectively) were termed A and the transition elements (copper and zinc Groups) were termed B. But in the other Groups conflicting usages exist; in one system, the Main Group was termed A throughout while in the other, more common one, it was termed B. Thus Group IVB might be the carbon Group of Main elements or the titanium Group of Transition elements depending on the author. The A/B usage is slowly disappearing but until it dies out completely, the reader has to check which convention any particular author is using.
M (Main) and T (Transition) Subgroups	This is an unambiguous nomenclature recently brought in to replace the A/B one. It is certainly to be preferred but is not yet widely used.

Trivial names for Groups of elements

Alkali metals*	Li, Na, K, Rb, Cs, Fr	
Alkaline earth metals*	Ca, Sr, Ba, Ra	
Chalcogens* or Calcogens	O, S, Se, Te, Po	
Halogens*	F, Cl, Br, I, At	
Inert gases* or rare or noble gases	He, Ne, Ar, Kr, Xe, Rn	
Rare earth elements*	Sc, Y, La, to Lu inclusive	These divisions are not usually strictly observed
Lanthanum series*	La to Lu inclusive	and these three names tend to be used inter-
Lanthanides*	Ce to Lu inclusive	changeably.
Actinium series*	Ac onwards	
Actinides*	Those elements where the $5f$ shell is being filled (see Chapter 11).	
Transuranium elements*	The elements following U	
Coinage metals	Cu, Ag, Au	
Platinum metals	Ru, Os, Rh, Ir, Pd, Pt	
Noble metals	An ill-defined term applied to the platinum metals, Au, and sometimes includes Ag, Re, and even Hg.	
Metal and non-metal	These two terms are widely used but it is not clear precisely where the boundary between them comes. The term metalloid or semi-metal is often applied to elements of intermediate properties such as B, Si, Ge, or As.	

---

*These names are approved by the 1957 International Union of Pure and Applied Chemistry Rules for Nomenclature of Inorganic Chemistry.

the valence shell and the underlying filled shell, which is reflected in the successive ionization potentials, puts an effective upper limit on the valency of an element and this, of course, follows from the electronic structures of the elements as derived from the wave equation in the early part of this chapter.

2.14 Electron affinity
The electron affinity is the energy of the reverse process to

ionization: the uniting of an electron with a gaseous atom or ion or molecule. The energy change in this process is the electron affinity of the species.* Electron affinities are difficult to measure experimentally and only a few have been directly

*The convention used here is the accepted thermodynamic one of endothermic changes being positive and exothermic changes negative, the opposite convention may be found in older texts.

TABLE 2.8 Ionization potentials of the elements

Element   Ionization potentials (electron volts, 1 eV = 96·484 kJ mol^{-1})

Element	1st	2nd	3rd	4th	5th	6th	7th	8th
H	13·6							
He	24·6	54·4						
Li	5·39	75·7						
Be	9·32	18·2	154					
B	8·30	25·1	37·9	259				
C	11·26	24·4	47·9	64·5	392			
N	14·5	29·6	47·4	77·5	97·9	552		
O	13·6	35·1	54·9	77·4	114	138	739	
F	17·4	35·0	62·6	87·2	114	157	185	954
Ne	21·6	41·0						
Na	5·14	47·3						
Mg	7·64	15·0	80·1					
Al	5·98	18·8	28·4	120				
Si	8·15	16·3	33·5	45·1	167			
P	11·0	19·7	30·2	51·4	65·0	220		
S	10·4	23·4	35·0	47·3	72·5	88·0	281	
Cl	13·0	23·8	39·9	53·5	67·8	96·7	114	348
Ar	15·8	27·6						
K	4·34	31·8						
Ca	6·11	11·9	51·2					
Sc	6·56	12·8	24·8	73·9				
Ti	6·83	13·6	29·5	43·2	99·8			
V	6·74	14·7	29·3	48·7	65·2	129		
Cr	6·76	16·5	31·0	49·6	73·2	90·3	161	
Mn	7·43	15·6	33·7	53·6	76·3	98	119	
Fe	7·90	16·2	30·6	57·1	78	102	128	151
Co	7·86	17·1	33·5	53	83·5	106	132	161
Ni	7·63	18·2	35·2	56	78	110	136	166
Cu	7·72	20·3	36·8	59	82	106	140	169
Zn	9·39	18·0	39·7	62	86	112	142	177
Ga	6·00	20·4	30·6	63·8				
Ge	7·88	15·9	34·2	45·7	93·0			
As	10·5	20·3	28·0	49·9	62·5	127		
Se	9·75	21·3	33·9	42·7	72·8	81·4		
Br	11·8	19·1	25·7	50·3				
Kr	14·0	26·4						
Rb	4·18	27·4						
Sr	5·69	11·0						
Y	6·6	12·3	20·4					
Zr	6·95	14·0	24·1	34·0	83			
Nb	6·77	13·5	28·1	38·3	49·5	103		
Mo	7·18	15·2	27·1	40·5	56	72	125	
Tc	7·45	15·3	29	43	59	76	94	162
Ru	7·36	16·4	28·5	46·5	63	81	100	119
Rh	7·46	18·1	31·0	45·6	67	85	105	126
Pd	8·33	19·9	33·4	49·0	66	90	110	132
Ag	7·57	22·0	39·7	52	70	89	116	139
Cd	8·99	16·8	38·0					
In	5·79	18·8	27·9	57·8				
Sn	7·33	14·6	30·5	39·6	80·7			
Sb	8·64	18·6	24·7	44·0	55·5			
Te	9·01	21·6	30·7	37·7	60·5	(72)		
I	10·4	19·1	31·5					
Xe	12·1	21·1						
Cs	3·89	23·4						
Ba	5·21	10·0						
La	5·61	11·4	19·2					
Hf	5·5	14·9	21	31				
Ta	7·88	16·2	22·3	33·1	45			
W	7·98	14·0	24·1	35·4	48	61		
Re	7·87	13·1	26·0	37·7	51	64	79	
Os	8·7	15	25	40	54	68	83	99
Ir	9·2	16	27	39	57	72	88	104
Pt	9·0	18·7	28·5	41·1	55	75	92	109
Au	9·22	20·1	30·5	43·5	58	73	96	114
Hg	10·4	18·7	34·3					
Tl	6·11	20·3	29·7	50·5				
Pb	7·41	15·0	32·0	42·3	69·4			
Bi	8·5	16·8	25·4	45·5	55·7			
Ra	5·28	10·1						
Th			29·4					

determined with accuracy. Zollweg has recently compiled a tabulation of first electron affinities corresponding to

$$M_{(g)} + e^- = M_{(g)}^-$$

by examining measured and interpolated values normalised to the accurately known first electron affinities of elements like the halogens and oxygen. These first electron affinities are listed in Table 2.9.

The stability of the rare gas configuration is reflected in the high electron affinities of the halogens which are forming the anion with the rare gas electronic structure. Conversely, the electron affinities for the rare gases, where the extra electron in the anion starts a new quantum shell, are all endothermic.

While most of the first electron affinities, shown in Table 2.9, are exothermic, the second electron affinities for elements forming doubly charged anions are always large and positive (i.e. energy has to be provided to add the second electron) with the result that the formation of doubly (or higher) charged anions requires the net addition of energy. For example:

$$O_{(gas)} + 2e^- = O^{2-}_{(gas)}$$

requires 703 kJ mol^{-1}. Such values are usually derived indirectly from the Born-Haber cycle (see section 4.4).

It should be noted that the largest exothermic electron affinity, that of chlorine, is smaller than the ionization potential of cesium which is the least endothermic of any atom. The result is that any electron transfer between a pair of atoms to form a pair of ions is an endothermic process:

$$Cl_{(gas)} + Cs_{(gas)} = Cs^+Cl^-_{(gas)} \quad \Delta H = +12\cdot1 \text{ kJ mol}^{-1}$$

and usually a considerable amount of energy is required:

$$e.g. \quad I_{(gas)} + Na_{(gas)} = Na^+I^-_{(gas)} \quad \Delta H = +180 \text{ kJ mol}^{-1}$$

It follows that the formation of an ionic compound from its component elements occurs only because of the additional energy provided by the electrostatic attractions between the ions in the solid. This is further discussed in Chapter 4.

2.15 Atomic and other radii

As the decrease in the probability density distribution of an electron on going outwards from the nucleus follows an exponential function (section 2.6), it never exactly equals zero and there is therefore no unambiguous definition of an atomic radius, even of an isolated atom. In a molecule or in a solid or other chemical environment, the electrons are subject to the fields of all the neighbouring atoms and their distribution will depend on the detailed chemical environment. This is to say that, even if it is possible to define the radius of an isolated atom (say as the radius of the 95 per cent contour), the radius of this atom in combination will be different and will differ from one compound to another, although in most cases this latter variation appears to be relatively small. Furthermore, only the distances between atoms (bond lengths) can be measured experimentally and

TABLE 2.9 Electron affinities of the elements (after Zollweg) (kJ mol^{-1})
(Exothermic changes are taken as negative)

1	2	3	4	5	6	7	8	9	10	11	12	13	14	15	16	17	18
H																	He
−74·5																	+21·2
Li	Be											B	C	N	O	F	Ne
−59·8	−36·7											−17·3	−122·3	+20·1	−141·3	−337·4	+28·9
Na	Mg											Al	Si	P	S	Cl	Ar
−52·2	+21·2											−19·3	−131	−68·5	−196·8	−349·2	+35·7
K	Ca	Sc	Ti	V	Cr	Mn	Fe	Co	Ni	Cu	Zn	Ga	Ge	As	Se	Br	Kr
−45·4	+186	+70·5	+1·93	−60·8	−93·5	+93·5	−44·5	−102	−156	−173	−8·7	−35·3	−139	−103	−203	−324·1	+40·5
Rb	Sr	Y	Zr	Nb	Mo	Tc	Ru	Rh	Pd	Ag	Cd	In	Sn	Sb	Te	I	Xe
(−37·6)	+145	+38·6	−43·5	−109	−114	−95·5	−145	−162	−98·5	−193	+26·1	−19·3	−99·5	−90·5	−189	−295·2	+43·5
Cs	Ba	La	Hf	Ta	W	Re	Os	Ir	Pt	Au	Hg	Tl	Pb	Bi	Po	At	Rn
(−36·7)	+46·4	−53·1	+60·8	−14·4	−119	−36·7	−139	−190	−247	−270	+18·6	−30·4	−99·5	−91·5	−127	−270	

the radii of the atoms forming the bond have to be deduced from these bond lengths.

Although the atomic radius is not an exact concept, it is possible to compile sets of atomic radii which reproduce most of the observed interatomic distances to within ten per cent or so. These sets of atomic radii are valuable, as any marked discrepancy between the observed bond length and that calculated from the atomic radii suggests that there is some change in the type of bond or some other effect which should be investigated further. Moreover, when working with closely related compounds the bond lengths should agree much more closely than to ten per cent so that quite small discrepancies are meaningful and worth further study. A number of sets of values for radii are required depending on the nature of the bonding—covalent, ionic, or metallic.

A set of covalent radii may be derived by starting from the experimentally-measured bond lengths in the elements. If these bond lengths are divided by two they give reasonable values for the radii of the atoms, and then the atomic radii of elements which do not form single bonds in the elemental state may be deduced from the bond lengths in suitable compounds with elements of known radii. A few simple examples of the process of building up a set of atomic radii are shown below.

Element	Bond length, pm	Atomic radius, pm
F_2	142	F = 71
Cl_2	199	Cl = 99
Br_2	228	Br = 114
I_2	267	I = 134
C (diamond)	154	C = 77

Molecule		Found	Bond length, pm Calculated from above	Difference
CF_4	C – F	132	148	16
CCl_4	C – Cl	177	176	1
CBr_4	C – Br	191	191	0
CI_4	C – I	214	211	3

The agreement between experimental and calculated values is excellent for the heavier halogens but less good for fluorine. It is found from a wide number of fluorine compounds that better general agreement with experiment is found if the

atomic radius of fluorine is taken as F = 64 pm. This is a purely empirical correction chosen to give the best fit with experimental data. In a similar way, the value used for the hydrogen radius is H = 29 pm although the bond length in the H_2 molecule is 74 pm. With these and similar empirical adjustments to the experimental values the Table of atomic radii shown in Table 2.10a was built up.

Most values calculated from Table 2.10a will agree to within 20–30 pm with the experimental bond lengths. The discrepancies are often wider when hydrogen or fluorine are involved as is illustrated by the measured values shown in Table 2.10b for the halides of the carbon Group elements.

TABLE 2.10a Atomic radii in covalent molecules, pm

Be	B	C	N	O	F	H
89	80	77	70	66	64	29
	Al	Si	P	S	Cl	
	126	117	110	104	99	
Zn	Ga	Ge	As	Se	Br	
131	126	122	121	117	114	
Cd	In	Sn	Sb	Te	I	
148	144	140	141	137	133	
Hg	Tl	Pb	Bi			
148	147	146	151			

	B	C	N	O
Double bond radii	71	67	62	62
Triple bond radii	64	60	55	

TABLE 2.10b Bond lengths of halides of the heavier elements of the carbon group, pm

	Silicon	Germanium	Tin
MF_4	154	167	?
MCl_4	201	208	231
MBr_4	215	231	244
MI_4	243	250	264

A self-consistent and semi-empirical set of values of this type is the best that can be done with a single set of figures for atomic radii. A number of suggestions have been made for modifying the calculated bond length to allow for environmental effects. One example is the Schomaker-Stevenson correction which allows for the polarity of the bond. This is:

$$r_{A-B} = r_A + r_B - 0 \cdot 09|x_A - x_B|$$

where r_{A-B} is the bond length, r_A and r_B are the covalent radii, and x_A and x_B are the electronegativities (see section 2.16). This formula does improve the agreement between calculated and experimental values in many cases, especially for fluorides, but the discrepancies are still significant and it is probably better to accept the purely empirical nature of the atomic radius and seek for other evidence to establish the existence of special effects within the bond.

The discussion above applies only to single bonds. When double or triple bonds are present, the bond length is shortened and appropriate values of the atomic radius must be used. For example, in ethylene the $C=C$ distance is 135 pm and in acetylene $C\equiv C$ is 120 pm, corresponding to a double bond radius for carbon of 67 pm and a triple bond radius of 60 pm. Approximate values for other double and triple bond distances may be calculated by using the radii which are given in the last two lines of Table 2.10a. It will be noticed that the variations in bond lengths due to multiple bonding are larger than the uncertainties associated with the empirical nature of the set of atomic radii, but not by a very great margin. This means that attempts to deduce the bond order from variations in the bond length are legitimate but should be treated with some reserve unless closely similar compounds are being discussed.

In addition to the covalent radius just discussed, a further, much larger radius called the Van der Waals' radius is characteristic of atoms in covalent compounds. This radius represents the shortest distance to which atoms which are not chemically bound to each other will approach before repulsions between the electron clouds come into play. The Van der Waals' radius therefore governs steric effects between different parts of a molecule. Some values of Van der Waals' radii are shown in Table 2.11.

TABLE 2.11 Van der Waals' radii, pm

H	N	O	F
120	150	140	135
	P	S	Cl
	180	190	185
	As	Se	Br
	200	200	195
	Sb	Te	I
	220	220	215

The ionic radii are a set of empirical self-consistent parameters which apply to ionic compounds in the same way as the covalent radii apply to molecular compounds. The ionic radii also vary slightly with the environment of the ion, and much larger variations are found for the hydride ion, H^-, which is fully discussed in Chapter 8. It is more difficult, however, to devise a set of ionic radii as there is no obvious way of dividing the observed internuclear distances, r_{MX}, in ionic compounds into cationic (r_+) and anionic (r_-) radii. This is in contrast to the case of covalent or metallic radii where there are distances between like atoms which can be divided into two.

One approach was to assume that in a compound with a large anion and a small cation, such as LiI, the anions would be in contact. Then half the I–I distance equals the radius of I^-. This value is then used, in compounds with larger cations such as NaI, KI, etc., to calculate cation radii for Na^+, K^+ etc., and these in turn allow the calculation of the radii of other anions such as Cl^- or O^{2-}. Finally, the radius of Li^+ is derived from some compound with a small anion such as LiCl or Li_2O. This method, with further refinements, was used to compile the set of empirical ionic radii shown in Table 2.12, due to Goldschmidt.

The value of 145 pm for oxygen is quoted in this set although the lower values of 140 pm or 135 pm are usually more compatible with transition metal values.

An alternative approach, used by Pauling, was to assume that the radii of isoelectronic ions, such as Cl^- and K^+, varied inversely as their effective nuclear charge. This then gave a way of dividing the experimentally observed M–X distances and allowed a different set of internally consistent ionic radii to be built up. Either the Pauling or the Goldschmidt radii allow a reasonable prediction of interatomic distances in crystals but, of course, the two sets of values must not be mixed.

In four cases, an experimental value for certain ionic radii has been derived from detailed x-ray analysis of solids. When the electron density arising from the inner electrons is subtracted out, analysis of the remaining electron distribution gives the position of minimum electron density along the line joining cation and anion, and thus the ionic radii. Unfortunately, this electron density map of the valency electrons also contains any experimental uncertainties (such as those arising from thermal motion) so that the precise values of the electron density minima may be open to challenge. The 'experimental' ionic radii which have so far been determined are

$$Li^+ = 92 \text{ pm}, \quad F^- = 109 \text{ pm in LiF}$$
$$Na^+ = 118 \text{ pm}, \quad Cl^- = 164 \text{ pm in NaCl}$$
$$Mg^{2+} = 102 \text{ pm}, \quad O^{2-} = 109 \text{ pm in MgO}$$
$$Ca^{2+} = 126 \text{ pm}, \quad F^- = 110 \text{ pm in CaF}_2$$

It will be seen that these values are internally self-consistent, and they add up to give good agreement with other experimental interatomic distances (e.g. $Ca^{2+} + O^{2-} = 240$ pm: experimental value in CaO = 240 pm). However, the values derived from the electron density maps make cations markedly larger, and anions smaller, than the values of either the Goldschmidt or the Pauling sets of ionic radii. Ladd (see references in Appendix A) has proposed a modified set of ionic radii giving weight to experimental electron densities and has shown that these may be used to calculate consistent heats of hydration of ions. Ladd's values are shown in Table 2.12.

TABLE 2.12 Ionic radii, pm

Ion	Radius (after Goldschmidt)	Radius (Ladd)
Li^+	68	86
Na^+	98	112
K^+	133	144
Rb^+	148	158
Cs^+	167	184
Mg^{2+}	65 (Be^{2+} = 30)	87
Ca^{2+}	94	118
Sr^{2+}	110	132
Ba^{2+}	129	149
O^{2-}	145 (135)	125
S^{2-}	190	170
Se^{2-}	202	181
Te^{2-}	222	197
F^-	133	119
Cl^-	181	170
Br^-	195	187
I^-	216	212
Cu^+	96	
Ag^+	126 (Au^+ = 137)	127
Zn^{2+}	83	
Cd^{2+}	103 (Hg^{2+} = 112)	114
Tl^+	149	154

(a) Goldschmidt values for first row transition elements are M^{2+} about 80 pm, M^{3+} about 55 pm.
(b) For lanthanide ions see Table 10.1.
(c) For eight-coordination multiply Goldschmidt values by 103 pm and for four-coordination, by 95 pm.
(d) Ladd gives Mn^{2+} = 93 pm, Fe^{2+} = 90 pm and Co^{2+} = 88 pm.

A set of metallic radii may be compiled in a similar manner.

When dealing with all these sets of radii it is essential to keep in mind that the experimentally determined quantities are the inter-atomic or inter-ionic distances which can be measured to high accuracy (usually to within a few tenths of a picometre). The values listed for the atomic or ionic radii are empirical and chosen to give the best fit over the widest range of experimental data. As a result, small deviations between calculated and measured values (of up to 5 pm or so) are significant only if very critically examined. Larger differences (of the order of 10 pm) suggest the presence of abnormal bonding, either multiple bonds or strong polarization effects in covalent compounds, or polarization and covalent contributions in ionic compounds.

2.16 Electronegativity

One parameter which is widely used in general discussion of the chemical character of an element is its electronegativity. This is defined as *the ability of an atom in a molecule to attract an electron to itself*. There is no direct way of measuring this ability though a number of indirect methods have been suggested, such as the proposal of Mulliken who defined the electronegativity of an atom as the average of its electron affinity and ionization potential (as the electron affinity is a measure of the tendency of the atom to gain an electron, and the ionization potential indicates its tendency to lose an electron). This is the most fundamental of a number of proposed definitions of electronegativity, but it can be applied only to those few elements whose electron affinities are known. Electronegativities have been defined in terms of a number of other parameters including bond energies, by Pauling, and in terms of nuclear magnetic resonance properties in a treatment by Allred and Rochow. Pauling's values have been widely used for many years but those proposed recently by Allred and Rochow are rather more extensive, and a rounded-off set of these values is shown in Table 2.13.

TABLE 2.13 Electronegativity values after Allred and Rochow

Values are rounded off to the nearest 0·05. The base is taken as H = 2·1 to fit with Pauling's values.

Li	Be											B	C	N	O	F
0·95	1·5											2·0	2·5	3·05	3·5	4·1
Na	Mg											Al	Si	P	S	Cl
1·0	1·25											1·45	1·75	2·05	2·45	2·85
K	Ca	Sc	Ti	V	Cr	Mn	Fe	Co	Ni	Cu	Zn	Ga	Ge	As	Se	Br
0·9	1·05	1·2	1·3	1·45	1·55	1·6	1·65	1·7	1·75	1·75	1·65	1·8	2·0	2·2	2·5	2·75
Rb	Sr	Y	Zr	Nb	Mo	Tc	Ru	Rh	Pd	Ag	Cd	In	Sn	Sb	Te	I
0·9	1·0	1·1	1·2	1·25	1·3	1·35	1·4	1·45	1·35	1·4	1·45	1·5	1·7	1·8	2·0	2·2
Cs	Ba	La	Hf	Ta	W	Re	Os	Ir	Pt	Au	Hg	Tl	Pb	Bi	Po	At
0·85	0·95	1·1	1·25	1·35	1·4	1·45	1·5	1·55	1·45	1·4	1·45	1·45	1·55	1·65	1·75	1·95
Fr	Ra	Ac														
0·85	0·95	1·0														

Lanthanides range from 1·0 to 1·15

Actinides range from 1·1 to 1·2

There has been a great deal of discussion, argument, and often confusion, about the significance of electronegativity values, largely because various authors have attempted to press the concept too far. The electronegativity is extremely valuable as a brief summary, within one parameter, of the general chemical behaviour of an atom but it must be used in a general way and little significance attaches to small differences in values between two atoms. The most electro-negative elements occur in the top right-hand corner of the Periodic Table and electronegativity falls on going down a Group towards the heavier elements or on going to the left along a Period towards the alkali metals.

Electronegativities are most useful in the guidance they give to the electron distribution in a bond. In a bond $A-B$ between two atoms, the electron density in the bond may lie evenly between the two atoms or be concentrated more towards one atom, say B, than towards the other, when the bond is said to be *polarized*. In the limiting case, when the electron density of the bonding electrons is entirely on B, an electron has been fully transferred from A to B and an ionic compound, A^+B^-, forms. The electron density distri-bution in the bond may be predicted from the electronega-tivities of A and B. If A and B have the same electronegativi-ties, it follows from the definition that A and B attract the electrons in the bond equally and no polarization results. If B is more electronegative than A, its attraction for the bond electrons is the stronger and polarization results, the degree of polarization being proportional to the difference in electronegativity. A large electronegativity difference favours the formation of ions and, as a rough guide, an ionic com-pound forms between A and B if they differ in electro-negativity by more than two units. Thus elements with very high or very low electronegativities are more likely to form ionic compounds than those with intermediate values.

The electronegativity of an element increases with its oxidation state—for example, manganese in permanganate is more electronegative than as the Mn^{2+} ion—and it also depends on the other atoms attached to the one in question. Thus, carbon in H_3C-X is less electronegative than carbon in F_3C-X, as the highly electronegative fluorine atoms in the trifluoromethyl compound remove more electron density from the carbon in the $C-F$ bonds than do the hydrogen atoms in the $C-H$ bonds of the methyl compounds. As a result, the carbon atom in F_3C-X has more tendency to attract the electrons in the $C-X$ bond than has the carbon atom in H_3C-X. It follows that the electronegativity values given in Table 2.13 represent the behaviour of the elements in an 'average' chemical environment and the effective electronegativity of an element in any particular compound depends in detail on its environment.

2.17 Coordination number, valency, and oxidation state

The three terms, coordination number, valency, and oxidation state, are used to describe the environment and chemical state of an atom in a compound. The three overlap somewhat in meaning and application, but the use of each has advantages in certain circumstances.

The simplest term to describe an atom in a compound is its coordination number, which is the number of nearest neighbours to the given atom, whatever the bonding between them. The coordination number is a purely empirical pro-perty of the element determined from the structure of the compound. This simplicity is the main advantage in the use of the term, as a compound may be described by the co-ordination numbers of its constituent atoms, however diffi-cult it may be to determine the bonding between these atoms.

When more information about the atom is required, the valency or the oxidation state must be determined. Valency is a familiar term and need not be described in detail. Basically, it describes the bonding of the atom and it is a theoretical term whose use demands more than the experi-mentally-determined properties of the compound in question. This problem is often disguised by familiarity, but it arises in an acute form in the many cases where a compound or class of compounds is discovered and the structures determined long before an adequate theoretical description of the bond-ing, and hence the valency, is available. One example is nickel carbonyl, $Ni(CO)_4$, which was known for many years before there was an adequate theory of its bonding. Its structure has been written at various times with the nickel-carbon monoxide bond as $Ni=C=O$, $Ni-C\equiv O$, $Ni\leftarrow C\equiv O$ and $Ni\rightleftharpoons C\equiv O$ implying that the nickel, respectively, eight-, four-, zero- and zero-valent.

Apart from this type of problem, the valency nomencla-ture is sometimes clumsy (just because it gives a more com-plete picture of the molecule). For example, cobalt in the ion $[Co(NH_3)_6]^{3+}$ has to be described as having a covalency of six and an electrovalency of three. There are also occasions when the term valency conceals differences in properties. An example is given by ammonia, NH_3, and nitrite ion, NO_2^-. In both compounds the nitrogen atom is properly described as trivalent and yet it has to be oxidized to pass from one compound to the other, and it is more useful in some con-texts to discuss ammonia and related compounds such as the amines, R_3N, separately from the nitrites and other trivalent oxy-compounds.

Considerations such as the above, led to the introduction of a narrower, more empirical term, oxidation number (or oxidation state). The oxidation number of an element in a compound may be simply determined from a number of empirical rules and it is quite independent of the nature of the bonding. Obviously, it gives less information about the chemical state of the element than does an accurate descrip-tion in terms of valency but it is useful and convenient when that extra information is not required or available.

The oxidation number of an atom in a compound is defined by the following rules:

(i) The oxidation number of an atom in the element is zero.

(ii) The oxidation number of an atom in an ionic compound is equal to the charge on that atom (with the sign).

(iii) The oxidation number of an atom in a covalent com-pound is equal to the charge which it would have in the most probable ionic formulation of the compound.

The first two rules are perfectly clear but a little experience is required to find the artificial ionic form required by rule (iii). The electronegativities of the elements in the compound usually serve to make the most probable ionic formulation clear, as illustrated by the examples given below:

Compound	More electro-negative element	Ionic formulation	Oxidation numbers
BCl_3	Cl	$B^{3+}Cl_3^-$	B = III, Cl = $-$I
SO_2	O	$S^{4+}O_2^{2-}$	S = IV, O = $-$II
NH_3	N	$N^{3-}H_3^+$	H = I, N = $-$III
NH_4^+	N	$[N^{3-}H_4^+]^+$	H = I, N = $-$III
NO_2^-	O	$[N^{3+}O_2^{2-}]^-$	N = III, O = $-$II
CrO_4^{2-}	O	$[Cr^{6+}O_4^{2-}]^{2-}$	Cr = VI, O = $-$II
$Cr_2O_7^{2-}$	O	$[Cr_2^{6+}O_7^{2-}]^{2-}$	Cr = VI, O = $-$II

Notice, in the last column, that the sum of the oxidation numbers of the atoms equals the overall charge on the species. Although atoms may be shown with large charges, e.g. Cr^{6+} or S^{4+}, this by no means implies the existence of such unlikely ions. To make this clear, it is usual to indicate the oxidation state by Roman numbers—Cr(VI) or S(IV).

In nearly all compounds, rules (i) to (iii) are equivalent to taking O = $-$II (except in peroxides), H = $+$I (except in ionic hydrides), and halogens = $-$I (except in their oxygen compounds).

Oxidation and reduction are very simple to define in terms of oxidation numbers. Oxidation is any process which increases the oxidation number of an element while reduction corresponds to a decrease in the oxidation number. For example, the conversion of ammonia to nitrogen involves an increase from $-$III to zero in the oxidation number of the nitrogen and is an oxidation by three steps, the conversion of ammonia to nitrite involves an oxidation by six steps, while the conversion to nitrate involves a change of eight steps to nitrogen (V). On the other hand, the change from ammonia, NH_3, to ammonium ion, NH_4^+, involves no change in the oxidation numbers and is not an oxidation. The same applies to the change from chromate to dichromate which sometimes causes trouble in analytical calculations. Further examples are provided by the range of nitrogen compounds below:

Oxidation number of the nitrogen	Examples
$-$III	NH_3 or NH_4^+
$-$II	N_2H_4
$-$I	NH_2OH
0	N_2
I	N_2O or $N_2O_2^{2-}$
II	NO
III	N_2O_3 or NO_2^-
IV	N_2O_4
V	N_2O_5 or NO_3^-

In complex ions, if the ligand is a neutral molecule like ammonia in $[Co(NH_3)_6]^{3+}$ or water in $[Cu(H_2O)_4]^{2+}$, the metal has an oxidation number equal to the charge, Co(III) and Cu(II) respectively. Similarly, nickel in nickel carbonyl, $Ni(CO)_4$, has an oxidation number of zero. If the ligand is charged, then the oxidation number of the metal must balance with the total charge on the ion: Fe(II) in ferrocyanide, $[Fe(CN)_6]^{4-}$, Fe(III) in ferricyanide, $[Fe(CN)_6]^{3-}$, or Co(III) in $[Co(NH_3)_3Cl_3]$ and in $[CoF_6]^{3-}$.

The use of oxidation numbers simplifies the calculations involved in oxidation-reduction titrations. In the overall reaction, the change in oxidation state of the reductant must balance that of the oxidant. The reaction stoichiometry is thus readily worked out from the oxidation state changes of the reactants. A full account of the method is given in the standard analytical textbooks but the following examples may provide some illustration of the approach.

The oxidation of arsenite by permanganate in acid solution

$$MnO_4^- + AsO_3^{3-} \text{ to } Mn^{2+} + AsO_4^{3-}$$

The manganese change is from MnO_4^-, where the Mn = VII to Mn^{2+} with Mn = II; change in manganese oxidation state = -5.

The arsenic change is from AsO_3^{3-}, where the As = III to AsO_4^{3-} with As = V; change in arsenic oxidation state = $+2$.

The reaction stoichiometry is therefore:

$$2MnO_4^- + 5AsO_3^{3-}$$

The equation may then be balanced by introducing hydrogen ions and water molecules in the usual way to give:

$$2MnO_4^- + 5AsO_3^{3-} + 6H^+ = 2Mn^{2+} + 5AsO_4^{3-} + 3H_2O$$

The reaction between iodate and iodide

$$IO_3^- + I^- \text{ to } I_2$$

In this case, the oxidant and the reductant end up in the same form. In iodate, the iodine is in the V oxidation state so that the change in going from iodate to iodine is by -5. The change in oxidation state from the $-$I in iodide to the element is by $+1$ and the reaction stoichiometry is therefore:

$$IO_3^- + 5I^-$$

The balanced equation is:

$$IO_3^- + 5I^- + 6H^+ = 3I_2 + 3H_2O$$

If this reaction is carried out in concentrated hydrochloric acid, instead of in dilute acid as above, the final product is not iodine but iodine monochloride, ICl, in which the iodine has an oxidation state of $+$I. In this case, the change from iodate to ICl is -4 and the change from iodide is $+2$ so that the balanced equation becomes:

$$IO_3^- + 2I^- + 6H^+ = 3I^+ + 3H_2O$$

As far as most calculations are concerned, only the reaction stoichiometry has to be known and the use of oxidation numbers in the calculation gives this very rapidly and easily.

The oxidation state concept breaks down in those cases where an ionic formulation is ambiguous. One example is in the case of the metal nitrosyls which contain groups, $M - NO$, which could quite validly be formulated in three ways—as $(NO)^+$, $(NO)^-$, or with neutral NO groups. Similar difficulties are encountered in, for example, the hydrides of boron or phosphorus where the electronegativities ($B = 2 \cdot 0$, $P = 2 \cdot 05$, $H = 2 \cdot 1$) are so close that doubts arise whether to write H^+ or H^-: in fact, the hydrogen is negatively polarized in most boron-hydrogen compounds and positively polarized in most phosphorus-hydrogen ones. In organic chemistry, also, the oxidation state concept is not very useful; it is more convenient to discuss reactions such as $CH_4 \rightarrow CH_3Cl$ in terms of substitution rather than in terms of a change in the carbon oxidation state. These three concepts, coordination number, oxidation state, and valency, become less empirical and convey increasing amounts of information in that order.

3 Covalent Molecules

3.1 Introduction

In the last chapter, a picture of the electron structure of atoms was built up from the theory of wave mechanics taken together with experimental data. This knowledge of atomic structure will now be used to examine the process of combining atoms to form molecules. The formation of covalent bonds is discussed in this chapter, while ionic compounds are the subject of Chapter 4.

The basic idea of a covalent bond is that due to Lewis, of two electrons shared between two atoms and binding them together. One way of translating this idea into terms of wave mechanics will be discussed here, though a number of other approaches are possible and these are presented in the references given on page 280.

The simplest of all molecules is the positive ion of the hydrogen molecule, H_2^+. As this molecule has only a single electron moving in the nuclear field, there is only one del-squared term in the kinetic energy part of the wave equation, while the potential energy part contains three terms—the attractions between each nucleus and the electron and the repulsion between the nuclei. The equation is similar to that of a hydrogen-like atom (see equation 2.4), but now the nuclear field is not spherically symmetrical. Although the problem is more difficult than that of the hydrogen atom, the wave equation for the hydrogen molecule ion can be solved exactly and the resulting energy levels and energy of formation agree with the experimental values. Just as with atoms, major difficulties arise in the case of molecules as soon as more than one electron is involved. For example, the presence of the second electron in the hydrogen molecule, H_2, makes it a molecular analogue of the helium atom. In the wave equation for the hydrogen molecule (compare equation 2.7) there will be two del-squared terms in the kinetic energy part describing the two electrons and there are six components of the potential energy term. These are the four attractions between each electron and each nucleus, the repulsion between the nuclei, and finally the inter-electron repulsion term which provides the main source of difficulty in these calculations.

It is clear that such complexities increase rapidly as the number of electrons rises, and methods of approximation

have to be found. Before turning to these, it is of interest to record the results of the more rigorous calculations of electronic energies for some simple molecules. These are shown in Table 3.1.

TABLE 3.1 Calculations of total electronic energies

Molecule	Calculated value, kJ mol^{-1}	Experimental value, kJ mol^{-1}
H_2^-	2 887·2	2 893·8
H_2	3 081·2	3 081·1
CH_4	104 850	106 210
Other diatomic molecules like CO, N_2, or HF	Differences between calculated and experimental values lie in the range 0·5—1·5%	

The total electronic energy of a species is the energy evolved when all its constituent particles (nuclei and electrons) are brought from infinite separation and combined to form the molecule or ion in its equilibrium configuration.

The agreement is very close in these cases and gives grounds for confidence in the general approach. The calculations are very long and complex, however, and many of them require the assistance of a computer. In addition, the quantity most readily derived from the calculation is the total electronic energy of the molecule, while the changes involved in a reaction are dependent on the small differences between such total energies. At present, these total energies cannot be calculated to the very high order of accuracy required for direct predictions of reaction paths or molecular structures. Simplifications must be introduced in order to solve the equations and the over-all theory becomes a semi-quantitative guide. Even in this relatively modest form, the wave mechanical approach to molecular structure has wrought an impressive change in the way in which chemists think about molecules.

In this text, the theory will be used as a general guide, and diagrams rather than calculations will be used to describe the processes of molecule formation. The reader is asked to keep in mind that these diagrams do mirror the calculations and that definite values may be found for the parameters—such as bond lengths and bond energies—which are qualitatively described here. Fuller accounts are given by Coulson or Cartmell and Fowles (see references on page 280).

Diatomic Molecules

One well-known approximation method for the wave equation for molecules is the method of molecular orbitals. In this approach, the aim is to construct orbitals analogous to the atomic orbitals of Chapter 2 but centred on both nuclei. Then the electrons are fed in—two into each orbital in order of increasing energy—to build up the electronic structure of the molecule, just as the electronic structures of the atoms were built up.

The first problem is to find a way of constructing these molecular orbitals. One starting point is to consider that, when the electron in a molecule is close to one nucleus, it is in almost the same environment as in the free atom. This suggests that the molecular orbitals may be derived from some combination of atomic orbitals. The simplest way of combining the orbitals is additively and a simple *linear combination of atomic orbitals* (LCAO) has been widely used.

3.2 The combination of s orbitals

In the case of H_2^+, the molecular orbitals are formed from the linear combinations (3.1) of the atomic orbitals ψ_X and ψ_Y on the two hydrogen atoms, X and Y:

$$\phi_B = \psi_X + \psi_Y$$
$$\phi_A = \psi_X - \psi_Y \quad \dots\dots\dots\dots (3.1)^*$$

Consider the case where the wave functions ψ_X and ψ_Y represent $1s$ orbitals on the two hydrogen atoms. The result of combining these two atomic orbitals into the molecular orbital ϕ_B may be shown diagrammatically, using the curves for the $1s$ orbital which were given in Chapter 2 in Figure 2.8.

In Figure 3.1, the curves giving the complete cross-section through each nucleus, X and Y, of their respective $1s$ orbitals are drawn out and the process of addition to give ϕ_B is indicated to the right. The two nuclei are brought together to a distance equal to the interatomic distance in the molecule, and the atomic electron density curves are summed to give the electron density curve of the molecule. This curve is a cross-section through the two nuclei of the electron distribution in the molecule. This distribution is symmetrical about the molecular axis so that the three-dimensional

---

*In these equations—and in all similar equations such as (3.2)—a fully rigorous discussion requires that the RHS is multiplied by a factor N, called the *normalizing constant*. This is a numerical factor which adjusts the equation so that $\int \phi^2 d\tau = 1$; i.e. so that the probability of finding the electron described by the molecular orbital ϕ somewhere in space is unity (compare the requirement for atomic orbitals in Chapter 2). In this particular case, as the atomic orbitals ψ_X and ψ_Y are also normalized (i.e. $\int \psi_X^2 d\tau = 1 = \int \psi_Y^2 d\tau$) $N = 1/\sqrt{(2)}$ and the full version of equation (3.1) is

$$\phi_B = \frac{1}{\sqrt{(2)}} (\psi_X + \psi_Y).$$

However, as N is a numerical constant, its actual value rarely affects the qualitative discussion of molecular orbitals which we are giving here and we shall normally omit it.

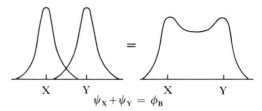

$$\psi_X + \psi_Y = \phi_B$$

FIGURE 3.1 *The formation of the molecular orbital ϕ_B (σ_s)*
The nuclei X and Y are placed at the measured internuclear distance in the molecule and the wave function curves are super-imposed to give the summation curve shown. Rotation of the diagram about the XY axis gives the complete molecular orbital in three dimensions.

electron density is obtained by rotating the figure about the XY axis.

The parts of ϕ_B outside the internuclear region follow an exponential curve and the electron density falls off rapidly on moving away from the nuclei. The orbital ϕ_B may be shown in projection as a contour map as in Figure 3.2a, or in the more convenient boundary diagram of Figure 3.2b where the contour line which encloses 90 or 95 per cent of the electron density is drawn (compare with atomic orbitals as in Figure 2.8c).

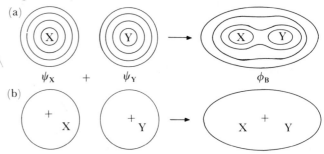

(a)

$$\psi_X \quad + \quad \psi_Y \quad \phi_B$$

(b)

FIGURE 3.2 (a) *Contour representation of ϕ_B (σ_s)*
(b) *Boundary contour representation of ϕ_B (σ_s)*
The contour line enclosing 90% of the electrons density is drawn. Such diagrams represent either the wave function or its square, the probability density function. The wave function diagram is the more useful and shows the sign of the wave function, as here.

It is clear from these diagrams that, in the molecular orbital ϕ_B, there is an accumulation of electron density in the region between the two nuclei. This markedly decreases the repulsion between the two nuclear charges, with the result that the presence of electrons in this molecular orbital holds the two atoms together and a bond is formed. The orbital ϕ_B is termed a *bonding molecular orbital*.

The molecular orbital ϕ_A may be treated in the same way. The combination of atomic $1s$ orbitals ($\psi_X - \psi_Y$) is shown in Figure 3.3, and the contour representation and the boundary line diagram are given in Figure 3.4. In this molecular orbital, ϕ_A changes sign at the mid-point of XY and the electron probability density, ϕ_A^2, falls to zero here across a *nodal plane* perpendicular to the internuclear axis. Thus, if an electron is placed in the orbital ϕ_A, electron density is removed from the region between the nuclei and accumulated on the remote side of the atoms. The internuclear repulsion has full effect and no bond results. ϕ_A is termed an *antibonding molecular orbital*.

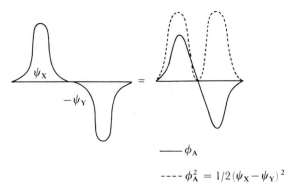

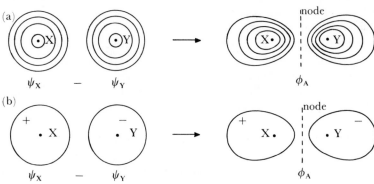

FIGURE 3.4 (a) *Contour representation of* ϕ_A (σ_s^*)
(b) *Boundary contour representation of* ϕ_A (σ_s^*)
(b) shows the molecular orbital wave function which changes sign across the nodal plane perpendicular to the internuclear axis.

FIGURE 3.3 *The formation of the molecular orbital* ϕ_A (σ_s^*)

----$\phi_A^2 = 1/2(\psi_X - \psi_Y)^2$

When the energies, E_B and E_A, of the molecular orbitals ϕ_B and ϕ_A are calculated, it is found that E_B is less than the energy of the constituent atomic 1s orbitals while E_A is greater than the atomic orbital energy by the same amount, ΔE. That is, the molecular orbital ϕ_B is stabilized, and the molecular orbital ϕ_A is destabilized, relative to ψ_{1s}, by the energy ΔE. This may be shown on an energy level diagram as in Figure 3.5. The two atomic orbitals thus combine to form two molecular orbitals, one of which is more stable than the

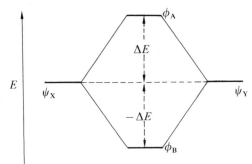

FIGURE 3.5 *Energy level diagram for* ϕ_B *and* ϕ_A (σ_s *and* σ_s^*)
This shows the energies of the molecular orbitals, ϕ_B and ϕ_A, relative to the energies of the constituent atomic orbitals. The atomic orbital energy levels are indicated by the horizontal lines to left and right while the molecular orbitals are shown in the centre of the diagram. Those atomic orbitals which contribute to a particular molecular orbital are connected to it by the finer sloping lines. This convention is followed in all diagrams of this type. The bonding orbital, ϕ_B, is more stable than the atomic orbitals by the same amount of energy as the antibonding orbital, ϕ_A, is less stable than the atomic orbitals.

atomic orbitals while the other is less stable by the same amount of energy. The process of deriving the electronic structure of a molecule can then be formalized in steps similar to those used for atoms. The nuclei are first placed together at the appropriate distance, then molecular orbitals are constructed from the atomic orbitals, and finally the electrons are fed into the molecular orbitals in order of increasing energy. Just as with atomic orbitals, a molecular orbital holds no more than two electrons and, when there are a number of molecular orbitals of equal energy, the electrons enter them singly with parallel spins. The number of mole-

cular orbitals formed must exactly equal the number of atomic orbitals used in their construction.

Although, in the above outline, it has been assumed that the inter-nuclear distance (the bond length) is a known factor—and it will usually be known experimentally—it should be noted that in a full theoretical treatment it is possible to derive the optimum bond length by finding the value which gives the minimum total energy of the system. This has been done for a number of simpler cases with results that agree with the experimentally determined distances.

The formation of molecules can now be followed by using the energy level diagram of Figure 3.5. For the hydrogen molecule ion, H_2^+, the electronic structure is shown in Figure 3.6. The orbital of lowest energy in the molecule is ϕ_B and the electron goes into this, gaining energy, ΔE, relative to its energy in the atom. The whole system, of two hydrogen

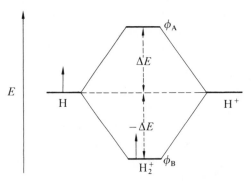

FIGURE 3.6 *The electronic structure of* H_2^+

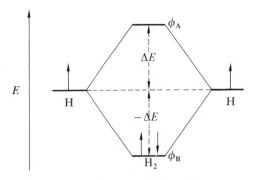

FIGURE 3.7 *The electronic structure of* H_2

nuclei and one electron, is thus more stable by ΔE as the molecule than as separate atoms.

In the hydrogen molecule, H_2, (Figure 3.7) there are two electrons to be considered. These both enter ϕ_B and will have their spins paired. The gain in energy of the molecule over the two isolated atoms is $2\Delta E$ (less a relatively small term due to the interelectron repulsion) and this bond energy, in the hydrogen molecule, is regarded as that of a normal single bond. It follows that the bond in the ion, H_2^+, which has about half the energy of formation, may be regarded as a 'half-bond'. This accords with experiment. H_2^+ exists as a transient species in electric discharges and its bond energy, which may be determined from its spectrum, is about half the energy of H_2.

Next, consider a three-electron molecule such as the positive ion of diatomic helium, He_2^+, whose energy level diagram is shown in Figure 3.8. Since ϕ_B is filled by the first

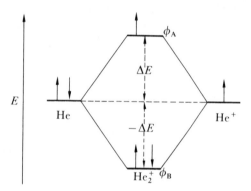

FIGURE 3.8 *The electronic structure of* He_2^+

two electrons, the third must be placed in the antibonding molecular orbital ϕ_A. The energy of formation of He_2^+ is therefore $(2\Delta E - \Delta E)$. The net gain in energy from re-arranging two helium nuclei and three electrons as a molecule is thus ΔE, corresponding to a 'half-bond' again. (It should be remarked that the actual energies both of the atomic orbitals and of the molecular orbitals in helium are different from the energies of the corresponding orbitals in hydrogen, because of the difference in the nuclear charge. The difference between the atomic and molecular orbital energies, ΔE, will however, be of the same order of size.)

Finally, in a four-electron molecule such as He_2, two electrons would be placed in the bonding orbital ϕ_B and two in the antibonding orbital ϕ_A. The energy of formation is $(2\Delta E - 2\Delta E)$ equal to zero and no bond results. In fact, helium exists as a monatomic gas.

An analysis such as this may be extended in an exactly similar way to all other s orbitals with higher n values, and also to the more general case of diatomic molecules where the two atoms are not the same. In this case, the two atomic orbitals which will combine to form the molecular orbital will be of different energies, and the most favourable combination will not be in the 1:1 ratio of equation 3.1, but some more general expression of the form:

$$\phi_B = \psi_1 + c\psi_2 \dots \dots \dots \dots (3.2)$$

will be needed. The value of the mixing coefficient c which

will give the optimum energy has to be found in the course of the calculation. The energy level diagram which corresponds to this more general case is shown in Figure 3.9. Here the bonding orbital is stabilized, and the antibonding orbital destabilized, by equal amounts of energy, ΔE, calculated from the mean energy of the contributing atomic orbitals ψ_1 and ψ_2. The gain in energy when an electron is taken from the more stable of the two atomic orbitals is the smaller amount labelled $\Delta E'$. Clearly, $\Delta E'$ decreases as the energy difference between the atomic orbitals ψ_1 and ψ_2 increases. If $\Delta E'$ is too small no molecule results. It follows that useful molecular orbitals are formed only when the combining atomic orbitals are of similar energy. As a general rule, this

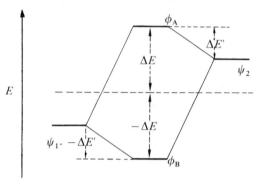

FIGURE 3.9 *Energy level diagram for the combination of s orbitals of unlike atoms*
Stabilization and destabilization is relative to the average energy level of the two atomic orbitals. Note that the differences $\Delta E'$ in energy from the atomic orbitals are also equal.

limitation implies that only orbitals in the valency shells of atoms will combine to form molecular orbitals. Thus, in hydrogen chloride, the hydrogen $1s$ orbital is of too high an energy to combine with $1s$ or $2s$ orbitals on the chlorine atom (whose energy levels are greatly stabilized relative to those of hydrogen by the attraction of the nuclear charge of 17), and it is too stable to interact with the chlorine $4s$, or higher, orbitals. The hydrogen $1s$ orbital is comparable in energy with the chlorine $3s$ or $3p$ orbitals and could form molecular orbitals with these.

3.3 The combination of *p* orbitals

If the molecular axis in a diatomic molecule is taken as the z-axis*, then the p_z orbitals on the two atoms may combine to form molecular orbitals similar in type to those formed by s orbitals. Just as with the s orbitals, the p_z orbitals on the two atoms X and Y combine to form two molecular orbitals, ϕ_1 and ϕ_2, which are similar to those given by equations 3.1:

$$\phi_1 = \psi_X + \psi_Y$$
$$\phi_2 = \psi_X - \psi_Y \quad \dots \dots \dots \dots (3.3)$$

where the atomic orbitals ψ_X and ψ_Y are here the p_z orbitals. The process of combination is illustrated using the boundary

---

*We are adopting here the usual convention that the *unique* direction in a molecule is taken as the z-axis. For example. the molecular axis in a linear molecule or the axis perpendicular to the molecule in a planar species would normally be labelled z.

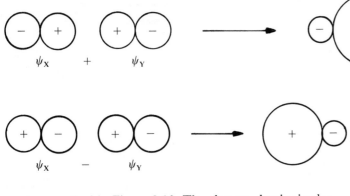

contour method in Figure 3.10. The electron density in the first orbital, ϕ_1^2, which results from the addition of the two atomic p_z orbitals, is concentrated in the region between the two nuclei and this molecular orbital is bonding. In the second orbital, there is a nodal plane between the nuclei and the electron density, ϕ_2^2, is concentrated in the regions remote from the internuclear region, falling to zero between the nuclei. This orbital is therefore antibonding. Thus the combination of the two atomic p_z orbitals gives two molecular orbitals, one of which is bonding and one antibonding, just as in the case of the s orbitals. These molecular orbitals, ϕ_B, ϕ_A, ϕ_1, and ϕ_2, which are formed from s or p_z orbitals are symmetrical around the molecular axis and are termed sigma (σ) molecular orbitals by analogy with the symmetrical s atomic orbitals. To form a systematic nomenclature, subscripts are used to indicate the type of atomic orbital which goes to form the molecular orbital, and the antibonding molecular orbitals are starred. Thus the molecular orbitals discussed so far are named systematically as follows:

$$\phi_B = \sigma_s : \phi_A = \sigma_s^* : \phi_1 = \sigma_p : \phi_2 = \sigma_p^*$$

The formation of sigma molecular orbitals is not restricted to combinations of like atomic orbitals. The only necessary condition is that the two components are symmetrical in sign about the bond axis. Then, if they are of similar energy, they may combine to form a molecular orbital. For example, an s and a p_z orbital may combine to form a bonding sigma molecular orbital as shown in Figure 3.11. The corresponding sigma antibonding orbital also exists.

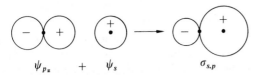

FIGURE 3.11 *The combination of an s and a p_z orbital*
Such an orbital is symmetrical around the internuclear axis and concentrates electron density between the nuclei. The corresponding antibonding orbital may be formed in a similar manner.

Orbitals with different symmetries about the bond axis cannot combine to form molecular orbitals. This is illustrated in Figure 3.12 for the cases of s with p_y and p_z with d_{yz}. It can be seen that the shaded 'plus-plus' and 'plus-minus' areas of overlap cancel each other out.

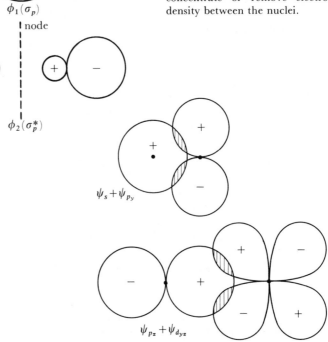

FIGURE 3.12 *Combinations where no bond results*
Orbitals can only combine to give molecular orbitals if both are of the same symmetry with respect to the molecular axis; that is, if the signs of the wave functions change across the axis in the same way.

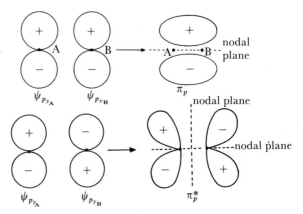

FIGURE 3.13 *The combination of atomic p_y orbitals*
When the z-axis is the molecular axis, the p_y orbitals are antisymmetrical with respect to this axis—that is, the wave function changes sign on crossing the axis. Such orbitals combine to give bonding, π, and antibonding, π^*, orbitals which are also antisymmetrical with respect to the molecular axis. The bonding orbital accumulates electron density between the nuclei, although less effectively than a σ orbital. The antibonding orbital has a nodal plane between the nuclei and perpendicular to the axis.

Although the p_y and p_x orbitals cannot enter into sigma bonding because they are antisymmetric (i.e. change in sign) across the bond axis, they can form a different type of bond by overlapping 'sideways-on' as shown in Figure 3.13.

Such molecular orbitals, which have one nodal plane containing the bond axis, are called pi (π) molecular orbitals—again by analogy with atomic p orbitals. The first combination shown in Figure 3.13 accumulates charge in the region between the nuclei and is therefore bonding. The second mode of combining the atomic p orbitals has a nodal plane lying between the nuclei and perpendicular to that containing the bond axis. This second molecular orbital is antibonding. These are named, systematically, as π_p and π_p^* respectively.

The bonding π_p molecular orbital is of lower energy than the contributing atomic p orbitals, while the antibonding π_p^* molecular orbital is of higher energy by an equal amount. Because of the existence of the nodal plane containing the nuclei in the π orbitals, the π_p interaction has rather less effect on the electron density between the nuclei than has the σ_p interaction, and the energy gap between the bonding and the antibonding π molecular orbitals is less than that between the bonding and antibonding σ orbitals.

Since the p_x and p_y atomic orbitals are both perpendicular to the molecular axis and identical with each other apart from their orientation, they both form π orbitals (with π_{p_x} perpendicular to π_{p_y}) in exactly the same way. The π_x and π_y levels are equal in energy as are the π_x^* and π_y^* levels. The combination of all the p atomic orbitals on two atoms to give the molecular orbitals of a diatomic molecule may therefore be shown on a composite energy level diagram, as in Figure 3.14.

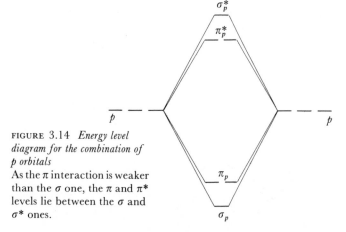

FIGURE 3.14 *Energy level diagram for the combination of p orbitals*
As the π interaction is weaker than the σ one, the π and π^* levels lie between the σ and σ^* ones.

This discussion may readily be extended to include d atomic orbitals which may combine with each other or with s or p orbitals to give σ or π molecular orbitals. It is also possible for two d orbitals to overlap each other with all four lobes to give a molecular orbital with two nodal planes containing the bond axis and mutually at right angles to each other. Such an orbital is termed a delta (δ) molecular orbital, but these are rarely encountered. Some examples of molecular orbitals involving atomic d orbitals are shown in Figure 3.15.

3.4 The formation of diatomic molecules

The discussion of covalent bonding between a pair of atoms, given in the previous sections, may be summarized:

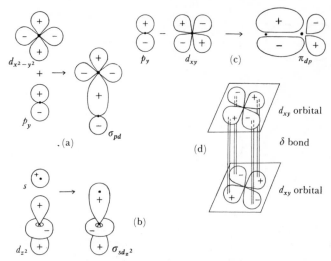

FIGURE 3.15 *Molecular orbitals involving d orbitals*
These diagrams indicate the formation of σ, π, and δ bonds, involving d orbitals. (a) sigma bond between p_y and $d_{x^2-y^2}$, (b) sigma bond between s and d_{z^2}, (c) pi bond between p_y and d_{xy}, (d) delta bond between two d_{xy} orbitals.

(a) Two atomic orbitals may combine to form two molecular orbitals centred on the nuclei, one of which concentrates electrons in the region between the nuclei and is bonding, while the other removes electron density from this region and is antibonding.

(b) Only those atomic orbitals which are of similar energy and of the same symmetry with respect to the inter-atomic axis may combine to form molecular orbitals.

(c) Any atomic orbitals which are symmetrical with respect to the bond axis combine to form sigma molecular orbitals.

(d) Any atomic orbitals which are anti-symmetrical with respect to the bond axis may combine to form pi molecular orbitals.

It is now possible to discuss the electronic structures of some diatomic molecules in the light of these points. The combination of the atomic orbitals in the valency shell may be displayed on a general energy level diagram, as in Figure 3.16, which is formed by combining the energy level diagrams of Figures 3.5 and 3.14.

Consider first the case of the fluorine molecule, F_2. The fluorine atom has the electronic configuration $(1s)^2(2s)^2(2p)^5$, and the inner $(1s)^2$ shell is too tightly held to the nucleus to play any part in the bonding. There are thus seven valency electrons from each atom to be fitted into the molecular orbitals of the fluorine molecule and, when these are filled in order of increasing energy, the arrangement shown in Figure 3.16 results. This may be written as an equation (where the atomic symbol should be taken to include the inner core electrons):

$$2F(2s)^2(2p)^5 = F_2(\sigma_s)^2(\sigma_s^*)^2(\sigma_p)^2(\pi_p)^4(\pi_p^*)^4$$

In the fluorine molecule there are eight electrons in bonding orbitals (σ_s, σ_p, π_p) and six in antibonding sigma and pi orbitals. This gives an excess of two bonding over antibonding electrons and corresponds to the single bond in the fluorine molecule.

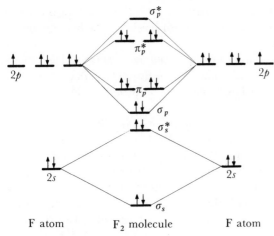

FIGURE 3.16 *Electronic structure of* F_2
An energy level diagram of the molecular orbitals is formed and all the valency electrons are placed in the molecular orbitals, according to Hund's Rules.

If the oxygen molecule is next examined, the advantage of using the more detailed molecular orbital approach will be seen. According to the valency octet theory the oxygen molecule is double-bonded and written $\ddot{:}O\ddot{:} \ \ddot{:}O\ddot{:}$. It is found experimentally, though, that the oxygen molecule has two electrons with unpaired spins and this property cannot be explained by the octet theory. The molecular orbital energy level diagram for oxygen is shown in Figure 3.17. The outer electronic structure of the oxygen atom is $(2s)^2(2p)^4$, so that the oxygen molecule has two fewer electrons than the fluorine molecule. These two electrons are removed from the highest level, the π^* one, leaving only two antibonding π electrons. As there are two electrons to enter the two π^* orbitals, which are of equal energy, they enter these orbitals singly, keeping their spins parallel in accordance with Hund's rules, giving the oxygen molecule two unpaired electrons. The structure is:

$$2O(2s)^2(2p)^4 = O_2(\sigma_s)^2(\sigma_s^*)^2(\sigma_p)^2(\pi_p)^4(\pi_p^*)^2$$

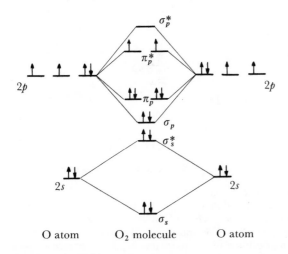

FIGURE 3.17 *Electronic structure of* O_2

There are eight electrons in bonding orbitals and four in the antibonding orbitals σ_s^* and π_p^*, giving a net count of four bonding electrons corresponding to the double bond. The molecular orbital theory therefore gives the same bond order as the electron pair theory but also readily accounts for the presence of the unpaired electrons.

If diatomic neon, Ne_2, were to form, two further electrons would have to be added to the F_2 structure and they could go only into the highest energy antibonding orbital, σ_p^*. The number of antibonding electrons and the number of bonding electrons both equal eight, no bond results, and neon exists as a monatomic gas.

The discussion so far has assumed that sigma molecular orbitals formed from atomic s orbitals are quite distinct from those formed from the p_z orbitals. This is a simplification: in principle, all orbitals of sigma symmetry on the atoms are expected to make some contribution to all the molecular sigma orbitals. That is, the two s and two p orbitals are, in general, combined in various proportions to give four molecular orbitals of sigma symmetry which we shall relabel σ_1, σ_2, σ_3, and σ_4, in order of increasing energy. The two extreme levels are similar to those in the simpler scheme with σ_1 close to σ_s and σ_4 close to σ_p^* in character. The major difference comes with σ_2 and σ_3, which both have substantial s and p contributions. The major effect of the mixing is to stabilize σ_2 below σ_s^* and to destabilize σ_3 relative to σ_p^*. The relation between the two schemes for the sigma molecular orbitals is indicated in Figure 3.18 parts (a) and (b).

The importance of this mixing between s and p orbitals depends inversely on the difference in energy between them. With atoms to the right of the Periodic Table, like O or F, the separation into σ_s and σ_p orbitals is a good approximation, and the schemes for F_2 or O_2 shown in Figures 3.16 and 3.17 are satisfactory. Further to the left, the atomic s and p orbitals are closer in energy and the interaction between them is significant. Thus the nitrogen molecule, N_2, has the energy level diagram shown in Figure 3.18c. There are five electrons for each atom, and these fill the orbitals up to σ_3. In this case, σ_3 is sufficiently destabilized that it lies at higher energy than the bonding π orbitals. Thus nitrogen is described:

$$2N(2s)^2(2p)^3 = N_2(\sigma_1)^2(\sigma_2)^2(\pi)^4(\sigma_3)^2$$

From Figure 3.18 it will be seen that the bonding character added to σ_2 relative to σ_s^* exactly equals the loss of bonding character in σ_3 relative to σ_p. Thus there is no nett bonding effect when σ_2 and σ_3 are both filled. Thus N_2 has six more bonding than antibonding electrons ($\sigma_1^2\pi^4$) making a triple bond. (Note that this result is also obtained on the simpler bonding scheme where N_2 would be like Figure 3.17 without the two π^* electrons.)

The separate σ_s and σ_p scheme, used as in Figure 3.16, always gives the correct bond order and usually gives the correct magnetic properties. Thus this system may be used with confidence when dealing with the electronic configuration of a new diatomic species. The modified σ_1 to σ_4 scheme, gives a more accurate picture of the relative energies of σ

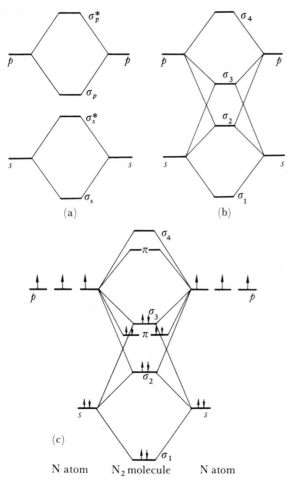

(a)　　　　　　(b)

(c)

N atom　　N₂ molecule　　N atom

FIGURE 3.18　　(a) *Sigma combinations without s-p mixing.*
　　　　　　(b) *Sigma combinations with s mixed with p.*
　　　　　　(c) *The electronic structure of* N_2.

This shows the relation between the $\sigma_s \sigma_s^* \sigma_p$ and σ_p^* levels (a) and the levels σ_1 to σ_4 (b) which result when s and p contributions are allowed to mix. The main effect is to stabilize σ_2 relative to σ_s^* and to destabilize σ_3 relative to σ_p. As a result, in the structure of N_2 (Figure 3.18c) the bonding π levels lie below σ_3. The bonding and antibonding contributions of σ_2 and σ_3 cancel, leaving σ_1 and the two π levels to make up a total bond order of three.

and π orbitals for N_2, and also for the transient diatomic molecules found in the vapour of the lighter elements, such as B_2 or C_2. The total energy of the four sigma orbitals is the same in both schemes, as it must equal that of the four contributing atomic orbitals, but the distribution differs, particularly between the two intermediate orbitals σ_2 and σ_3 compared with σ_s^* and σ_p.

This type of analysis also applies to molecules where the atoms have higher n values than two for their valency levels. For example, the electronic structures of the other halogen molecules are exactly the same as for fluorine—

$$X_2 (\sigma_s)^2 (\sigma_s^*)^2 (\sigma_p)^2 (\pi_p)^4 (\pi_p^*)^4$$

but here the atomic symbol X is to indicate that all the inner non-valency electron shells remain held by the nuclear attraction and play no part in the bonding. For example, for bromine which has the configuration $(1s)^2 (2s)^2 (2p)^6 (3s)^2 (3p)^6 (3d)^{10} (4s)^2 (4p)^5$, the molecular orbitals are constructed

from the $4s$ and $4p$ atomic orbitals and all the inner shells of the first, second, and third levels remain held by the atomic nuclei.

The analysis may be extended to heteronuclear molecules where the two atoms are in different periods. For example, in ClF, the larger nuclear charge of the chlorine atom lowers the energy of all its orbitals compared with the corresponding fluorine ones. As a result, the $3s$ and $3p$ levels of chlorine are approximately equal to the $2s$ and $2p$ levels of fluorine (and the Cl $2s$ and $2p$ levels are so tightly bound that they play no part in the bonding). The electronic diagram of ClF is then very similar to that of F_2 shown in Figure 3.16, except that the molecular orbitals are formed by overlap of F orbitals of the second shell with the corresponding Cl orbitals of the third shell.

The structure of hydrogen halides follows similarly. For example, in HCl, the H $1s$ orbital is approximately equal to the Cl $3s$ and $3p$ orbitals in energy and is of correct symmetry to form a sigma bond either with the $3s$ or the $3p_z$ orbital (taking z as the H-Cl direction). The overlap is best with the $3p_z$ orbital, and a sigma bonding orbital is formed as in Figure 3.11, together with the corresponding antibonding orbital. The remaining valency orbitals of the chlorine are non-bonding and hold electron pairs. Thus HCl is described

$$\text{H}(1s)^1 + \text{Cl}(3s)^2 (3p)^5 = \text{HCl}(\sigma_{s,p})^2 (\sigma_{s,p}^*)^0 (3s)^2 (3p_x)^2 (3p_y)^2.$$

If we now turn to consider heteronuclear species where the two atoms are in the same period, the only difference is in the relative levels of the atomic orbitals with those of the atom with highest nuclear charge lying lowest in energy. Consider nitric oxide and its cation, NO^+ (the nitrosonium ion). The

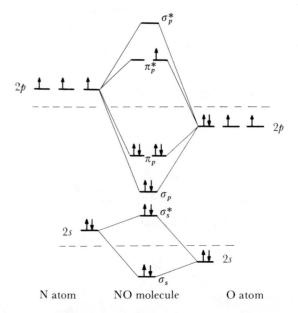

N atom　　NO molecule　　O atom

FIGURE 3.19 *Electronic structure of* NO
The energy level diagram differs from that of a homonuclear diatomic molecule, such as O_2, only in having the atomic levels at different energies. The splitting of the bonding and antibonding molecular orbitals of any particular type is symmetrical about the average energy of the constituent atomic orbitals.

energy level diagram for nitric oxide is given in Figure 3.19. The atomic orbital energy levels are no longer equal because of the difference in nuclear charges. The nitric oxide molecule has an odd number of electrons and the one of highest energy occupies the antibonding π orbital. There are eight bonding electrons, in the sigma s and p levels and in the pi p level and three antibonding electrons, $(\sigma_s^*)^2$ plus $(\pi_p^*)^1$, leaving a net excess of five bonding electrons corresponding to a bond order of two and a half. If this molecule is ionized, the highest energy electron is the one removed and this is in the antibonding pi orbital. The excess of bonding over anti-bonding electrons goes up to six and the bond order rises from two and a half to three. (NO^+ is isoelectronic with N_2.) The cation is therefore more strongly bonded than the parent molecule. This can be seen experimentally in the infrared spectrum where the $N-O$ stretching frequency occurs at higher energy for NO^+ than for NO.

In carbon monoxide and in the cyanide ion there are the same number of valency electrons as in the nitrogen molecule. (Such species are termed *isoelectronic*.) They have the same molecular electronic structure and the σ_3 level lies above the bonding π orbitals as in Figure 3.18c. The species are triplybonded and the formation equations may be written:

$$C(2s)^2(2p)^2 + O(2s)^2(2p)^4 = CO(\sigma_1)^2(\sigma_2)^2(\pi)^4(\sigma_3)^2$$
and $C(2s)^2(2p)^2 + N(2s)^2(2p)^3 + e-$
$$= CN^-(\sigma_1)^2(\sigma_2)^2(\pi)^4(\sigma_3)^2$$

The energy level diagrams will differ from that for N_2 only in the relative levels of the atomic orbitals (compare also Figure 3.19).

The molecular energy level diagram is useful for explaining bond orders and magnetic properties in a qualitative manner, as in the above examples, but it is also used more quantitatively in the discussion of the excited states of the molecules. If the molecule absorbs energy equal to the difference between the energy of an occupied orbital and that of one of the higher, empty orbitals, an electron will undergo a transition into the upper orbital, e.g. $\pi \rightarrow \pi^*$ in N_2. The energy differences between molecular orbitals can thus be observed in the electronic spectrum of a diatomic molecule and such experimental values compared with the calculated energy levels, just as in the case of atomic orbitals. The electronic transitions for the light diatomic molecules such as nitrogen are of relatively high energy and occur in the far ultraviolet region of the spectrum.

Polyatomic Molecules

The molecular orbital approach may be extended to cover molecules of more than two atoms in two different ways. One way would be to take the n nuclei of an n-atomic molecule, together with their inner non-bonding electron shells, place them in their positions in the molecule, and then combine all the valency shell atomic orbitals on all the atoms into poly-centred molecular orbitals embracing all the nuclei. The total number of such *delocalized* molecular orbitals equals the total number of atomic orbitals used in their construction and each may hold up to two electrons. The valency electrons would then be fed into these n-centred molecular orbitals, in order of increasing energy, to form the molecule in a process which is exactly analogous to the building-up of atomic and diatomic molecular structures which has just been discussed.

An alternative approach is to think of each pair of bonded atoms separately as forming a two-centred bond just as in the diatomic molecules. The whole molecule is built up from such two-centred *localized* molecular orbitals. This latter approach corresponds to the long-familiar ideas of electron pair bonds and it is the one which is mainly used to describe sigma bonding in polyatomic molecules. The picture of delocalized many-centred bonds is less familiar but it is very useful in the discussion of π-bonded species and also finds application in the discussion of certain σ-bonded molecules, especially the 'electron-deficient' molecules typified by the boron hydrides.

In a complete treatment, both the method of localized bonds and the method of delocalized ones give exactly the same description of the electron density distribution in the molecule, and they are often used interchangeably in advanced work.

The most important property required from any theory of polyatomic molecules is the shape of the molecule. It will be shown in the next sections that molecular shapes may be derived from fairly simple ideas on electron pair repulsion and that more sophisticated descriptions of the bonding may then be applied to fit the shapes which have been so determined.

3.5 The shapes of molecules and ions containing sigma bonds only

The case of sigma-bonded species is the simplest to deal with and is discussed first.

When two atoms are bound together, electron density is concentrated in the region of space between them (sections 3.2 and 3.3). If a central atom is bonded to a number of others, it is reasonable to expect the bonds from the central atom to be as far apart as possible in order to reduce the electrostatic repulsions between the electron-dense regions in the bonds. For a triatomic molecule, such as $BeCl_2$, a linear configuration $Cl-Be-Cl$ will minimize the repulsions between the electrons in the two $Be-Cl$ bonds. Similarly, when there are three attached atoms, as in BCl_3, an equilateral triangle with the $Cl-B-Cl$ angles all $120°$ is the expected form for the molecule. Table 3.2 shows the expected configurations for molecules of the types AB_n. In each case the configuration is the one of highest symmetry.

Most examples where the coordination number of the central atom is greater than six are transition element complexes. Iodine is seven-coordinate in the pentagonal bipyramidal IF_7 and tellurium, for example, is seven- and eight-coordinate in complexes derived from the hexafluoride. There are many examples of the cases shown in Table 3.2,

TABLE 3.2 Shapes of AB_n molecules

n	Formula	Shape	Angles	Examples
2	AB_2	linear	$B-A-B = 180°$	$BeCl_2$, $HgCl_2$
3	AB_3	triangular	$B-A-B = 120°$	BF_3
4	AB_4	tetrahedral	$B-A-B = 109°\ 28'$	$SiCl_4$, BH_4^-, NH_4^+
5*	AB_5 or $AB_3B'_2$	trigonal bipyramidal	$\begin{cases} B-A-B = 120° \\ B'-A-B' = 180° \\ B-A-B' = 90° \end{cases}$	PCl_5
6	AB_6	octahedral	$B-A-B = 90°$	SF_6, PCl_6^-

*All the B positions in all these configurations are equivalent, except in the trigonal bipyramid where the two apical positions (B') are not equivalent to the three equatorial (B) ones.

among ions as well, as among molecules, and the tetrahedron and octahedron are especially common shapes.

If more than one type of atom is bonded to the central one, the configuration becomes less symmetrical although retaining the basic shape shown in the Table. (The tetrahedral configuration of all kinds of organic molecules is an obvious example.) In a molecule AB_rX_s, if the $A-X$ bond is shorter and stronger than the $A-B$ bond the electron density near A in the X directions will be greater than in the B directions. As a result, the BAB angles close up and the XAX angles open out relative to the values given in Table 3.2 for symmetrical AB_n cases. In addition, the AX bonds will tend to be as far apart in the molecule as possible; for example an octahedral AB_4X_2 species would be expected to be most stable in the *trans* configuration, compare Figure 12.18.

It is not always obvious, in a particular compound, in which direction away from the symmetrical configuration a distortion will occur. Bond lengths, bond polarities, and steric effects may all come into play and tend in opposite directions. However, most changes in bond angles due to unsymmetrical substitution are relatively small—c.f. the parameters of the silyl halides below—and the basic shape remains the one of Table 3.2.

$H-Si-H$ angle in silyl halides, SiH_3X:

X = F, angle = $109\frac{1}{2}°$: X = Cl, angle = $110\frac{1}{2}°$:
X = Br, angle = $111\frac{1}{2}°$

The most important case of distortion by unsymmetrical substitution comes when a substituting atom is replaced by an unshared pair of electrons on the central atom. For example, ammonia NH_3, is not a plane triangular molecule as a casual glance at Table 3.2 would suggest, but is pyramidal with an $H-N-H$ angle of $107°$. This angle is near the tetrahedral value and a count of the electrons on the nitrogen shows that there is a lone or unshared pair of electrons in addition to the three bond pairs. These four pairs of electrons adopt the tetrahedral arrangement, as in CH_4, showing that the lone pair can be regarded as localized in a particular direction in space. Thus ammonia is an extreme case of an unsymmetrically substituted tetrahedron, AB_3X, rather than a triangular AB_3 species. Since such a lone pair is attracted

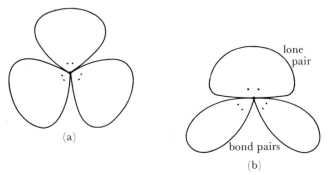

FIGURE 3.20 *Three electron pairs on a central atom:* (a) *three equal pairs,* (b) *two bond pairs and a lone pair*
As the lone pair is influenced only by the central nucleus, its density is concentrated much more closely to the central atom than the bond pairs and it thus dominates the stereochemistry.

only by the central atom nucleus instead being subject to the fields of two nuclei as is a bonding pair, the lone pair electron density is concentrated close to the central nucleus as in Figure 3.20. The lone pair electrons thus exert a greater repulsion effect than bonding electron pairs and the bond angles tend to close up when a lone pair is present. For example, the $X-M-X$ angles of all the nitrogen group trihalides such as NF_3 or SbI_3 all lie within the range $97°-104°$ compared with the tetrahedral angle of $109\frac{1}{2}°$. Again, where geometrical isomerism is possible, lone pairs will tend to be as far apart as possible. Thus a central atom with two lone pairs and four bond pairs around it, is expected to form a square planar molecule, with the two lone pairs in the *trans* positions of Figure 3.24c.

When lone pairs are taken into account, the structure of any sigma-bonded species may be formulated in terms of two simple rules:

1. The basic shape of a molecule or ion depends on the number of electron pairs surrounding the central atom (lone pairs plus bond pairs) and is that which follows from Table 3.2, assuming that lone pairs occupy positions in space.

2. Repulsions decrease in the order:— lone pair—lone pair > lone pair—bond pair > bond pair—bond pair, with the results that (a) lone pairs tend to be as far apart from each other as possible and (b) bond angles close up compared

TABLE 3.3 Shapes of species with lone pairs of electrons

Total number of electron pairs in the valence shell	Basic shape	Number of lone pairs	Formula	Shape	Examples
3	triangle	1	AB_2E	V-shape	$SnCl_2$ (gaseous)
4	tetrahedron	1	AB_3E	trigonal pyramid	NH_3, PF_3
		2	AB_2E_2	V-shape	H_2O, SCl_2
5	trigonal bipyramid	1	AB_4E	see Figure 3.23	$TeCl_4$
		2	AB_3E_2	T-shape	ClF_3
		3	AB_2E_3	linear	$(ICl_2)^-$
6	octahedron	1	AB_5E	square pyramid	IF_5
		2	AB_4E_2	square plane	$(ICl_4)^-$

with those given in Table 3.2 for the regular structure of the same total number of electron pairs.

It is convenient to use the symbol E for a lone pair of electrons and some cases are listed in Table 3.3 and discussed below. In all these cases the ligands (a general term for any atom or group attached to the central atom) may be radicals like alkyl groups, or ions like cyanide, or neutral groups donating lone pairs like water or ammonia, as well as atoms as in the Table. Thus, $Zn(H_2O)_4^{2+}$ is a zinc ion (with no valency electrons of its own) surrounded by four electron pairs donated by the four-co-ordinated water molecules and the shape is therefore tetrahedral.

Figure 3.21 shows the case where there are three electron pairs in the valence shell. When one of these pairs is unshared, the V-shaped molecule of Figure 3.21b results. (It

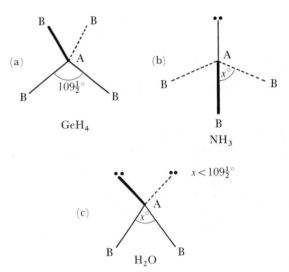

FIGURE 3.22 *Shapes of species with four electron pairs around the central atom:* (a) *four bonds*, (b) *three bonds and one lone pair*, (c) *two bonds and two lone pairs*

carbon Group elements are regular tetrahedra, as in Figure 3.22a, with bond angles of 109° 28′. In ammonia, one position is occupied by the lone pair of electrons, and the molecular shape is the trigonal pyramid of Figure 3.22b, with the H−N−H angle reduced to 107° 18′ by the repulsion of the lone pair. Water has two unshared electron pairs on the oxygen atom, so that the molecule is V-shaped as shown in Figure 3.22c and the increased repulsion reduces the H−O−H angle to 104° 30′.

The bond angles in the heavy element analogues of ammonia and water all show considerable reduction from the tetrahedral angle. Table 3.4 lists some values. The angles of the compounds of the oxygen Group are all smaller than those of the corresponding compounds of the nitrogen Group elements, reflecting the enhanced repulsion effect of the two lone pairs.

The variations in angle shown in Table 3.4 may be rationalized in terms of the electron densities in the bonds of these compounds: the higher the bond electron density at the central atom, the more resistance there is to the bond-

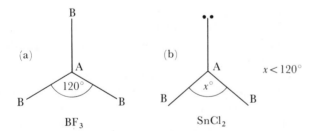

FIGURE 3.21 *Shapes of species with three electron pairs around the central atom:* (a) *three bonds*, (b) *two bonds and one lone pair*

must be noted, of course, that there is no experimental method of determining the position of lone pairs. The atomic positions can be determined but the electron pair positions must be deduced from the resulting structure. In this and the following Figures, the lone pairs are indicated schematically to show the relation to the basic structure.) Because of the greater repulsion of the lone pair, the bond angle is reduced from the value of 120° for the trisubstituted molecule of Figure 3.21a.

The shapes derived from the tetrahedron are shown in Figure 3.22. These species are well illustrated by the cases of methane, ammonia, water, and their heavy atom analogues. Methane, CH_4, and the other tetrahydrides of the

TABLE 3.4 Bond angles in some AX_3E and AX_2E_2
compounds

	(XAX angles)		
NH_3 107·3°	NF_3 102·5°		
PH_3 93·5°	PF_3 (104°)	PCl_3 100·1°	
AsH_3 91·8°	AsF_3 (102°)	$AsCl_3$ (103°)	
SbH_3 91·3°		$SbCl_3$ 99·5°	
H_2O 104·5°	OF_2 101·5°	OCl_2 (112°)	
H_2S 92·2°		SCl_2 (102°)	
H_2Se 91·0°			
H_2Te 89·5°			

All values measured on molecules in the gas phase. The values are
accurate to $\pm 0.5°$ except for those in brackets which are known within 3°.

closing effect of the lone pairs. Thus, comparing NH_3 and
PH_3, the P—H bond is longer than the N—H bond and
therefore the electron density is less. Furthermore, as
phosphorus is less electronegative than nitrogen, the electron
density is concentrated nearer the hydrogen in P—H than in
N—H. Both these factors mean that there is less electron
density at the phosphorus atom in the P—H bonds, than is
the case at nitrogen in NH_3, and therefore the $\widehat{HPH}$ angle
in PH_3 is much more drastically reduced by the lone pair
repulsion. A similar effect occurs if the phosphorus, in turn,
is replaced by the rather larger and less electronegative
atoms, As or Sb, and the $\widehat{HAsH}$ and $\widehat{HSbH}$ angles show
further contraction. However, the biggest relative change in
size and electronegativity comes between nitrogen and
phosphorus (compare Tables 2.9 and 2.13) and this is the
point where the biggest change in bond angle occurs.
Parallel changes are found for H_2O, H_2S, H_2Se and H_2Te.

If NH_3 and NF_3 are now compared, changes in bond
length play only a minor role as fluorine is only a little bigger
than hydrogen. However, the electronegativities fall in the
order F > N > H so that the electron density in the N—F
bond is concentrated towards the fluorine—the reverse of the
polarization of the N—H bond. Thus the lone pair repulsion
closes up the $\widehat{FNF}$ angles more than the $\widehat{HNH}$ ones. Again, a
parallel effect is observed in H_2O and OF_2.

Unfortunately, the angles in the remaining halogen compounds of these elements are less accurately known. It is
clear, however, that they are all around 100° and therefore
larger than in the hydrides while electronegativity effects
would lead us to predict angles smaller than in the hydrides.
This may be a result of steric effects with the large halogen
atoms coming into contact, or it may be an indication that
π bonding is occurring between the halogen and the central
atom (see section 3.7) which returns electron density to the
central atom.

When there are five electron pairs around the central
atom, the basic shape is the trigonal bipyramid in Figure
3.23. When some of these are lone pairs a number of alternatives become possible, as the apical and equatorial positions
in the bipyramid are not equivalent. If one lone pair is
present, it could be in an apical (B′) position giving $AB_3B'E$

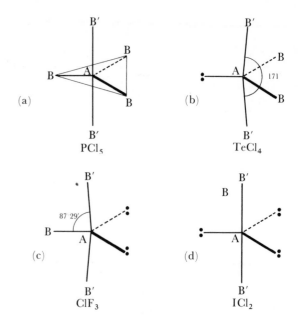

FIGURE 3.23 *Shapes of species with five electron pairs around the central
atom:* (a) *five bonds,* (b) *four bonds and one lone pair,* (c) *three bonds and
two lone pairs,* (d) *two bonds and three lone pairs*
The axial and equatorial positions are not equivalent in the trigonal
bipyramid. All examples known so far with lone pairs have these
lone pairs in equatorial positions.

or in an equatorial (B) position giving the $AB_2B'_2E$ configuration. If the lone pair is apical, it makes an angle of 90°
to the three A—B bonds in the equatorial plane, together
with an angle of 180° to the apical A—B′ bond. In the
$AB_2B'_2E$ arrangement, the equatorial lone pair makes only
two 90° angles (to the two AB′ bonds) and also two angles of
120° to the two A—B bonds. The reduction in repulsions due
to the drop from three to two 90° interactions is the dominating factor, and the equatorial position of the lone pair is the
preferred one as illustrated in Figure 3.23b. Thus, tellurium
tetrachloride adopts the distorted tetrahedron of the $AB_2B'_2E$
configuration.

When two lone pairs are present in the trigonal bipyramid,
there are three possible configurations. Both lone pairs may
be apical giving AB_3E_2, or both may be equatorial giving
ABB'_2E_2, or one might be apical and one equatorial in
$AB_2B'E_2$. The choice of which of these configurations is
adopted is not unambiguous. In the only case so far decided
experimentally, that of ClF_3, the lone pairs are equatorial in
the ABB'_2E_2 arrangement to give a T-shaped molecule. In
this configuration, the angle between the lone pairs is 120°
and there are four lone pair-bond pair angles of 90° (to the
A—B′ bonds) plus two of 120° (to A—B bonds). In the configuration AB_3E_2 the angle between the two apical lone pairs
is increased to 180° at the price of having six 90° angles
between lone pairs and bond pairs. As all the electron
densities fall off in an exponential manner, the interaction
between electrons will rise sharply with decreasing angle and
it appears that the increased number of lone pair—bond
pair interactions at 90° outweigh the decrease in the lone
pair—lone pair angle, in chlorine trifluoride at least. The
molecular shape is shown in Figure 3.23c. It may well be that

other molecules with two lone pairs and three bond pairs will be found to adopt one of the other possible configurations. In chlorine trifluoride, the $F-Cl-F$ angles are reduced to $87\frac{1}{2}°$ by the repulsion of the two lone pairs.

When three lone pairs are present, a number of alternative configurations is again possible but the only one found to date is the $AB_2'E_3$ arrangement with all the lone pairs equatorial, giving a linear molecule as in Figure 3.23d. In this case, the lone pairs are arranged symmetrically around the bonds so the angle is exactly $180°$.

When six electron pairs are present around the central atom, the basic shape is the regular octahedron shown in Figure 3.24a. In this case, unlike the trigonal bipyramid, all the positions are equivalent and all the BAB angles are $90°$. (In the conventional diagram of the octahedron the equatorial positions are commonly linked up to clarify the figure, and this tends to suggest that the apical and equatorial positions are non-equivalent, so care must be taken until such diagrams become familiar.)

If one unshared pair is present, the structure becomes the square pyramid of Figure 3.24b. When there are two lone pairs, the configuration which minimizes the lone pair—lone pair interaction is the square planar configuration in which the lone pairs are *trans* to each other at an angle of $180°$, as in Figure 3.24c. There are no examples of more than two lone pairs in the octahedral case.

With these ideas in mind, it is possible to put the process of determining the shape of any singly-bonded species on a formal basis in terms of two simple rules. These are:

1. Determine the number of electron pairs on the central atom by adding up the number of valency electrons on the central atom plus one electron for each bond (this is the electron contributed by the second element to the bond) plus one electron for each negative charge, or minus one for each positive charge.

2. The shape then follows from the cases listed in Tables 3.2 and 3.3. Two other rules may be added to allow for special classes of compound:

3. When a co-ordinate bond is present, two electrons are added to the total around the central atom since the donor atom or group is providing both the electrons of the bond.

4. Transition elements follow the rules above but their valency shell *d* electrons are not always included when determining the number of electrons around the central atom. The shapes of species with partly-filled *d* shells is discussed in Chapter 12; here we need only note that the electron pair analysis applies rigorously to those cases, d^0, d^5 and d^{10}, where the *d* orbitals are symmetrically occupied and also, in practice, to all configurations except d^4, d^7 and d^9 (which are distorted) and d^8, which is commonly square planar.

This approach is illustrated in the examples below.

(a) The two coordinated species, $PbCl_2$, SCl_2, ICl_2^- and $Ag(CN)_2^-$.

(i) $PbCl_2$: lead has the configuration $[Xe] (4f)^{14} 5(d)^{10} (6s)^2 (6p)^2$. Thus the electron count is:

$$\text{valency electrons Pb} = 4 \text{ electrons}$$
$$add \text{ bond electrons from 2Cl} = 2 \text{ electrons}$$
$$\text{total} = 6 \text{ electrons} = 3 \text{ pairs}$$

$PbCl_2$ has therefore the V-shaped structure of the AB_2E type:

Notice that the analysis applies only to an isolated molecule of lead chloride and such V-shaped molecules are found in the vapour. In the solid, chlorine atoms are shared between two lead atoms and a condensed structure results.

(ii) SCl_2: in a similar manner, the electron count is:

$$S = 6 \text{ electrons}$$
$$2Cl = 2 \text{ electrons}$$

SCl_2 has therefore four electron pairs and a V-shaped AB_2E_2 structure:

$$\overset{\displaystyle S}{\underset{\displaystyle Cl \quad Cl}{\diagdown}}$$

(iii) ICl_2^-:
$$I = 7 \text{ electrons}$$
$$2Cl = 2 \text{ electrons}$$
$$add \text{ for negative charge} = 1 \text{ electron}$$

ICl_2^- has thus five electron pairs and the structure is the linear $AB_2'E_3$ one: $(Cl-I-Cl)^-$.

(iv) $Ag(CN)_2^-$: silver has the electronic configuration $[Kr](d)^{10}(s)^1$ of which only the *s* electron is used. The

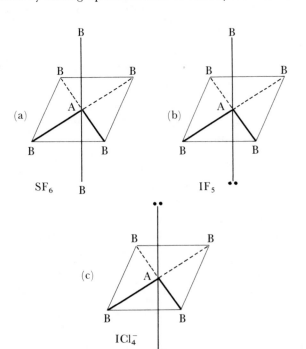

FIGURE 3.24 *Shapes of species with six electron pairs around the central atom: (a) six bonds, (b) five bonds and one lone pair, (c) four bonds and two lone pairs*

number of electrons around the silver is thus:

$$Ag \ s \ electron = 1$$
$$add \ \text{bond electron from each CN} = 2$$
$$add \ \text{negative charge electron} = 1$$

$$total = 4 \ electrons$$

Thus there are two electron pairs on the Ag and the structure is linear AB_2: $(NC-Ag-CN)^-$.

(b) The rare gas fluorides, XeF_2, XeF_4, and XeF_6. The structures derived here are based on the above electron pair scheme: a different approach has also been used and this is discussed in Chapter 15. The xenon atom is regarded as using all eight s^2p^6 electrons in this scheme. The electron count then proceeds as follows.

XeF_2:
$$Xe = 8 \ electrons$$
$$\text{bond electrons from } 2F = 2 \ electrons$$

XeF_2 has thus five electron pairs around the xenon atom and the structure is the linear AB_2E_3 type.

XeF_4:
$$Xe = 8 \ electrons$$
$$4F = 4 \ electrons$$

XeF_4 is therefore the square planar AB_4E_2 structure derived from an octahedral arrangement of six electron pairs.

XeF_6:
$$Xe = 8 \ electrons$$
$$6F = 6 \ electrons$$

In XeF_6 there are thus seven electron pairs around the xenon. The only seven-pair structure known is that of IF_7 which is a pentagonal bipyramid: the most likely structure for XeF_6 would thus be a distorted octahedron formed by replacing one bond pair in IF_7 by an unshared pair. A second possibility would be a pentagonal pyramid, if the unshared pair occupied an apical position.

The alternative description of the xenon fluorides described in Chapter 15 agrees with the above in giving XeF_2 as linear and XeF_4 as square planar (and this also agrees with the experimentally-determined structures). The theories differ in the case of XeF_6, the alternative description being as a regular octahedron. The XeF_6 structure has recently been reported to be a distorted octahedron: i.e. the lone pair does have a steric effect.

(c) PCl_5 and its dissociation ions, PCl_4^+ and PCl_6^-. Phosphorus pentachloride exists as monomeric PCl_5 units in the gas phase but forms $PCl_4^+PCl_6^-$ in the solid by transfer of a chloride ion.

PCl_5:
$$\text{valency electrons P} = 5 \ electrons$$
$$\text{bond electrons from } 5Cl = 5 \ electrons$$

PCl_5 thus forms a trigonal bipyramid in the gas phase.

PCl_4^+:
$$P = 5 \ electrons$$
$$4Cl = 4 \ electrons$$
$$\text{Charge} = -1 \ electron \ \text{(for the plus charge)}$$

Thus there are four electron pairs around the phosphorus in PCl_4^+ and the shape is tetrahedral.

PCl_6^-:
$$P = 5 \ electrons$$
$$6Cl = 6 \ electrons$$
$$\text{Charge} = 1 \ electron$$

There are thus six electron pairs in PCl_6^- and the shape is a regular octahedron.

The ionization in the solid state probably derives from the impossibility of packing the five-coordinate structure regularly.

(d) As an example of coordinate bonds, consider the adduct $BF_3.NH_3$.

Boron has three valency electrons, therefore BF_3 has three electron pairs around the boron and is a planar molecule with FBF angles of 120°. Ammonia is a trigonal pyramid with a lone pair on the nitrogen as already seen. As the boron has four valency shell orbitals, one remains vacant in BF_3 and this can accept the electron pair of the ammonia to form the adduct.

$F_3B \leftarrow NH_3$: The electron count at the boron atom in the adduct is

$$\text{valency electrons B} = 3 \ electrons$$
$$\text{one bond electron from } 3F = 3 \ electrons$$
$$\text{two bond electrons from N} = 2 \ electrons \ \text{(as the electron}$$
$$\text{pair is donated)}$$

The boron has therefore four electron pairs around it and the configuration (of the three F atoms and the N) is tetrahedral: the FBF angle is decreased towards the tetrahedral value of $109\frac{1}{2}°$.

The electron count at the nitrogen atom remains at four electron pairs as the two donated electrons remain in the nitrogen valence shell. The nitrogen is thus surrounded tetrahedrally by the three H atoms and the B atom.

(e) Transition metal compounds.
(i) $TiCl_4$. Titanium has four valency electrons, d^2s^2. The electron count is

$$\text{valency electrons Ti} = 4 \ electrons$$
$$\text{bond electrons from } 4Cl = 4 \ electrons$$
Thus $TiCl_4$ is tetrahedral.

(ii) $CoCl_4^{2-}$. Cobalt has the configuration d^7s^2. The electron count is most easily derived in the following way. Consider the complex as derived by four Cl^- ions donating an electron pair each to a Co^{2+} ion—which does not use its seven d electrons. Then the electron count at the cobalt is

$$Co^{2+} = 0 \ electrons$$
$$\text{two bond electrons from } 4Cl^- = 8 \ electrons$$
Thus the complex is tetrahedral.

In a similar way, any ML_4 or ML_6 complexes of transition elements, M, and any ligand, L (such as halide, cyanide, water, ammonia) are regarded as formed from an ion M^{n+} and four, or six, ligands donating a pair of electrons each. Any d electrons on M^{n+} are not used and the shapes are ML_4 = tetrahedral and ML_6 = octahedral. (Notice that $TiCl_4$ can be treated in this way as Ti^{4+} and $4Cl^-$: this gives the correct shape and electron count, although it is misleading in implying ionic bonding.)

Any transition metal compound may be treated in this way, and any d electrons remaining on the central metal ion are neglected in the electron count, but if this number of electrons is d^8 in an ML_4 compound, the shape may be square planar rather than tetrahedral and if the number of electrons is d^4, d^7, or d^9, the shapes will be distorted.

Coordination numbers other than four or six are less common for transition metal compounds but, where they are found, the shape of highest symmetry is most likely to be formed. The whole question of shapes of transition metal compounds is treated in detail in Chapter 12.

3.6 The shapes of species containing pi bonds

The extension of the discussion of the preceding section to molecules and ions containing π bonds is basically simple. As the π electrons in a double bond follow the same direction in space as the σ electrons of that bond, the electrons used in pi bonding have no major effect on the shape of the molecule. The molecular shape is determined mainly by the sigma bonds and lone pairs and the presence of pi bonding causes only minor changes due to the additional electron density. Thus, to determine the shape of a molecule or ion which contains one or more π bonds, the electrons which the central atom uses to form the pi bonds must be subtracted from the total of electrons. Then the shape of the species is determined by the number of lone pairs and sigma bond pairs around the central atom. This is best understood in terms of an example. In phosphorus oxychloride, $POCl_3$, the central atom is phosphorus which forms a double bond to the oxygen, $Cl_3P=O$. The two electrons of this π bond come, one from the phosphorus and one from the oxygen. Thus, when counting the number of electrons around the phosphorus atom, this electron used in the π bond must be subtracted as it has no steric effect. The calculation proceeds

valency electrons at P = 5 electrons
sigma bond electrons from O + 3Cl = 4 electrons
less electron used by P in π bond = −1 electron

Therefore there are four electron pairs around the phosphorus atom which are shape determining. $POCl_3$ is therefore tetrahedral. (The effect on the bond angles of the π bond is discussed later.)

Next consider the carbonate ion, CO_3^{2-}. The most reasonable formulation for this ion in the normal valency form is

$$O=C\diagdown_{O^-}^{O^-}$$

(note that in charged species involving divalent ligand atoms it is usually convenient to place the negative charge on the ligand atom and not on the central atom as in the simple sigma systems). Then the electron count at the carbon atom is

valency electrons at C = 4 electrons
sigma bond electrons from 3O = 3 electrons
less electron used in 1 π bond = −1 electron

thus there are three electron pairs to determine the shape and the carbonate ion is planar.

As a final example, take the sulphite ion, SO_3^{2-}. This would be formulated as $O=S\diagdown_{O^-}^{O^-}$ and therefore:

valency electrons at S = 6 electrons
sigma bond electrons from 3O = 3 electrons
less electron used in 1 π bond = −1 electron

hence, four electron pairs and the sulphite ion has an unshared pair on the S atom and is a trigonal pyramid.

A list of cases involving π bonds is given in Table 3.5.

The main difficulty in dealing with π-bonded species lies in deciding how many electrons the central atom is using for π bonds. It will usually be found satisfactory to write down a formula with single and double bonds which satisfies normal ideas of valency, as in the examples above, and then count up the number of double bonds from the central atom.

In a few cases, notably ozone, O_3, and the nitrate ion, NO_3^-, it will be found that it is necessary to write some of the bonds as coordinate bonds to keep normal valencies or to avoid violating the octet rule. These two are best written as

$$O=O\rightarrow O \text{ and } O=N\diagdown_{O}^{O^-}$$

. All the bonds are, despite this formalism, equivalent. The electron count then proceeds:

O_3	NO_3^-
O = 6 electrons (the central O)	N = 5 electrons
1O = 1 electron (the covalently bound O)	2O = 2 electrons (the two covalently bound Os)
1 π bond = −1 electron	1 π bond = −1 electron
total = 3 pairs	total = 3 pairs

Thus O_3 is a V-shaped molecule and NO_3^- is planar. Notice that, as the coordinate bonds originate from the central atom, the oxygen bound by the coordinate bond contributes no electrons to the count of those around the central atom.

Although a formula with localized double bonds is used to assist these calculations, it must not be thought that this type of formula gives a true description of the electron distribution in the species. If the ligand atoms are all equivalent, double bonds and charges are delocalized over the whole molecule or ion. For example, the three configurations:

$$O=C\diagdown_{O^-}^{O^-} \text{ , } O=C\diagdown_{O^-}^{O} \text{ , and } O=C\diagdown_{O}^{O^-}$$

are all equally likely for the carbonate ion, and it is found experimentally that all the C−O distances are identical and that there is a charge of $-\frac{2}{3}e$ on each oxygen. This delocalization is discussed in the next section.

Where π bonding is delocalized, as in this case, its only effect on the stereochemistry is to shorten all the bonds compared with the single bond lengths. If, however, the π bond is localized, as in $Cl_3P=O$ or $Cl_2S=O$ for example, then it exerts a steric effect. As the double bond region con-

TABLE 3.5 Structures of species with π-bonds

Number of valency electrons on the central atom	Number of σ-bonds	Number of π-bonds	Shape	Examples
Case 1: no lone pairs				
(Shape follows from column 2)				
4	2	2	linear	CO_2, HCN
	3	1	triangular	CO_3^{2-}
5	3	2	triangular	NO_3^-
	4	1	tetrahedral	POX_3, PO_4^{3-}, VO_4^{3-}
6	3	3	triangular	SO_3
	4	2	tetrahedral	CrO_4^{2-}, SO_4^{2-}
7	4	3	tetrahedral	IO_4^-, MnO_4^-
	6	1	octahedral	$IO(OH)_5$
Case 2: one lone pair				
(Shape is that due to the lone pair plus the number of bonds in column 2)				
5	2	1	V-shaped	NOCl, NO_2^-
6	3	1	trigonal pyramidal	$SOCl_2$, SO_3^{2-}
	2	2	V-shaped	SO_2
7	3	2	trigonal pyramidal	IO_3^-, XeO_3
	4	1	distorted (see Figure 3.25)	$IO_2F_2^-$
Case 3: two lone pairs				
(Shape is that due to the two lone pairs plus the bonds)				
7	2	1	V-shaped	ClO_2^-

tains two electron pairs and the bond is also shorter than a single bond, the electron density is high and so the repulsion between two double bonds, or between a double bond and a single bond is greater than that between two single bonds. The π bond thus has a steric effect similar to that of a lone pair and the bond angles between single-bonded ligands will be closed up because of the greater repulsion. For example, in $POCl_3$ the $Cl-P-Cl$ angle is reduced from the tetrahedral value to $103\frac{1}{2}°$.

Although this method, presented in the last two sections, of determining molecular shapes by considering the repulsions of electron-dense regions has proceeded from simple ideas of electrostatic repulsions, there is also some justification for it in wave mechanics. Basically, the property of electrons that leads to Hund's Rules—that electrons with parallel spins avoid being in the same region of space—is taken into account. For example, the most probable arrangement for a set of four electrons with parallel spins is at the corners of a tetrahedron. Four pairs of electrons may be regarded as two sets of four parallel electrons with the spins of one set of four anti-parallel to the spins of the second set. The arrangement of four electron pairs around a central atom may then be regarded as that of these two, anti-parallel, tetrahedral, sets of four with the tetrahedra brought into alignment by the presence of the ligands. This theory is termed *electron correlation* and is discussed in depth in the references by Linnett given on page 280.

The discussion of the last two sections may be summarized:

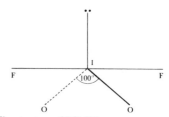

FIGURE 3.25 *The structure of* $IO_2F_2^-$

(i) Empirical ideas of the repulsion of the electron-dense regions represented by chemical bonds leads to the principle that bonds around a central atom will be arranged in the most symmetrical manner possible for the coordination number. These are the configurations given in Table 3.2.

(ii) Lone or unshared pairs of electrons on the central atom must be taken into account when deciding the shapes. These lone pairs fill coordination positions in the basic configurations to give the shapes detailed in Table 3.3 and in the commentary thereon.

(iii) The direction of distortion from the basic shapes in unsymmetrically substituted species may be predicted by considering the electron density and polarization within the bonds. When lone pairs are present, they have a dominating effect and the order of repulsions is lone pair—lone pair > lone pair—bond pair > bond pair—bond pair.

(iv) The same analysis applies to species with π bonds. The electrons contributed by the central atom to the π bonds are subtracted and the configuration depends on the number of

FIGURE 3.26 *The combination of the phosphorus $3p_x$ orbital with the $1s$ orbital of the hydrogen atom on the x-axis to give the molecular orbitals of a $P-H$ bond in PH_3*
The H atom in the x direction combines with a suitable phosphorus orbital to give bonding and antibonding two-centred molecular orbitals.

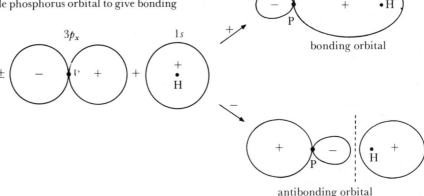

bonding orbital

antibonding orbital

bonds plus lone pairs left on the central atom as in the cases in Table 3.5. Delocalized π bonds have little steric effect but localized ones are regions of high electron density and cause distortion.

Although it is at first necessary to work out each structure systematically using the rules above, it will be found that the process is readily short-circuited with practice. For example, a major question is always whether there are lone pairs present, and these are nearly always found in compounds where the central element is showing an oxidation state less than its Group maximum one. Similarly, structures of iso-electronic species are usually the same, e.g., CO_3^{2-} and NO_3^-, or SO_2 and O_3, and this often helps to link up with known compounds.

Finally, one warning: the methods above apply on the assumption that the molecular formula is known and does not allow for polymerization. Thus, beryllium dichloride exists as $BeCl_2$ only in the gas phase, in the solid it is a chlorine-bridged polymer

$$
\begin{array}{ccccc}
Cl & & Cl & & Cl \\
\diagdown\diagup & & \diagdown\diagup & & \diagup \\
& Be & & Be & & Be \\
\diagup\diagdown & & \diagup\diagdown & & \\
Cl & & Cl & & Cl
\end{array}
$$

Similarly, aluminium tribromide exists as the dimer Al_2Br_6, and SiO_2 is a three-dimensional polymer instead of a discrete molecule like CO_2. However, as long as the molecular formula is known, the methods given above lead to the structures of most simple molecules and ions.

3.7 Bonding in polyatomic species

How are the shapes derived in the last two sections to be described in terms of the atomic orbitals in the molecule? There are already in the atomic orbitals some steric pro-perties, for example, the three p orbitals are at $90°$ to each other. It is thus possible to describe bond angles of $90°$ in terms of overlap with p orbitals of the central atom and suitable orbitals on the ligands. Thus phosphine, PH_3, where the $H-P-H$ angle is just above $90°$, could be de-scribed as having each $P-H$ bond formed by overlap of a phosphorus $3p$ orbital with the $1s$ orbital of the hydrogen to give a situation very similar to that in a diatomic molecule. The phosphorus atom has the valency shell configuration

$(3s)^2(3p_x)^1(3p_y)^1(3p_z)^1$. In the x direction, for example, the phosphorus $3p_x$ orbital (which will be written ψ_{p_x}) is of the correct symmetry to form a sigma overlap with the $1s$ orbital on that hydrogen atom lying in the x direction (written ψ_s). The resulting two-centred molecular orbitals are of the form given in equation 3.2:

$$
\sigma_{s,\,p_x} = \psi_s + c\psi_{p_x}
$$
$$
\sigma^*_{s,\,p_x} = \psi_s - c\psi_{p_x}
$$

where the value of c which gives the most favourable energy of combination would have to be determined in the course of the calculation (compare section 3.2). There are two electrons available, the hydrogen one and the one from the phosphorus p_x orbital, and these fill $\sigma_{s,\,p_x}$ leaving the anti-bonding orbital empty, and giving a single $P-H$ bond. This step is represented in Figure 3.26 and may be written as a formation equation:

$$
P(3p_x)^1 + H(1s)^1 = P-H(\sigma_{s,\,p_x})^2(\sigma^*_{s,\,p_x})^0
$$

Thus the whole process of forming this $P-H$ bond is just the same as forming the bond in a heteronuclear diatomic molecule (Figure 3.27).

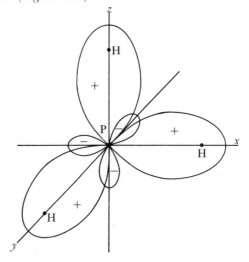

FIGURE 3.27 *The bonding orbitals in the phosphine molecule*
The overlap of the three phosphorus p orbitals with hydrogen s orbitals gives three $P-H$ bonds at $90°$ to each other, reflecting the angle between the p orbitals.

The P−H bonds in the y and z directions are formed in the same way giving the phosphine molecule:

$$P(3s)^2(3p_x)^1(3p_y)^1(3p_z)^1 + 3H(1s)^1$$
$$= PH_3(3s)^2(\sigma_{s,\,p_x})^2(\sigma_{s,\,p_y})^2(\sigma_{s,\,p_z})^2(\sigma^*_{s,\,p_x})^0(\sigma^*_{s,\,p_y})^0(\sigma^*_{s,\,p_z})^0$$

The molecule may be built up in this way by considering each pair of bonded atoms in turn as if they were in a diatomic molecule and using the methods of section 3.4. The lone pair of electrons is accommodated in the s orbital (and they will tend to be concentrated in that spatial segment of the s orbital remote from the bond directions) and the H−P−H bond angle follows as a consequence of the 90° angle between the p orbitals.

A similar discussion applies to the other hydrides of the heavier nitrogen Group elements and to the hydrides of sulphur and its heavier congeners. In the hydrogen sulphide molecule (Figure 3.28) the bonds are formed using two of the

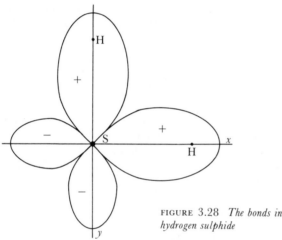

FIGURE 3.28 *The bonds in hydrogen sulphide*

two bonds from $s+p$ overlap

sulphur $3p$ orbitals, giving an H−S−H angle of 90°, and the third p orbital and the $3s$ orbital hold the two lone pairs. This analysis does not explain the small departures from the 90° angle in these molecules and cannot account for the much larger bond angles in water and ammonia.

3.8 Hybridization

The use of pure atomic s and p orbitals fails to account for the shape of most of the molecules mentioned in the previous sections. The linear, triangular, or tetrahedral shapes, for example, cannot be explained using simple atomic orbitals and the concept of mixed or *hybrid* orbitals has to be introduced. Consider first beryllium dichloride: the electronic configuration of the beryllium atom is Be = $[He](2s)^2(2p)^0$. The two electrons in the valency shell are paired in the s orbital, and the p orbitals are empty. In order to obtain divalency, the s orbital and a p orbital must be used and, to have two bonds at 180°, this $s+p$ configuration must be rearranged to form two equivalent orbitals. This is done by combining the two into two sp hybrids as shown in Figure 3.29. If the s orbital is superimposed on the p one as shown, the positive lobe of the p wave function is reinforced and the

negative lobe is diminished. This description reflects diagrammatically the mathematical process:

$$sp = \frac{1}{\sqrt{(2)}}(s+p)$$

which gives the orbital shown in Figure 3.29a. The second hybrid, which is directed in the reverse sense, is $1/\sqrt{2}(s-p)$ and is given in Figure 3.29b. The factors $1/\sqrt{2}$ arise as the

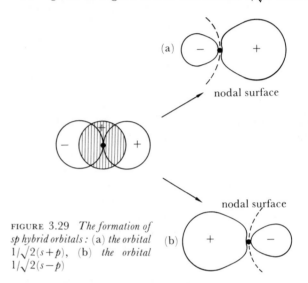

FIGURE 3.29 *The formation of sp hybrid orbitals:* (a) *the orbital* $1/\sqrt{2}(s+p)$, (b) *the orbital* $1/\sqrt{2}(s-p)$

total electron density in the two sp hybrids (found by squaring the two wave functions above and adding) must equal s^2+p^2, the electron density in the two constituent atomic orbitals. The total process is: $s+p = 2sp$ hybrids, two atomic orbitals giving two hybrid orbitals on the central atom. (Note that this process of mixing orbitals on the same atom, to give hybrid atomic orbitals, must be distinguished from the process of combining an s and a p orbital on different atoms to form σ molecular orbitals which was shown in Figure 3.11.) The two valency electrons are then placed one into each sp hybrid and each of these may overlap with a ligand orbital to form molecular orbitals.

The two sp hybrid orbitals are equivalent to each other and directed at 180°. They are used to form the bonds in a linear molecule. For example, the description of $BeCl_2$, given below, resembles the description of PH_3 apart from the use of hybrid orbitals.

BeCl$_2$. The chlorine atom has seven valency electrons and four valence shell orbitals. Six of the electrons are paired up in three of these orbitals, leaving the fourth singly-occupied. Let the Cl−Be−Cl axis be taken as the z-direction. The singly-occupied chlorine orbital must be of sigma symmetry with respect to the Be−Cl bond, and could be the chlorine s or p_z atomic orbitals or a hybrid such as sp pointing in the z direction. Molecular orbitals are formed by this chlorine σ orbital and the sp hybrid on the beryllium to give bonding and antibonding molecular orbitals, and the same happens for the second chlorine atom and the other sp beryllium hybrid. The complete process is first:

$$Be(s)^2(p)^0 = Be(sp_A)^1(sp_B)^1 \text{ (designating the two}$$
chlorine atoms A and B)

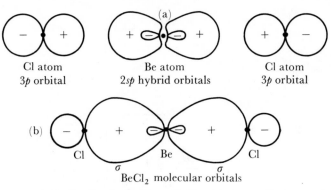

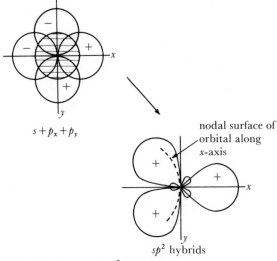

FIGURE 3.30 *Bonding in beryllium chloride:* (a) *the contributing atomic orbitals,* (b) *the bonding molecular orbitals*

that is, the beryllium valency electrons are placed one into each hybrid orbital.

Then $Be(sp_A)^1 + Cl_A(\sigma)^1 = Be-Cl_A(\sigma_{Be-Cl})^2(\sigma^*_{Be-Cl})^0$

where σ_{Be-Cl} is of the form $\psi_{sp} + \psi_p$ and similarly for the antibonding orbital σ^*_{Be-Cl}.

Similarly for the bond to the second chlorine

$Be(sp_B)^1 + Cl_B(\sigma)^1 = Be-Cl_B(\sigma)^2(\sigma^*)^0$

These are shown in Figure 3.30.

Each beryllium-chlorine bond is a normal single bond, just as in the diatomic molecules or in phosphine, except that it is formed using a hybrid atomic orbital on the beryllium instead of a simple one. The $Cl-Be-Cl$ angle of $180°$ follows from the angle between two sp hybrids.

Not only do these sp hybrid orbitals lead to the correct shape of the molecule but, in fact, they give more effective overlap than either of the simple atomic orbitals used in their construction, and the bond strength increases with the overlap of the constituent atomic orbitals. If it is assumed that the strength of a bond formed by an s orbital is unity, then the relative strengths of bonds formed by other atomic orbitals are as shown in Table 3.6.

TABLE 3.6 *Approximate strengths of bonds formed by various atomic orbitals*

Orbital	s	p	sp	sp^2	sp^3
Relative strength	1·0	1·73	1·93	1·99	2·00

A similar mixing process can be carried out with two p orbitals and the s orbital, to give three equivalent sp^2 hybrid orbitals directed at $120°$. These are shown in Figure 3.31. If one of the three orbitals is directed along the x-axis, this has contributions from the s and p_x orbitals. The other two hybrids are composed of the s, p_x, and p_y orbitals. The expressions for the three hybrids are, respectively:

$$\sqrt{(\tfrac{1}{3})}s + \sqrt{(\tfrac{2}{3})}p_x$$
$$\sqrt{(\tfrac{1}{3})}s - \sqrt{(\tfrac{1}{6})}p_x + \sqrt{(\tfrac{1}{2})}p_y$$
$$\sqrt{(\tfrac{1}{3})}s - \sqrt{(\tfrac{1}{6})}p_x - \sqrt{(\tfrac{1}{2})}p_y$$

The coefficients are again chosen so that the total electron density in the three hybrid orbitals adds up to $s^2 + p_x^2 + p_y^2$, the electron density in the three constituent orbitals.

The sp^2 hybrids are then used to form the bonds in a plane

FIGURE 3.31 *The formation of sp^2 hybrid orbitals*

triangular molecule, such as BCl_3. The process may be described as:

(i) Formation of the sp^2 hybrids and placing one valency electron in each

$$B(2s^2)(2p)^1 = B(sp^2)^1(sp^2)^1(sp^2)^1.$$

(ii) Each sp^2 hybrid is combined with a chlorine orbital of sigma symmetry to give one bonding and one antibonding two centre molecular orbital. These are of the form $sp_B^2 + c\sigma_{Cl}$ (bonding) and $sp_B^2 - c\sigma_{Cl}$ (antibonding). The electrons go into the bonding orbitals:

$$B3(sp^2)^3 + 3Cl(\sigma)^1 = BCl_3 3(\sigma_{B-Cl})^6 3(\sigma^*_{B-Cl})^0$$

The whole process is that three atomic orbitals on the boron, combined into three atomic hybrid orbitals, combine with three atomic orbitals on the three chlorines to give six molecular orbitals, three bonding and three antibonding. The six valency electrons fill the bonding orbitals, giving the boron trichloride molecule as shown in Figure 3.32.

In tetrahedral configurations, all three p orbitals and the s orbital combine to form four sp^3 hybrids, which are shown in Figure 3.33. When five- or six-coordinated structures have to be formed, it becomes necessary to use d orbitals as well as s and p orbitals. Five-coordination arises from sp^3d hybridization and the appropriate d orbital is the d_{z^2} orbital. (This can be seen qualitatively; the three equatorial positions in the trigonal bipyramid correspond to three sp^2 hybrids and the two orbitals in the $\pm z$-direction can arise only by mixing the atomic orbitals lying on the z-axis, that is p_z and d_{z^2}.) In a similar manner, octahedral hybridization involves six sp^3d^2 orbitals and, here, the appropriate d orbitals are d_{z^2} and $d_{x^2-y^2}$. The d orbitals may also contribute, in appropriate cases, to other shapes. For example, sd^3 (using d_{xy}, d_{yz} and d_{zx}) is also tetrahedral.

In each of the cases above, all the hybrid orbitals in a set are equivalent, have equal angles between them, and extend equally in space (except in the case of five-coordination as already discussed). For example, each sp^3 hybrid is equivalent to the other three and has one-quarter s character and three-quarters p character. In a molecule such as methane,

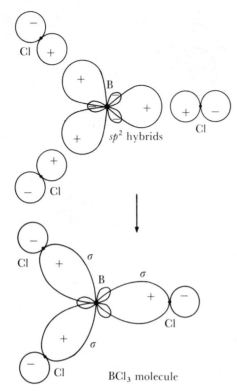

FIGURE 3.32 *The formation of boron trichloride*

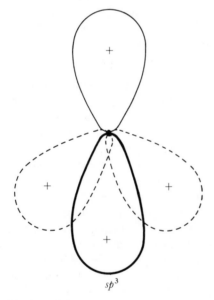

FIGURE 3.33 *The four sp^3 hybrid orbitals*
The four small negative lobes (compare Figure 3.31) have been omitted for clarity. The sp^3 hybrids are tetrahedrally directed.

the process of molecule formation involves a number of steps and, if the energies of each of these can be calculated, the stability of the product can be predicted. Such an analysis is often not possible for lack of data, but the case of the reaction of carbon and hydrogen to form the CH_4 molecule has been worked out and it may be compared with the possible alternative reaction to form CH_2. The energies involved in the formation of CH_2 and CH_4 are shown in Figure 3.34. It will be noted that the difference in heats of formation

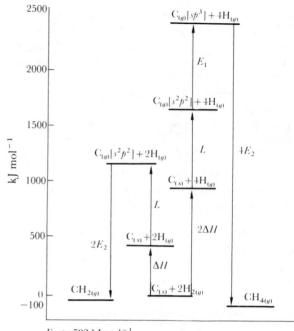

FIGURE 3.34 *Energy changes involved in the formation of CH_2 and CH_4*

$E_1 \approx 782$ kJ mol^{-1}
ΔH = heat of dissociation of gaseous $H_2 \approx 431$ kJ mol^{-1}
L = heat of vaporization of $C_{(s)} \rightarrow C_{(g)} \approx 711$ kJ mol^{-1}
E_2 = energy of $C-H$ bond ≈ 607 kJ mol^{-1}

favours CH_4 by a factor of 4/3 compared with CH_2 but the difference is only about one per cent of the total energies involved, which again highlights the difficulties of predicting chemical behaviour by direct calculation.

The carbon atom thus reacts in such a way that all its valency electrons and all its valency orbitals are involved in molecule formation, and this is generally the favoured mode of reaction, at least for lighter atoms. Thus, where there are empty valence orbitals as in $BeCl_2$ or BCl_3, further reaction to use these orbitals is likely and such molecules act as electron pair acceptors: similarly, where there are one or more lone pairs, as in ammonia or water, these tend to form donor bonds to suitable acceptors. Many examples of such behaviour will be found in the later chapters.

In all the cases discussed above, a given number of simple atomic orbitals were combined to give the same number of hybrid orbitals which were all equivalent to each other. These modes of hybridization correspond to the basic shapes of Table 3.2. It is quite possible, however, to form non-equivalent hybrids. For example, an unsymmetrical beryllium compound, BeXY, would almost certainly have a more favourable bonding structure if non-equivalent '$s+p$' hybrids were used with, say, the Be$-$X bond formed by a hybrid with rather more s character and the Be$-$Y bond from one with more p character.

In the same way, trigonal or tetrahedral hybrids may be non-equivalent. For example, in chloromethane, CH_3Cl, the $C-Cl$ bond will be formed by an '$s+p^3$' hybrid which has a different amount of s character from the other three '$s+p^3$' hybrids which form the $C-H$ bonds. The optimum amount of s character in such hybrids will vary from molecule to molecule, although the shape and basic hybrid structure is still tetrahedral.

Such non-equivalent hybrids become even more necessary when molecules containing lone pairs are under discussion. To get the bond angles of $107°$ in ammonia, the p character of the $N-H$ bonds has to be significantly increased over the $3/4\ p$ of the equivalent sp^3 hybrid, and the orbital holding the lone pair has proportionately more s character. Similar remarks apply to the case of water with a bond angle of $105°$.

There is indeed a complete range of possible hybrids from the four equivalent sp^3 hybrids at $109\frac{1}{2}°$ in methane, through the non-equivalent '$s+p^3$' hybrids in CH_3Cl, NH_3 and H_2O, down to the hybrids in PH_3 and H_2S which correspond to bond angles of just over $90°$ and are mainly p orbitals with a little s character for the bonds and s plus a little p for the lone pair. Moving the other way, towards larger angles, there is the same continuous variation from equivalent sp^3 hybrids at $109\frac{1}{2}°$, through non-equivalent arrangements, to the case of equivalent sp^2 hybrids at $120°$ plus an unused atomic p orbital.

In general terms, any of the structures, symmetrical or not, predicted by the methods of the preceding two sections, may have their electronic structures explained in terms of molecular orbitals formed by appropriate equivalent or non-equivalent hybrids on the central atom together with appropriate ligand orbitals. Hybrids can be constructed to fit any set of bond angles. It is also possible, in a more detailed treatment, to derive theoretical heats of formation by calculations involving hybridization schemes, although the results are too inaccurate in general to give a direct guide to the course of a reaction. It is also possible to be much more elaborate and introduce further small contributions from higher energy orbitals of suitable symmetry into all the hybridization schemes mentioned (for example, to add some d orbital character to the description of PH_3). Such elaborations usually give improved values for the calculated heats of formation and the like, but at the expense of considerable complexity in the calculations and with the loss of the simplicity of the basic scheme.

It should be clear from the above discussion that the idea of hybridization is very valuable in providing a *description* of the electron density in a molecule but that it does not provide any *explanation* of the electron arrangements or of the shapes of molecules. It is common to see statements such as 'methane is tetrahedral because it is sp^3 hybridized' but this description is incorrect.

As with other wave-mechanical calculations, hybridization presents a potential method of predicting molecular shapes. It is possible, in principle, to find the optimum hybridization scheme for a molecule such as water which will give the most favourable energy of formation and hence the bond lengths and angles. However, just as in the cases discussed earlier, the necessity of introducing approximate methods to solve the wave equations adds so much uncertainty to the values which result that the predictions are rather crude in the present state of the art.

3.9 Pi bonding in polyatomic molecules

Polyatomic molecules which contain isolated localized π bonds, such as the $P=O$ bond in Cl_3PO or R_3PO or the $S=O$ bond in sulphoxides R_2SO, may be treated similarly to the π-bonded diatomic molecules of section 3.4. These π bonds which are localized between two atoms are formed by sideways overlap of atomic orbitals of suitable symmetry. In the case of phosphorus oxychloride or the phosphine oxides, the σ bonds are formed from sp^3 hybrids on the phosphorus atom (using all the $3p$ orbitals on the phosphorus), so that the π bond requires the use of a suitable phosphorus $3d$ orbital to overlap with a p orbital on the oxygen. This bond is shown in Figure 3.35. The sulphoxide is similar with a lone pair on the sulphur and a $d_\pi-p_\pi$ bond (Figure 3.35b).

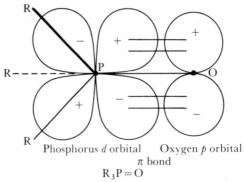

Phosphorus d orbital   Oxygen p orbital
π bond
$R_3P=O$

FIGURE 3.35a *Localized π bonds in a phosphine oxide*

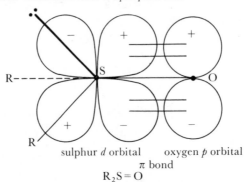

sulphur d orbital   oxygen p orbital
π bond
$R_2S=O$

FIGURE 3.35b *Localized π bonds in a sulphoxide*

Similarly, the triple bond in hydrogen cyanide, $H-C\equiv N$, is localized between the carbon and nitrogen atoms and formed by the overlap of the p_y and p_z orbitals of the carbon and nitrogen atoms.

Such localized π bonds are obviously similar to those in diatomic molecules, but they are relatively uncommon in inorganic chemistry. Most π bonds are *delocalized* over a number of atoms and are best discussed in terms of many-centred atomic orbitals. For example, in the carbonate ion, CO_3^{2-}, all the experimental evidence shows that all three oxygen atoms are equivalent, and so a structure with a localized π bond, such as $O=C{\overset{O^-}{\underset{O^-}{\big\langle}}}$, is incorrect and a description of the ion must be found which keeps the equivalence of the oxygens. This is done by first deriving the shape of the molecule by the methods of section 3.6 and thus accounting for the electrons and orbitals used in sigma bonding and in lone pairs. The remaining orbitals and electrons are then involved in the π system. The carbonate ion is planar with no unshared pair on the carbon. Let the molecular plane be the xy plane. Then the carbon uses its $2s$, $2p_x$ and $2p_y$ orbitals to form the sigma bonds (using sp^2 hybrids), and the oxygen atoms must also use their $2s$, $2p_x$ and $2p_y$ orbitals either in the sigma bonds or in accommodating non-bonding electrons (there is no chance of sideways overlap between oxygen orbitals in the plane of the molecule as the distances are too large). This accounts for all the s, p_x and p_y orbitals and for eighteen electrons, six in the three sigma bonds and four unshared electrons on each oxygen atom. The twelve s and p orbitals form the three bonding sigma orbitals and two orbitals on each oxygen to hold the nonbonding electrons, and all these are filled. The remaining three orbitals are the antibonding sigma orbitals and these remain empty and are of very high energy (see Figure 3.36). The ion has a total of

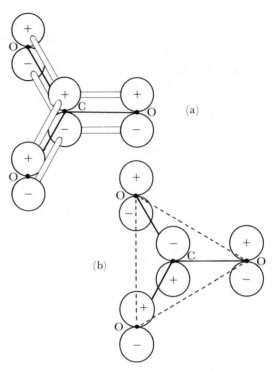

FIGURE 3.37 *Delocalized π orbitals in the carbonate ion:* (a) *the most strongly bonding orbital*, (b) *the most strongly antibonding orbital*

atom and each oxygen and is antibonding. This orbital is of the form:

$$\psi_4 = (p_C - p_1 - p_2 - p_3)$$

As there are four atomic p_z orbitals, there must be four molecular orbitals formed by them. The remaining two orbitals ψ_2 and ψ_3 lie between ψ_1 and ψ_4 in energy. One is weakly bonding and the second is weakly antibonding (see Figure 3.36).

The six remaining electrons enter the three π orbitals of lowest energy leaving ψ_4 empty. The weakly bonding and weakly antibonding effects of the two intermediate orbitals cancel each other, and the carbonate ion is left with one effective π bond over the whole molecule with a resulting C−O bond order of $1\frac{1}{3}$ (one for the σ bond and $\frac{1}{3}$ for the π bond), which corresponds with that implied by the simple formula.

A more detailed discussion of π orbitals, especially of the orbitals of intermediate energy such as ψ_2 and ψ_3, is beyond the scope of this text, but the general properties of such π systems are readily recognized. Four generalizations about polycentred π orbitals are possible.

1. The number of many-centred molecular orbitals equals the number of component atomic orbitals, in the case of the carbonate ion this is four. Each molecular orbital may hold up to two electrons.

2. The molecular π orbital of lowest energy is that obtained by combining all the atomic orbitals with the same sign so that there are no nodes between pairs of atoms in the resulting π orbital. In the carbonate ion, this orbital is ψ_1.

3. The molecular π orbital of highest energy is generally one where there is a node between each pair of atoms, that is, where the sign of the wave function is reversed between each

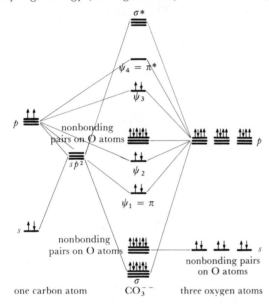

FIGURE 3.36 *Schematic energy level diagram for the carbonate ion* For clarity in drawing, any hybridization of the oxygen orbitals (e.g. to give sp^2 hybrids) has been neglected. Each oxygen atom has two pairs of non-bonded electrons—shown as the pair in the s orbitals and the pair of non-bonding p electrons. The delocalized π orbitals are ψ_1, ψ_2, ψ_3, and ψ_4. The bonding and antibonding sigma C−O orbitals are shown as σ and σ*.

twenty-four electrons (including two for the charge) so that six have still to be accommodated and the p_z orbitals on the four atoms have yet to be used. These four p_z orbitals are combined to form four, four-centred, π orbitals each holding two electrons. The pi orbital of lowest energy is of the form:

$$\psi_1 = (p_C + p_1 + p_2 + p_3)$$

where the constants have been omitted and the orbitals referred to are the carbon $p_z(p_C)$ and the three oxygen p_z orbitals respectively. This orbital is shown in Figure 3.37a. It has no nodes, concentrates electron density in the regions between the atomic nuclei, and is bonding. The orbital of highest energy, by contrast, has a node between the carbon pair of atoms as in ψ_4 of the carbonate ion. Such an orbital is strongly antibonding and no π bonding can occur in a case where electrons have to be placed in this type of orbital.

4. The remaining molecular π orbitals are intermediate in energy and their energies fall symmetrically about the mean energy of the strongly bonding and the strongly antibonding orbitals. Thus, in the carbonate case, ψ_2 is weakly bonding and ψ_3 is weakly antibonding by the same amount. In other cases, especially where there is an odd number of contributing orbitals, one or more of the polycentred π orbitals will be nonbonding. Degenerate pairs of orbitals may also be found in this group.

The form of the intermediate molecular orbitals is not always clear from simple considerations, but these generalizations make it possible to work out whether there will be any π bonding in a molecule without knowing any more about the formation of the intermediate orbitals. The one necessary condition is that it should be possible to leave the highest antibonding orbital empty and there will then be a net π bonding effect. If other, more weakly antibonding π orbitals also remain empty, the bonding effect is enhanced.

The classical example of a delocalized π system is, of course, the case of benzene. The π orbitals here are enumerated below and illustrated in Figure 3.38 and the reader can see how they fit the generalizations above.

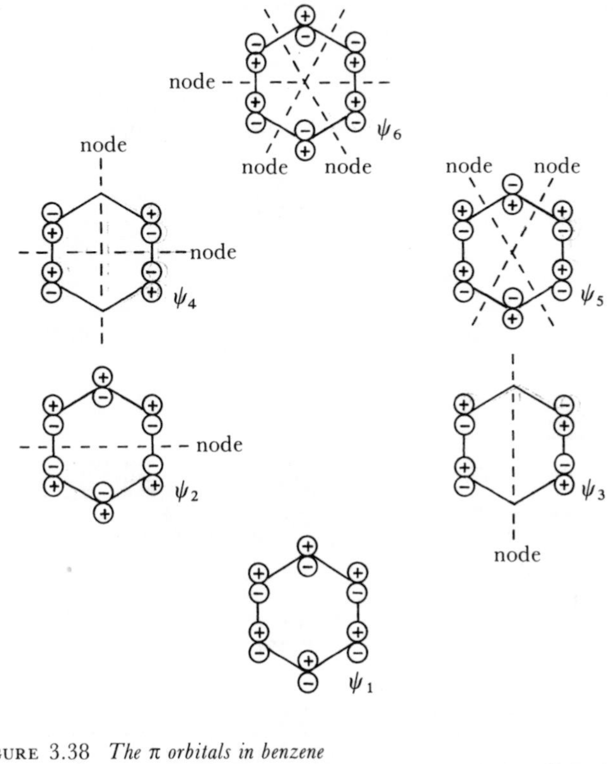

1. There are six component atomic orbitals and six π orbitals result.

2. The orbital of lowest energy has the form:

$$\psi_1 = k(p_1 + p_2 + p_3 + p_4 + p_5 + p_6)$$

This has no nodes between atoms and is strongly bonding.

3. The orbital of highest energy has the form:

$$\psi_6 = k(p_1 - p_2 + p_3 - p_4 + p_5 - p_6)$$

Each pair of atoms is separated by a node and the orbital is strongly antibonding.

4. The other four molecular π orbitals fall into two sets of degenerate pairs. One pair has one node cutting the ring and is weakly bonding while the other pair of orbitals has two nodes cutting the ring and is weakly antibonding. The six π electrons fill the three bonding levels, leaving the three antibonding orbitals empty.

A few more inorganic examples will be discussed to illustrate how far these relatively simple principles will suffice to carry a discussion of π systems.

The nitrite ion, NO_2^-

In this ion, the two oxygen atoms are equivalent and the methods of section 3.6 show that the ion is bent with a lone pair on the nitrogen atom. Again, let the molecule lie in the xy plane, then only the p_z orbitals on each atom will be involved in π bonding. Of the eighteen electrons in the valency shells of the component atoms, four will form the sigma bonds, two give the nitrogen lone pair and there are four nonbonding electrons on each oxygen atom leaving four electrons to go into π bonds. There are three atomic p_z orbitals and thus there are three molecular, three-centre, π orbitals. The lowest energy orbital extends evenly over the molecule and is bonding, while the highest energy π orbital has a node between the nitrogen atom and each oxygen and is antibonding. The third π orbital is nonbonding and has its electron density largely on the oxygens. The four electrons enter the bonding and nonbonding π orbitals, giving a net effect of one π bond over the molecule or an $N-O$ bond order of $1\frac{1}{2}$.

Carbon dioxide, CO_2

This is a linear molecule with no lone pairs on the carbon, by the method of section 3.6. If the molecular axis is taken as the z-axis, the s and the p_z orbitals on the three atoms will form the sigma bonds or hold nonbonding electrons on the oxygen atoms. The molecule has a total of sixteen valency electrons, and four of these form the sigma bonds while there is one nonbonding pair on each oxygen atom leaving eight electrons for π bonding. The p_y and p_x orbitals on each atom are available to form π bonds. Consider first the p_y orbitals. There are three of these and therefore three delocalized molecular π orbitals will be formed. The lowest energy orbital will have no interatomic nodes and be of the form (omitting constants):

$$\psi_1 = (p_O + p_C + p_O)$$

FIGURE 3.38 *The π orbitals in benzene*
ψ_1 is bonding, ψ_2 and ψ_3 are degenerate and weakly bonding, ψ_4 and ψ_5 are degenerate and weakly antibonding and ψ_6 is strongly antibonding. The component p orbitals are shown as of uniform, small size for clarity.

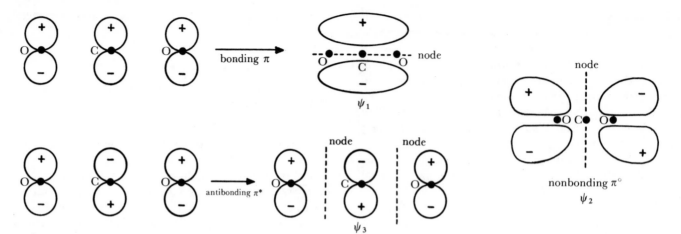

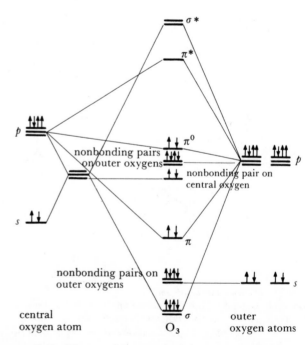

FIGURE 3.39   *The π orbitals in carbon dioxide*

while the highest energy orbital will have nodes between the carbon and each oxygen and have the form:

$$\psi_3 = (p_O - p_C + p_O)$$

The third molecular orbital, ψ_2, will be nonbonding. In the case of the p_x orbitals, exactly the same combinations will occur, as these orbitals are identical with the p_y ones apart from their direction in space. The resulting molecular orbitals are also identical; that is, the six atomic p_y and p_x orbitals combine to form six molecular orbitals, two of which are bonding and of the form of ψ_1, two nonbonding like ψ_2, and two antibonding like ψ_3. The eight valency electrons fill the bonding and nonbonding pairs giving a net effect of two π bonds over the molecule and a total C–O bond order of two. The orbitals in carbon dioxide are shown in Figure 3.39 and the energy level diagram for the molecule is shown schematically in Figure 3.40.

Ozone, O₃

It is interesting to compare the cases of ozone and carbon dioxide. There are two more valency electrons in ozone and, if these are added to the energy level diagram of carbon dioxide in Figure 3.40, they would have to be placed in the antibonding π orbitals, reducing the net π bonding over the molecule from two to one and, if both electrons were placed in one of the π* orbitals, say the one in the y direction, the whole π system in the y direction would have zero bonding effect. All this implies that the equivalence of the y and x directions in CO₂ will disappear in O₃, i.e., the molecule is no longer linear. This conclusion may also, of course, be derived by the methods of section 3.6. Ozone is isostructural and isoelectronic with the nitrite ion and has a bent structure, with a nonbonding pair on the central oxygen atom and two nonbonding pairs on each of the two terminal oxygens. There are three, three-centre, π orbitals of which the bonding

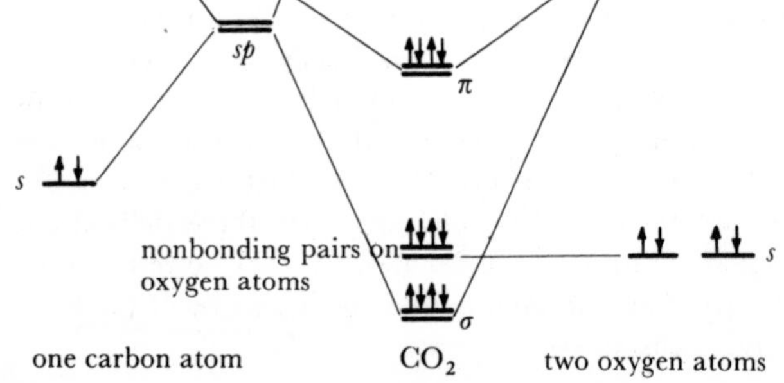

FIGURE 3.40   *Diagram of the energy levels in carbon dioxide*

FIGURE 3.41   *Diagram of the energy levels in ozone*

and nonbonding ones are occupied giving an $O-O$ bond order of $1\frac{1}{2}$. The schematic energy level diagram of ozone is shown in Figure 3.41 for comparison with carbon dioxide.

The nitrate ion, NO_3^-

This ion is planar and there are no unshared electrons on the nitrogen atom. The total number of electrons is twenty-four. The ion is isoelectronic and isostructural with carbonate and has the same π electron configuration, with two electrons in a strongly-bonding π orbital delocalized over the whole molecule and four electrons in weakly-bonding and weakly-antibonding orbitals. There is a net effect of one π bond in the ion and a $N-O$ bond order of $1\frac{1}{3}$.

Sulphur trioxide, SO_3

This is a plane triangular molecule with twenty-four valency electrons like the nitrate ion. It might therefore be described in the same way if the sulphur atom used only its $3s$ and $3p$ orbitals. There would then be six electrons in π orbitals but with a net bonding effect of only one π bond. If, however, the sulphur atom makes use of its $3d$ orbitals in addition, it is possible to construct π orbitals from six atomic orbitals (S $3p_z$, $3d_{xz}$, $3d_{yz}$, plus three O $2p_z$) instead of four. There would then be three bonding π orbitals and three anti-bonding ones, and the six π electrons can all be placed in bonding orbitals, giving three delocalized π bonding orbitals and a total $S-O$ bond order of about two (the approximation comes in as all three π orbitals are not equally stable).

In a similar way, SO_2 is isoelectronic and isostructural with O_3 but the possibility exists of the sulphur using d orbitals to increase the number of bonding π orbitals.

Whenever d orbitals of relatively low energy are available, there is the possibility that they will contribute to the bonding. This applies especially to molecules containing very electronegative elements bonded to elements in the third and higher periods and thus to all such oxygen compounds. The extent of d orbital participation will depend on the relative energies and the number of electrons to be accommodated and may be decided only by detailed calculation in each case.

Summary

Shapes of polyatomic molecules may be treated as follows:
1. Where only sigma bonds and unshared pairs of electrons are involved, the shape of simple molecules is determined by the number of electron pairs around the central atom and the bonding is best treated in terms of localized two-centred molecular orbitals formed between each pair of bonded atoms. This treatment may often be extended quite considerably to more complex species (for example, those with more than one 'central atom') but the predictions become less secure as the complexity rises.
2. The basic shapes of simple species with pi bonds again depend only on the number of sigma and lone pairs on the central atom and this can be calculated by the methods of section 3.6. Localized pi bonds between pairs of atoms are treated in terms of two-centred molecular orbitals just as in the case of diatomic molecules.
3. Where a number of equivalent atoms occur in a species with pi bonding, the pi electrons must be treated as delocalized over the whole molecule. Polycentred molecular pi orbitals are constructed from atomic orbitals of suitable symmetry, and these have the property that the molecular orbital of lowest energy and that of highest energy are readily distinguishable, and, as long as the latter remains unoccupied, a net pi bonding effect results. The treatment of delocalized pi bonding by relatively simple ideas is less complete that that of sigma bonding, and problems arise particularly in the cases where d orbital participation is possible. These more complicated cases can only be fully treated by detailed calculation which is beyond the scope of this text. However, the relatively simple ideas outlined in the earlier parts of this chapter allow a reasonably accurate description of the shapes of the large majority of simple polyatomic molecules and ions.

4 The Solid State

Simple Ionic Crystals

4.1 The formation of ions

In the last chapter, the Lewis electron pair theory was extended on the basis of wave mechanics to give a full description of the covalent bond. In this section, ionic bonding will be examined and the various factors which determine the formation and stability of ionic compounds will be discussed.

The basic process in the formation of an ionic compound is the transfer of one or more electrons from one type of atom to another: the resulting ions are then held together by electrostatic attraction. The arrangement of the ions in the solid is the one which gives the highest electrostatic energy. To see what factors determine this arrangement, consider the process of bringing up successive anions around a given cation. If there are already n anions surrounding the cation, the addition of a further anion produces an extra attraction between its charge and the cation charge, and also produces a number of repulsions between its charge and the charges on the n anions already present. There are thus two opposing tendencies. One is to increase the attractive forces by making the coordination number of the cation as large as possible and this is balanced by the increase in the repulsive forces as more and more anions are added. When the two tendencies balance, the final structure results. An exactly similar argument holds, of course, for the number of cations to be found around an anion. The repulsions are at a minimum if the distribution of ions is as symmetrical as possible. Thus ions which are three-coordinated have their neighbours at 120° in a triangular arrangement, four-coordinated ions have a tetrahedral, six-coordinated ions an octahedral, and eight-coordinated ions a cubic arrangement. In addition, some coordination numbers, such as five, which do not pack regularly in a solid are not observed in ionic crystals.

The coordination numbers in a solid of given formula, such as AB, depend on the number of the larger ions which may be packed around the smaller one. The stoichiometry—in this example, 1:1—then determines the coordination number of the larger ion. As the formation of a cation involves the removal of electrons, cations are always smaller

than the parent atoms (for example, the atomic radius of K is 203 pm, while the radius of K^+ is 133 pm). Conversely, the addition of electrons to atoms to form anions involves an increase in radius (for example, F = 72 pm but F^- = 136 pm in radius). As a result, anions are generally larger than cations and it is the number of anions which can pack around a cation which usually determines the coordination numbers and the structures. For example, sodium chloride crystallizes in a structure where the sodium ion is surrounded by six chloride ions (Figure 4.1a), and the chloride ion has, of course, six sodium ions around it. The larger cesium cation allows a coordination number of eight in the structure of cesium chloride (Figure 4.1b).

The number of anions which can pack around a given cation may be determined from the ratio of the radii of the cation and anion. This *radius ratio*, r_+/r_-, may thus be used to give an indication of the likely coordination number for a salt of given formula type. (Notice that the argument is exactly the same if the anion is smaller than the cation, except that it is the ratio of anion radius to cation radius, r_-/r_+, that is the important one.) If it is assumed (i) that ions are charged, incompressible spheres of definite radius (the validity of this assumption is discussed in section 4.5), (ii) that the central ion adopts the highest coordination number which allows it to remain in contact with each neighbour (this is a good approximation for the balance of attractive and repulsive forces referred to above), then the radius ratio limits corresponding to different coordination numbers may be calculated from purely geometrical considerations.

For example, in six-coordination, a cross-section through a site in the lattice appears as in Figure 4.2.

When the anions just touch BAB' = 90°.

Then,

$$AB = AB' = (r_+ + r_-) \text{ and } BB' = 2r_-$$

$$\frac{AB}{BB'} = \frac{r_+ + r_-}{2r_-} = \cos 45° = 1/\sqrt{2}$$

$$\therefore r_+ = \sqrt{2}r_- - r_-$$

$$\text{or } r_+/r_- = \sqrt{2} - 1 = 0.41$$

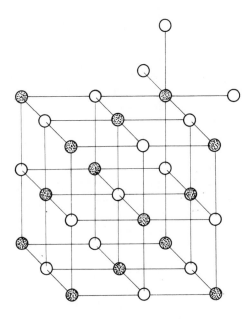

(a) sodium chloride

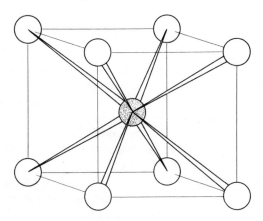

(b) cesium chloride

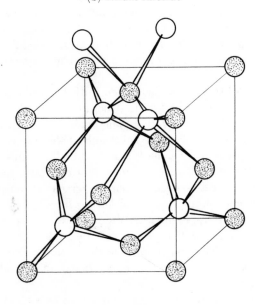

(c) zinc blende

FIGURE 4.1 *Structures of* AB *solids,* (a) *sodium chloride,* (b) *cesium chloride,* (c) *zinc blende*

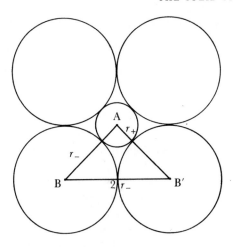

FIGURE 4.2 *Cross-section through an octahedral site*

Note on crystal structure diagrams

It follows from the discussion in section 4.1 that ions, or atoms, in crystals are expected to be as close together as possible and that the available space will be filled as completely as possible. However, if such a 'spacefilling' situation is represented directly in a diagram it is very difficult to see the arrangement of the atoms. As a result, most diagrams give an exploded view of the crystal. This convention has been adopted here. In accurate diagrams, the centres of the atoms are positioned exactly but their diameters would be reduced. In most of the diagrams in this chapter, further slight distortions or alterations in perspective have been introduced to make the structural arrangement easier to interpret. In a number of cases, further conventions are used to assist understanding. Atoms or ions in the top layer may be distinguished from more distant layers by the thickness of the circles used to represent them or, as for example in Figure 4.3, by using darts to represent the bonds joining them with the broad end of the dart on the atom nearest the front of the unit cell. Thus in the rutile diagram, Figure 4.3a, the Ti atom at the body centre has three O atoms in front of it and three behind it.

In more elaborate structures, there is a conflict between showing clearly the coordination of each atom and avoiding an extremely elaborate diagram showing many repeat units. Compare the diagram of CdI_2 (Figure 4.10a), which shows the coordination but does not readily give the metal layers, and the photograph of a model of the same structure (Plate I) which gives a better impression of the metal occupying every second layer between I sheets by showing a larger portion of the structure. Any diagram is necessarily a compromise and a formalized representation of a three-dimensional, space-filling structure and the reader should search for as many representations as possible of difficult structures in the references given. The study of models in three dimensions greatly clarifies the more complex structures and every opportunity of examining such models should be taken.

Similar calculations may be carried out for all the coordination numbers. The results indicate the range of values for the radius ratio within which different coordination numbers should be stable. These are shown below.

r_+/r_- : 0·155 to 0·23 to 0·41 to 0·73 to higher values
C.N. :   3   4   6   8

The validity of this simple method of predicting the coordination number may be assessed by examining the structures of

TABLE 4.1 Radius ratios and structures of some AB and AB$_2$ solids

Compound	r_+/r_-	AB Structure	Compound	r_+/r_-	AB$_2$ Structure
KF	0·98	sodium chloride	CaF$_2$	0·79	fluorite
KCl	0·73	sodium chloride	SrF$_2$	0·91	fluorite
KBr	0·68	sodium chloride	BaF$_2$	1·08	fluorite
KI	0·62	sodium chloride	MgF$_2$	0·52	rutile
RbF	0·92*	sodium chloride	TiO$_2$	0·48	rutile
RbCl	0·82	sodium chloride (and cesium chloride)	SnO$_2$	0·51	rutile
RbBr	0·76	sodium chloride	GeO$_2$	0·38	rutile and also a 4:2 form as in SiO$_2$
RbI	0·69	sodium chloride	CeO$_2$	0·72	fluorite
CsF	0·81*	sodium chloride			
CsCl	0·93	cesium chloride (and sodium chloride)			
CsBr	0·87	cesium chloride			
CsI	0·76	cesium chloride			

*These values are r_-/r_+ as the cations are larger than the anion.

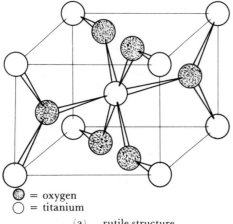

= oxygen
◯ = titanium

(a)   rutile structure

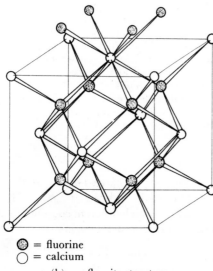

= fluorine
◯ = calcium

(b)   fluorite structure

FIGURE 4.3 *The structures of* AB$_2$ *solids*, (a) *rutile*, TiO$_2$, (b) *fluorite*, CaF$_2$

some AB and AB$_2$ compounds. Ionic solids of formula AB generally adopt either the sodium chloride or the cesium chloride structures of Figure 4.1 while the common AB$_2$ structures are those of rutile (titanium dioxide) in which the coordination is 6:3, or of fluorite (calcium fluoride) where the coordination is 8:4. These structures are shown in Figure 4.3. The structures and radius ratios of some ionic halides and oxides are listed in Table 4.1 calculated using the ionic radii of Table 2.12.

Among the AB$_2$ structures, where the eight-coordinated fluorite structure is expected to be replaced by the six-coordinated rutile structure at a radius ratio of 0·73, the agreement with prediction is remarkably good. The lower limit for the stability of six-coordination comes at 0·41, and germanium dioxide, with a ratio of 0·38, occurs both in a rutile form and in a form isomorphous with one of the silica structures where the germanium coordination number is four.

In the AB structures, the agreement is less good and the sodium chloride (or rock salt) structure persists through a wider range of radius ratios than predicted, both at the upper end as shown in the Table and at the lower end of the range as shown by the lithium halides of radius ratios down to 0·28 for LiI, all of which have the rock salt structure. However, the reasonably good general agreement shows that even this very simple model (of ions as hard spheres) is a reasonable first approximation and the radius ratio may be used with some confidence in predicting the structures of ionic crystals. The uncertainties in the values of ionic radii (see section 2.15) must be remembered in considering borderline cases.

A number of other structures adopted by crystals of more complex stoichiometry are shown in Figure 4.4. Less symmetrical configurations adopted by compounds in which the bonding is not purely ionic are discussed later.

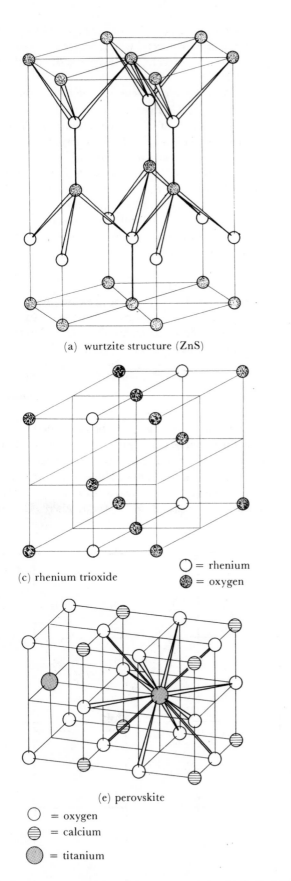

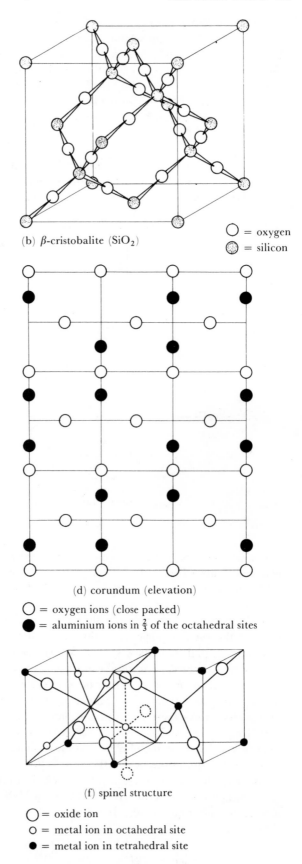

FIGURE 4.4 *Further examples of structures*, (a) *wurtzite*, (b) *β-cristobalite* (SiO₂), (c) *rhenium trioxide*, (d) *corundum* (Al₂O₃), (e) *perovskite* (CaTiO₃), (f) *spinel* (Al₂MgO₄).

These structures all have the ions in positions of maximum sym-metry with neighbours disposed regularly around them, with the exception of corundum and spinel where the oxygen positions are in a close-packed arrangement but the nearest neighbours are not of highest symmetry.

4.2 The Born-Haber cycle

What factors determine whether given elements combine to form an ionic solid? These may be found by considering the energy changes involved in the formation of an ionic solid from the elements and we shall use sodium chloride as an example. This has a measured heat of formation (H_f) of $-410\cdot9$ kJ mol^{-1}; i.e. for the reaction

$$Na_{(metal)} + \tfrac{1}{2}Cl_{2(gas)} = Na^+Cl^-_{(solid)}, \; H_f = -410\cdot9 \text{ kJ mol}^{-1}$$

This reaction may be broken down into simpler steps whose energies are known, as in the diagram of Figure 4.5, by a method due to Born and Haber.

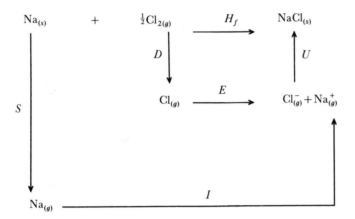

	energy changes		kJ mol^{-1}
$Na_{(metal)} = Na_{(gas)}$	heat of sublimation	S	$+109\cdot0$
$Na_{(gas)} = Na^+_{(gas)} + e^-$	ionization potential	I	$+493\cdot7$
$\tfrac{1}{2}Cl_{2(gas)} = Cl_{(gas)}$	$\tfrac{1}{2}\times$ heat of dissociation	D	$+120\cdot9$
$e^- + Cl_{(gas)} = Cl^-_{(gas)}$	electron affinity	E	$-359\cdot8$
$Na^+_{(gas)} + Cl^-_{(gas)} = Na^+Cl^-_{(solid)}$	lattice energy	U	$-764\cdot4$

FIGURE 4.5 *Born-Haber cycle for the formation of sodium chloride*

Known as the Born-Haber Cycle, this gives the net energy change calculated from the five simpler steps shown in the table below the diagram, and hence a calculated heat of formation.

In the case of sodium chloride, the values of all the quantities in the cycle are known so that a calculated heat of formation (equal to $-400\cdot8$ kJ mol^{-1}) may be found and compared with the experimental value. The close agreement confirms that the model of ions as incompressible spheres is a reasonable one in this case. Similar agreement between cycle values and experimental ones is found in many cases and this gives grounds for confidence in using the cycle in other ways. The most useful of these is the evaluation of one of the cycle quantities when the others are known. In particular, the electron affinity E which is difficult to measure is usually determined from the Born-Haber cycles of a series of appropriate salts. It can be seen from Figure 4.5 that:

$$H_f = S + I + D + E + U$$
$$\text{hence } E = H_f - U - S - I - D$$

The Born-Haber cycle may also be used to determine whether or not the bonding in a compound is purely ionic. If the calculated lattice energy and that derived experi-

mentally do not agree, this is a strong indication that the assumption of ionic forces, which is made in order to do the calculation, is incorrect. This aspect is discussed further in section 4.5.

Our main reason for discussing the formation of an ionic solid in terms of the Born-Haber cycle is to allow a more detailed assessment of the contribution of each component of the energy cycle to the heat of formation of the ionic solid. The factors which determine whether a compound is ionic or covalent may thus be isolated and discussed. These factors will be examined in turn starting with the lattice energy U which is the most important exothermic term in the cycle. Then the endothermic terms—the ionization potential, the electron affinity, and the heats of atomization—will be examined.

4.3 The lattice energy

The energy change when the gaseous ions are brought together from infinite separation to their equilibrium distances in the solid is the lattice energy, U. That is, U is the heat of the reaction (for an AB solid)

$$A^{z^+}_{(gas)} + B^{z^-}_{(gas)} = A^{z^+}B^{z^-}_{(solid)} \; \dots\dots\dots\dots U$$

This energy arises from electrostatic interactions between the ions. When the geometry of the solid array of ions is known, the lattice energy may be calculated. The method may be illustrated by considering the square two-dimensional array of ions shown in Figure 4.6. Any one cation in a square array of interatomic distance r has four anions at a distance r as nearest neighbours. The electrostatic potential energy between these ions is $4Z^+Z^-e^2/4\pi\epsilon_0 r$, where Z^+ and Z^- are

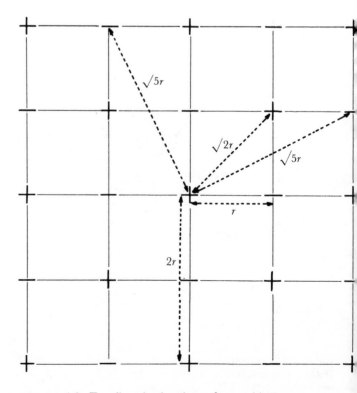

FIGURE 4.6 *Two-dimensional analogue of a crystal lattice*

the positive and negative charges on the ions (assuming an AB stoichiometry), e is the charge on the proton, and ϵ_0 is the permittivity of a vacuum. The next-nearest neighbours to the given cation are four cations at a distance $\sqrt{2}r$ and a repulsion exists between the given cation and these ions of $4(Z^+)^2 e^2/4\pi\epsilon_0\sqrt{2}r$. Then come four more cations at $2r$, eight anions at $\sqrt{5}r$ and so forth. The total electrostatic energy of a cation in this square array is thus:

$$E = 4e^2(Z^+)(Z^-)/4\pi\epsilon_0 r + 4e^2(Z^+)^2/4\pi\epsilon_0\sqrt{2}r$$
$$+ 4e^2(Z^+)^2/8\pi\epsilon_0 r + 8e^2(Z^+)(Z^-)/4\pi\epsilon_0\sqrt{5}r + \ldots$$
$$= -Z^2 e^2/4\pi\epsilon_0 r(4 - 2\sqrt{2} - 2 + 8/\sqrt{5}\ldots)$$

where $Z^+ = -Z^-$ has been put equal to Z for an AB crystal.

The convergent infinite series in the bracket may be evaluated from the geometrical properties of the array and its sum may be found. This sum is called the *Madelung constant*, A, after its first evaluator. A different geometrical array, for example a rectangular one, would have a different value of the Madelung constant and the whole analysis may be extended to three dimensions. For example, for a rock salt lattice the Madelung constant

$$A_{\text{NaCl}} = (6 - 12/\sqrt{2} + 8/\sqrt{3} - 6/2 + 24/\sqrt{5}\ldots)$$
$$= 1 \cdot 748 \ldots$$

The Madelung constant is the factor relating the electrostatic forces and the spatial arrangement of the ions in a crystal. It depends only on the geometry of the crystal and is independent of the nature or charge of the ions.

The electrostatic energy of a cation in an AB lattice may therefore be written:

$$E = \frac{-Z^2 e^2 A}{4\pi\epsilon_0 r}$$

The electrostatic energy of the anion in an AB crystal is the same as that for the cation. If a mole is considered, the electrostatic energy is E times Avogadro's constant, N. (In detail, the energy is $N \times (E_{\text{cation}} + E_{\text{anion}}) \div 2$: the division by two is to avoid counting each attraction twice over.) Thus, the electrostatic energy of a mole of an AB ionic compound is

$$NE = \frac{-Z^2 e^2 AN}{4\pi\epsilon_0 r}$$

So far, only the electrostatic forces between the ions have been considered but Born introduced a second, repulsive, energy term into the equation to take account of the repulsion which arises at very short interionic distances when the electron clouds of the ions start to interpenetrate. This force must be taken into account, as the expression above for NE tends to infinity as r tends to zero: the repulsive force rises steeply as the interionic distance decreases and is represented by a term $E_{\text{rep}} = B/r^n$ where B is a constant similar to the Madelung constant and n is also a constant, of the order of

nine for sodium chloride. The total energy of the crystal, which is the lattice energy U, is then:

$$U = NE + NE_{\text{rep}} = \frac{-Z^2 e^2 AN}{4\pi\epsilon_0 r} + \frac{NB}{r^n}$$

The constant B can be eliminated, since the lattice energy is a minimum at the equilibrium value of $r = r_0$, so that by setting $dU/dr = 0$ for $r = r_0$, it is found that B can be expressed in terms of the other constants. Then:

$$U = \frac{-Z^2 e^2 NA}{4\pi\epsilon_0 r_0}(1 - 1/n)$$

All the quantities on the right-hand side of this expression are known or may be found; r_0 comes from direct experimental evidence and n from calculation or experiment. The lattice energy U can thus be found by a combination of calculation and experimental determination of parameters—though difficulties arise in the determination of the Madelung constant for lattices where the symmetry is low.

It can be seen that the properties of an ionic solid which determine the lattice energy are the geometry, as reflected in the Madelung constant, the interionic distance, and the charges on the ions. The most important effect is that of the charge which appears as a squared term. The lattice energy of an $A^{2+}B^{2-}$ solid is four times that of an A^+B^- solid of the same geometry.

Of the geometrical factors, the Madelung constant has only a minor effect on the lattice energy as the values of the constant for different structures of the same stoicheiometry are similar. For example, in AB structures the Madelung constants are as follows:

8-coordinated, CsCl structure, $A = 1 \cdot 763$
6-coordinated, NaCl structure, $A = 1 \cdot 748$
4-coordinated, wurtzite structure, $A = 1 \cdot 641$

When the stoichiometry changes, the Madelung constant changes to a greater extent, for example, $A = 5 \cdot 04$ for the fluorite structure; but changes in the stoichiometry imply changes in the charges and it is difficult to make comparisons between different formula types.

Variations in the second geometrical factor, the interionic distance, have a more important effect on the size of the lattice energy. The lattice energy depends inversely on r_0 so that, within a given structure type, the lattice energy will fall as the ionic sizes increase. This is illustrated for compounds of the sodium chloride structure in Figure 4.7. It will be clear from the discussion of section 4.1, that the ratio of the anion to cation sizes should not vary too widely as r_0 varies or some other structure will become the more stable one. The effect of variations in charge and interionic distance is best seen by comparing the lattice energies of a series of compounds of the same structure formed by related elements.

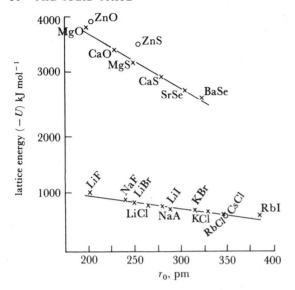

FIGURE 4.7 *Lattice energies* (plotted as $-U$)
Lattice energies are plotted as a function of r_0 for a number of cases of the sodium chloride structure and for CsCl and ZnS. The marked effect of the ionic charge is clear: it will also be noted that the variation in structure to 8:8 or 4:4 coordination has only a minor effect.

This is done in Figure 4.7 where the lattice energies of the alkali halides and the alkaline earth chalcogenides of the sodium chloride structure are plotted. The marked effect of changes in charge is very obvious, as well as the linear fall in lattice energy as the ions (and therefore the interionic separation) increase in size. The values for two other compounds of the AB type but of different structures, CsCl and ZnS (the latter having the wurtzite structure) are also shown in the Figure to show the effect of the Madelung constant.

To summarize, the lattice energy, and therefore the probability that the formation of an ionic solid will be energetically favourable, is increased by (i) increasing the charges on the ions, (ii) decreasing the interionic separation, i.e. by having small ions, and (iii) by changes in the Madelung constant, though this effect is relatively small.

4.4 The endothermic terms in the formation of an ionic solid

In general, the largest of the energy contributions which have to be added to the system before an ionic solid is formed is the ionization potential of the cation, I. If the formation of an ionic solid is to be favoured, the ionization energy should be as small as possible. From the table of ionization potentials (Table 2.8), it will be seen that the ionization potentials of the elements have the following characteristics:
(i) The ionization potentials increase from left to right across a Period.
(ii) In any Group, the ionization potential decreases with increasing size down the Group.
(iii) For a given element, the first ionization potential is less than the second, which in turn is less than the third, etc.
It thus requires least energy to form cations of large atoms on the left of the Periodic Table, that is of the larger elements of the alkali and alkaline earth elements. The higher the

charge on the ion, the greater is the energy required in its formation.

The energy of formation of the anion is the electron affinity. The electron affinities (compare section 2.14) may represent an endothermic or an exothermic contribution to the heat of formation of an ionic solid, but even the most exothermic electron affinity is less than the smallest ionization potential so that the formation of gaseous cation plus anion from the gaseous atoms is endothermic for any pair of elements. Only a few electron affinities are exothermic and these are all for the formation of singly-charged anions. The formation of doubly-charged anions is always a strongly endothermic process and this true *a fortiori* for anions of higher charge.

The heats of atomization of the elements in their standard states depend very much on the form in which the element exists. Little in the way of generalization can be said except that where elements in a Group occur in the same form—as for example, the halogens—there is a tendency for the heat of atomization to decrease with increasing atomic weight, but often with the lightest element anomalous. As the values for sodium chloride show, the heats of atomization commonly represent only a minor contribution to the energy balance.

The effect of the endothermic terms on the heat of formation of an ionic compound may be summarized by saying that least energy is required, and therefore formation of an ionic solid is most favourable, when the ions are (i) of low charge, (ii) large, so that the interionic distance in the solid is large, and (iii) formed from elements at the extremes of the Periodic Table.

If these factors are compared with those which lead to a high lattice energy, it is seen that the two main requirements for low endothermic energies—large ions of low charge—are exactly opposed to those which favour high lattice energies. The small ions of high charge which give the highest exothermic contribution are precisely those which require the highest endothermic energies of formation from the elements in their standard states. The formation of an ionic solid therefore depends on the detailed balance of all the energy contributions in each individual case. Once again, the direction of the chemical change depends on small differences between large values and is difficult to predict *a priori*. In general, ionic solids are formed by the s metals and the transition elements in the II or III oxidation states (though often as complexes) with anions from the halogen or chalcogen Groups. Simple ions with charges greater than two are less common, but the lanthanide elements form M^{3+} ions, and the existence of Th^{4+} is well established.

If the values of the ionization potentials are examined in detail, it will be seen that successive ionization potentials for an element rise in a fairly regular manner as long as only electrons in the valence shell are removed. Since lattice energies vary as the square of the charge on the ions, it might seem that, on balance, ion formation would be most favoured in cases where the charges are high. Unfortunately, a further complication appears which upsets this conclusion, in that highly-charged ions are those most likely to cause polarization and a departure from purely ionic bonding.

4.5 Bonding which is not purely ionic

In the discussion above, the assumption was made that ions were hard spheres and that the only forces (except at very short distances) were electrostatic ones between the charges. That this is a reasonable approximation in many cases is shown by the agreement between the calculated and measured heats of formation of the solids. However, in a solid such as lithium iodide where the cation is very small and the anion is large (Figure 4.8) the high charge density on the cation distorts the rather diffuse electron cloud of the anion as indicated in the figure. The centre of negative charge in the anion no longer coincides with the centre of positive charge and an induced dipole is present in the anion. In such a case, the anion is termed *polarizable* and the cation,

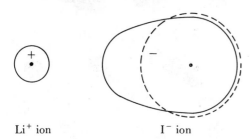

Li$^+$ ion   I$^-$ ion

FIGURE 4.8 *Diagrammatic representation of polarization in* LiI

polarizing. This polarization represents a departure from purely ionic bonding in the compound. Thus the elementary idea of ions as hard spheres has to be modified by allowing for distortions of the spherical electron clouds where ions have high charge densities.

In a similar way, the purely covalent bond discussed in the last chapter is uncommon. The pair of electrons which form a covalent bond are equally shared between the constituent atoms only if these are identical or, by coincidence, of the same electronegativity. If the two atoms differ in electronegativity, the bonding electrons are more strongly attracted by the more electronegative and a dipole is created in the bond (Figure 4.9). The pure covalent bond with the electrons equally shared between the two atoms and the pure ionic bond with the electron completely transferred from one atom to the other are the two extreme cases. There is a complete range of bond types lying between the two, the ionic bond distorting in the manner of Figure 4.8, and the covalent

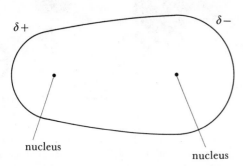

$\delta+$ $\delta-$

nucleus   nucleus

FIGURE 4.9 *Polarization in a covalent bond*

bond distorting in the manner of Figure 4.9, till the polarized covalent bond and the polarized ionic bond become indistinguishable.

There is one reservation about this picture: the pure covalent bond does exist (between two atoms of the same element) but the pure ionic bond is an abstraction as there is bound to be some polarization between any pair of ions. However, since the environment of an anion in a crystal is symmetrical, the slight polarizations in a strongly ionic compound have only a minor effect on the bond energy and can be regarded as a normal attribute of ionic solids.

The polarizing power of the smallest ion (which is generally the cation) depends on the density of charge on it. The polarizing power is thus greatest for small, highly-charged ions like Mg^{2+} or Al^{3+}: so much so that the smaller congeners of these ions, Be^{2+} and B^{3+}, do not exist and beryllium and boron are covalently bound in their compounds. Polarizability is greatest for large ions like I$^-$, and especially for those with diffuse electron clouds like H$^-$ or N^{3-}.

A marked degree of polarization shows up in the lattice energy values. When a dipole is induced in one ion, there is an ion-dipole attraction to be added to the ion-ion attraction which is used in the calculation of lattice energies on the purely ionic model. The actual lattice energy of a polarized solid should thus be higher than that calculated on a purely ionic model. The values in Table 4.2 illustrate this.

TABLE 4.2 Calculated and experimental lattice energies, $-U$

| Compound | Lattice energy (kJ mol^{-1}) | | Difference |
	Experimental	Calculated	
LiF	1009·2	1019·2	−2·4
NaF	903·9	900·8	+3·1
LiH	905·4	979·1	−73·7
NaH	810·9	845·2	−34·3
KH	714·2	741·4	−27·2
AgF	954	920	+34
AgCl	904	833	+71
AgBr	895	816	+79
AgI	883	778	+105
KF	801·2	805·4	−4·2
KCl	697·9	702·5	−4·6
KBr	672·4	674·9	−2·5
KI	631·8	637·6	−5·8
TlCl	732	686	+46
TlBr	720	665	+55
TlI	695	636	+59
RbCl	677·8	677·8	0
RbBr	649·4	653·1	−3·7
RbI	613·0	619·2	−6·2

The differences for the alkali halides are less than one per cent and the direction of deviation is random. The alkali hydrides have experimental values which differ from the values calculated on an ionic model by up to eight per cent and the experimental values are all low. The difference is greatest for the lithium compound, showing the large polarizing effect of the small lithium ion. The sign of the deviation probably results from the unusual compressibility of the hydride ion (compare section 8.1).

The values of the silver and thallium(I) halides are given for comparison with those of the alkali metals of most similar radius, e.g. potassium and rubidium. The figures illustrate the effect of the filled d shell in such ions. The values for the silver and thallous halides (which have the sodium or cesium chloride structures) show marked discrepancies between the calculated and experimental lattice energies and these differences increase from the fluoride to the iodide. In these cases the experimental values are all higher than the calculated ones, showing the presence of the additional energy term due to polarization, which is largest for the iodides. It is interesting to note that gold(I) iodide, AuI, which has a markedly higher lattice energy $(-1050 \text{ kJ mol}^{-1})$ than silver iodide, has had its structure determined. This consists of chains . . . $Au-I-Au-I-$ . . . and is clearly not ionic at all. The silver salt thus provides an intermediate case between the ionic alkali metal iodides and the covalent aurous iodide. This example illustrates that the structure of a solid may provide an additional criterion for a departure from ionic bonding.

It is characteristic of an ionic solid that the forces are equal in all directions so that a symmetrical arrangement of ions results. At the other extreme, in a molecular solid formed by a covalent compound, there are two quite different kinds of force: very strong interactions in the bonds between the atoms of the molecule, and very weak interactions between molecules. When the solid is intermediate between these extreme types, there is often stronger bonding in some directions than in others and this shows up in a lowered symmetry of the structure. Thus chains may be formed as in gold(I) iodide above, and another common deviation from ionic bonding is the formation of layer lattices as in CdI_2 (Figure 4.10a and Plate I). In general, departure from purely ionic bonding is shown by the adoption of lattices where ions do not have their neighbours in the most symmetrical possible environments. One example is nickel arsenide which is an AB structure with 6:6 coordination but, although the six neighbours of the nickel atom are disposed octahedrally, the six neighbours of the arsenic atom lie at the corners of a trigonal prism. Some of the commonly adopted structures of this less regular type are shown in Figure 4.10.

It must be noted that, although structures where the atoms are in environments of low symmetry indicate the presence of non-ionic contributions to the bonding, the converse is not true. Many compounds with symmetrical structures are not ionic. For example, many compounds of the transition metals with small atoms, the so-called 'interstitial' compounds, such as TiC or CrN, have the sodium

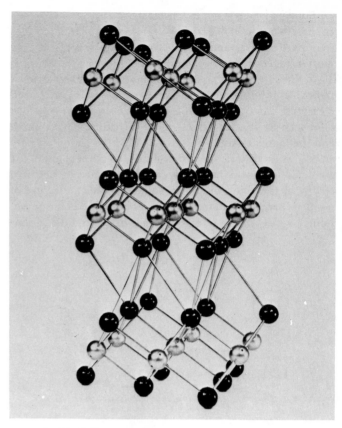

PLATE I   *The* CdI_2 *structure*

Photograph of a model of the CdI_2 structure showing the layers of metal atoms (silver) sandwiched between layers of iodines (black). Compare with Figure 4.10a, which shows a smaller portion of the structure in a perspective that illustrates the hexagonal cell.

PLATE II   *The nickel arsenide structure*

A model of the nickel arsenide structure showing the trigonal prism of Ni atoms around the black As atom. The octahedral coordination of the Ni is best seen for the atom at the centre of the model.

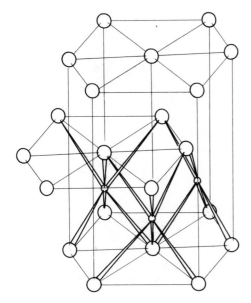

O = iodine atom

o = cadmium atom

(a)

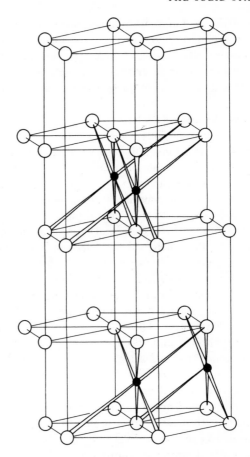

● = bismuth atom

O = iodine atom

(b)

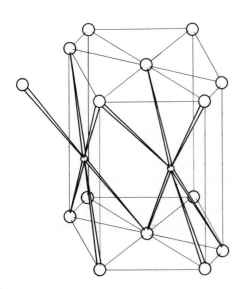

O = As atom

Ō = Ni atom

(c)

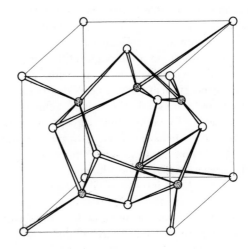

FIGURE 4.10 *Less regular structures:* (a) *cadmium iodide* (see also Plate I which shows three I—Cd—I layers), (b) *bismuth triiodide*, (c) *nickel arsenide* (see also Plate II which shows a larger portion of the structure), (d) Mn_2O_3

In these structures some or all the ions have their neighbours in coordination positions which are not of the highest possible symmetry and this reflects the directional, non-ionic character of part or all of the bonding forces.

⊛ = oxygen atom

O = manganese atom

(d)

chloride structure but there is no question of these compounds containing C^{4-} or N^{3-} ions and their bonding has appreciable metallic character. Similarly, the silver halides have the sodium chloride structure although there is appreciable covalent character in the bonding as just discussed.

4.5 Metallic bonding

It has been possible, so far, to present bonding in compounds in terms of a linear variation between pure ionic bonding and pure covalent bonding, but this is an over-simplification and a third type of bond must be considered, though only briefly. This is metallic bonding. While elements to the right of the Periodic Table favour electron pair sharing and are covalently bound in their elemental form, those to the left, which readily lose their valency electrons, are metallic elements. A metal has been picturesquely described as 'an array of cations in a sea of electrons'. The cations usually assume one of three simple arrangements described below and the valency electrons become completely delocalized over the whole structure. The electrons are mobile, accounting for the typical metallic properties of high electrical and thermal conductivity, and there are no underlying directed bonds. From one point of view, metallic bonding is the limit of the process of delocalizing electrons, only here the electrons are σ ones and not π electrons as in the examples of last chapter. If (instead of constructing two-centre orbitals as in covalent compounds) delocalized orbitals with an infinite number of centres are constructed and then electrons are placed in these orbitals so that they are evenly shared by the infinite number of cations, the result is a wave-mechanical description of a metal where the electrons are standing waves over the whole crystal.

It is not possible to go into further detail here, but the point to grasp is that there are three basic types of bonding: ionic, involving complete transfer of an electron from one atom to another; covalent, with the sharing of a pair of electrons between two atoms; and metallic, where the electrons are completely delocalized over the crystal. If these three extreme types are thought of as being placed at the corners of a triangle, then some compounds will be represented by points near the vertices of this 'bond triangle', with bonding predominantly of one type. Some compounds will be represented by points along an edge of the triangle, with bonding intermediate between two types. Finally, the majority of compounds would be represented by points within the area of the triangle, showing that the bonding had some of the characteristics of all three types. This idea is illustrated schematically in Figure 4.11.

There are three common metallic structures which are illustrated in Figure 4.12, and nearly all metals adopt one or other of these. Two are based on close-packing of spheres, i.e. the metal ions (assumed to be spherical) are arranged to fill the space as closely as possible. For close packing, a layer of spheres can be arranged in only one way, as shown in Figure 4.12a, in which each sphere is in contact with six neighbours. A second layer can be arranged on top of the first, again in only one way if close packing is to be preserved,

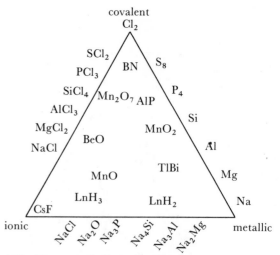

FIGURE 4.11 *Diagrammatic illustration of bond types*
Few bonds are purely ionic, covalent, or metallic and most have some characteristics of all three types and would lie within a triangular plot of the type shown. This presentation also emphasizes that there is no sharp boundary between bonds of different types.

and this is shown in Figure 4.12b. There are two possible ways of adding the third layer, both preserving close packing. These are either directly above the first layer (Figure 4.12c) or in a third position illustrated in Figure 4.12d. The first case then gives a repetition of the second layer for the fourth one, the first and third layers for the fifth one, and so on in an ABABAB . . . arrangement. This gives rise to *hexagonal close packing* (*hcp*) which is shown in Figure 4.12e and in Plate IIIa where the ABABA arrangement is more obvious. Plate IIIb shows another view of the hcp model illustrating the close-packed layer.

In the second case, the arrangement of the first three layers may be labelled ABC and the repeated pattern is then ABCABCABC . . . , to give *cubic close packing* (*ccp*). This is shown in Figure 4.12f and in Plate IV. In this structure, the close-packed layer is not parallel to the base of the cube, as it was in hcp, but lies along the body diagonal of the cubic array as indicated by the shaded plane in Figure 4.12f. Put another way, the close-packed layers of Figure 4.12d have to be turned through 45° to give the unit cube. Plate IVa shows a model of the ccp array while Plate IVb shows the model with spheres removed to illustrate the close-packed plane parallel to the body diagonal. An alternative name for ccp is *face-centred cube* (*fcc*) describing the orientation of Figure 4.12f.

The third common metal structure is the *body-centred cube* (*bcc*), shown in Figure 4.12g, which is not close-packed and in which the coordination number is eight. These high coordination numbers are typical of metal structures.

A number of common crystal structures are closely related to one or other of the close-packed structures above. In cubic close packing, there are two different kinds of interstitial sites, see Figure 4.12b. One of these is a tetrahedral site between four of the spheres, and the other is an octahedral site between six of the spheres. There are as many octahedral sites as there are spheres, and twice as many tetrahedral sites as spheres. Now suppose that a small atom was inserted into

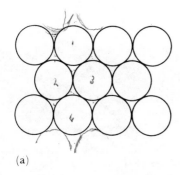

(a)

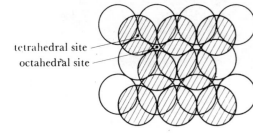

tetrahedral site
octahedral site

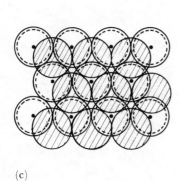

(b)

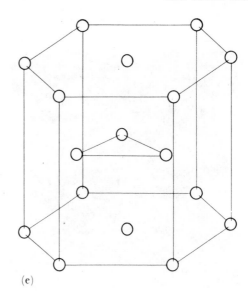

(e)

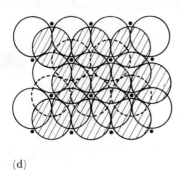

(c)

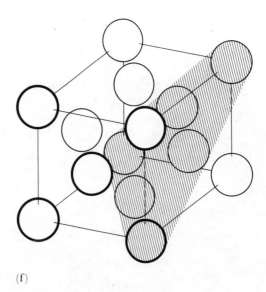

(f)

(d)

FIGURE 4.12 *Common metal structures and their construction:* (a) *a close-packed layer of spheres,* (b) *addition of a second layer to* (a) *in close-packing,* (c) *the addition of a third layer to* (a) *and* (b) *directly above* (a), (d) *alternative way of adding third layer,* (e) *hexagonal close-packing,* (f) *cubic close packing (fcc), close-packed layer shaded,* (g) *the body-centred cube*

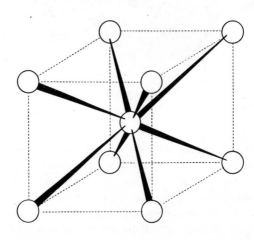

(g)

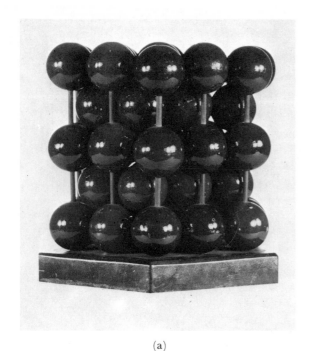

(a)

(b)

(a)

(b)

PLATE III (a) *Model illustrating hexagonal close packing. The ABABA arrangement is clearly seen.*
(b) *Another view of the model in (a) showing the close-packed layers.*

PLATE IV (a) *Model illustrating cubic close packing, in the face-centred cube orientation.*
(b) *The model of (a) with close-packed layer shown at 45° to fcc edges.*

each octahedral site in a close-packed structure, leaving the large atoms in contact, the structure which results is of formula AB. The two kinds of atoms are six-coordinated and this is the same as the sodium chloride structure. Thus the sodium chloride structure may be described as derived from cubic close packing with all the octahedral sites occupied. Of course, in some compounds with the sodium chloride struc-

ture, the ions are of such relative sizes that the smaller ones force the larger ones out of contact with each other so that they are no longer close packed, but their relative positions remain the same and the structure is often described in terms of the close-packed form.

Many of the other common structures may be described in similar terms, with the larger ions—usually the anions—in cubic or hexagonal close packing, and the smaller ions occupying some fraction of the octahedral or tetrahedral sites in the close-packed structure. Table 4.3, at the end of the Chapter, summarizes the common structures in terms both of the coordination numbers and of their relation to the close-packed structures.

Apart from metals and alloys, metallic bonding is found in a number of other types of compound. One class consists of the so-called 'interstitial' hydrides, carbides, nitrides and borides of the transition metals. These compounds are formed with a variety of compositions—W_2C, TiN, ZrH_2 etc.—which do not commonly fit any normal ideas of valency. The compounds have metallic properties, such as conductivity and magnetic ordering but have different structures from the parent metals. The bonding is still the subject of some controversy but most theories agree in leaving some valency electrons in conduction bands to give metallic properties. Thus these compounds may be pictured as, for example, ionic structures but with additional electrons providing metallic properties (see section 8.5 on the metallic hydrides). Many low oxidation state halides of the transition metals may belong to this class. These compounds, which exist only in the solid state, like ThI_2 or $NbCl_2$ may be regarded as containing M^{v+} cations (v = normal Group oxidation state of the metal) and v electrons which form anions and provide conduction electrons. Thus, ThI_2 would contain Th^{4+} ions, two I^- ions and two conduction electrons per formula unit.

A more restricted form of metallic bonding, which emphasizes the similarity between metallic and delocalized covalent bonding, is found in the 'cluster' molecules and ions which are being explored. Examples include the $M_6X_8^{4+}$ ions of molybdenum and tungsten (section 14.4), the $M_6X_{12}^{2+}$ ions of niobium and tantalum (section 14.3) both of which contain octahedra of metal atoms held together both by halogen bridges and by delocalized metal–metal bonding within the cluster. Other clusters include the triangle of rhenium atoms in $Re_3Cl_{12}^{4-}$ and the Bi_9^{5+} cluster in lower bismuth halides (section 15.6). The latter is interesting as the metal atoms are held together by delocalized metallic bonding alone, without the addition of bridging groups. The bonding in boron hydrides and their ions is also similar. Thus there is no clear division between metallic and delocalized covalent bonding, the more metallic species being those where the electrons are the more mobile.

Finally, it must be remarked that there exists a wide variety of simple bonds between two metal atoms, many examples of which have long been known. For example, a variety of compounds R_6M_2 for M = Pb, Sn, Ge, and R a variety of organic groups; $(CO)_5Mn-Mn(CO)_5$; a number of transition metals bonded to Hg, Ge, Sn as in $Pt(SnCl_3)ClL_2$ (L = π-bonding ligand) and so on. All these cases involve a direct metal–metal bond and a number, especially the transition metal non-transition metal bonds, probably involve π as well as sigma bonding.

This field of metal–metal bonding in clusters or small molecules has attracted a lot of attention recently. One hope is that these studies may throw some light on the action of metallic catalysts by presenting, as it were, a small portion of metallic structure which can be studied in solution.

4.6 Complex ions

Although the discussion so far has been concerned with simple ions, most of the points made apply to complex ions as well. Many of the structures of compounds containing complex ions are simply related to those discussed for simple ions. For example, the relationship between the calcium carbide structure of Figure 4.13, and the sodium chloride

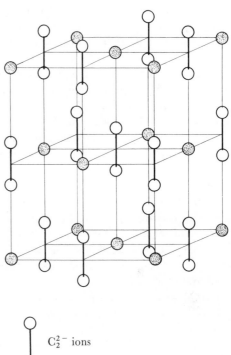

C_2^{2-} ions

Ca ions

FIGURE 4.13 *The calcium carbide structure*

structure is obvious. Similarly, sodium nitrate, calcium carbonate, and potassium bromate all have structures in which the anion occupies the chloride ion position of the sodium chloride structure. Sodium iodate has a cesium chloride structure while potassium nitrate and lead carbonate have nickel arsenide structures. In most cases of complex ion salts, as in all the above examples, the actual symmetry of the lattice is lower than that of the sodium chloride lattice as the complex ions are not spherical. The calcium carbide lattice, for example, is elongated in the direction parallel to the axes of the carbide ions to give a tetragonal rather than a cubic lattice. However, there are a number of cases where complex ionic lattices are of high symmetry. This occurs if the complex ion is able to rotate in its lattice position at room temperature. The resulting, averaged-out, configuration is spherical as in the alkali metal borohydrides, MBH_4. These compounds have actual sodium or cesium chloride configurations at room temperatures as the BH_4^- ions are freely rotating. An interesting intermediate case is provided by the alkali metal hydroxides. These have lattices of low symmetry at room temperature, in which the OH^- groups are not rotating. When the temperature is raised, enough energy is present to allow the hydroxide ions to rotate and the alkali hydroxides undergo a transition to a high temperature form with the sodium chloride structure. Complex cations behave in exactly the same way; most ammonium salts, for example, have sodium chloride or cesium chloride structures as the NH_4^+ ion is freely rotating. Hydrated and other complex cations also typically form lattices of high symmetry with large anions. Thus $[Co(NH_3)_6][TlCl_6]$ has the sodium chloride structure while $[Ni(H_2O)_6][SnCl_6]$ crystallizes in the cesium chloride structure.

Compounds of more complicated formula type also mirror the structures of the simple ionic types. One example is provided by the many compounds which crystallize with the K_2PtCl_6 structure. This is related to the structure of calcium fluoride but with the cations in the fluoride positions and the large anion in the calcium site (called the *anti-fluorite* structure).

Compounds containing complex ions may also be found with the less symmetrical structures associated with significant non-ionic contribution to the bonding. Lead carbonate and one form of calcium carbonate have the ions in the same positions as in nickel arsenide while a number of complex fluorides, such as K_2GeF_6, have layer structures related to that of cadmium iodide.

When complex ions are present, a permanent dipole often exists, and the interaction with this dipole has to be added to the interactions between ions, and between induced dipoles and ions, which have already been discussed. One example is provided by the hydroxide ion where there is a permanent dipole $O^{\delta-} - H^{\delta+}$ in addition to the negative charge. In the high-temperature form where the hydroxyl group is freely rotating, the ion-dipole forces are equally directed, but when the hydroxyl groups become fixed in orientation, the existence of the dipole means that the forces between anion and cation differ in the direction of the dipole from those in directions perpendicular to the dipole. Salts containing small complex ions may thus be equivalent to those with simple ions when the complex ions can rotate freely, or the presence of permanent dipoles may introduce a directional element into the bonding in the crystal. Just as simple ions may occur in compounds which are ionic or which have non-ionic contributions to the forces in the crystal, so do small complex ions occur in symmetrical crystals which are ionic to a high degree of approximation and also in less symmetrical compounds with layer or other 'non-ionic' structures.

In addition to compounds with small, discrete, complex ions, there are very extensive series of compounds of large condensed ions, especially those containing condensed oxyanions such as polyphosphates or silicates. Such compounds form a vast topic of their own and the structural problems involved have often been very difficult to study. A number of cases will be met later in this book but a brief survey of the silicates is given here to illustrate some of the general structural characteristics of such compounds.

4.7 Silicates

The $-Si-O-Si-$ linkage forms very readily by elimination of water between two $Si(OH)$ groups and a very wide variety of silicates is found. As far as is known, the coordination number of silicon with respect to oxygen is always four, and all the silicates are built up from SiO_4 tetrahedra.

Simple silicate anions, SiO_4^{4-}

This type of silicate is represented by Mg_2SiO_4 where the magnesium ions are surrounded octahedrally by oxygen atoms. The structure is a general one found in a wide class of minerals. The magnesium ions may be replaced by any other divalent cations of about 80 pm in diameter, such as Fe^{2+} or Mn^{2+}. The mineral *olivine* is a naturally occurring example of such a magnesium silicate where about one in ten of the magnesium ions is replaced by a ferrous ion.

A less simple silicate which still contains discrete silicate ions, is the scandium compound, $Sc_2Si_2O_7$. The scandium ions lie within an octahedra of oxygen atoms and the anions are $Si_2O_7^{6-}$ ions formed by joining two SiO_4 tetrahedra through an oxygen atom. These two anions are shown in Figures 4.14a and b. Although the simple silicate ion, SiO_4^{4-}, occurs in a fairly common class of minerals, examples of the disilicate anion, $Si_2O_7^{6-}$, are rare and the scandium silicate appears to be the only one with a well-established structure.

Rings and chains, SiO_3^{2-}

If each SiO_4 tetrahedron shares an oxygen with each of two neighbouring tetrahedra, ring anions (Figure 4.14c) or chain anions (Figure 4.14d) result. Two different ring sizes are known. The smaller contains three SiO_4 tetrahedra linked together as $Si_3O_9^{6-}$ in, for example, $BaTiSi_3O_9$, while the larger has six linked tetrahedra in the ion $Si_6O_{18}^{12-}$. The latter is found in *beryl* (emerald), $Be_3Al_2Si_6O_{18}$. In the structure of these minerals, the rings are arranged in sheets with their planes parallel and the metal ions lie between the sheets, binding the parallel rings together by electrostatic forces. The barium, titanium, and aluminium atoms are coordinated to six oxygen atoms while the smaller beryllium atom is four-coordinate to oxygen.

The chain structure is found in many minerals and these are known collectively as *pyroxenes*. Examples are $MgSiO_3$ and $CaMg(SiO_3)_2$. In these compounds, the silicate chains lie parallel to each other and the cations lie between the chains and bind them together. The magnesium ions are six-coordinate to oxygen while the larger calcium ions are eight-coordinated.

Two chains may be linked together to form double chains which correspond to the formula, $Si_4O_{11}^{6-}$ (Figure 4.14e). These double chain structures are also common and are represented by the class of minerals called *amphiboles*, which includes most asbestos minerals. One example is

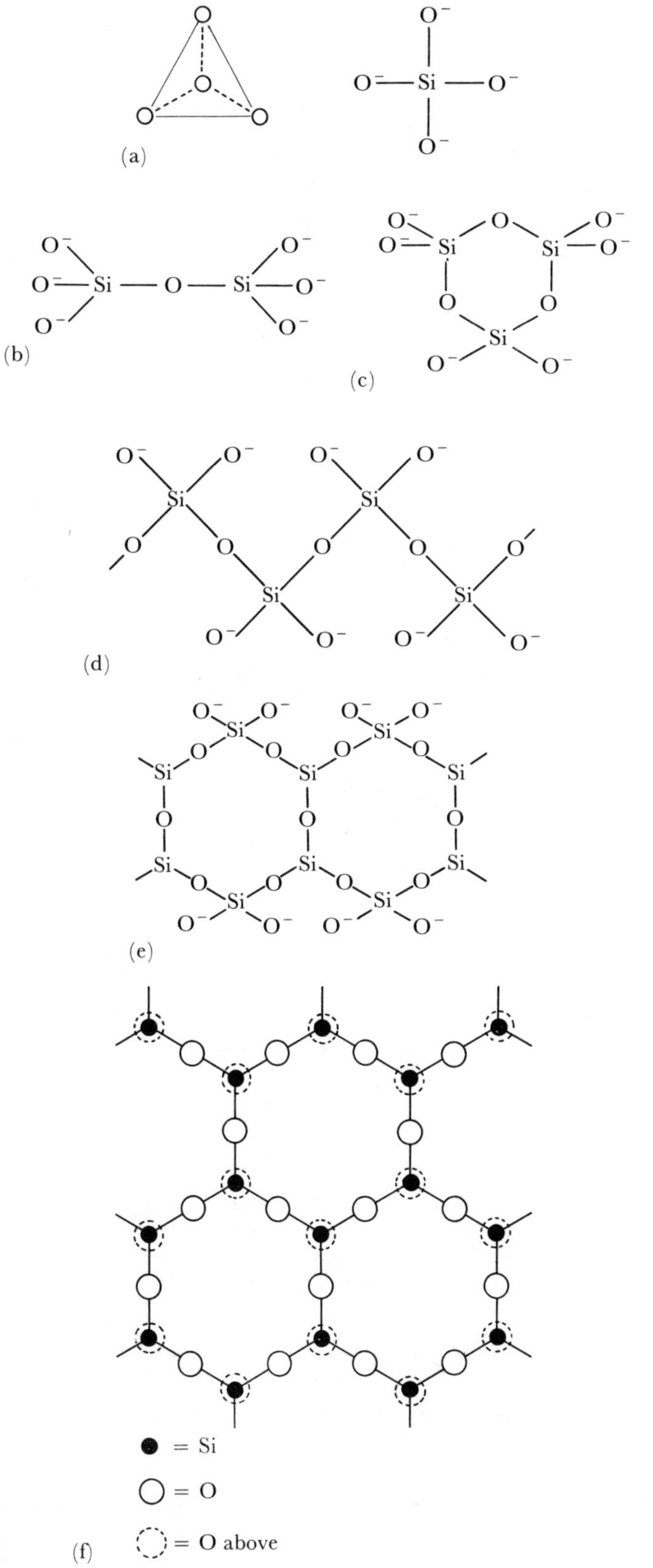

FIGURE 4.14 *Structural units in silicates:* (a) *simple* SiO_4^{4-} *tetrahedron,*
(b) *linked tetrahedra in* $Si_2O_7^{6-}$, (c) SiO_3^{2-} *rings,* (d) SiO_3^{2-} *chains,*
(e) *double chains,* (f) *sheet structures*
The SiO_4 coordination unit in all these structures is always tetrahedral, hence the chains, rings and sheets are puckered structures.
In diagrams of these complex structures, it is convenient, as here,
to draw the projection on the base plane of the tetrahedral units.

tremolite, $Ca_2Mg_5(Si_4O_{11})_2(OH)_2$, where the magnesium
ions are coordinated to six oxygens, three from the silicate
and three from the hydroxyl groups. Tremolite is the mineral
to which the name asbestos was first given, although the
term is now applied more widely. As with all minerals, the
above composition is an idealized one and other ions of
similar size may be present. In particular, some of the magnesium may be replaced by iron, and when the iron content
rises to about 2 per cent the mineral is termed *actinolite*.

Tremolite is also characteristic of the amphiboles in containing OH groups in the structure.

Sheet structures, $Si_4O_{10}^{4-}$
If the SiO_4 tetrahedra have three corners in common with
adjacent tetrahedra, the sheet structure of Figure 4.14f
results. A wide variety of sheet silicate minerals is known
and these usually contain hydroxyl groups, e.g. *talc*,
$Mg_3(OH)_2Si_4O_{10}$. It is also common to find that some of the
silicon atoms in the sheets have been replaced by aluminium
atoms which are of the same size. This happens, for example,
in the *micas* such as *phlogopite*, $KMg_3(OH)_2Si_3AlO_{10}$. In
addition to the micas, the main minerals in the group of
sheet structures are the *clay minerals* such as the kaolins,
used in ceramics, vermiculite, used as a soil conditioner, and
bentonite which is used as a binder and adsorbent.

Three-dimensional structures
If SiO_4 tetrahedra share all four oxygen atoms with adjacent
tetrahedra, the infinite three-dimensional structure of silica,
SiO_2, results (see Figure 4.4b). It is possible to replace some
of the silicon atoms by aluminium atoms, which are very
similar in size, and then cations must be introduced to
balance the charges in these three-dimensional aluminosilicates. These minerals are represented by the important
class of *felspars* which are the most abundant of all the rock-
forming minerals. Examples include *orthoclase*, $KAlSi_3O_8$,
and *anorthite*, $CaAl_2Si_2O_8$.

A further important set of aluminosilicates are the *zeolites*.
These have structures with very open lattices with channels
running through them, which can take up small molecules
such as water or carbon dioxide. In these minerals, the
positive ions are readily exchanged for others by treatment
with a strong solution of an appropriate salt and the zeolites
have found extensive uses as ion exchangers, especially in
water softeners.

This brief outline of silicate structures gives only a slight
indication of the rich variety of structures which are known.
The number of structures which exist arise mainly from the
different ways in which the SiO_4 tetrahedra can link up,
and from the possibility of replacing some of the silicon atoms
by others of similar size, especially aluminium. A similar
proliferation of structures based on linked MO_4 tetrahedra
are found in other polyanions, especially the polyphosphates
and polyvanadates although these are less rich in variety.
There are also many examples of polyanions built up from
units other than tetrahedra. Thus, polyborates comprise
planar BO_3 units as well as tetrahedral BO_4 ones, while
polyanions of larger elements, like the very extensive series

of polymolybdates and polytungstates, are built up from MO_6 octahedra.

4.8 The crystal structures of covalent compounds

In the largely ionic or metallic compounds discussed above there is a reasonable uniformity of bond strength throughout the crystal. Either all the bonds are identical, as in salts of simple ions and in elemental metals, or, even when there are strongly bonded units within the solid, as in the salts of complex ions, these units are bonded together by strong ionic forces. Such relative uniformity of bond strength is reflected in the hardness and fairly high melting points which are common among the compounds discussed above. When compounds in which the bonding is largely covalent are examined, these properties are no longer found except in relatively few examples, like diamond or silica, which are very hard and high melting. The vast majority of covalent compounds form soft, low-melting solids of low crystal symmetry, and the actual structure of the solid form of these compounds is usually of relatively minor importance for an understanding of their chemistry.

The hard, high-melting covalent compounds are those where covalent, electron pair bonding extends throughout the crystal so that the whole crystal is a 'giant molecule'. An obvious example is diamond in which each carbon atom is bonded tetrahedrally to four others and this structure continues throughout. In order to melt or fracture such a crystal, strong covalent bonds must be broken and this requires considerable energy. Similar examples are silica, with four-coordinate silicon and two-coordinate oxygen throughout, silicon and silicon carbide and the tetrahedral form of boron nitride—all with the diamond structure—and oxides of other elements of moderate electronegativity, such as aluminium.

The structures of the elements in the carbon Group give a good illustration of this type of giant molecule and the effect of deviations from it. Carbon, silicon, germanium, and tin (in the grey form) all occur with the diamond structure but the valency electrons become increasingly mobile with increasing atomic weight so that silicon, germanium, and grey tin have conductivities increasing in that order. These conductivities are much higher than those of insulators, such as diamond, but many times less than the conductivities of true metals. Such compounds are called *semi-conductors*. The valency electrons are largely fixed in the bonds but have a small mobility, in other words, these elements are at the start of the trend from covalent to metallic properties.

White tin and lead have different structures from the diamond one (Figure 4.15) and tend much more towards metallic forms. White tin has the atoms in a distorted octahedral configuration with four nearest neighbours and two more a little further away, while lead has an approximately close-packed structure but with an abnormally large interatomic distance. These two elements represent further steps away from the covalent, low-coordinate structures towards the high-coordinate structures with close-packing which are typical of metals. Both have conductivities in the metallic

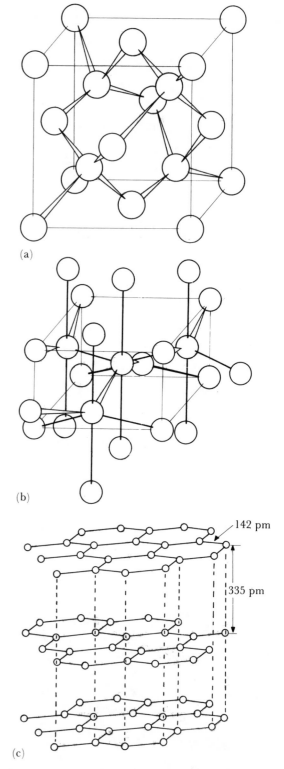

(a)

(b)

(c)

142 pm

335 pm

FIGURE 4.15 *Structures of Group IV elements:* (a) *diamond*, (b) *white tin*, (c) *graphite*

range, and thus mobile electrons. However, even lead differs a little from the typical metal in its high interatomic spacing. These elements would all lie along the covalent-metal edge of the ionic-covalent-metallic triangle of bond types with diamond at the covalent apex, silicon, germanium and grey tin near the covalent end of the edge, and white tin and lead nearer the metallic end of the edge.

The other structure shown by an element in this Group is the graphite form of carbon. This is a much softer solid than diamond with a pronounced horizontal cleavage so that it readily flakes. These properties reflect the bonding as the carbons are three-coordinated in a planar sheet, and the fourth electron and the fourth orbital form delocalized π bonds extending over the sheet. The sheets of carbon atoms are very strongly bound but the forces bonding the sheets together are relatively weak (shown, for example, by the $C-C$ distances of 142 pm within the sheets and 335 pm between them), so the sheets readily slide over one another, giving graphite its lubricating properties. Thus the strength and external properties of the solid as a whole reflect the bonding, especially the weakest links in the solid.

When ordinary covalent compounds are considered, it is the weak intermolecular forces which are reflected in the crystal properties. In a simple molecule such as methane, the atoms are strongly linked together by directed bonds in the molecule, but the only forces between molecules are the very weak 'van der Waals' interactions due to induced dipoles. Thus the compound melts readily and the solid is soft and readily fractured as only these weak interactions have to be overcome. Naturally, in such a case, properties of the solid give no information about the bonds in the compounds.

TABLE 4.3 Summary of common structures and their relation to close-packing

Structure	Figure	Coordination	Description in terms of close-packing
		FORMULA TYPE AB	
zinc blende ZnS	4.1c	4:4 both tetrahedral	S atoms ccp with Zn in half the tetrahedral sites (every alternate site occupied)
wurtzite ZnS	4.4a	4:4 both tetrahedral	S atoms hcp with Zn in half the tetrahedral sites
sodium chloride NaCl	4.1a	6:6 both octahedral	Cl atoms ccp with Na in all the octahedral sites
nickel arsenide NiAs	4.10c	6:6 Ni octahedral As trigonal prism	As atoms hcp with Ni in all the octahedral sites (note that the two types of position cannot be equivalent in hcp)
cesium chloride CsCl	4.1b	8:8 both cubic	Not close-packed. The AB_8 and A_8B arrangements are like bcc
		FORMULA TYPE AB_2	
β-cristobalite SiO_2	4.4b	4:2 tetrahedral and linear	The Si atoms occupy both the Zn and S positions in zinc blende (this is equivalent to two inter-penetrating ccp lattices) and the O atoms are midway between pairs of Si
rutile TiO_2	4.3a	6:3 octahedral and triangular	Not close-packed. The Ti atoms lie in a considerably distorted bcc
fluorite, CaF_2 and anti-fluorite, Li_2O	4.3b	8:4 cubic and tetrahedral	Ca atoms ccp with F in all the tetrahedral sites O atoms ccp with Li in all the tetrahedral sites
cadmium iodide CdI_2	4.10a	6:3 layer lattice octahedral and the 3:coordination is irregular	I atoms are hcp and Cd atoms are in octahedral sites between every second layer
			The $CdCl_2$ structure is similar but the Cl atoms are ccp
		FORMULA TYPE AB_3	
rhenium trioxide ReO_3	4.4c	6:2 octahedral and linear	O atoms are in $\frac{3}{4}$ of the ccp sites and Re atoms are in $\frac{1}{4}$ of the octahedral sites

Structure	Figure	Coordination	Description in terms of close-packing
bismuth triiodide BiI_3	4.10b	6:2 layer lattice octahedral and the 2:coordination is non-linear	I atoms are hcp and Bi atoms occupy $\frac{2}{3}$ of the octahedral sites between every second layer The $CrCl_3$ structure is similar but the Cl atoms are ccp
		FORMULA TYPE M_2O_3	
corundum Al_2O_3	4.4d	6:4 octahedral, and at four of the six corners of a trigonal prism	O atoms are hcp with Al atoms in $\frac{2}{3}$ of the octahedral sites. (cf. NiAs with $\frac{1}{3}$ Ni missing). Ilmenite ($FeTiO_3$) is the same structure with Fe and Ti atoms in the Al sites
manganese(III) oxide Mn_2O_3	4.10d	6:4 six of eight cube corners, and tetrahedral	Mn atoms ccp with O atoms in $\frac{3}{4}$ of the tetrahedral sites (cf. fluorite)
		OTHER TYPES	
perovskite $CaTiO_3$	4.4e	Ca—12O (ccp) Ti—6O (octahedral) O—2Ti (linear) and 4Ca (square) giving distorted octahedron	Ca and O atoms together are ccp, with Ca in $\frac{1}{4}$ of the positions in a regular manner. The Ti atoms are in $\frac{1}{4}$ of the octahedral sites
spinel $M_2^{III}M^{II}O_4$	4.4f	M^{II}—tetrahedral M^{III}—octahedral	O atoms are ccp and $\frac{1}{8}$ of the tetrahedral sites and $\frac{1}{2}$ the octahedral sites are occupied by metal atoms. In a *normal* spinel, the M^{II} ions are tetrahedral and M^{III} octahedral. In *inverse* spinels, M^{II} ions are octahedral and half M^{III} are octahedral and half are tetrahedral

Notes: (i) Cubic close-packed = ccp: hexagonal close-packed = hcp: body-centred cube = bcc.
    (ii) In both ccp and hcp, there are two tetrahedral sites and one octahedral site for each atom in the close-packed lattice.

5 Solution Chemistry

Aqueous Solutions

Most work in inorganic chemistry is carried out in solution and the observed results depend to a large extent on the properties of the solvent. These will be examined in this chapter. Water is still by far the commonest solvent, but an ever-increasing number of reactions are carried out in other solvent media. These range from solvents like liquid ammonia, which have much in common with water, through more exotic media like anhydrous hydrogen fluoride or liquid bromine trifluoride, to molten salts and even molten metals. The major features of aqueous chemistry are discussed first and then follows the extension of these principles to nonaqueous solvents, with detailed discussion of the properties of a small number of representative solvent systems.

It is convenient to divide the discussion of solution behaviour into three sections;
(i) solubility and solvolysis, which depend directly on the interaction between solvent molecules and the solute,
(ii) acid-base behaviour,
(iii) oxidation-reduction behaviour.
None of these types of behaviour is independent of the others but it is convenient to make these broad divisions. They will be discussed in turn, first of all as they apply to solutions in water.

5.1 Solubility

The energy changes which govern solubility may be discussed by first considering the process of dissolving an ionic solid in water. The principal enthalpy changes may be related by a simple cycle diagram such as Figure 5.1. The process of solution is treated as occurring in two steps: the ions in the solid are separated to infinity as gaseous ions, which requires the input of the lattice energy, U; the separated gaseous ions are hydrated by the water molecules with the evolution of the heats of hydration, H_{aq}, of the cation and anion. The heat of solution, H_s, is the difference between the lattice energy and the heats of hydration.

The factors affecting the lattice energy were discussed in section 4.3. It will be recalled that the lattice energy is

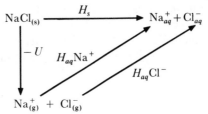

U = lattice energy of NaCl
H_s = heat of solution of NaCl
$H_{aq}\text{Na}^+$ = heat of hydration of Na^+
$H_{aq}\text{Cl}^-$ = heat of hydration of Cl^-

FIGURE 5.1 *Enthalpy changes in the solution of sodium chloride*
The process of solution may be split up into simple steps, whose heat changes are known or measurable, in the same way as the formation of an ionic solid was treated in the Born-Haber cycle.

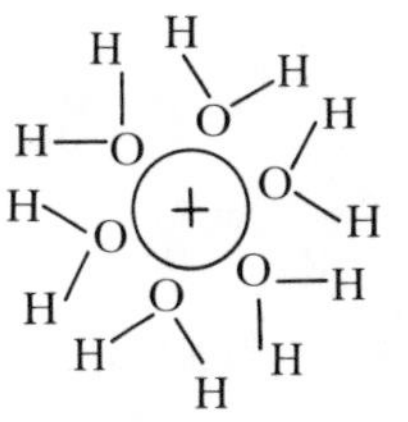

hydration of cation by water

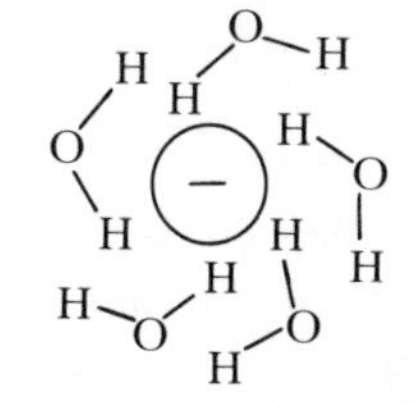

hydration of anion by water

FIGURE 5.2 *Diagrammatic representation of the solvation of ions*

greatest for small ions of high charge and is increased if polarization effects are present to add the attractions between the ionic charges and induced dipoles to those between the ions themselves.

The heats of hydration of the gaseous ions arise from the electrostatic attractions between the ionic charges (and dipoles if any) and the dipole of the water molecules. These interactions are shown schematically in Figure 5.2. In the water molecule, each $O-H$ bond is polarized in the sense $O^{\delta-}-H^{\delta+}$ by the uneven sharing of the bonding electrons between the very electronegative oxygen and the less electronegative hydrogen. In addition, the effective negative charge on the oxygen atom is increased by the two unshared pairs of electrons. In the case of an anion, the relatively positive hydrogen atoms interact with the negative charge to give the anionic heat of hydration. The cations interact with the relatively negative oxygen atoms of the water molecules and the major effect is probably that due to the lone pairs. The energy of hydration of the cation is usually the most important exothermic term as the cation-lone pair interaction is strong and is enhanced by the general small size and consequent high charge density of cations. The strength of the anion and cation interactions with the water molecules increases with the charge on the ions and is inversely proportional to their sizes. The effect of ion size on the hydration energies is seen in Figure 5.3 where the heats of hydration of the alkali halides are plotted against the anion radius.

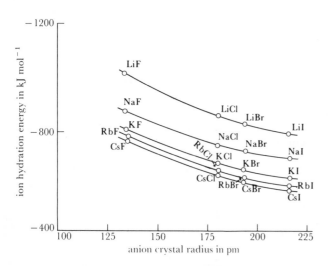

FIGURE 5.3 *Heats of hydration of the alkali halides*
The heats of hydration are plotted against the anion radius to illustrate the effect of ion size on the hydration energy.

It can be seen that the task of predicting the solubilities of ionic solids is very similar to that of predicting the formation of an ionic lattice which was discussed in the last chapter. The heat of solution, and therefore the probable solubility, is the result of the balance between the lattice energy of the solid and the heats of hydration of the gaseous ions. Both these factors are increased if the ions are small and of high charge, so that the difference between them may be expected to vary in a random manner and a general correlation of solubility with ion properties is impossible, each individual case having to be calculated separately. Solubility is yet another example of a phenomenon which depends on the small difference between large energies. There are, however, two factors which appear to have an over-riding effect on

the balance of energies involved in solution. One is the effect of the ionic charge. When this is increased, the ionic sizes and the crystal structure remaining constant, the lattice energy increases more than the heats of solvation of the ions. Thus, A^+B^- solids, for example, are much more soluble than $A^{2+}B^{2-}$ solids of the same structure. Highly-charged salts are generally of low solubility. A second effect which usually tends to decrease the solubility is the presence of polarization in the solid. A polarizing cation increases the lattice energy by the additional ion-dipole force which its presence introduces, and such a cation would also polarize the solvent molecules and increase the hydration energy in a similar manner. In this case, the effect of these forces appears to be largest in the solid and the presence of polarizing cations usually leads to low solubility, at least in cases like the silver and thallous salts which were discussed in Chapter 4.

The discussion above has been centred on heat changes whereas the true driving force is the free energy of solution which includes the entropy of solution as well as the heat. In general, entropy changes would be expected to favour the formation of solutions as the solute is much less ordered than the solid. (It will be recollected that a high degree of order corresponds to low entropy.) However, in a coordinating solvent such as water, the disorder of the solvent is reduced when ordered hydration sheaths are formed around the ions. If the ions are of high charge density, this increase in the ordering of the solvent may outweigh the decreased order of the solute. Thus, entropy changes should favour solution processes in general, but may oppose them for ions of high charge density which would be strongly solvated.

Another property of water, in addition to its power of hydrating ions, is important in determining its solvent power for ionic compounds. This is its very high dielectric constant, $D = 81.1$. The attraction between opposite charges e_1 and e_2 at a distance r in a medium of dielectric constant D is $e_1e_2/4\pi\epsilon_0 Dr^2$. Thus the interposition of a medium of high dielectric constant reduces the forces between ions in solution, hindering their precipitation, and assists the passage into solution of ions in the solid. Water is among the best of all solvents for ions and this is a result of the strong solvating forces for ions combined with the high dielectric constant.

By contrast, water is a poor solvent for covalent compounds, especially if these have no dipole or only a relatively weak one. To return to the heat cycle of Figure 5.1, a covalent compound may go into solution as single molecules or as dimers or polymers, and the energy required to separate the appropriate species is analogous to the lattice energy U of an ionic compound. In the case of the covalent 'giant molecule' like diamond or silica, the heat required to separate out any particle from the solid is too large to be compensated by the solvation energy—because strong covalent bonds would have to be broken—and such compounds are completely insoluble in water. In the case of a molecular solid which will dissolve as molecules, the heat required is the very small amount of energy needed to overcome the weak van der Waals' forces and this is readily available. Such solids are not usually soluble in water although the heat of hydration, which would arise from the interaction of the water dipoles with the dipoles induced by them in the covalent molecules, should be greater than the heat required to break up the solid. This low solubility arises because there is a third energy term to be added to the cycle. This is the heat required to overcome the attractions between the water molecules themselves and allow the solute molecules to enter between them. This energy of the 'water structure', which arises from the attractions between the partial charges of the water dipoles, is negligible compared with the energies involved in the solution of an ionic solid and was neglected in the earlier discussion, but it becomes important when the much weaker interactions with covalent molecules are considered. Thus, before a covalent solid dissolves, enough energy must be available to separate the solute molecules and to separate the water molecules, and this can only be provided by the heat of hydration of the solute molecules. For non-polar covalent molecules, the energy of the water structure is the dominant term and such compounds are of low solubility. When polar molecules are involved, both the forces within the solid and the forces between the solute molecules and the water are increased, and the water structure energy no longer dominates the energy balance. Solubilities in such cases depend on the detailed balance of enthalpies but they tend to be higher than those of non-polar compounds. For example, the solubility of the non-polar molecule methane in water at room temperature is about 0·004 mole per litre, while the polar methyl iodide dissolves to the extent of 0·110 mole per litre.

This discussion may be summarized as follows:

(a) In the case of ionic solids strong forces are involved, those between ions in the solid and those between ions and dipoles in solution. Solubility depends on the detailed balance in each case, but salts with balanced numbers of ions with charges greater than one are of low solubility and compounds where there are appreciable polarization effects are also of low solubility.

(b) In the case of 'giant molecules' the strength of the forces binding the solid predominates and such compounds are insoluble in water.

(c) Covalent nonpolar solids are weakly bound and would be somewhat more strongly bonded to water molecules than to themselves, but there is insufficient energy to break down the 'water structure' and such compounds are insoluble in water.

(d) Polar covalent compounds present an intermediate range of interactions between those of (a) and (c). Solubilities vary but tend to be higher than those in class (c).

With some solutes, interaction with the water molecule goes further than simple coordination and new chemical species are formed. Such a reaction is termed *hydrolysis*. The distinction between hydration and hydrolysis is not completely clear-cut but hydrolysis implies a more extensive and less reversible interaction of the solute with the water molecule.

Some examples are given below:

$$SiCl_4 + 4H_2O = Si(OH)_4 + 4HCl*$$
$$O^{2-} + H_2O = 2OH^-$$
$$HCl + H_2O = H_3O^+ + Cl^-$$
$$BF_3 + 2H_2O = H_3O^+ (HOBF_3)^-$$

5.2 Acids and bases

The qualitative properties of acids—sharp taste, solvent power, effect on the colours of dyes, etc.—and the corresponding properties of bases were first listed by Boyle and the succeeding centuries have seen a number of theories for acid-base properties developed and discarded. At present, three concepts are used to deal with acid-base phenomena in various solvents. These three theories overlap considerably although each has special uses and weaknesses. That of most value when dealing with aqueous solutions is the *protonic concept* of Brönsted and Lowry who characterized and related acids and bases by the equation:

$$A \text{ (acid)} \rightleftharpoons B \text{ (base)} + H^+ \text{ (proton)} \ldots \ldots (5.1)$$

An acid is a proton donor and a base is a proton acceptor. Since the proton in this theory is the hydrogen nucleus which has such a high charge density that it is never obtained free in the condensed state (see Chapter 8), it follows that this is a 'half-equation' and the acidic properties of a molecule are observed only when it is in contact with the basic form of a second species. That is, the observed equation is always:

$$A_1 + B_2 \rightleftharpoons A_2 + B_1 \quad \ldots \ldots \ldots \ldots \ldots (5.2)$$

Such an equilibrium is displaced in the direction of the weaker acid and base, since the stronger acid is a stronger proton donor than the weaker acid and thus more of its molecules are in the basic form. The base corresponding to a given acid is called the conjugate base; clearly a strong acid has a weak conjugate base, and *vice versa*. The acid strength can be expressed in terms of the equilibrium constant of the above reaction:

$$K = \frac{[A_2][B_1]}{[A_1][B_2]} \quad \ldots \ldots \ldots \ldots \ldots (5.3)$$

This gives a method of expressing relative acid strengths. If one acid-base pair is taken as the standard, then the strengths of all other pairs may be expressed in terms of the equilibrium constants involving the standard acid. It is most convenient to choose as the standard acid-base pair, the one involving the solvent. Thus in water, the acid strength is defined with respect to the pair $H_3O^+ - H_2O$ and the equilibrium used to define acid strengths is:

$$A + H_2O \rightleftharpoons B + H_3O^+ \text{ with } K' = \frac{[B][H_3O^+]}{[A][H_2O]} \ldots (5.4a)$$

*These products undergo further reaction.

TABLE 5.1 Strengths of acids in water

Acid	Conjugate base	K	pK	
$HClO_4$	ClO_4^-	?	?	unknown K but very high
HCl	Cl^-	ca 10^7	ca -7	HBr and HI greater
HNO_3	NO_3^-	?	?	unknown K but very high
H_2SO_4	HSO_4^-	?	?	unknown K but very high
H_3O^+	H_2O	55·5	$-1·74$	
HSO_4^-	SO_4^{2-}	2×10^{-2}	1·70	compare with H_2SO_4
H_3PO_4	$H_2PO_4^-$	$7·5 \times 10^{-3}$	2·12	
HF	F^-	$7·2 \times 10^{-4}$	3·14	compare with HCl
$H_2PO_4^-$	HPO_4^{2-}	$5·9 \times 10^{-8}$	7·23	
NH_4^+	NH_3	$3·3 \times 10^{-10}$	9·24	typical weak base
HPO_4^{2-}	PO_4^{3-}	$3·6 \times 10^{-13}$	12·44	compare with H_3PO_4 and $H_2PO_4^-$
H_2O	OH^-	$1·07 \times 10^{-16}$	15·97	strong base
OH^-	O^{2-}	below 10^{-36}	above 36	

Providing dilute solutions are being used, the concentration of water $[H_2O]$ is essentially constant, so that the strength of an acid can be defined by:

$$K = [B][H_3O^+]/[A] \ldots \ldots \ldots (5.4b)$$

The strengths of bases are quite naturally expressed by means of the K values and there is no need for a separate scale of base strengths. Since strong acids have weak conjugate bases and *vice versa*, the order of base strengths is the inverse of the order of acid strengths. Some typical K values for acids and bases in water are shown in Table 5.1. Also included are pK values which are the negative logarithms, analogous to the well-known pH values.

The relative strengths of the acids and bases which lie between the two acid pairs of water

$$H_3O^+ \rightleftharpoons H_2O + H^+$$
$$\text{and} \quad H_2O \rightleftharpoons OH^- + H^+ \quad \cdots \cdots \cdots (5.5)$$

are well defined, but it is impossible to measure the acid strengths of acids stronger than H_3O^+ (or of bases stronger than OH^-) as they are completely converted to the water acid (or base), that is, the equilibrium in (5.4) lies completely to the right. No acid stronger than the hydrated proton can exist in water. One way of determining the relative strengths of acids such as nitric acid or perchloric acid, which are completely dissociated in water, is to compare their catalytic powers in an acid-catalysed reaction. The catalysis of the inversion of sucrose gives an order of strengths as follows: $HClO_4 > HBr > HCl > HNO_3$. An alternative method is to measure the dissociation in a solvent which is more acidic than water.

An alternative definition of acids and bases was proposed by Cady and Elsey and is often termed the *solvent system* definition. They suggested that the acid and base in a particular solvent should be defined in terms of the ions formed in the self-dissociation of the solvent. For example, water

dissociates slightly in the sense:

$$2H_2O \rightleftharpoons H_3O^+ + OH^- \ldots \ldots \ldots (5.6)$$
$$\text{acid} \qquad \text{base}$$

and an acid in water is any substance which enhances the concentration of the solvent cation, H_3O^+, while a base is any substance which enhances the concentration of the solvent anion, OH^-.

This definition is closely related to that of Lowry and Brönsted, the dissociation of equation (5.6) corresponding to the two Brönsted acid pairs of water as given in equation (5.5). A similar correlation between the solvent system and Lowry-Brönsted definitions holds for any protonic solvent, but the solvent system definition has the advantage that it is readily extended to solvents which do not contain dissociable hydrogen atoms.

The third definition of acids and bases is that due to Lewis, who defines an acid as a lone pair acceptor and a base as a lone pair donor. This definition is of little advantage in discussing reactions in water so an account of it is deferred to section 5.5.

Strengths of oxyacids

Many compounds which show acidic properties fall into the class of oxy-acids, that is, they contain the Group $X-O-H$. Two general observations may be made about the strengths of these acids.

First, where there are a number of OH groups attached to the central atom, the pK values for the removal of the successive ionizable hydrogens increase by about five each time (compare the values for phosphoric acid and the two phosphate ions in Table 5.1).

Second, the strength of the acid depends on the difference $(x-y)$ between the number of oxygen atoms and the number of hydrogen atoms in the molecule H_yXO_x. When x is equal to y, pK is about $8·5 \pm 1$. If $(x-y) = 1$, pK is about $2·8 \pm 1$. If $(x-y)$ is two or more, the pK value is markedly less than zero. Examples of the last case are provided by the first three oxy-

TABLE 5·2 Strengths of oxyacids, H_yXO_x

$(x-y) = 1$	acid	HNO_2	H_2SO_3	H_3AsO_4	H_5IO_6
	pK	3·3	1·90	3·5	3·29
$(x-y) = 0$	acid	$HClO$	H_3BO_3	H_4GeO_4	H_6TeO_6
	pK	7·50	9·22	8·59	8·80

acids in Table 5.1 while some examples of the first two cases are given below in Table 5.2.

As oxygen is the second most electronegative element, when the central atom X has an oxygen atom attached to it as well as the OH group, that is in $O=X-O-H$, electron density will be withdrawn from X by the oxygen and this effect will be transmitted to the $O-H$ bond, making it easier for the hydrogen to dissociate as the positive ion. The acid strength should therefore rise with the number of X=O groups, that is with $(x-y)$ as the Tables show. Two cases are known where simple acids do not fit this pattern. One is provided by the lower acids of phosphorus, phosphorous acid H_3PO_3 where $(x-y) = 0$ but $pK = 1·8$, and hypophosphorous acid H_3PO_2 with $(x-y)$ equal to -1 and $pK = 2$. In both these molecules, the value of $(x-y)$ does not correspond to the number of X=O bonds as there are direct $P-H$ bonds present, one in phosphorous acid and two in hypophosphorous acid. Allowing for these, both acids have one $P=O$ bond, which should correspond to pK values in the range $2·8 \pm 1$, as is found.

The second exception is provided by carbonic acid, H_2CO_3, which is expected to have a pK value of about $2·8$ and has an actual value of $6·4$. In this case, as well as the acid dissociation equilibrium, there is a further non-protonic equilibrium in solution:

$$H_2CO_3 \rightleftharpoons H_2O + CO_{2(aq)}$$

When allowance is made for this dissolved carbon dioxide, the effective pK value of carbonic acid is about $3·6$, which just falls within the expected range.

5.3 Oxidation and reduction

Oxidation, originally defined in terms of combination with oxygen and later generalized to include combination with other electronegative elements, is nowadays commonly defined as the *removal of electrons* from the element or compound which is oxidized. *Reduction*, similarly, has been defined in terms of removal of oxygen and electronegative elements, addition of hydrogen or electropositive elements and now in terms of *gain of electrons* by the element or compound in question. In cases involving ions, for example in:

$$Na + \tfrac{1}{2}Cl_2 = Na^+Cl^-,$$
$$\text{or} \quad 2Fe^{3+} + Sn = 2Fe^{2+} + Sn^{2+},$$

the direction of electron transfer is clear from the equations. When only covalent species are involved, it is usually clear that some electron rearrangement has taken place but there is often no obvious electron transfer, as in:

$$\tfrac{1}{2}H_2 + \tfrac{1}{2}Cl_2 = HCl$$

where the difference in electron density at the hydrogen atom, say, is the relatively small one due to the polarization of the $H-Cl$ bond compared with the unpolarized $H-H$ bond. It is in these cases that the definitions of oxidation and reduction in terms of oxidation numbers, using electronegativity values, discussed in Chapter 2 are so useful.

Some quantitative measure of oxidizing or reducing power is necessary and this is known as the *redox potential* of the reactant. The tendency to gain or lose an electron may be measured as an electrical potential under standard conditions and expressed relative to a suitable standard value (as only the relative values of potentials may be measured). The standard conditions used are a temperature of 25 °C, unit activity (which is usually taken as unit concentration) of the ions concerned and, for gases, one atmosphere pressure. The standard potential is provided by the hydrogen electrode (Figure 5.4), which consists of a platinum plate, coated with

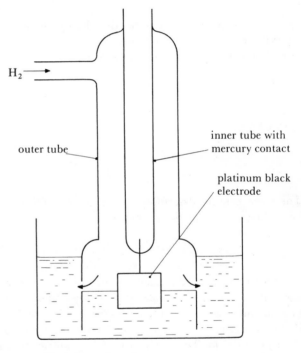

FIGURE 5.4 *The hydrogen electrode*

platinum black, partly dipping into a solution containing hydrogen ions at unit activity. Hydrogen gas at one atmosphere pressure is passed over the platinum plate and bubbled through the solution so that the platinum is in contact both with the gas and the hydrogen ions in solution and catalyses the attainment of equilibrium between them:

$$H^+ \text{ (aq, } a = 1) + e^- \rightleftharpoons \tfrac{1}{2}H_2 \text{ (gas, } p = 1 \text{ atm)}$$
$$\quad (a \text{ is activity})$$

The potential of this electrode is defined to be zero at 25 °C (298 K) and this is written $E°(298 \text{ K}) = 0·000$ V.

The potential of any other oxidation-reduction or redox system is measured by immersing a platinum wire or other inert conductor in a solution which contains both the oxidized and the reduced form of the system at unit activity, and measuring the potential of this half-cell against the hydrogen electrode. For example, the standard potential of the ferric/ferrous system is measured from the cell:

$$\text{Pt, H}_2(p = 1 \text{ atm}) \left| \text{H}^+ (a = 1) \right\| \begin{array}{l} \text{Fe}^{3+} (a = 1) \\ \text{Fe}^{2+} (a = 1) \end{array} \right| \text{Pt}$$

(It is usually more convenient to use some other electrode of accurately known potential in place of the hydrogen electrode which is awkward to handle.) If the potential of a gaseous element is to be determined, an electrode similar to the hydrogen electrode is used, while the half-cell used to determine the potential of a metallic element consists simply of a rod of the metal (which is defined to have unit activity) immersed in a solution of the metal ions at unit activity. For example, the oxidation potential of iron going to ferrous ions is measured as the standard potential of the cell:

$$\text{hydrogen electrode} \left\| \text{Fe}^{2+} \right| \text{Fe} \ (a = 1)$$

In the half-equation describing the change in the redox system, the oxidized form is always written on the left hand side, that is the equations are in the form:

$$\text{Ox} + n \text{ e}^- = \text{Red} \dots\dots\dots(5.7)$$

where Ox stands for the oxidized form and Red for the reduced form,

$$\begin{array}{ll} \text{e.g.} & \text{Fe}^{3+} + \text{e}^- = \text{Fe}^{2+} \\ \text{or} & \text{Fe}^{2+} + 2\text{e}^- = \text{Fe} \end{array}$$

The short form for describing the electrode is written similarly: Ox|Red or, in the examples above, $\text{Fe}^{3+}|\text{Fe}^{2+}$ and $\text{Fe}^{2+}|\text{Fe}$.

The sign convention which applies for these potentials takes a negative sign for the potential Ox|Red to mean that the reduced form of the redox system is a better reducing agent than is hydrogen under standard conditions. This is tantamount to measuring the sign of the metal in the $\text{M}^{z+}|\text{M}$ electrode relative to the hydrogen electrode. Thus the standard potential for $\text{Fe}^{2+}|\text{Fe}$ is -0.41 V, showing that iron going to ferrous ions is a better reducing agent than hydrogen going to hydrogen ions. The potential for $\text{Fe}^{3+}|\text{Fe}^{2+}$ is $+0.77$ V so that ferrous ions going to ferric ions provide a much worse reducing agent than hydrogen. This is, of course, equivalent to saying that ferric going to ferrous is a useful oxidizing agent. The above sign convention is the one recommended by the International Union of Pure and Applied Chemistry (IUPAC) but the opposite convention (reversing the equation and the sign) is also found.*

The full range of redox potentials extends over about six volts, from -3 V for the alkali metals which are the strongest reducing agents to $+3$ V for fluorine, the strongest

oxidizing agent. The values at these extremes cannot be measured directly as these very reactive elements decompose water, but they can be derived by calculation. The range of useful redox reagents in an aqueous medium is about half of the total range, from about 1.7 V for strong oxidizing agents like ceric or permanganate to about -0.4 V for a strong reducing agent like chromous, or about -0.8 V for reduction by fairly active metals such as zinc. The values for many common oxidation-reduction couples are given in Table 5.3.

The potential of a redox system where the oxidized and reduced forms are not at unit activities is related to the standard potential, E^0, by the equation:

$$E = E^0 + \frac{RT}{nF} \ln \frac{[\text{Ox}]}{[\text{Red}]} \dots\dots\dots(5.8)$$

where R is the gas constant, T the absolute temperature, F the Faraday constant, and n the number of electrons transferred in the oxidation process as in equation (5.7). Transferring to logarithms to the base ten and introducing the values of the constants

$$E = E^0 + \frac{0.059}{n} \log \frac{[\text{Ox}]}{[\text{Red}]} \text{ at } 25 \text{ °C} \dots\dots(5.9)$$

This equation permits the extent of reaction between two redox systems to be calculated, and can be used, for example, to decide whether a particular reaction will go to completion or not. For the general reaction:

$$a \text{ Ox}_1 + b \text{ Red}_2 \rightleftharpoons b \text{ Ox}_2 + a \text{ Red}_1, \dots\dots(5.10)$$

$$E_1 = E_1^0 + \frac{RT}{nF} \ln \frac{[\text{Ox}_1]^a}{[\text{Red}_1]^a} \text{ and } E_2 = E_2^0 + \frac{RT}{nF} \ln \frac{[\text{Ox}_2]^b}{[\text{Red}_2]^b}.$$

At equilibrium, $E_1 = E_2$ so that:

$$\log \frac{[\text{Ox}_2]^b [\text{Red}_1]^a}{[\text{Ox}_1]^a [\text{Red}_2]^b} = \log K = \frac{n}{0.059} (E_1^0 - E_2^0) \dots(5.11)$$

where K is the equilibrium constant for the reaction. Consider as examples the oxidations of iodide ion, first by ceric ion and secondly by ferric ion. In the first case:

$$\text{Ce}^{4+} + \text{I}^- = \text{Ce}^{3+} + \tfrac{1}{2}\text{I}_2$$

E^0 for I_2/I^- is 0.54 V and E^0 for $\text{Ce}^{4+}/\text{Ce}^{3+}$ is 1.61 V hence:

$$\log K = \frac{1}{0.059} (1.61 - 0.54) = 18.48,$$

$$\text{i.e. } K \simeq 10^{18}$$

Thus iodide is completely oxidized to iodine by ceric ion.

In the second case, E^0 for $\text{Fe}^{3+}/\text{Fe}^{2+}$ is 0.77 V so that for the reaction:

$$\text{Fe}^{3+} + \text{I}^- = \text{Fe}^{2+} + \tfrac{1}{2}\text{I}_2,$$
$$\log K = 1/0.059 (0.76 - 0.52) = 4.068$$

Hence K is approximately 10^4 so that a small but significant

---

*Equation (5.7) is a reduction from left to right, which is the normal way of reading, but it is an oxidation from right to left. It has been suggested that these standard electrode potentials may be called reduction potentials. The term redox is, however, recommended to illustrate the reversible nature of the reactions.

TABLE 5.3 Examples of standard redox potentials in acid solution

Couple	Reaction equation	E^0 (volts)
Li^+/Li	$Li^+ + e^- = Li$	$-3{\cdot}05$
M^+/M (M = K, Rb, Cs)	$M^+ + e^- = M$	$-2{\cdot}93$
Ba^{2+}/Ba	$Ba^{2+} + 2e^- = Ba$	$-2{\cdot}90$
Sr^{2+}/Sr	$Sr^{2+} + 2e^- = Sr$	$-2{\cdot}89$
Ca^{2+}/Ca	$Ca^{2+} + 2e^- = Ca$	$-2{\cdot}76$
Na^+/Na	$Na^+ + e^- = Na$	$-2{\cdot}71$
Mg^{2+}/Mg	$Mg^{2+} + 2e^- = Mg$	$-2{\cdot}38$
La^{3+}/La	$La^{3+} + 3e^- = La$	$-2{\cdot}37$
$\frac{1}{2}H_2/H^-$	$\frac{1}{2}H_2 + e^- = H^-$	$-2{\cdot}23$
Be^{2+}/Be	$Be^{2+} + 2e^- = Be$	$-1{\cdot}70$
Al^{3+}/Al	$Al^{3+} + 3e^- = Al$	$-1{\cdot}67$
Zn^{2+}/Zn	$Zn^{2+} + 2e^- = Zn$	$-0{\cdot}76$
Cr^{3+}/Cr	$Cr^{3+} + 3e^- = Cr$	$-0{\cdot}74$
S/S^{2-}	$S + 2e^- = S^{2-}$	$-0{\cdot}51$
H_3PO_3/H_3PO_2	$H_3PO_3 + 2H^+ + 2e^- = H_3PO_2 + H_2O$	$-0{\cdot}50$
$CO_2/H_2C_2O_4$	$2CO_2 + 2H^+ + 2e^- = H_2C_2O_4$	$-0{\cdot}49$
Fe^{2+}/Fe	$Fe^{2+} + 2e^- = Fe$	$-0{\cdot}41$
Cr^{3+}/Cr^{2+}	$Cr^{3+} + e^- = Cr^{2+}$	$-0{\cdot}41$
H_3PO_4/H_3PO_3	$H_3PO_4 + 2H^+ + 2e^- = H_3PO_3 + H_2O$	$-0{\cdot}20$
Sn^{2+}/Sn	$Sn^{2+} + 2e^- = Sn$	$-0{\cdot}14$
$H^+/\frac{1}{2}H_2$	$H^+ + e^- = \frac{1}{2}H_2$	$0{\cdot}00$
$S_4O_6^{2-}/S_2O_3^{2-}$	$S_4O_6^{2-} + 2e^- = 2S_2O_3^{2-}$	$0{\cdot}09$
Sn^{4+}/Sn^{2+}	$Sn^{4+} + 2e^- = Sn^{2+}$	$0{\cdot}15$
Cu^{2+}/Cu^+	$Cu^{2+} + e^- = Cu^+$	$0{\cdot}15$
Cu^+/Cu	$Cu^+ + e^- = Cu$	$0{\cdot}52$
$\frac{1}{2}I_2/I^-$	$\frac{1}{2}I_2 + e^- = I^-$	$0{\cdot}54$
H_3AsO_4/H_3AsO_3	$H_3AsO_4 + 2H^+ + 2e^- = H_3AsO_3 + H_2O$	$0{\cdot}56$
O_2/H_2O_2	$O_2 + 2H^+ + 2e^- = H_2O_2$	$0{\cdot}68$
Fe^{3+}/Fe^{2+}	$Fe^{3+} + e^- = Fe^{2+}$	$0{\cdot}77$
Hg_2^{2+}/Hg	$\frac{1}{2}Hg_2^{2+} + e^- = Hg$	$0{\cdot}80$
Hg^{2+}/Hg	$Hg^{2+} + 2e^- = Hg$	$0{\cdot}85$
$\frac{1}{2}Br_2/Br^-$	$\frac{1}{2}Br_2 + e^- = Br^-$	$1{\cdot}09$
IO_3^-/I^-	$IO_3^- + 6H^+ + 6e^- = I^- + 3H_2O$	$1{\cdot}09$
$IO_3^-/\frac{1}{2}I_2$	$IO_3^- + 6H^+ + 5e^- = \frac{1}{2}I_2 + 3H_2O$	$1{\cdot}20$
$\frac{1}{2}Cr_2O_7^{2-}/Cr^{3+}$	$\frac{1}{2}Cr_2O_7^{2-} + 7H^+ + 3e^- = Cr^{3+} + 7/2H_2O$	$1{\cdot}33$
$\frac{1}{2}Cl_2/Cl^-$	$\frac{1}{2}Cl_2 + e^- = Cl^-$	$1{\cdot}36$
$HIO/\frac{1}{2}I$	$HIO + H^+ + e^- = \frac{1}{2}I_2 + H_2O$	$1{\cdot}45$
$BrO_3^-/\frac{1}{2}Br_2$	$BrO_3^- + 6H^+ + 5e^- = \frac{1}{2}Br_2 + 3H_2O$	$1{\cdot}52$
MnO_4^-/Mn^{2+}	$MnO_4^- + 8H^+ + 5e^- = Mn^{2+} + 4H_2O$	$1{\cdot}52$
$HBrO/Br_2$	$HBrO + H^+ + e^- = Br_2 + H_2O$	$1{\cdot}59$
H_5IO_6/IO_3^-	$H_5IO_6 + H^+ + 2e^- = IO_3^- + 3H_2O$	$1{\cdot}60$
Ce^{4+}/Ce^{3+}	$Ce^{4+} + e^- = Ce^{3+}$	$1{\cdot}61$
$HClO/\frac{1}{2}Cl_2$	$HClO + H^+ + e^- = \frac{1}{2}Cl_2 + H_2O$	$1{\cdot}64$
$HClO_2/HClO$	$HClO_2 + 2H^+ + 2e^- = HClO + H_2O$	$1{\cdot}64$
H_2O_2/H_2O	$H_2O_2 + 2H^+ + 2e^- = 2H_2O$	$1{\cdot}77$
$\frac{1}{2}S_2O_8^{2-}/SO_4^{2-}$	$\frac{1}{2}S_2O_8^{2-} + e^- = SO_4^{2-}$	$2{\cdot}01$
O_3/O_2	$O_3 + 2H^+ + 2e^- = O_2 + H_2O$	$2{\cdot}07$
$\frac{1}{2}F_2/F^-$	$\frac{1}{2}F_2 + e^- = F^-$	$2{\cdot}87$
$\frac{1}{2}F_2/HF$	$\frac{1}{2}F_2 + H^+ + e^- = HF$	$3{\cdot}06$

proportion of iodide would remain in equilibrium with the iodine. For a one electron change, an equilibrium constant of one million requires a difference in the standard potentials of the two redox systems of $0·354$ V, so this is about the minimum difference necessary for a complete reaction.

The values of the standard potentials may also be used to determine the stability of different oxidation states in solution. If the potentials for the various oxidation states of iron in Table 5.3 are examined, it will be seen that iron going to ferrous ions is a much better reducing agent than ferrous ions going to ferric ions. Oxidation of metallic iron in aqueous solution therefore gives ferrous ions first and stronger oxidation is required to get to ferric ions. By contrast, the values for the different oxidation states of copper show that copper going to cuprous ions is a worse reducing agent than copper going to cupric ions (that is, an oxidizing agent strong enough to oxidize copper to cuprous, $E^0 = 0.52$ V, is more than strong enough to oxidize cuprous to cupric, $E^0 = 0·15$ V). The cuprous state is therefore avoided in aqueous solution. Put alternatively, cuprous ions in water disproportionate to cupric ions and metallic copper:

$$2\,Cu^+ \rightarrow Cu^{2+} + Cu$$

In this reaction, the cuprous ion may be regarded as Ox_1 (with copper as the corresponding reduced species, Red_1) and also as Red_2 (with cupric ion as Ox_2) in the general equation (5.10). Substituting the standard potentials in equation (5.11) gives

$$K = \frac{[Cu^{2+}]\,[Cu]}{[Cu^+]^2} = approx.\ 10^6$$

so that only one part in a million of the original cuprous copper remains in solution as cuprous copper at equilibrium. This disproportionation of the copper oxidation states may be reversed by adding some reagent which forms a very stable complex with the cuprous ion, so stable that there is significantly less than one part per million of cuprous ion in equilibrium with the cuprous complex. The cuprous ion is thus removed from the cuprous-cupric equilibrium by complexing and the disproportionation reaction reverses. One example of such a complexing agent is cyanide ion which gives the very stable $Cu(CN)_2^-$ complex ion.

The above examples give some indication of the ways in which the redox potentials may be used. Let us examine the potential in more detail and analyze its components.

The type of cell used for determining these potentials, as on page 77.

$$Pt, H_2 \mid H^+ \mid\mid M^+ \mid M$$

implies the reaction:

$$\tfrac{1}{2}H_2 + M^+ = H^+ + M$$

The electrode potentials of the left and right electrode are E_H^0 and E_M^0 respectively and the potential difference, the e.m.f., of the cell is $(E_M^0 - E_H^0)$. By convention we set $E_H^0 = 0$.

The electrode potential E^0 of the cell is related to the free energy change ΔG^0 of the reaction by the equation:

$$\Delta G^0 = -nFE^0$$

and ΔG^0 is related to the heat of reaction ΔH^0 and entropy of reaction ΔS^0 at temperature T by

$$\Delta G^0 = \Delta H^0 - T\,\Delta S^0$$

For an approximate treatment, and especially for comparative purposes, the entropy changes may be neglected and the approximate relation

$$-nFE^0 \approx \Delta H^0 \quad \dots\dots\dots\dots (5.12)$$

may be used.

Let us now examine ΔH^0 in detail. This means the enthalpy of the products minus the enthalpy of the reactants which we may write

$$\Delta H^0 = H^0(M) + H^0(H^+) - H^0(M^+) - H^0(\tfrac{1}{2}H_2)$$

which can be regrouped as

$$\Delta H^0 = \{H^0(H^+) - H^0(\tfrac{1}{2}H_2)\} - \{H^0(M^+) - H^0(M)\}$$

and abbreviated as

$$\Delta H^0 = \Delta H^0(H) - \Delta H^0(M) \quad \dots\dots (5.13)$$

The terms $\Delta H^0(H)$ and $\Delta H^0(M)$ may be analyzed by means of a cycle, similar to those used to discuss lattice energies or solubilities. For the formation of a cation from the element in aqueous solution we may examine the contribution of various terms to $\Delta H^0(M)$.

where H_A is the heat of atomization of the element, I is the ionization potential (or the sum of the first z ionization potentials if a Z^+ cation is formed) and H_{aq} is the heat of hydration of the gaseous ion. An exactly similar cycle may be constructed for the formation of an anion in solution except that the ionization potential must be replaced by the electron affinity when the gaseous atom goes to the gaseous anion.

Thus, for a cation,

$$\Delta H^0(M) = H_A + I + H_{aq}$$

and for an anion,

$$\Delta H^0(X) = H_A + E + H_{aq}$$

The values of H_A, I (or E) and H_{aq} are known from experiment and, for hydrogen, equal respectively $+218$, $+1310$, and -1070 kJ mol^{-1}, giving $\Delta H^0(H) = 452$ kJ mol^{-1}.

Thus we may now write from equations (5.12) and (5.13):

$$-nFE_M^0 \approx 452\ \text{kJ mol}^{-1} - \Delta H^0(M)$$

and E_M^0 will be negative if $\Delta H^0(M)$ is less than 452 kJ mol^{-1}.

The alkali metals have ΔH^0 values of about 188 kJ mol^{-1} reflecting their low ionization potentials and sublimation energies, so that their standard potentials are strongly negative. The fall in standard redox potentials from cesium or barium to sodium or magnesium shown in Table 5.3 reflects the regular decrease in ionization potential and sublimation energy with increasing size, while the anomalous position of lithium is mainly the result of the greater hydration energy of the small cation. In the case of the halogens, the sum of the atomization energy (half the bond energy of X_2) and the electron affinity is approximately constant, so that the fall of the redox potential from fluorine to iodine again reflects the fall in hydration energy as the size of the ion increases.

This treatment need not be confined to the case of elements, for example the Fe^{3+}/Fe^{2+} system depends on the energy changes:

$$
\begin{array}{ccc}
Fe^{3+} + e^- & \longrightarrow & Fe^{2+} \\
\text{(gas)} & & \text{(gas)} \\
\downarrow & & \downarrow \\
Fe^{3+} + e^- & \longleftarrow & Fe^{2+} \\
\text{(hydrated)} & & \text{(hydrated)}
\end{array}
$$

where the overall change depends on the hydration energies of the Fe(II) and Fe(III) ions and on the third ionization potential of iron.

These cycles provide one method of calculating potentials which are unobtainable experimentally, either because the species are not stable in water or because the attainment of equilibrium is too slow.

It must be noted that all the above treatment of standard redox potentials gives no indication of the rates of reaction. Although it may seem from the values of the potentials that a certain molecule should be readily oxidized by another, the rate of reaction might be so slow that nothing is observed. For example, the values given in Table 5.3 show that iodine should oxidize thiosulphate to tetrathionate as in the standard volumetric method and that oxidation of thiosulphate to sulphate should also be significant. However, the reaction to tetrathionate is quantitative, showing that the oxidation to sulphate must be extremely slow. In other words, the quantitative nature of oxidation to tetrathionate depends on a kinetic factor and is not revealed by the thermodynamic approach involved in redox potentials.

The potentials which appear in Table 5.3 as involving simple ions do in fact apply to the hydrated forms of these ions, as the potentials are determined in aqueous solution. In some of these cases, the water molecules are strongly held and the heat of hydration is an important factor in the potential. If these coordinated water molecules are replaced by other ligands, the potential changes. For example, the potential of 0·77 V for hydrated iron(II) and (III) changes to 0·36 V for the hexacyanide complexes and rises as high as 1·2 V for the dipyridyl complexes, $Fe(dipy)_3^{3+}/Fe(dipy)_3^{2+}$. The potential also changes with the acidity when, for example, hydrated species in dilute acid change to hydroxy complexes in more alkaline media.

If the redox potential in a solvent other than water is in question, it will clearly differ from that of the hydrated species in water. In the case of a simple ion, the difference lies in the heat of solvation compared with the heat of hydration, the contribution of the heat of atomization and the ionization potential (or electron affinity) remaining the same as in water. Although the solvation energy of any set of ions will neither be the same nor in the same order as in water, gross differences are less likely than small ones. In other words, ions which are strongly hydrated are likely to be strongly solvated by another ionizing solvent. This means that the general order of oxidizing powers should be similar in all solvents to that obtaining in water, although there may be marked differences in detail from one solvent to the next. Since, in general, thermodynamic measurements in nonaqueous solvents are much more sparse than those made in water, this generalization is usually the best that can be obtained. Similar remarks apply to oxidation-reduction reactions in systems which differ even more from that of aqueous solution, such as fused salts and other high temperature melts. In such systems, considerable changes take place in the relative stability and redox powers of chemical species, and, although the values derived from aqueous measurements may provide some guide, it is often better to go back to the fundamental parameters, such as ionization potentials, and base predictions on these.

Non-aqueous Solvents

Water is so familiar and accessible a solvent that its properties are usually taken for granted and underlie most general statements about inorganic compounds. Such general ideas as 'stability' usually imply 'stability in an environment containing air and water', and many inaccessible compounds or highly reactive oxidation states in this type of environment become stable and manageable if handled in a non-aqueous medium. In fact, one of the advantages of studying nonaqueous solvents has been the increased attention to, and wider insight into, chemistry in water that has resulted.

The practical reasons for using solvents other than water lie partly in the extended range of experimental conditions which are available and partly in the exclusion of water as a reactant. Many anhydrous compounds, for example, cannot be prepared from the hydrates and were unknown until the use of other solvents was introduced. Such compounds often have unusual and unexpected properties. The non-aqueous solvents range widely in intrinsic reactivity, from solvents such as sulphur dioxide which commonly act only as inert media for the reaction, to solvents such as anhydrous hydrogen fluoride which react with nearly all non-fluoride solutes. A number of selected examples of such solvent systems are discussed individually later but first the discussion of the early part of this chapter on solubility and solvent interaction and on acid-base behaviour will be extended to include non-aqueous solvents. It has already been indicated how the treatment of redox behaviour may be extended and further illustrations occur in the course of the discussion of specific solvents.

5.4 Solubility and solvent interaction in non-aqueous solvents

The extension of the discussion of solubility in section 5.1 to non-aqueous solvents is fairly easy. A very generalized energy diagram is shown in Figure 5.5. E_1 is the energy required to

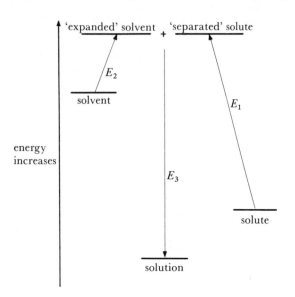

FIGURE 5.5 *Energy changes in the solution of a solid in a solvent*
This very generalized version of Figure 5.1 shows the energy changes for the solution of any solid in any solute, whether an ionic or a covalent system is involved.

change the solute into the form in which it will exist in solution. In the case of an ionic solid, this means providing the lattice energy to separate the ions, while for a solute which dissolves as molecules it means providing the energy required to separate the molecules or to form dimers or other species. E_2 is the energy required to separate the solvent molecules, breaking up any solvent 'structure' to allow the admission of the solute particles. E_3 is the energy given out when the separate solute particles associate with the solvent molecules. (It may be noted that the energy of hydration as commonly measured is not E_3 but the difference $[E_3 - E_2]$.) Two general cases exist. One, the generalization of the 'strong forces' case of section 5.1, is the case of ions or strongly polar molecules dissolving in a polar solvent. Here, the important balance of energies involves strong attractions between ions and permanent dipoles. The other extreme is the 'weak forces' case of weakly bound, nonpolar, covalent molecules dissolving in nonpolar solvents. Here, the balance of forces involves weak attractions between induced dipoles, and the order of magnitude of the effects is less than that in the strong forces case by a factor of about ten. The major conclusion is that 'mixed cases' will all represent situations where solubility is low. Thus, ions will not dissolve in non-polar solvents because, although E_2 is low, E_3 is also low and

the energetics are dominated by E_1. Conversely, nonpolar molecules are insoluble in polar solvents as E_1 and E_3 are both low and there is insufficient energy to supply E_2. This analysis is at the base of the very old generalization about solubilities that 'like dissolves like'.

Although a clear answer can be given in the mixed cases, it must be noted that the detailed variation of solubilities within the 'strong forces' or within the 'weak forces' class depends on the detailed balance of energies of similar magnitudes, so that a knowledge of solubilities in one solvent gives only a very general guide to the solubilities of similar compounds in similar solvents. This is illustrated by the solubilities given in Table 5.4.

TABLE 5.4 Solubilities in liquid ammonia and sulphur dioxide

| Compound | Solubility (millimole/litre at 0 °C) | | |
	Water	Ammonia	Sulphur dioxide
NaCl	6 100	2 200	insoluble
NaBr	7 710	6 210	1·4
NaI	10 720	8 800	1 000
KCl	3 760	18	5·5
KBr	4 490	2 260	40
KI	7 720	11 060	2 490
AgCl	0·005	20	0·05
AgBr	0·003	135	0·16
AgI	10^{-6}	4 000	0·68

For the alkali halides, the values in ammonia and sulphur dioxide follow the same order as in water, which is, in turn, the inverse order of the lattice energies, and most of the solubilities fall going from water to ammonia to sulphur dioxide but the relative sizes of the solubilities change enormously. The much greater solubilities of the iodides and the silver salts in ammonia and in sulphur dioxide show that polarization effects producing induced dipoles are more important in these two solvents than in water. It is worth noting that the solubility of the silver halides in liquid ammonia is in the reverse order to their solubility in aqueous ammonia.

The dielectric constants of non-aqueous solvents have an important effect on their power to dissolve ions. The general drop in solubilities of the alkali halides from H_2O to SO_2 reflects the drop in dielectric constant from 81 for water to 22 for ammonia and 12 for SO_2. Although important, the dielectric constant alone is a poor guide to solvent power for ions. The value for liquid hydrogen cyanide of 123 is one of the highest measured and yet this is a poor solvent for ions as the energy of solvation is low.

The general interaction between solvent molecules and ions in solution, which is analogous to hydration in water, is termed *solvation* and follows similar mechanisms. In liquid ammonia, for example, cations are ammonated by interaction

TABLE 5.5 Examples of solvolysis and other interactions

$SnCl_4 + 4H_2O$	$= Sn(OH)_4$		hydrolysis*
$+ 8NH_3$	$= Sn(NH_2)_4 + 4NH_4Cl$		ammonolysis*
$+ 4HF$	$= SnF_4 + 4HCl$		solvolysis
$+ 2SeOCl_2$	$= SnCl_4.2SeOCl_2$ (or $2SeOCl^+ + SnCl_6^{2-}$)		solvation(?)
$LiBr + NH_3$	dissolves: $Li(NH_3)_4Br$ recovered		solvation
$+ SO_2$	$= Li_2SO_3 + SOBr_2$		solvolysis*
$KNO_3 + 4HF$	$= K^+ + H_2NO_3^+ + 2HF_2^-$		solvolysis

*(= further reaction of the product takes place)

with the lone pair of electrons on the nitrogen atom plus the interaction with the negative end of the $N^{\delta-} - H^{\delta+}$ dipoles. The anions interact with the relatively positive hydrogens.

In addition to solvation, the solute may react with the solvent molecules to break up the solvent molecule, the solute species, or both, in the reaction called *solvolysis*—analogous to hydrolysis in water. The distinction between solvation and solvolysis is not completely clear-cut but solvolysis implies a more extensive and less reversible interaction with the solvent molecule. Some examples are given in Table 5.5.

5.5 Acid-base behaviour in non-aqueous solvents

Of the two theories of acids and bases discussed in section 5.2, the protonic theory of Lowry and Brönsted is restricted to solvents containing ionizable hydrogen atoms. The application of the theory is then exactly the same as for aqueous systems, except that the strengths of different acids are most conveniently measured relative to the acid-base pairs characteristic of the solvent. For example, K values in liquid ammonia may be measured relative to:

$$NH_4^+ = NH_3 + H^+.$$

Such values would fall in a similar order to those obtained in water or in any other solvent, although there might be minor inversions.

The solvent system definition was developed for use in non-aqueous solvents and provides a definition of acid and base in non-protonic solvents. It is very convenient to use if work is being carried out on one particular solvent and it provides a guide to the properties to be expected for a new system, but it does suffer the disadvantage that the definition of acid or base changes from one solvent to another. A more important reservation which must be made is that it only suggests an acid-base system for a particular solvent and does not prove that the solvent ions do in fact behave as acid and base. This point must be proved experimentally. These remarks are extended and illustrated in the discussion of the sulphur dioxide solvent system (section 5.10).

As in the case of water (section 5.2), the solvent's acid-base properties place a limit on the range of the acid strengths of solutes in it. The strongest acid existing in a solvent is the solvent acid and all solutes which are more acidic are converted into the solvent acid—and similarly for bases—so that the number of solutes which are strong acids (i.e. those which

are completely converted to the solvent acid) varies with the intrinsic acid strength of the solvent. In an acidic solvent such as glacial acetic acid, some solutes which are strong acids in water are not completely dissociated to the solvent acid, that is, they are weak acids in this medium. This is the case for HNO_3 or HCl in acetic acid. Indeed some molecules which are weakly acidic in water are basic in acetic acid. This latter effect is even more marked in strongly acidic solvents such as absolute sulphuric acid where even HNO_3 acts as a base and accepts a proton from the solvent. The relative strengths of those solutes which are strong acids in water may thus be differentiated by examining their properties in more acidic solvents. On the other hand, a basic solvent like liquid ammonia has a very weak solvent acid so that all solutes which are acids, strong or weak, in water are strong acids in ammonia. Base strengths in basic and acidic solvents are affected in an analogous manner.

An acid solute which *is* completely dissociated in an acidic solvent, such as perchloric acid in glacial acetic acid, is a much stronger acid than in water and finds valuable application in analysis for determining very weakly basic functions. In a similar way, the solution of a strong base in a basic solvent may be used for the estimation of very weakly acidic groups. Examples of these applications are given in sections 5.7 and 5.8.

Both the proton-transfer and solvent system definitions of acids and bases are in general use for non-aqueous solvents and they are practically equivalent when applied to a protonic solvent. The solvent system definition—for example:

	acid	base
$2 NH_3$	$\rightleftharpoons NH_4^+$	$+ NH_2^-$
$2 H_2SO_4$	$\rightleftharpoons H_3SO_4^+$	$+ HSO_4^-$
$2 HCN$	$\rightleftharpoons H_2CN^+$	$+ CN^-$

is equivalent in each case to a pair of Brönsted acids,

Acid $\rightleftharpoons$	conjugate base + H^+
NH_4^+	NH_3
NH_3	NH_2^-
$H_3SO_4^+$	H_2SO_4
H_2SO_4	HSO_4^-
H_2CN^+	HCN
HCN	CN^-

A third, much more extensive, theory of acids and bases was proposed by Lewis, who defined an acid as an electron pair acceptor and a base as an electron pair donor. This definition includes the Brönsted and solvent system definitions as special cases. For example, the proton is an electron pair acceptor and H_2O or NH_3 or H_2SO_4 or HCN, among our examples, act as donors. This definition emphasizes the similarity between all coordination reactions, for example:

Lewis acid	Lewis base	product
H^+	H_2O	H_3O^+
	OH^-	H_2O
	F^-	HF
BF_3	H_2O	$H_2O:BF_3$
	OH^-	$(HOBF_3)^-$
	F^-	BF_4^-

but it is usually regarded as too wide in scope to be useful. It includes solvation reactions:

Lewis acid	Lewis base	
M^{z+}	$x\ H_2O$	$[M(H_2O)_x]^{z+}$
	$x\ NH_3$	$[M(NH_3)_x]^{z+}$

and at least the initial steps of solvolysis reactions, for example:

$SnCl_4$	NH_3	$(SnCl_4:NH_3)$

but also includes many other types of reaction, including oxidation-reduction reactions. The Lewis theory is more extensively applied in America than in Europe but use of the general term 'Lewis acid' for acceptor molecules like BF_3 is very common, even by authors who make no other use of the theory.

5.6 General uses of non-aqueous solvents

It is now possible to outline some general reasons for choosing to work in a solvent other than water and to suggest points which will guide the choice of solvent. These are best seen by comparison with the properties of water itself.

The effect of the acid-base properties has already been discussed. In water, the strongest acid is H_3O^+ and the strongest base is OH^-, all solutes of higher acidity or basicity being completely converted to these. If the effect of more acidic or more basic media is to be investigated, clearly a suitable non-aqueous solvent must be chosen. The most fully-investigated solvents from this point of view are glacial acetic acid, anhydrous hydrogen fluoride, and anhydrous sulphuric acid for acidic media, while ammonia and the amines have been the most-studied basic media. At higher temperatures, fused salts provide suitable media. Acidic systems include ammonium salts of strong acids, acidic anion salts such as HSO_4^- or HF_2^-, and acidic oxides such as silica or phosphorus pentoxide. Basic systems include the oxides or hydroxides of the alkali metals.

In a rather similar way, the oxidation-reduction properties of the solvent place limits to the reactions which may be carried out in it. Water is attacked rapidly by oxidizing agents with standard potentials greater than about $1 \cdot 7$ V, and by reducing agents with oxidation potentials below about $-0 \cdot 4$ V. Outside these limits, use of water as the solvent is undesirable. Non-aqueous solvents which are

useful for carrying out oxidizing reactions include sulphuri acid and higher valency oxides such as dinitrogen tetroxide The most important solvent for strong reductions is liqui ammonia, which dissolves the alkali and alkaline eart metals to give a very strongly reducing system.

The choice of a solvent is clearly also dependent on it solvent powers. Solvents of high polarity and coordinating power are good ionizing media, and the range runs from these down to solvents such as the hydrocarbons which tak up only molecular solids. In preparation reactions, a solven is often chosen because of the insolubility of a certain com pound in it. For example, the reaction in water between KCN and $NiSO_4$ gives $Ni(CN)_2$ by direct precipitation bu the product is very finely divided and difficult to filter an wash. The same reaction in liquid ammonia gives an easil handled precipitate of K_2SO_4, not $Ni(CN)_2$, and the potas sium salt is readily removed and the nickel cyanide recovered by evaporation.

A solvent may also be chosen because of the way in which it reacts with solutes. One example is in the preparation o anhydrous compounds. Since water coordinates strongly, i is often impossible to remove it without decomposing the compound, and many anhydrous compounds are availabl only by syntheses which avoid water, or other strongly coordinating solvents. One example is anhydrous coppe nitrate, $Cu(NO_3)_2$, which is a volatile solid prepared by reaction in N_2O_4. When solvolytic reactions are possible the solvent may be chosen to avoid these, as in the use o liquid sulphur dioxide for reactions with non-metal halides On the other hand, a solvent is often used because it doe react with the solute, and the major example of this latte case is anhydrous hydrogen fluoride which solvolyses almos every compound placed in it. This is a standard method o preparing fluorine-containing compounds.

A fairly wide variety of compounds have been used as non aqueous solvents and the more important ones are listed in Table 5.6 with notes on their special uses. The general points made above are illustrated by more detailed discussion o four systems. These are glacial acetic acid and liquid am monia, representing protonic acid and base, and two non protonic solvents, bromine trifluoride and sulphur dioxide The solvent system definitions of acid and base appear to hold in BrF_3, but they break down in SO_2. Table 5.7 shows the more important physical properties of these solvents.

5.7 Liquid ammonia

Liquid ammonia was one of the first non-aqueous ionizing solvents to be studied and is now the one which is most extensively used and best known. It is readily available in reasonable purity and is easily dried by the action of sodium. The liquid range is from -77 °C to -33 °C, and this presents some handling problems, but these are partly compensated by the relatively high latent heat of vaporization. Liquid ammonia is normally handled in cooled vessels—and it is often used as its own coolant—or at room temperature under pressure. A wide range of special techniques have been

TABLE 5.6 Typical non-aqueous solvents

Solvent	Postulated self-ionization	Acids	Bases	Comments
NH_3	$NH_4^+ + NH_2^-$	ammonium salts	amides, e.g. KNH_2	See section 5.7 for discussion. A number of similar solvent systems have been studied including N_2H_4, NH_2OH, and organic amines such as $C_2H_4(NH_2)_2$ and CH_3NH_2.
HCN	$H_2CN^+ + CN^-$	protonic acids	alkali cyanides	Very poor solvating power therefore a poor solvent for ions despite the high dielectric constant.
CH_3CN				Similar to HCN but a better ionizing solvent as it coordinates moderately well.
$HCONH_2$ and CH_3CONH_2				Little-investigated solvents of high dielectric constant which may well be interesting solvents. Acetamide is said to be a good solvent for ionic compounds.
HF	$H_2F^+ + HF_2^-$	F^- acceptors, e.g. BF_3 $HF + BF_3 =$ $H_2F^+BF_4^-$	alkali fluorides	Good ionizing solvent: many non-fluorine species react, e.g. $H_2SO_4 \rightarrow HSO_3F$. Used in the preparation of fluorine compounds, e.g. $AgNO_3 \rightarrow Ag^+ + H_2NO_3^+ + 2F^-$ $BF_3 + F^- \rightarrow BF_4^-$ then, $Ag^+ + BF_4^- \rightarrow AgBF_4$.
CH_3COOH	$CH_3COOH_2^+ + CH_3COO^-$	protonic acids	ionizable acetates	See section 5.8. Other carboxylic acids behave similarly, especially HCOOH.
H_2SO_4	$H_3SO_4^+ + HSO_4^-$	$H_2S_2O_3$ ($HClO_4$ weak: most common acids are bases in H_2SO_4)	soluble bisulphates H_2O, HNO_3, H_3PO_4	
N_2O_4	$NO^+ + NO_3^-$	NOCl	alkali nitrates	Used to prepare anhydrous nitrates, nitrato- and nitro-complexes and nitrosyl compounds: commonly used in admixture with organic compounds such as ethyl acetate.
SO_2	Conflicting evidence, see section 5.10.			Useful as an inert reaction medium.
$SeOCl_2$	$SeOCl^+ + Cl^-$ (or $SeOCl_3^-$)	Cl^- acceptors e.g. $SnCl_4$	organic bases like pyridine ($\rightarrow C_5H_5NSeOCl^+Cl^-$)	Dissolves many elements with reaction and dissolves a variety of metal chlorides. Other salts are converted to the chloride.

Other oxyhalides, NOCl and $POCl_3$ for example, are similar. Some conductiometric and potentiometric titration data exist to support the self-ionization mode and acid-base behaviour which is postulated.

Solvent	Postulated self-ionization	Acids	Bases	Comments
BrF_3	$BrF_2^+ + BrF_4^-$	F^- acceptors	ionizable fluorides	Discussed in section 5.9. IF_5 behaves similarly.
$AsCl_3$	$AsCl_2^+ + AsCl_4^-$	Cl^- acceptors	Cl^-	$AsBr_3$, $SbCl_3$ and ICl are analogous

TABLE 5.7 Physical properties of non-aqueous solvents

	H_2O	NH_3	CH_3COOH	SO_2	BrF_3
m.p. (°C)	0	-77.7	16.6	-75.5	9.0
b.p. (°C)	100	-33.4	118.1	-10.2	126
dielectric constant	78.5 (18 °C)	23 (b.p.)	9.7 (18 °C)	17.3 (-16 °C)	—
specific conductivity ($\Omega\,m^{-1}$)	6×10^{-8} (25°C)	5×10^{-9} (b.p.)	0.5—0.8×10^{-8} (25 °C)	4×10^{-8} (b.p.)	8×10^{-3} (m.p.)
heat of vaporization ($kJ\,mol^{-1}$)	40.7	23.6	24.3	25.0	42

developed for handling ammonia so that few problems are now met.

The dielectric constant and the self-ionization are both lower than for water, indicating that liquid ammonia will be a poorer ionizing solvent than water. Soluble salts include most ammonium salts, nitrates, thiocyanates and iodides. Fluorides and most oxy-salts are insoluble, and solubility among the halides increases $F^- < Cl^- < Br^- < I^-$. Calcium and zinc chlorides, which are extremely soluble in water, are quite insoluble in liquid ammonia, but they do take up a large amount of ammonia forming solid ammoniates with eight and ten molecules of ammonia respectively. Calcium chloride is thus a useful absorbent for traces of ammonia. Since ammonia is less highly associated than water, it is a better solvent for organic compounds, especially for those with fairly small carbon radicals. Unsaturated hydrocarbons, alcohols, esters, ammonium salts of acids, and most nitrogen compounds are soluble.

The self-ionization of ammonia is written:

$$2\,NH_3 \rightleftharpoons NH_4^+ + NH_2^-$$

and ammonium compounds behave as acids while amides are bases. Ammonium iodide, nitrate, or thiocyanate are very soluble and concentrated solutions will slowly dissolve metals with the evolution of hydrogen:

$$Mg + 2NH_4^+ = Mg^{2+} + H_2 + 2NH_3$$

This reaction is rapid with the active alkali and alkaline earth metals but slow with less active metals such as magnesium or iron. As ammonia is more basic than water, these acids are weaker, so the solvent power for metals does not extend to the less active metals. The commonest base is potassamide, KNH_2, which is much more soluble than sodium amide. Acid-base titrations between potassamide and ammonium salts may be carried out and followed conductiometrically or by using phenolphthalein. The ammonia system brings out very weakly acidic functions in molecules. Thus, urea, $CO(NH_2)_2$, which is a weak base in water, acts as a weak acid in liquid ammonia and may be neutralized by amide.

This enhancement of weakly acidic functions by the basic solvent finds application in analysis. Ammonia itself is rarely used because of its low boiling point, but simple organic derivatives such as ethylamine, $CH_3CH_2NH_2$ (which is related to ammonia as ethanol, CH_3CH_2OH, is to water), or ethylenediamine, $H_2NCH_2CH_2NH_2$, are used as solvents for the determination of weak acids. Titrants used include methoxides of the alkali metals and soluble hydroxides which are easier to purify and standardize than amides. Applica-

tions include the determination of phenols and related compounds in ethylenediamine, and the determination of carbon dioxide which may be separated from other gases by solution in acetone and determined as a weak acid with sodium methoxide, CH_3ONa. Suitable indicators or potentiometric methods are used to determine the end points.

There are many examples of amphoteric behaviour in liquid ammonia. For example:

$$Zn^{2+} \underset{NH_4^+}{\overset{NH_2^-}{\rightleftharpoons}} Zn(NH_2)_2 \downarrow \underset{NH_4^+}{\overset{NH_2^-}{\rightleftharpoons}} Zn(NH_2)_4^{2-}$$

compare with

$$Zn^{2+} \underset{H_3O^+}{\overset{OH^-}{\rightleftharpoons}} Zn(OH)_2 \downarrow \underset{H_3O^+}{\overset{OH^-}{\rightleftharpoons}} Zn(OH)_4^{2-}$$

Amides and imides of the less-reactive metals dissolve in excess potassamide:

e.g. $$PbNH + NH_2^- + NH_3 = Pb(NH_2)_3^-$$

and even sodamide, which is insoluble in ammonia, dissolves in potassamide to give the ammonosodiate:

$$NaNH_2 + 2\,NH_2^- = Na(NH_2)_3^{2-}.$$

Hydrolysis of heavy metal salts to basic salts (oxy- and hydroxy-compounds) is paralleled by reactions between ammonia and such salts. The analogues of the oxy- and hydroxy-compounds in the aqueous system are compounds containing the groups, amide $-NH_2$, imide $=NH$, and nitride $\equiv N$. For example, lead nitrate dissolves in liquid ammonia and the compound $PbNH.NH_2.PbNO_3$ may be isolated from the solution. If an ammonium salt is added, this compound dissolves, while the addition of potassamide precipitates it.

The most striking property of liquid ammonia is its ability to dissolve the active metals to give fairly stable, blue solutions. As ammonia is more resistant to reduction than is water, the reaction:

$$M + NH_3 = MNH_2 + \tfrac{1}{2}H_2$$

is much slower than the reaction:

$$M + H_2O = MOH + \tfrac{1}{2}H_2$$

If the reagents are pure and dry, sodium solutions in liquid ammonia may be preserved for several weeks and even the much more reactive cesium gives solutions which may be kept overnight. These solutions are formed by all the alkali metals, by the alkaline earth metals, and by the reducible

TABLE 5.8 Reduction reactions by metal solutions in liquid ammonia

Reactant	Reduction products
O_2	metal peroxides, O_2^{2-}, and superoxides, O_2^-
S, Se, Te, As, Sb, Bi, Sn, Pb and their oxides or halides	white binary compounds such as M_2S or M_4Pb and highly coloured poly-anions such as M_2S_x $x = 2$ to 7) (deep red), M_3Bi_3 (violet) or M_4Pb_9 (deep green)
metal oxides, halides, and dissociable complexes	metal
complexes which do $\quad$ $Ni(CN)_4^{2-}$ not dissociate $\quad$ e.g. $Co(CN)_6^{2-}$ $\qquad\qquad$ $Ni(C\equiv CH)_4^{2-}$	$Ni_2(CN)_6^{4-}$; $Ni(CN)_4^{4-}$ $Co(CN)_4^{4-}$ $Ni(C\equiv CH)_4^{4-}$
hydrides of Groups IV and V	ions formed by removal of one or two hydrogens PH_2^- or SnH_2^{2-} and SnH_3^-
Ge_3H_8	$2GeH_3^- + GeH_2^{2-}$ (i.e. each $Ge-Ge$ bond is broken)

lanthanide elements such as samarium. In addition, very dilute solutions may be formed from less reactive elements, such as magnesium, beryllium, aluminium, and the other lanthanides, by electrolytic means. The alkali metals are also soluble in amines and dilute solutions are formed in certain ethers.

Dilute metal solutions are coloured a very deep blue and concentrated ones have a metallic, coppery appearance. The solutions are conducting and strongly reducing. In these solutions, the valency electrons are ionized off the metal and become solvated in a manner which is not yet fully understood:

$$Na \rightarrow Na^+_{(ammoniated)} + e^-_{(ammoniated)},$$

and the unique properties of the solutions are associated with the presence of these readily available electrons.

The major application of metal solutions is in reduction reactions. In organic chemistry, the skeletal single bonds, $C-C$, $C-O$, $C-N$, are stable in these solutions as are isolated double bonds and single benzene rings, but nearly all other functional groups and most unsaturated compounds are reduced. In inorganic chemistry, the most interesting reductions have been the formation of polyanions, the reductions of hydrides, and the formation of transition metal complexes in unusually low oxidation states. Some examples are given in Table 5.8.

5.8 Anhydrous acetic acid

To some extent, glacial acetic acid is the converse solvent to liquid ammonia. It is readily available, easily purified, and differs from water in undergoing less self-ionization, and in having a markedly lower dielectric constant. It is a moderately strong acid and is used to investigate weakly basic functions, just as ammonia and the amines are used for weakly acid functions. The lower dielectric constant makes acetic acid a poorer solvent for ions, and most compounds which are insoluble in water are insoluble in acetic acid. Soluble compounds include most acetates, nitrates, halides, cyanides, and thiocyanates. The strong acids all dissolve in glacial acetic acid as do basic compounds like water and ammonia. A wide range of polar organic compounds are also soluble.

The self-ionization of acetic acid is:

$$2\,CH_3COOH = CH_3COOH_2^+ + CH_3COO^-$$

so that acetates of the reactive metals are bases. The normal protonic acids are acids in acetic acid by virtue of reactions such as:

$$HClO_4 + CH_3COOH = CH_3COOH_2^+ + ClO_4^-$$

(compare $HClO_4 + H_2O = H_3O^+ + ClO_4^-$), while ammonia, say, is a base as the acetate ion is produced by the reaction:

$$NH_3 + CH_3COOH = CH_3COO^- + NH_4^+$$

Of the mineral acids which are strong acids in water (that is are completely dissociated to the H_3O^+ ion), only perchloric acid is strongly dissociated in glacial acetic acid. The relative strengths of the common acids in acetic acid are $HNO_3 = 1 < HCl = 9 < H_2SO_4 = 30 < HBr = 160 < HClO_4 = 400$. The acetates which are strongest bases are those of the alkali metals and ammonium, while bismuth, lead, and mercuric acetates are ten to a hundred times weaker. Even the strongest bases and acids are poor electrolytes in acetic acid, largely as a result of the low dielectric constant of the solvent.

Acid-base titrations between the acetates and the mineral acids in acetic acid may be demonstrated by indicators or electrometrically. In addition, perchloric acid in glacial acetic acid is a widely used reagent for the estimation of the weakly basic functions of amines, amino-acids, metal salts of organic acids, and the like.

Amphoteric behaviour has also been demonstrated in acetic acid. For example, the addition of sodium acetate to a solution of a zinc salt precipitates zinc acetate, which redissolves in excess acetate:

$$Zn^{2+} \underset{CH_3COOH_2^+}{\overset{CH_3COO^-}{\rightleftharpoons}} Zn(CH_3COO)_2 \downarrow$$

$$\underset{CH_3COOH_2^+}{\overset{CH_3COO^-}{\rightleftharpoons}} Zn(CH_3COO)_4^{2-}$$

Copper and lead(II) acetates also show amphoteric behaviour, while lead(IV) tetra-acetate decreases in solubility as sodium acetate is added to its solution, which corresponds to 'salting out' a poor electrolyte.

Acetic acid is a convenient reaction medium for the preparation of covalent hydrolysable compounds. For example, tin reacts smoothly and controllably with halogens in acetic acid:

$$Sn + 2X_2 = SnX_4$$

and the stannic halide is readily isolated by distillation or crystallization.

Other weak acids which have been studied as solvent systems include formic acid and hydrogen cyanide, but investigations in these systems have been largely confined to studies of solubility and acid-base relationships. Of the strong acids, attention has been concentrated on hydrogen fluoride and sulphuric acid, in which the principal type of reaction is solvolysis. This is also true of bromine trifluoride.

5.9 Bromine trifluoride

Bromine trifluoride is a much more restricted solvent than the two discussed above but it is one to which the ideas of the solvent system theory appear to apply and it has been quite widely used as a preparative medium. It must be handled by special techniques which require experience but these are now highly developed and present few problems in a well-equipped laboratory. The main requirements are a rigorous exclusion of moisture and of materials such as tap greases which can be fluorinated or oxidized.

The alkali metal fluorides are soluble but most other ionic fluorides are rather insoluble. The more covalent fluorides of elements in higher oxidation states are also soluble. Most other compounds are either insoluble or converted to fluoro-compounds.

The self-ionization of bromine trifluoride is much more extensive than that of the other solvents in Table 5.6 and is presumed to follow the equation:

$$2\,BrF_3 = BrF_2^+ + BrF_4^-$$

By the solvent system definition, solutes which increase the concentration of BrF_2^+ are acids and those which increase the concentration of BrF_4^- are bases. The bromofluorides of the alkali metals and a number of other elements, e.g. $AgBrF_4$, are known and act as bases, while the alkali metal fluorides add on a molecule of the solvent and may also be regarded as bases in the system. No solute is known which produces the solvent cation by dissociation (as the mineral acids produce H_3O^+ in water) that is, there are no *donor acids* known, but a number of solutes are known which react with the solvent molecule by removing a fluoride ion and these are termed *acceptor acids*. This is true of most covalent fluorides, which form fluoro-complexes, for example:

$$VF_5 + BrF_3 = BrF_2^+ + VF_6^-$$
$$SnF_4 + 2BrF_3 = 2BrF_2^+ + SnF_6^{2-}$$

Neutralization reactions are represented by the interaction of these acids and bases and may be followed conductiometrically:

$$BrF_2^+VF_6^- + KBrF_4 = KVF_6 + 2BrF_3$$
$$\text{acid} \qquad \text{base} \qquad \text{salt} \quad \text{solvent} \cdot$$

Solvolysis of some of these complex fluorides is also observed. For example, if hexafluorotitanate is treated with bromine trifluoride the reversible reaction

$$K_2TiF_6 + 4BrF_3 \rightleftharpoons (BrF_2)_2TiF_6 + 2KBrF_4$$

is observed leading to the formation of the solvent base in equilibrium.

The main use of bromine trifluoride is in the preparation of fluorides and fluorocomplexes of elements in the higher oxidation states. This is complemented by the use of iodine pentafluoride or selenium tetrafluoride with reducing agents as solvents for the preparation of lower oxidation state fluorides. Among the transition metal compounds which have been prepared in bromine trifluoride are the hexafluoro-complexes of Ti(IV), V(V), Nb(V), Ta(V), Mn(IV), Ru(IV & V), Os(IV & V), Rh(IV), Ir(IV & V), Pd(IV), Pt(IV), and Au(III). The first preparation of gold trifluoride was also made in this solvent. Another interesting series of reactions involves the reactions of non-metal oxides with fluorocomplexes in bromine trifluoride to give fluosulphonates, nitronium complexes, and nitrosonium complexes, such as:

$$SO_3 + K_2SO_4 \rightarrow KSO_3F$$
$$NO_2 + Au \rightarrow (NO_2)AuF_4$$
$$NO_2 + As_2O_3 \rightarrow (NO_2)AsF_6$$
$$NOCl + GeO_2 \rightarrow (NO)_2GeF_6$$

(Note that several of these reactions involve oxidations as well as fluorination.)

5.10 Sulphur dioxide

The study of sulphur dioxide as a solvent extends back as far as that of liquid ammonia, with the early definitive work being done at the turn of the century. Sulphur dioxide was found to be conducting and the postulated self-ionization was:

$$2SO_2 \rightleftharpoons SO^{2+} + SO_3^{2-}$$

According to this, ionizable sulphites, such as Cs_2SO_3, were bases, and acids were species which produced SO^{2+}. The thionyl halides were found to be ionized and the equilibrium:

$$SOX_2 \rightleftharpoons SO^{2+} + 2X^-$$

was postulated, so that these molecules were solvent acids. It was indeed found that thionyl chloride reacted with soluble sulphites in a 1:1 ratio and the titration curve was reasonably like the expected one. On the basis of these observations and others on solvate formation, solvolysis, and amphoteric behaviour, the above self-ionization was accepted as the basis for interpreting reactions in sulphur dioxide.

The existence of these sulphur dioxide results was an important reason for the general acceptance of the solvent system definition of acids and bases, since this was not confined to protonic solvents and appeared to fit for sulphur dioxide, the most extensively studied non-protonic system of the time. As other non-protonic solvents came to be studied, the self-ionization theory was applied and found in every case to give a satisfactory basis for systematizing reactions and to be a useful guide in the study of the solvent. This was the case for all the non-protonic solvents listed in Table 5.6.

However, if the self-ionizations in that list are studied, it will be seen that only in the case of sulphur dioxide is the separation of doubly-charged ions postulated. Table 5.7 shows that sulphur dioxide has a fairly low dielectric constant and the self-conductivity is also low. This indicates that the self-ionization theory might be on a rather weak basis in this case. This suspicion has been confirmed by recent work on radio-isotope exchange between sulphur dioxide and thionyl compounds. It was found that there was negligible exchange of either ^{18}O or ^{35}S between thionyl chloride and sulphur dioxide, showing that the concentration of any ion common to both compounds was negligible (compare the instantaneous exchange of hydrogen isotopes between protonic acids and water). Therefore ionization to SO^{2+} cannot be occurring. The 'neutralization' reactions between sulphites and thionyl compounds were probably not simple reactions between solvent ions but involved two-step reactions via SOX^+ intermediates.

These results for sulphur dioxide must throw some doubt on the whole solvent system concept of acids and bases in non-protonic solvents and some workers have suggested its abandonment. It must be granted that there is no absolute proof of its validity in other systems. All the evidence depends on indirect observations, such as 'neutralization' reactions between solvent ions. Furthermore, the interpretation, in terms of the solvent system theory, of the work on oxychlorides like $POCl_3$ and $SeOCl_2$ has been heavily criticized recently and a rationalization of the observations in terms of the Lewis theory has been proposed.* However, the interpretation of the oxychloride systems is arguable, and the sulphur dioxide system is a particularly unfavourable example for the reasons outlined above. Thus, the solvent system theory will probably continue to be used to interpret reactions in protonic and non-protonic systems, but with more marked reservations in the latter cases until the existence of the solvent self-ionization has been proved by radio-

*See the article by Drago and Purcell in the references quoted on page 282.

chemical or other methods. At present, the odd situation exists where the solvent system concept may still be applied to most solvents except sulphur dioxide, the very one that led to its adoption.

As far as the status of acids and bases in sulphur dioxide is concerned, the present picture is confused. It has been shown that Brönsted acids, such as HCl, react with Brönsted bases in liquid sulphur dioxide, and the solvent is acting as a more or less inert reaction medium (just as a solvent like benzene does). The possibility of solvent ionization is not completely excluded, however, until the possibility of self-ionization under suitable conditions has been studied, for example, in the presence of an inert co-solvent which raises the dielectric constant of the system. At present, however, it is best to regard sulphur dioxide, not as a self-ionizing solvent which is the parent of an acid-base system, but as a relatively inert reaction medium. This has always been the main use of sulphur dioxide and it is extremely valuable as a solvent for reactions between weakly ionic or fairly polar molecules.

Among the compounds which are very soluble in sulphur dioxide are the alkali metal, or ammonium, iodides, thiocyanates, and carboxylic acid salts. Many classes of organic compounds are soluble, as are the more covalent halides and pseudohalides of the Main Group elements. Solvates of soluble salts are often recovered when the solvent is removed. Examples include $NaI.4SO_2$, $KSCN.SO_2$, or $AlCl_3.SO_2$. The sulphur dioxide is usually much less tightly bound than is water or ammonia in hydrates or ammoniates.

Amphoteric behaviour is also observed (using for the moment the solvent system terminology). Thus aluminium chloride reacts with a solution of sulphite with the precipitation of aluminium sulphite, $Al_2(SO_3)_3$, and this redissolves in excess sulphite to form the trisulphito-aluminate, $Al(SO_3)_3^{3-}$.

Sulphur dioxide finds its main use as a solvent for the reaction of readily hydrolysable halides and related compounds. Thus thionyl compounds may be prepared:

$$2SCN^- + SOCl_2 = SO(SCN)_2 + 2Cl^-$$

Chloro-complexes are readily prepared:

$$NOCl + SbCl_5 = (NO)^+(SbCl_6)^-,$$
$$3SOCl_2 + 2SbCl_3 = (SO)_3^{2+}(SbCl_6)_2^{3-} \text{ (in solution only)}$$

and a few solvolytic reactions are observed:

$$PCl_5 + SO_2 = POCl_3 + SOCl_2$$

A considerable range of organic reactions has also been conducted in sulphur dioxide.

6 Experimental Methods

In the discussion of the chemistry of the elements which makes up the later part of this book, the structures of a number of compounds are described. It is the aim of this chapter to give a brief indication of the more important experimental methods of determining these structures, and a description of some of the more recently developed methods of separating compounds is also given. Fuller accounts of the individual methods are given in the references.

Separation Methods

6.1 Ion exchange

The use of clay minerals and zeolites for base exchange and water treatment has been established for many years, but a major advance was made when the use of synthetic resins containing specific functional groups was introduced. The materials in present use are cross-linked polystyrene resins and similar types containing active groups. These are:

sulphonic groups	$-SO_2OH$	giving strongly acid cation exchangers
carboxylic groups	$-COOH$	giving weakly acid cation exchangers
quaternary ammonium groups	$-NR_3^+$	giving strongly basic anion exchangers
amine groups	$-NR_2$; $-NH_2$; $-NHR$	giving weakly basic anion exchangers,

where the bond is to the polymer skeleton. The polymer skeleton of the resin acts as an inert, unreactive framework to support the functional groups and has a relatively porous structure with about ten per cent cross-linking.

The sulphonic acid resins* are strong acids and the hydrogen atoms ionize and are readily replaced by cations. If a solution of a salt, say NaX, is passed down a column containing the resin, the Na^+ is replaced:

$$c-ResinH + NaX \rightarrow c-ResinNa + HX$$

In a similar way, the strongly basic resins* containing quaternary ammonium groups have hydroxyl groups which ionize completely and are exchangeable with anions:

$$a-ResinOH + NaX \rightarrow a-ResinX + NaOH,$$
$$or\ a-ResinOH + HX \rightarrow a-ResinX + H_2O$$

This illustrates one important use of these resins, a strong acid cation exchanger used in series with a strong base anion exchanger will remove all dissolved salts from water and is widely used for water-treatment, especially in boilers to prevent scale formation.

Strongly acid cation exchangers in the sodium form will exchange the sodium ions for other cations:

$$c-ResinNa + M^{n+} = c-ResinM + nNa^+$$

These reactions are equilibrium reactions and the affinity of the resin for the cation in solution depends on the charge, the size of the ion, and the concentration. In 0·1 M solutions, the series $Th^{4+} > Fe^{3+} > Al^{3+} > Ba^{2+} > Pb^{2+} > Sr^{2+} > Ca^{2+} > Fe^{2+} > Co^{2+} > Mg^{2+} > Ag^+ > Cs^+ > Rb^+ > NH_4^+ = K^+ > Na^+ > H^+ > Li^+$ has been established, but in concentrated solutions the effect of valency is reversed and univalent ions are favoured over multi-valent ones. Thus, the sodium form of the resin may be used to remove, say, calcium ions from a dilute solution and the calcium can be recovered by treating the resin with a concentrated solution of a sodium salt. The use of a strongly basic anion exchanger for a similar purpose is fairly obvious.

The weakly acid cation exchangers behave as insoluble weak acids. They may be buffered so that exchange takes place at a controlled pH. In acid solution they are undissociated and have little exchange capacity, but in neutral or alkaline media they behave similarly to the strong acid resins as cation exchangers, being rather more selective for divalent cations. The hydrogen form is useful to produce a weak acid from one of its salts:

$$ResinH + CH_3COONa \rightarrow ResinNa + CH_3COOH$$

*Cation-exchange resins are abbreviated as c-resins and anion exchange resins as a-resins.

Similar remarks apply to the weakly basic anion exchange resins.

The uses of the ion-exchange materials will be fairly obvious from the above outline of their properties. In preparations or analysis, any specific ion may be replaced by another one or by a hydrogen or hydroxyl ion. In the latter cases the liberated acid or alkali may be titrated to determine the quantity of cation or anion respectively in the original solution. An example of a preparative application is the use of a strongly basic anion exchanger to prepare carbonate-free alkali metal hydroxides. A solution of, say, sodium hydroxide contaminated with carbonate is run down a column containing a strong base anion exchanger. This has a higher affinity for the doubly-charged carbonate ion which is then replaced by hydroxyl ion from the column. Pure sodium hydroxide solution is recovered.

An obvious extension of this technique is the removal of interfering ions in analysis, and the resins may also be used to concentrate trace constituents. The resins also find use as catalysts, especially as acid or base catalysts, and in the determination of dissociation constants and activity coefficients. As the exchange process is an equilibrium, the resins may be used in the separation of isotopes. For example, a separation of ^{14}N and ^{15}N has been effected by the exchange between a cation exchange resin and ammonium hydroxide:

$$\text{Resin}^{14}NH_4 + {}^{15}NH_4OH = \text{Resin}^{15}NH_4 + {}^{14}NH_4OH$$

A band of NH_4^+ travelling down a long column of the resin by means of a large number of absorption-desorption steps gradually becomes enriched with $^{15}NH_4^+$ at the trailing edge and with $^{14}NH_4^+$ at the leading edge. By the time the band is extended to about forty times its original width, the tail fraction contains 99 per cent ^{15}N.

6.2 Chromatography

Chromatography is a process for separating a mixture which depends on the redistribution of the components between a stationary phase and a mobile one. The components may be adsorbed on the stationary phase or be held by more specific chemical bonds. The typical arrangement has the stationary phase in a column and the mixture is passed up or down the column in the gas or liquid phases. One form of chromatography involves the use of ion exchangers and follows on from the discussion in the previous section.

Suppose that a mixture of cations in solution have similar chemical properties, they will not be separated from each other by the simple ion exchange processes discussed above. If the cations are absorbed at the top of a cation exchange column and then treated with a weak complexing agent—one with an affinity for the cation similar to that between the cation and the column material—then the equilibrium,

cation on ion exchanger + complexing agent
⇌ cation in complex in solution,

will depend sensitively on the nature of the cation. If the band of mixed cations on the column is washed down (eluted) by a solution of the complexing agent, those cations which form the strongest complexes will spend more time in solution than on the column, and will travel down the column faster than the cations whose complexes are weaker and which remain longer on the ion exchanger. As the cation band travels down the column, it starts to separate into its components and, if the column is long enough, the different components are recovered separately. The process is shown diagrammatically in Figure 6.1. As there are also random

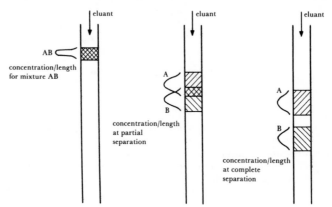

FIGURE 6.1 *Separation of a mixture by column chromatography*
The figure shows schematically the concentration gradients of the two components, expressed as Gaussian distributions, at different stages in the separation.

processes of diffusion affecting the ions, the distribution of cation concentration down the column is gaussian. The classic example of this method of separation was the separation of the lanthanide elements, using citrate as the weak complexing agent. A similar use of anion exchange resins allows a separation of anions. If the ions are not too closely similar, the separation may be effected without the use of a complexing agent. For example, the halides may be separated on an anion exchange resin by eluting with a sodium nitrate solution, and an additional element of control is introduced by varying the concentration of the eluant.

Many other systems have been devised for chromatographic separation. Materials such as silica gel or cellulose which act by adsorption, or by a mixture of adsorption and chemical interaction with adsorbed water, are widely used, and one common application is in paper chromatography. A wide variety of ions and molecules may be separated by using suitable solvents moving over paper. For example, the alkali metals may be separated by using a mixture of alcohols as the solvent, and the various phosphate anions may be separated using a two-dimensional method with a basic solvent flowing in one direction followed by an acidic one in a direction at right angles, Figure 6.2.

Another modification of this method is in the separation of gas mixtures by passing them in a stream of nitrogen, or helium, through a column containing a high-boiling liquid supported on an inert material. For example, the silicon or germanium hydrides may be separated by this vapour phase chromatography by passing the hydride mixture through a column of silicone oil on powdered brick.

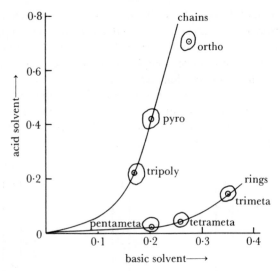

FIGURE 6.2 *Separation of phosphate anions by two-dimensional paper chromatography*
The ring anions fall on one curve and the chain ones on another. Higher members of each type would lie on the prolongations of these curves towards the origin.

6.3 Solvent extraction

A mixture may be separated into its components by treating it with two or more immiscible liquids, most commonly with an aqueous phase and an organic phase. Ionic and strongly polar species remain in the aqueous phase while less polar species dissolve into the organic liquid. For example, germanium may be separated by forming the tetrachloride in contact with a solvent, such as carbon tetrachloride, in which it is soluble. As the hydrolysis of the tetrachloride is reversible in presence of hydrochloric acid, all the germanium ends up in the organic phase:

Ge(IV) in aqueous solution $+ HCl \rightleftharpoons GeCl_4 + H_2O$
(as oxy- or hydroxy-species)

solution in CCl_4

The nature of the solute may be altered by changing the pH, by adding complexing agents, or by adding common ions, while considerable variation in the organic phase is also possible. There is thus a fair chance that any given mixture may be separated by suitably varying the conditions.

Perhaps the classic example of the application of this technique is the separation of uranium and plutonium from each other and from the fission products that accumulate in an atomic pile. One method involves the solution of the material in nitric acid, in which the uranium is in the VI state and is more covalent than the plutonium, which remains in the IV state. The uranium is extracted into ether and then the plutonium and fission products are precipitated and strongly oxidized to convert the plutonium to Pu(VI). This can then be extracted from the fission products in turn. A variation, which particularly lends itself to a cyclic process of repeated extraction, is to separate the plutonium and uranium in the VI state into an organic solvent which is then carried into a second vessel where it is subject to mild reduction, in contact with water, which leaves the U(VI)

unchanged but converts the plutonium to Pu(III), which is washed back into the aqueous phase by dilute nitric acid.

All these solvent extraction processes may be adapted for the separation of very similar solutes by using a continuous flow method, when fresh organic phase is brought into contact with the partially extracted aqueous phase and *vice versa*. This 'counter-current' method corresponds basically to a chromatographic separation involving two mobile liquid phases instead of one mobile phase and one held stationary on an absorbing phase. The close similarity of all these methods means that they may often be used interchangeably and the problem of separating chemically similar elements or compounds can now be treated by very powerful and versatile methods.

Structure Determination

6.4 Diffraction methods

The most accurate and powerful method of determining the structure of solids is by X-ray diffraction. Just as light can be diffracted by a grating of suitably spaced lines, so can the much shorter wavelength X-rays be diffracted. The regular array of atoms or ions in a crystalline solid provides a suitably spaced three-dimensional grating for the diffraction of a beam of X-rays, and the diffraction pattern can be detected by using a photographic film or other detector. When a beam of X-rays passes through a crystal it meets various sets of parallel planes of atoms. The diffracted beams from atoms in successive planes cancel unless they are in phase, and the condition for this is given by the Bragg relationship:

$$n\lambda = 2d\sin\theta$$

where λ is the wavelength of the X-rays, d is the distance between successive planes, and θ is the angle of incidence of the X-ray beam on the plane. From a set of distances d for different sets of planes in the crystal, the positions of the atoms can be derived. The angles θ can be measured and the values of d can be calculated if the wavelength of the X-rays is known. If the intensities of the diffracted beam in each direction can also be measured, complete structure determinations are possible. As each atom in the lattice acts as a scattering centre, the total intensity in a given direction of the diffracted beam depends on how far the contributions from individual atoms are in phase. The essential feature of complete X-ray structure determinations is a trial-and-error search for that arrangement of atoms which best accounts for the observed intensities of reflections. For large molecules, or structures of low symmetry, this process involves numerous calculations and is best carried out by computer.

There are a number of experimental methods for studying X-ray diffraction but two are widely used. One uses a single crystal of the compound which is mounted so that it can be rotated about a crystal axis. A monochromatic beam of X-rays (i.e. of a single wavelength λ) shines on the crystal. As it rotates, successive sets of planes are brought into reflecting positions and the reflected beams are recorded as a function of the crystal rotation. The positions and intensities of the reflected beams are measured and the complete structure can be derived from the resulting data.

A second method is used for the many cases when substances are available only as a crystalline powder and no single crystal is obtainable. The powder is packed in a thin capillary tube and illuminated by a monochromatic beam inside a circular camera. All possible crystal orientations occur at random, so that some crystallites will be correctly orientated to fulfil the Bragg condition for each value of d. The reflections are recorded as lines on a film placed round the inside of the camera and these positions give the d values. A complete analysis is possible only for crystals of high symmetry.

Although X-ray methods give the fullest and most reliable data, they are restricted to solids and are not very suitable for determining the positions of very light atoms. Two other diffraction methods are available, electron diffraction and neutron diffraction. By virtue of the wave properties of electrons, a beam of electrons may be diffracted. Electron diffraction is used to determine the structures of gaseous molecules and gives very accurate values for bond lengths and angles. The process of interpretation depends on matching the observed diffraction pattern with ones calculated for different model structures, so that it is not fully reliable as a method of structure determination but does give accurate parameters for molecules whose overall structure is known. Electron diffraction is a function of the atomic number and hydrogen atoms are, as a rule, not detectable.

A beam of neutrons behaves in a similar manner to a beam of electrons and gives information about the positions of light atoms and is thus complementary to the other two techniques. Neutron diffraction is particularly useful for locating light atoms, such as H, in solids after the heavy atom positions have been found by X-ray diffraction.

It is also possible to carry out X-ray diffraction experiments on gases but the technique is more difficult than electron diffraction and gives similar information.

6.5 Spectroscopic methods

If an atom, molecule, or ion, with two energy states differing in energy by ΔE is irradiated with continuous electromagnetic radiation, the radiation of frequency corresponding to ΔE will be absorbed and the species raised to the upper energy state. This absorption, or the consequent emission of radiation as the species returns to the ground state, may be detected and provides information about the energy states. The energy change is related to the frequency v' of the radiation by the relation

$$\Delta E = hv' \dots\dots\dots\dots\dots (6.1)$$

where h is Planck's constant equal to $6.625\,6 \times 10^{-34}$ J s. The frequency is related to the wavelength, λ, as their product equals the speed of light

$$v'\lambda = c = 2.998 \times 10^{10} \text{ cm s}^{-1} \dots\dots (6.2)$$

Thus the energy change per mole obeys the relations

$$\Delta E = Nhv' = Nhc/\lambda = Nhc\tilde{v} \dots\dots (6.3)$$

where $\tilde{v}$ is called the *wave number* or Kayser and is the number of cycles per centimetre, hence the unit is cm^{-1}. Equation (6.3) leads to the relation between different units given on page xiv.

The electromagnetic spectrum spans a very wide range, from gamma and cosmic rays with energies in excess of 10^6 kJ mol^{-1} down to radiowaves corresponding to small fractions of a joule per mole. In different regions of the spectrum, the energy corresponds to differences in two states of a system spanning many kinds of transition. Thus a nuclear transformation involves very high energies, in the gamma ray region, while the reversal of an electron spin in a magnetic field of 0·1 tesla would involve a tiny energy corresponding to the shorter radiowaves. Nevertheless, both these processes, and many others of intermediate energy, may be used to yield information of value to the chemist. The major regions of the electromagnetic spectrum, and the transitions corresponding to the interaction with them, are listed in Table 6.1. There is no clear boundary between any two regions and the energy ranges and types of transition overlap: we give arbitrary round-figure ranges (which are thus slightly different for different units) whose boundaries are often set by arbitrary experimental factors. As there is such a wide range of energies, and so many different types of transition are involved, there have grown up a number of separate areas of study which have developed fairly independently and with their own conventions. In particular, this means that a variety of units have been used and some of the traditional ones in the important regions are indicated in Table 6.1. Notice that these include units of energy, wavelength, and frequency.

The gamma and X-ray regions of the spectrum give information about the nucleus and the inner, closed shell, electrons and are therefore of less direct interest to the chemist although the Mossbauer effect which involves gamma resonance absorption, and the derivation of atomic energy levels from X-ray spectra, as in Moseley's experiments, are important. The average chemist is more likely, however, to make direct measurements of *electronic spectra* in the ultraviolet and visible region, of *vibrational spectra* in the infrared region and in the Raman effect, and of *nuclear magnetic resonance (nmr) spectra* in the radiofrequency region. These methods are therefore discussed in more detail below. Pure rotational spectra, studied in the microwave region, give accurate values for the moments of inertia and thus provide values for bond lengths and angles for sufficiently small or symmetric molecules which must also have a permanent dipole moment to interact with the radiation. By using isotopic substitution, a number of independent parameters may be determined, but even so, only a limited number of molecules are simple enough to be analyzed. Thus microwave spectroscopy gives information complementary to that derived from electron diffraction for small molecules in the gas phase. Electron spin resonance which is observed for

TABLE 6.1 Principal regions of the electromagnetic spectrum

Region	Approximate range in			Transition excited
	Wavelength (m)	Energy		
		(SI)	(Commonly-used)	
gamma rays	$<10^{-10}$	$>10^6$ kJ mol^{-1}	$>10^4$ eV	nuclear transformations
X-rays	10^{-8} to 10^{-10}	10^4 to 10^6 kJ mol^{-1}	100 to 10^4 eV	transitions of inner shell electrons
ultraviolet	4×10^{-7} to 10^{-8} }	10^2 to 10^4 kJ mol^{-1}	1 to 100 eV or }	transitions of valence shell
visible	8×10^{-7} to 4×10^{-7} }		10^4 to 10^6 cm^{-1} }	electrons including $d \rightarrow d$ and $f \rightarrow f$
infrared	10^{-4} to $2 \cdot 5 \times 10^{-6}$	1 to 50 kJ mol^{-1}	100 to 4000 cm^{-1}	molecular vibrations
microwave and far infrared	10^{-2} to 10^{-4}	10 to 1000 J mol^{-1}	1 to 100 cm^{-1}	molecular rotations
radio frequency	$\sim 10^{-2}$	~ 10 J mol^{-1}	3×10^4 MHz	electron spin reversal in magnetic field of 1 A m^{-1}
	~ 10	$\sim 0 \cdot 01$ J mol^{-1}	10 to 100 MHz	nuclear spin reversal in magnetic field of 1 A m^{-1}

1 eV = $96 \cdot 49$ kJ mol^{-1} = $23 \cdot 06$ kcal mol^{-1} and is equivalent to 8068 cm^{-1} and $2 \cdot 419 \times 10^8$ MHz: 1 Am^{-1} = 10^4 gauss
Ranges are rounded-off and are not converted exactly from one unit to the next.

species with unpaired electrons is also a tool available for only a restricted number of species. Rotational changes often accompany vibrational changes and are seen as a fine structure to the vibrational absorptions, and similarly, electronic transitions may have fine structures due to concomitant rotational and vibrational transitions.

Study of ultraviolet visible and infrared spectra all require the same basic equipment. The sample must be placed in a beam of radiation which can be continuously varied in frequency and a detector is required to show absorption of energy. For example, visible spectra require a source of white light which is scanned in frequency by using a rotating prism, while the detector is a photocell whose output may be converted to a movement of a pen on a recorder. An infra-red spectrum may be recorded using a heated element as source, an alkali halide prism or a grating to change the frequency, and a thermocouple as detector.

Vibrational transitions may also be detected in the Raman effect. In this, the sample is illuminated with strong mono-chromatic radiation, and some of the re-emitted quanta are found to have gained or lost a (smaller) quantum corresponding to the energy of one of the fundamental vibrational modes. The spectrum of the scattered radiation thus consists of a very strong line corresponding to the incident radiation and a number of other lines whose energy differences from the primary line give the energies of vibrational transitions of the molecule. Up till now, the main limitation in Raman spectroscopy has been in finding a primary source of illumination with a sufficiently intense beam of radiation spread over a narrow frequency range. The advent of lasers has greatly improved the potential of Raman spectroscopy and considerable developments are under way in laser Raman work.

Electronic spectra. Transitions of outer shell electrons fall in the general wave number range of 100 000 cm^{-1} to 10 000 cm^{-1}, that is in the ultraviolet, visible, and near infrared

regions of the electromagnetic spectrum. The transition involved are those between sigma, pi, or nonbonding, n orbitals in the valence shell, such as those involved in Figures 3.36 or 3.40. Not all transitions are allowed, only

$$\sigma \rightarrow \sigma*$$
$$n \rightarrow \sigma*$$
$$\pi \rightarrow \pi*$$
$$n \rightarrow \pi*$$

while sigma to pi or pi to sigma transitions are forbidden. It is clear from an energy level diagram like Figure 3.40 that this is the approximate order of decreasing energy. The largest energy is required for $\sigma \rightarrow \sigma*$ transitions and these are found in the far ultraviolet regions in which the molecules of the atmosphere also absorb. Evacuated spectrometers have to be used and this region is less studied.

Most work is done in the 50 000 to 10 000 cm^{-1} region 200 nm to 1 μm, where transitions involving nonbonding and pi electrons are involved. For lighter atoms, such transitions are usually in the ultraviolet, giving colourless compounds, but extended conjugation, giving rise to a large number of closely spaced pi orbitals, may lower the energy of the transition and give a coloured compound. In addition heavy atoms have their outer orbitals closer together in energy so that the transitions again fall into the visible region. Such transitions account for the colours of many iodides, for example. In addition to these transitions, transition elements with partly occupied d or f orbitals show bands, usually in the visible region, due to $d \rightarrow d$ or $f \rightarrow f$ transitions. Such transitions are formally forbidden and give rise to weak bands; they are discussed in more detail in Chapters 10 and 12.

While the position of an absorption band corresponds to the energy of the transition, its intensity depends on the nature and quantity of the absorbing material. The relation between the *absorbance*, A, the pathlength l and the concentration C is given by the Beer-Lambert law

$$A = kCl \qquad (6.4$$

where k is a constant characteristic of the material. For a pathlength of 1 cm, and with the concentration expressed in moles per unit volume, the constant k becomes ϵ, the *molar extinction coefficient*. Allowed transitions generally have molar extinction coefficients in the range 10^3 to 10^5 l mol^{-1} cm^{-1} while d–d or f–f bands are much weaker with ϵ typically 10^{-1} to 10 l mol^{-1} cm^{-1}. The absorbance, formerly termed the optical density, is defined as

$$A = \log\,(I_0/I) \dots\dots\dots\dots (6.5)$$

where I_0 is the intensity of the incident light and I the intensity after the light has passed through the sample. Most modern spectrometers record the absorbance directly.

Electronic spectra are used primarily to give information about the energy levels of the valency orbitals, and for inorganic chemistry this is particularly widely studied for transition elements. Use may also be made, through Beer's law, of absorbance measurements in quantitative analysis. Here, measurements are preferably made on strong bands and at, or near, maxima in the absorption. Thus, for example, the strong allowed bands of CrO_4^{2-} would be used for estimating Cr, rather than the weak d–d transitions of Cr^{3+} ions. (Note that species like chromate or permanganate, which are in the Group oxidation state, have no d electrons not involved in bonding so that these colours are not due to d–d transitions.) In a more qualitative way, electronic spectra may also be used to indicate the presence of particular groupings of atoms. For example, π–π^* transitions in a benzene ring will be relatively constant in position and intensity as long as π bonding substituents are absent: thus the presence of a phenyl group would be indicated by such bands in the electronic spectrum. Such an application is of more value in organic chemistry where such a chromophore is likely to occur in a fairly constant environment, but similar applications are of value particularly in organometallic chemistry.

Vibrational spectra. Vibrational modes of a molecule are excited by the absorption of quanta whose energy lies in the infrared region of the spectrum, from about 4000 cm^{-1} downwards. Vibrational transitions are also detected in Raman scattering. As the selection rules for infrared absorption differ from those governing Raman scattering, the two techniques are complementary and both infrared and Raman spectra need to be measured to obtain the maximum amount of information.

The information obtainable from vibrational spectroscopy depends on the size and symmetry of the molecule. For a diatomic molecule, assuming simple harmonic motion, the wavenumber is given by

$$\tilde{v} = \frac{1}{2\pi c}\sqrt{\frac{k}{u}} \quad \dots\dots\dots\dots (6.6)$$

where k is the force constant (the proportionality between the extension of the bond and the restoring force) in N m^{-1} and u is the reduced mass ($1/u = 1/m_1 + 1/m_2$) of the two atoms. Thus the vibrational frequency is directly related to the force constant, which in turn is related to the bond

strength. Absorptions of successive quanta of vibrational energy will continue until the molecule dissociates, and the frequency at which this occurs gives the bond energy. From the rotational fine structure, the moment of inertia may be derived and hence the bond length (if the atomic masses are known). The existence of isotopes of an element may be proved by observing different moments of inertia for the same compound. For example, hydrogen chloride is found to have a bond length of 128·1 pm and two moments of inertia corresponding to $H^{35}Cl$ ($I = 2\cdot649 \times 10^{-40}$ g cm^2) and $H^{37}Cl$ ($I = 2\cdot653 \times 10^{-40}$ g cm^2).

For polyatomic molecules, the position is more complicated. Vibrations involve not only bond stretching, but angle deformation and often twisting modes as well. An n-atom species has $3n-6$ degrees of vibrational freedom ($3n-5$ for a linear species) and a corresponding number of force constants are required to describe the vibrations. However, it is unlikely that all $3n-6$ vibrations will be observable in the majority of cases. This is partly due to practical difficulties of detecting weak bands and resolving closely overlapping ones, and partly due to degeneracy. For example, a tetrahedral molecule like GeH_4 has $3 \times 5 - 6 = 9$ modes of vibration but the maximum number of bands observable in the infrared is two and in the Raman, four. This is because two of the modes are degenerate, giving only one fundamental, and two further groups of three are triply degenerate. Thus there are only four observable bands, two triply degenerate, one doubly degenerate and one non-degenerate and these are found in the Raman spectrum. Of these, only the triply degenerate modes involve a dipole change so that only these two are also observed in the infrared. Analysis in terms of force constants is thus difficult, though isotopic substitution may help.

The use of vibrational studies for small and moderate-sized molecules is more limited. However, the number of bands expected in the spectrum is predictable from the symmetry of the molecule by the methods of *group theory* (this leads to the prediction above for GeH_4) and thus the spectrum may be used to determine which one out of a number of structures is the correct one. For example, a square planar AB_4 species has 3 infrared active bands (two of which are doubly-degenerate) and three Raman bands which do not coincide with the infrared modes. The ninth vibration is inactive in both the infrared and Raman. If this is compared with the prediction for a tetrahedral AB_4 species given above, it will be seen that these two shapes could readily be distinguished unless the intrinsic intensities or spacings were very unfavourable. In a similar way, the two possible structures for a species like perchloryl fluoride could be distinguished, as the $FClO_3$ structure, based on a tetrahedron, would have six bands active in both infrared and Raman (as three are doubly-degenerate) while the hypofluorite form, $FOClO_2$, has no degenerate bands and all nine vibrations would be seen in both infrared and Raman.

It should be noted that structural evidence of this sort may be used to support a structure but does not offer absolute proof. As some bands are inevitably difficult to observe, the assignment of six bands for perchloryl fluoride does not prove the first structure, as it is quite possible that the remaining three would be too weak or too close to others to be observed. The negative argument is much more definite —observation of seven or more fundamentals *would* disprove the first structure. Similarly, a species AB_4 showing three or more fundamentals in the infrared could not be a tetrahedron. These examples show the type of evidence that may be obtained by applying arguments based on the molecular symmetry, using group theory. The full discussion of these methods is beyond our scope, but the first step is to determine the molecular symmetry and this is discussed in the third Appendix. It will also be seen from these examples how valuable it is to be able to observe vibrational spectra both in the infrared and in the Raman. Often different modes are active in different effects, the extreme example being provided by species such as square planar AB_4 which have a centre of symmetry where no modes are both Raman and infrared active. Even where all the expected modes are active in both effects, it is likely that bands which are weak in one effect will be strong in the other. One example is offered by stretching modes involving similar heavy atoms which are often weak in the infrared, as the dipole change is small, but are strong in the Raman because a fairly extended electron cloud is moving. Thus the Si—Ge stretching mode in H_3SiGeH_3, which is allowed in both the infrared and the Raman, is too weak to be observed in the infrared but gives a strong Raman band at about 350 cm^{-1}. A similar effect is found for many metal-metal stretching modes.

In more complicated species, this approach breaks down because the number of predicted bands becomes so large that detailed assignment is impossible. The vibrational spectrum is still useful, but in a more qualitative way. First, certain groups in a molecule may absorb in fairly constant regions of the spectrum and can thus be identified. For example, a CN group, with its triple bond, absorbs at about 2000 cm^{-1}, a far higher wave number than modes involving single bonds and heavier atoms. Thus the presence of cyanide as a ligand may always be detected, and similarly of CO in metal carbonyls. Indeed, symmetry information may be derived from the number of bands in the 2000 cm^{-1} region, to distinguish *cis* and *trans* $ML_3(CN)_3$ for example, as these vibrations are little affected by the presence of other groups (compare also Figure 6.3). From the mass effect, modes involving hydrogen are also found at higher frequencies than those involving any other substituent and these also are readily distinguished. Such partial analyses are often valuable, and much information may be derived by comparing related species. Thus, the Cl—F and Cl—O—F alternatives for ClO_3F above, might be further distinguished by comparison with chlorine fluorides and other hypofluorite species to see if bands appear in regions characteristic of Cl—F stretching or of O—F stretching.

Finally, a purely qualitative approach may be made in which a complex species may be identified with a known compound if their spectra are identical. In this case, the

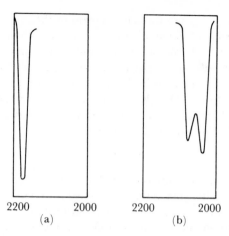

FIGURE 6.3 C—N *stretching bands in the infra-red spectra of* (a) $Ni(CN)_4^{2-}$, *and* (b) $Ni_2(CN)_6^{4-}$ *ions*
The CN absorption occurs in a similar region for both compounds (a) at 2124 cm^{-1} and (b) at 2045 cm^{-1} and 2075 cm^{-1}. The difference in symmetry between the two ions is reflected in the number of absorptions.

more complex the spectrum, the more definite would be the identification.

Nuclear magnetic resonance. When an atom, such as hydrogen, with a nuclear spin of $\frac{1}{2}$ is placed in a magnetic field, the spin may take up one of two orientations, either parallel or antiparallel to the field vector. These two correspond to different energy states and, for hydrogen in a field of about one tesla (10^4 gauss) the energy difference is about 10^{-2} J mol^{-1}. This corresponds to a frequency of about 40 MHz. Recently developed magnets allow larger fields and modern instruments use 60 or 100 MHz for proton resonance, or even 220 MHz using a superconducting coil. In the experimental arrangement commonly used, the sample is placed in a cylindrical tube in the field and irradiated with a fixed frequency. The field is slightly modulated by passing a current through coils and energy absorption is detected at resonance when

$$h\nu = \Delta E = g_N\beta_N H \dots\dots\dots\dots (6.7)$$

Here H is the magnetic field, β_N is the nuclear magneton $= eh/2\pi Mc = 5{\cdot}050\ 5 \times 10^{-27}$ J T^{-1} and g_N is a constant, called the nuclear g factor, which is a characteristic of the element.

The great value of nmr to the chemist is that the magnetic field actually experienced by a particular nucleus is the sum of the applied field and fields induced in the electrons around the magnetic nucleus. Thus atoms of the same element which are in different chemical environments resonate at slightly different values of the external field and these differences in *chemical shift* may be detected and yield structural information. In other words, the position of the absorption depends on the atom to which the hydrogen is bonded and on the other bonds nearby. The classic case is that of ethanol, CH_3CH_2OH, where the absorption due to the hydrogen bonded to oxygen is in a different position to the absorptions of the carbon-bonded hydrogens. Furthermore, the hydrogens in the methyl group are in a different electronic environment to those in the methylene (CH_2) group, so the absorptions due to the two types of carbon-bonded hydrogens are separated. The three methyl-protons are in identical environments, as are the two methylene ones, and the resulting spectrum consists of three peaks, in the ratio of 3:2:1,

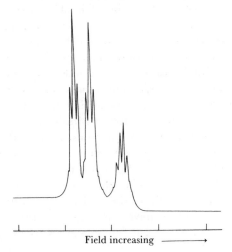

FIGURE 6.4 *The ^1H nuclear magnetic resonance spectrum of* $SiH_3GeH_2GeH_3$

The resonances are, in order of increasing field (1) the triplet due to the SiH_3 protons split by coupling to the two GeH_2 protons, (2) the triplet due to the GeH_3 protons, also split by the GeH_2 coupling, and (3) the multiplet due to the GeH_2 protons split by coupling both to the SiH_3 and to the GeH_3 protons. Notice that the coupling constant, $\mathcal{J}(SiH_3GeH_2)$ is almost identical to the constant $\mathcal{J}(GeH_3GeH_2)$, shown by the near-identity of the splittings in the two triplets, and therefore the signal from the GeH_2 protons is very close to the septet which would arise from coupling to six identical protons.

for the three types of hydrogen atoms, and in positions typical of methyl, methylene, and hydroxyl hydrogens. An example of the ^1H nmr spectrum of an inorganic hydride, $SiH_3GeH_2GeH_3$, is shown in Figure 6.4. There are three main envelopes of absorption, each showing fine structure—see below—in intensity ratio 3:3:2 which arise from the SiH_3, GeH_3 and GeH_2 protons respectively.

Such detailed information about the environment of an atom may be sufficient to determine the structure, or at least goes a long way towards this. For example, diborane B_2H_6, has had a number of structures proposed for it, including an ethane-like one (A) and a bridged structure (B).

$$(A) \quad \begin{matrix} H & & H \\ \diagdown & & \diagup \\ H{-}B{-}B{-}H \\ \diagup & & \diagdown \\ H & & H \end{matrix} \qquad (B) \quad \begin{matrix} H & H & H \\ \diagdown & \vdots & \diagup \\ B & & B \\ \diagup & \vdots & \diagdown \\ H & H & H \end{matrix}$$

The hydrogen magnetic resonance spectrum of (A) would consist of only one line, as all the hydrogen atoms are equivalent, whereas the hydrogen spectrum of (B) would have two lines, in the ratio of 4:2, corresponding to the terminal and bridge hydrogens respectively. The latter spectrum is observed, supporting the bridge structure.

Nuclear resonance may be observed for atoms other than hydrogen which have a spin of one half, and also for atoms with higher nuclear spins, although the spectra are more complicated in the latter cases. No resonance is possible when atoms have zero spin and this includes some common atoms such as ^{12}C and ^{16}O. So far, studies have been mainly on hydrogen resonance but other atoms studied include ^2H (deuterium), ^{11}B, ^{13}C, ^{14}N, ^{19}F, ^{29}Si, ^{31}P, and also sulphur, the halogens, and tin. Of these, fluorine is the most widely studied after hydrogen, and then phosphorus. One example of fluorine resonance is in the confirmation of the square pyramid structure of BrF_5 and IF_5 by the observation of two lines in the fluorine resonance, with intensities in the ratio 4:1, corresponding to the basal and apical fluorines respectively.

Further information may be obtained from the fine structure of the nmr bands which arises from the effects of spin-spin coupling. If an atom which has a nuclear spin, like H, is bonded to a second one which also has a nuclear spin, the local magnetic field will be affected by the orientation of the spin of the second nucleus. For example, in PH_3 the phosphorus-31 nucleus, which has a spin of $\frac{1}{2}$, may be aligned with or against the field, giving two different local resultant fields. As the energy difference between the two orientations is so small, there will be essentially equal numbers of molecules with each P spin alignment. Thus, in the proton resonance signal there will be two components corresponding to hydrogens bonded to P atoms with spins parallel or antiparallel to the external field. Thus the proton resonance signal is a doublet with components of equal intensities. The phosphorus atom also shows a resonance signal, though at much lower frequency, and this is split by the spins of the three H atoms into four components. There are four possible arrangements of the proton spins—all parallel, which we can label $+\frac{1}{2}$, $+\frac{1}{2}$, $+\frac{1}{2}$, or with two, one or no spins parallel to the field and respectively one, two or three antiparallel spins, labelled $+\frac{1}{2}$, $+\frac{1}{2}$, $-\frac{1}{2}$: $+\frac{1}{2}$, $-\frac{1}{2}$, $-\frac{1}{2}$: $-\frac{1}{2}$, $-\frac{1}{2}$, $-\frac{1}{2}$. There are thus four different net fields and four components to the signal. In addition, while the $+\frac{1}{2}$, $+\frac{1}{2}$, $+\frac{1}{2}$ and $-\frac{1}{2}$, $-\frac{1}{2}$, $-\frac{1}{2}$ arrangements can only result in one way, any one of the three protons may be the antiparallel one in the other two combinations. Thus, a net spin of $+\frac{1}{2}$ results from any one of the three sets $+\frac{1}{2}$, $+\frac{1}{2}$, $-\frac{1}{2}$ or $+\frac{1}{2}$, $-\frac{1}{2}$, $+\frac{1}{2}$ or $-\frac{1}{2}$, $+\frac{1}{2}$, $+\frac{1}{2}$, and similarly for a net spin of $-\frac{1}{2}$. As all possible spins combinations are equally probable, there will be three times as many molecules in the sample with net spin $+\frac{1}{2}$ or $-\frac{1}{2}$ as with net spin 3/2 or $-3/2$. Thus the phosphorus signal becomes a quartet with relative intensities 1:3:3:1. In a similar way, the signal of any atom bonded to n equivalent atoms of spin $\frac{1}{2}$ becomes an $(n+1)$ multiplet with intensities in the ratio of the binomial coefficients. Such spin-spin coupling is not limited to atoms which are directly bonded, as it is transmitted via the bonding electrons, and it may be observed for atoms separated by several bonds. Thus the CH_3 and CH_2 protons in ethanol couple to make the methyl signal a triplet (1:2:1 intensity ratio) and the methylene signal a quartet. This latter may be further split by coupling to the OH proton. Similarly, the SiH_3 signal in $SiH_3GeH_2GeH_3$ is split into a 1:2:1 triplet by coupling to the GeH_2 protons and the GeH_3 signal is also a triplet for the same reason. The GeH_2 signal is a more complex multiplet as these protons are coupled both to the SiH_3 ones and to the GeH_3 ones. The resultant fine structure is seen in Figure 6.4.

If the nucleus has a spin of more than $\frac{1}{2}$, the coupling splittings follow different rules. A nucleus of spin $n/2$ has $n+1$ orientations each equally likely, and this gives an $(n+1)$ multiplet of equal intensities. Thus the proton signal of the terminal hydrogens in diborane, above, when the coupling to ^{11}B with spin $= 3/2$ is taken into account, is a quartet with all components of equal intensity.

Thus the nmr investigation gives information about the relative numbers of magnetic nuclei of each type in a molecule, from the position and intensities of the signals, and shows something of the structure of the molecule from the spin-spin coupling. If more than one type of magnetic nucleus is present, the resonance of each may be studied although experimental difficulties are greater as the intrinsic intensity is lower than for ^1H. The resonance frequencies are so different that the study of each different nucleus is a separate process. At present, studies on ^1H are much the most common, but work on ^{19}F, ^{31}P, ^{11}B and ^{14}N is quite extensive.

Apart from its use in identifying compounds, information about reaction kinetics and exchange processes may be derived by studying the nmr signals over a range of temperatures.

Electron spin resonance, Mossbauer, and mass spectroscopy

While nmr is concerned with reversal of nuclear spin, *electron spin resonance* involves a very similar phenomenon, the reversal of the spin of an electron in a magnetic field. This involves a higher energy than the nuclear spin reversal, about 10 J mol^{-1} and thus a higher frequency of around 28 000 MHz. In an electron pair, the spins are already opposed and any reversal of one spin would be cancelled by that of the other (the energies involved are much lower than those required for excitation to the triplet state where the spins are parallel). Thus esr measurements can only be made on species with unpaired electrons like radicals and transition element compounds. For these, changes in the electron g factor (compare equation (6.7)), analogous to the nmr chemical shift, and spin-spin coupling to magnetic nuclei may be observed. One application is in the study of molecular orbitals. For example, if an electron is added to benzene to give the anion, $C_6H_6^-$, this electron will enter the lowest available orbital, which is one of the antibonding orbitals. The detailed structure of the electron resonance absorption then yields information about the interaction of the electron in this delocalized π-orbital with the atomic nuclei, and hence the distribution of this π-orbital over the atoms of the molecule may be determined and compared with the calculated one. In particular, an atom lying on a nodal plane should not interact with the electron, and this may be checked.

If an isotope, such as ^{119}Sn, has a metastable excited state which transforms to the normal ground state of the nucleus by emitting a gamma ray, the excited state can be used as a source and the ground state as target for the emission and resonant reabsorption of the gamma ray. This is the Mossbauer effect and is made use of in *Mossbauer spectroscopy*. The striking characteristic of the effect, which makes it so valuable, is the extremely well-defined energy of the transition. The line width is only about 10^{-12} of the energy, compared with about 1 cm^{-1} in 1000 cm^{-1} for a sharp infrared line for example. This means that the very small effect on the nuclear energy arising from the chemical environment may be detected in Mossbauer spectroscopy. Two interactions are important to us. First, as the nucleus has a slightly different radius in its ground and excited states, the resonance energy changes slightly with the electron density at the nucleus. This gives rise to a chemical shift whose magnitude reflects the s electron density as only electrons in s orbitals have any density at the nucleus. A good example is provided by tin species where chemical shifts lie in the sequence Sn(IV) < Sn(0) < Sn(II). In metallic tin (α-form) the element has the diamond structure with four more tin atoms surrounding each atom tetrahedrally. The configuration is thus s^1p^3 giving one s electron per tin atom. In tin (IV) compounds, all the valency electrons tend to be used in bonding so that the s electron population is less than one (and becomes 0 for the Sn^{4+} ion). On the other hand, in Sn(II) compounds, there is one unshared pair of electrons whose configuration approximates to s^2. Thus the order of chemical shifts in the Mossbauer follows from the configurations $s^0 < s^1p^3 < s^2$. Clearly, intermediate shifts give information about the s electron density, for example shifts between those for Sn^{4+} and for the metal show the direction of s electron drift in bonds to covalent tin(IV).

In the second phenomenon, the Mossbauer resonance may be split by interaction with the nuclear quadrupole moment. This arises where the nucleus has a spin of more than $\frac{1}{2}$, either in the ground state or in the metastable state. The nuclear quadrupole moment interacts with electric field gradients at the nucleus, and thus the quadrupole coupling indicates the degree of departure from spherical symmetry at the nucleus. That is, the quadrupole coupling gives information about the p and d electron densities. A further interaction which may be detected in the Mossbauer effect is the splitting of nuclear energy levels in a magnetic field. This may be imposed externally or arise internally from ferromagnetic or paramagnetic interactions and gives information about these.

The main limitation of Mossbauer spectroscopy is that only a limited number of elements have a suitable metastable nuclear state. All these have very short lifetimes and occur in the course of some decay sequence, so that the work requires an irradiating source and a suitable sequence of nuclear decay processes. The Mossbauer effect has been observed, or is predicted, for 49 elements but all with $Z \geqq 26$ (iron), except for potassium. It is experimentally easiest to study iron and tin but a fair amount of work has been done on others including Te, I, Xe, Au and several of the lanthanides. It has the advantage that the sample need only be a powder, so that it provides a method of studying insoluble, poorly crystalline materials of the heavier elements which are difficult or impossible to study in any other way.

One further spectroscopic technique which is making an impact on modern inorganic chemistry is *mass spectroscopy*. This is the refinement of Aston's method of determining isotope weights and has been extensively used by organic chemists in the last decade. More recently, extensive studies of metal carbonyls and organometallic compounds have appeared and other inorganic applications are becoming common.

In the experiment, a stream of the vapour of the substance to be studied is passed through a beam of electrons of energy usually in the region of 70 eV. These ionize the molecules, M,

$$M + e^- = M^+ + 2e^-$$

and the resultant ions may fragment to daughter products, radicals, and ions,

$$M^+ = M' + M''^+$$

The ions are passed through a magnetic field, and in many cases an electric field as well, and resolved into species with the same m/e ratio. The resolved peaks are detected to give the mass spectrum. If the *parent ion* M^+ can be detected (in some cases it is not) its mass may be measured with high accuracy and yields the molecular weight of the molecule. The use of double focusing (magnetic and electrical fields) yields masses accurate to about 1 ppm and thus allows analysis. For example, CO may be distinguished from $^{14}N_2$ or from ^{28}Si. The fragmentation path may also give useful information, and related compounds often have similar fragmentations. For example, metal carbonyls $M(CO)_n$ lose CO groups stepwise so that ions $M(CO)_x^+$ are observed for all values of x from n to 0. By varying the energy of the electron beam, the minimum energy required to ionize the molecule, or one of its fragments, may be determined, leading to values of the ionization potentials and information on bond energies.

A vapour pressure of only a fraction of a millimetre of mercury is required, and provision is made for heated samples, so that any species with a significant vapour pressure up to about 300 °C may be examined. The technique may thus be applied to a wide range of inorganic molecules, although the high resolution spectrometers and those useful for high masses, are elaborate and expensive.

6.6 Other methods

Other methods of investigating inorganic compounds include magnetic measurements and the measurement of dipole moments. If a sample of a compound is weighed in a magnetic field and then in absence of the field, a weight change will be observed. Most compounds are repelled by the field and show a decrease in weight; these are termed *dia-*

magnetic. The diamagnetism arises from the repulsion between the applied field and induced magnetic fields in the compound, and is a very small effect which occurs for all compounds. However, some compounds show a net attraction to a magnetic field and an increase in weight; these are termed *paramagnetic*. The paramagnetism arises where there are one or more unpaired electrons in the compound, and is a much larger effect than diamagnetism. An unpaired electron corresponds to an electric current, and hence to a magnetic field, by virtue of two effects, its spin, and its orbital motion. In most compounds, the effect of the orbital contribution is quenched out by the electric fields of surrounding atoms, and the spin-only magnetic moment is observed. This is given by:

$$\mu = 2\sqrt{[S(S+1)]}$$

where μ is the magnetic moment in units of Bohr Magnetons, and $S = \frac{1}{2}n$ equals the number of unpaired spins multiplied by the spin quantum number. This formula holds, to within ten per cent, for most compounds, allowing a direct determination of the number of unpaired electrons. In some cases, particularly when the unpaired electrons are in an f orbital, the orbital contribution is not quenched out and a more complex formula:

$$\mu = \sqrt{[4S(S+1) + L(L+1)]}$$

which involves the orbital quantum number, L, holds (see Figure 10.5). In some cases, the determination of the number of unpaired spins gives direct structural information. For example, consider a nickel(II) compound $NiL_4.2S$, where L is any ligand and S is a molecule of solvent. If the solvent molecules are not coordinated to the nickel, the NiL_4 species could well be square planar and have no unpaired electrons, while coordinated solvent would mean an octahedral NiL_4S_2 species with two unpaired electrons (compare section 13.8).

It is possible to gain much more detailed information from magnetic measurements than is indicated above. Other effects such as ferromagnetism and anti-ferromagnetism are observed, and much valuable information results from studying the variation of magnetic moment with temperature, concentration, and field strength. However, the simple Gouy method of weighing the sample in a magnetic field is readily carried out and yields considerable information, especially in transition metal chemistry.

The measurement of the dipole moment of a compound may also yield useful structural information. As any bond between atoms with different electronegativities is polarized, any molecule will have a dipole moment unless such bond dipoles are so arranged as to cancel out. As a simple example, if CO_2 is linear, the two C−O dipoles oppose each other and no resultant moment is observed, while if the molecule is bent a resultant dipole is observed (Figure 6.5). The figure also illustrates, as a further example, how *cis* and *trans* isomers may be distinguished by dipole moment measurements. Care, however, must be exercised in interpreting dipole moments: thus, NF_3 has an almost zero dipole moment (whereas NH_3 has a marked moment), not because the molecule is planar as once thought, but because the bond and lone pair dipoles cancel. However, with care in interpretation, dipole moments have proved a very useful adjunct to structural determinations, and the method—like the magnetic measurements above—has the advantage that it does not destroy the sample, and adaptations are available which require only a small amount of material.

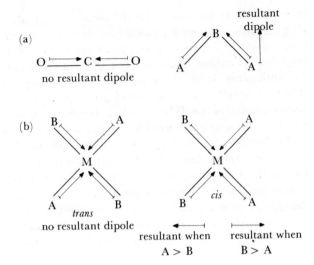

FIGURE 6.5 *The use of dipole moments to yield structural information*
Figure (a) shows how the two C−O moments cancel in linear O−C−O but would give rise to a resultant moment if the molecule was V-shaped. As CO_2 is observed to have no dipole moment, the linear structure is the correct one. Figure (b) shows how a *cis* square planar complex MA_2B_2 will have a resultant moment while the *trans* form will not.

7 General Properties of the Elements in Relation to the Periodic Table

7.1 Variation in energies of atomic orbitals with atomic number

In Chapter 2 the derivation of the existence, shapes, and energies of the atomic orbitals from the wave equation was described and it was shown how the structure of the Periodic Table could be derived by filling the atomic orbitals in order of increasing energy. A more detailed discussion of the properties of the elements demands a closer look at the variation in energy of the atomic orbitals as the atomic number increases.

Consider first the s orbitals. The energy of an electron in the $1s$ orbital of a hydrogen-like atom of atomic number Z, is $-Ze^2/8\pi\epsilon_0 a_0$; so that the energy decreases as Z increases. As the nuclear charge and the number of electrons in the atom increase, account must be taken of the repulsive effect of the extra electrons, as well as that of the increased nuclear charge. This is done by replacing the actual nuclear charge Z, by the effective nuclear charge, Z^*, which is the resultant of the nuclear charge and the electron charges as experienced by an electron in a particular orbital. For example, the $2s$ electron in lithium experiences an effective charge which is the resultant of the nuclear charge of $+3$ and the charges of the two $1s$ electrons. The effect of the inner electrons in reducing the effective charge experienced by the outer one is termed *shielding*. If the shielding effect of the two $1s$ electrons in lithium were perfect, the outer electron would experience an effective nuclear charge, Z^*, of $1\cdot0$ $(3-2)$, but the $2s$ orbital has finite electron density at the nucleus (see Figure 2.9b) so that an electron in the $2s$ orbital penetrates the $1s$ shell and thus experiences a greater nuclear charge than that calculated from perfect shielding. The result of shielding by inner electron shells is that the effective nuclear charge experienced by the outer electrons in an atom is always markedly less than the actual nuclear charge, Z, but, as the shielding is not perfect, the effective nuclear charge increases as Z increases, but more slowly. A useful indication of the shielding effects is given by Slater's rules which are summarised in Table 7.1. Application of these rules show that the effective nuclear charge experienced by one $1s$ electron in, say, carbon is $5\cdot7$ and in nitrogen, $6\cdot7$. Similarly, a $2s$ electron in carbon experiences an effective charge of $3\cdot25$, while a nitrogen $2s$ electron experiences a charge of $3\cdot9$. Such calculations, or direct experimental determination of energy levels by electron spectroscopy in one of its forms, allows the construction of diagrams showing the variation in energy levels with atomic number, such as Figure 7.1.

TABLE 7.1 Slater's rules for shielding contributions

The effective nuclear charge, Z^*, is given by $Z-\sigma$, where σ is the sum of the shielding contributions of all the other electrons in the atom, as follows

Principal quantum number, n, of shielding electrons	*Shielding contribution, σ*
n higher than principal quantum number of the electron under consideration	zero
n equal to principal quantum number of the electron under consideration	0·35 (for each electron) except that 0·30 is used for σ for a $1s$ electron acting on the second $1s$ electron
n is one less than the principal quantum number (a) for an s or p electron under consideration (b) for a d or f electron under consideration	0·85 1·00
n is less by two, or more, than the principal quantum number of the electron under consideration	1·00

In hydrogen-like atoms, the p orbital has the same energy as the s orbital with the same value of n, but in all other atoms where there is more than one electron and shielding effects come into play, the p orbital is more shielded than the corresponding s orbital as it does not penetrate so far towards the nucleus. It accordingly experiences a smaller effective nuclear charge and is of higher energy. The p orbitals also

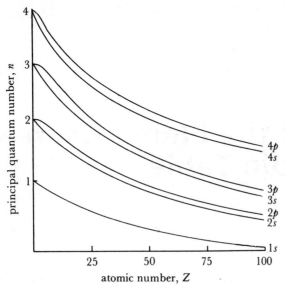

FIGURE 7.1 *Energies of the lower s and p orbitals as functions of Z*

experience an increasing charge as Z increases and the curves of their energies *versus* atomic number run roughly parallel with the curves of the corresponding s orbitals as Z increases. This is shown in Figure 7.1. The gap in energy between the s and p orbital of a given n value is much smaller than that separating the p orbital and the s orbital with the next higher n value.

The case of the d orbitals is more complicated. Figure 7.2 shows the variations in energy with increasing Z of the $3d$ orbital with respect to the $3s$, $3p$, $4s$ and $4p$ orbitals. The $3d$ orbital has the same energy as the $3s$ and $3p$ orbitals in the hydrogen atom and, since it scarcely penetrates the first and second quantum shells at all, it is perfectly shielded from the increase of nuclear charge as these two atomic levels are filled. Thus the $3d$ level is subject to the same effective nuclear charge (about unity) for Z values up to $Z = 10$, and the plot of the $3d$ energy against Z remains level. On the other hand, the $4s$ and $4p$ orbitals—which are of consider-

ably higher energy than the $3d$ orbital in the lightest elements —do penetrate the inner electron shells significantly, are less shielded, and drop steeply in energy as Z increases. When the $3s$ and $3p$ levels are filling, the $3d$ level still remains almost unaffected and the energy of the $4s$ level falls below it at about $Z = 15$. As a result of this, when the $3p$ shell is filled at argon, the next lowest energy level is $4s$ and not $3d$. The nineteenth and twentieth electrons therefore enter the $4s$ level, into which the $3d$ level does strongly penetrate. It experiences a marked increase in effective nuclear charge, and its energy falls from being nearly equal to that of the $4p$ level towards that of the $4s$ level. The twenty-first electron and the next nine enter the $3d$ level whose energy falls below that of the $4s$ level. As these two remain very close in energy, electrons readily switch between them; for example, copper, which might be $3d^9 4s^2$, is actually $3d^{10} 4s^1$, and gains the extra stability of the filled d shell by transferring an s electron. When the $3d$ shell is filled, the level next in energy is the $4p$ orbital. This filled d shell introduces an extra shielding effect on the higher orbitals, and the energy gap between the $4p$ and $5s$ levels at $Z = 36$, where the $4p$ level is filled, is larger than that expected by simple extrapolation from the $2p$–$3s$ and the $3p$–$4s$ differences.

A similar effect occurs for the $4d$ level relative to the $5s$ one. The effective nuclear charge experienced by the $5s$ level increases more rapidly with Z than does that of the $4d$ level, and the order of filling these levels is $4p$ then $5s$ then $4d$ and then $5p$.

The effect on the $4f$ level of increasing nuclear charge is rather similar to that on the $3d$ level at lower charges. The f orbitals penetrate the inner electron shells even less than do the d orbitals, and the $4f$ level is perfectly shielded as all the inner shells fill. The $5s$, $5p$, $6s$, $6p$, and $5d$ levels all drop below

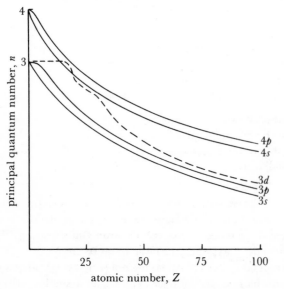

FIGURE 7.2 *Energies of the 3d and neighbouring orbitals as functions of Z*

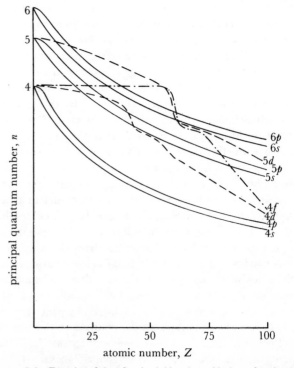

FIGURE 7.3 *Energies of the 4f and neighbouring orbitals as functions of Z*

the $4f$ level in energy as Figure 7.3 shows. The $5s$, $4d$, and $5p$ levels fill, and then the next level in energy is not the $4f$ nor the $5d$ but the $6s$ level. Both the $4f$ and $5d$ levels are strongly affected by the filling of the $6s$ level and drop very steeply in energy below the $6p$ level, but remain almost equal in energy to each other. In the event, the element lanthanum, which comes after the $6s$ level is filled, has its outer electron in the $5d$ shell but the following element cerium has the outer configuration $4f^2 5d^0 6s^2$, and the next twelve electrons fill the $4f$ shell, then the $5d$ and $6p$ levels are filled in turn to give the configuration of xenon. The $5f$ and $6d$ levels are very close in energy for similar reasons, and in this case the $6d$ level is partly filled first and then all the d and f electrons appear to change into the f shell, which is being filled in the heaviest of the artificial elements.

The complete diagram of the variation of all the orbital energies with Z is given in Figure 7.4. The energies in Figure 2.4 correspond to the orbital energies at the value of Z at which they start filling.

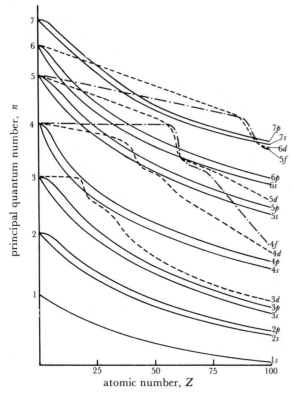

FIGURE 7.4 *The variation of the energies of the atomic orbitals with atomic number*

7.2 Exchange energy

The energy of an electron in an orbital depends on other factors besides the attraction of the nuclear charge and the electrostatic interaction between the electrons. There is a second, quantum-mechanical, interaction between electrons which is known as the *exchange energy*. There is no classical analogue to this energy which derives from the indistinguishability of electrons and the arrangement of their spins. The exchange energy is a function of the number of pairs of electrons with parallel spins, i.e. $E_{ex} = K \times P$, where K is a

constant and P is the number of pairs of parallel electrons. P is equal to the combination $_nC_2$ where n is the number of parallel spins, i.e. P has the following values:

n	1	2	3	4	5	6	7
$P = \dfrac{n(n-1)}{2}$	0	1	3	6	10	15	21

This energy is at the basis of Hund's Rule that electrons which enter orbitals of equal energy have parallel spins as far as possible. For example, the exchange energies for the various possible configurations of the electrons in the three p orbitals are shown below (note that the parallel and antiparallel electrons act as independent sets).

Number of electrons	Exchange energy if Hund's Rule is followed ($\times P$)		Exchange energy for maximum pairing ($\times P$)		Loss of energy in latter case ($\times P$)
1	↑	0	↑	0	0
2	↑ ↑	1	↑↓	0+0	1
3	↑ ↑ ↑	3	↑↓ ↑	1+0	2
4	↑↓ ↑ ↑	3+0	↑↓ ↑↓	1+1	1
5	↑↓ ↑↓ ↑	3+1	↑↓ ↑↓ ↑	3+1	0
6	↑↓ ↑↓ ↑↓	3+3	↑↓ ↑↓ ↑↓	3+3	0
	due to spins ↑ ↓		due to spins ↑ ↓		

An example of the stabilization due to exchange energy has already been noted in the case of the ground state of copper. The exchange energy of the actual configuration $d^{10} s^1$ is $20K$, from the two sets of five parallel electrons in the d shell. The energy of the alternative configuration $d^9 s^2$ is $16K$ (from a set of five plus a set of four in the d shell; the two s electrons are antiparallel of course). The exchange energy gain thus favours the d^{10} configuration, but against this must be set the loss in orbital energy in moving the electron from the $4s$ orbital to the $3d$ one. In the case of copper, the gain in exchange energy more than balances this loss, while in the case of nickel, which has the configuration $d^8 s^2$ but could be $d^{10} s^0$, the balance appears to lie the other way and the former configuration is the ground state. That the balance of energies is very close is shown by the configurations in the nickel and copper Groups:

nickel	$3d^8 4s^2$	palladium	$4d^{10} 5s^0$	platinum	$5d^9 6s^1$
copper	$3d^{10} 4s^1$	silver	$4d^{10} 5s^1$	gold	$5d^{10} 6s^1$

The exchange energy is responsible for the stability of the filled shell configurations, and also accounts for the stability of the half-filled shell with the maximum number of parallel electrons. Illustrations are to be found in the ground state configurations of transition and inner transition atoms where the d^5 and f^7 half-filled shells are favoured. Examples can be found in Table 2.5, including the configurations of chromium and gadolinium and their neighbours. This preference for half-filled and filled shell configurations is general in the Periodic Table, although the nice balance of energies means that configurations are not readily predictable (compare the ground state electronic configurations of the second transition series from ytterbium to cadmium in Table 2.5). The examples quoted so far have been confined to the ground states of atoms, but the stability of these special arrangements also shows up in the general chemistry of the elements. Manganese, for example, is particularly stable in the $+II$ state which is a d^5 configuration.

The interelectronic forces and the changes in nuclear charge play an important part in determining the stability and configurations of ions, but it must be noted that it is not possible to determine the detailed chemistry of an ion from the ground state configuration of its parent atom. For example, in most of the transition metals the $(n+1) s$ shell is filled while the nd level is only partly occupied. That is, in the atom the s shell is more stable than the d shell. However, when any of the transition elements form ions, it is always the s electrons which are lost first. Further, once one or more electrons are lost from an atom, the order of orbital stabilities is not necessarily the same as in the undisturbed atom. Thus, europium has the configuration, $4f^7 5d^0 6s^2$, while the next two elements have the configurations $f^7 d^1 s^2$ and $f^9 d^0 s^2$, yet all three lose three electrons to give a stable trivalent cation of valency configuration f^6, f^7 and f^8 respectively, just as if each element had had the outer configuration $f^n d^1 s^2$.

7.3 Stable configurations

With the reservations expressed above in mind, it is possible to generalize about stable electronic configurations by considering the interplay of orbital and exchange energies. In Figure 7.4, the line joining the orbital energies at the Z values at which these orbitals fill has to leap a number of wide gaps. The largest of these energy jumps come between the p orbitals of the nth quantum shell and the s orbitals of the $(n+1)^{th}$ level (and that between the $1s$ and $2s$ levels at the beginning of the Periodic Table). The electron arrangements which correspond to the filled levels just before these large energy gaps should be particularly stable, because the addition of an electron to the filled configuration yields a particularly small energy increment, while the removal of an electron from the filled level involves a marked loss of exchange energy. These particular configurations, $ns^2 np^6$, are the most stable arrangements found in the Periodic Table and, as atomic configurations, are those of the rare gases (together with the $1s^2$ configuration of helium).

The energy gaps between successive levels with the same l value decrease as the n values increase, so that all the atomic orbitals get closer in energy as the atomic number increases. This trend is not completely regular, however, and larger than average energy gaps occur, particularly between the $4p$ and $5s$ levels where the first set of d orbitals has been filled, and between the $6p$ and $7s$ levels where the first of the f levels comes. These energy jumps reflect the poorer-than-average shielding powers of d and f electrons.

Apart from the major discontinuities at the rare gases, there is also a gap in energy wherever the outermost electron enters a new atomic orbital. These gaps correspond to stabilization of the filled shell configurations, s^2, $d^{10} s^2$, and $f^{14} d^{10} s^2$, before the p orbitals are occupied, and also suggest

the possibility of transfer of s electrons into the d shell to give the d^{10} configuration, or of d electrons into the f shell to give the f^{14} arrangement, which was discussed in the previous section.

In addition to the rare gas configurations and to these other filled-shell configurations, whose stabilities follow from the discontinuities in orbital energy which accompany the start of a new shell, there are also some configurations which should have some additional stability from exchange forces. These are the half-filled shell configurations, p^3, d^5 and f^7. All these effects show up in the chemistry of the elements and they are neatly reflected by the ionization potentials. As the ionization potential measures the energy required to remove the least tightly bound electron from an atom or ion, its value will reflect the stability of the configuration from which the electron is being removed. Table 2.8 gives the ionization potentials of the elements.

The stability of the rare gas configurations can be seen, both from the high energies required to remove an electron from the rare gases themselves, and from the leap in the values of the potential when the rare gas configuration has to be broken (i.e. when the second electron is removed from an alkali metal, the third electron is removed from an alkaline earth, the fourth electron from a boron Group element, etc.). The very low first ionization potentials of the alkali metals and, to a lesser extent, the low first and second potentials of the alkaline earths, show how loosely held are the first one or two electrons outside the rare gas configuration.

The stabilities of filled and half-filled shells show up when the variation of the first ionization potentials across a Period is plotted. Figures 7.5 and 7.6 show such plots across a short and a long Period respectively. The stabilities of the s^2, d^{10} and p^3 configurations show up very clearly.

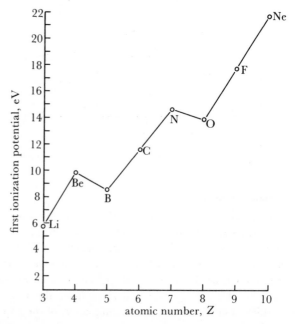

FIGURE 7.5 *Variation of first ionization potential across the First Short Period*

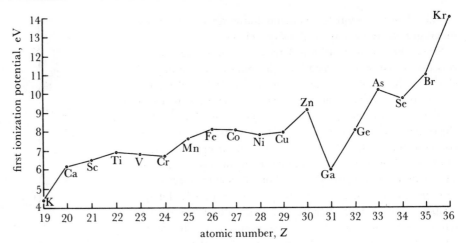

FIGURE 7.6 *Variation of the first ioniza-tion potential across the First Long Period*

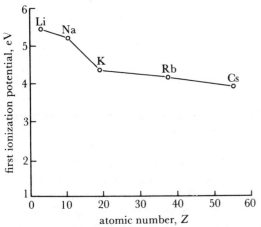

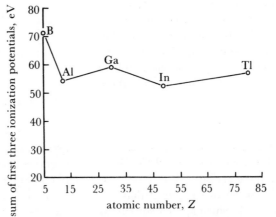

FIGURE 7.7 *Variation in ionization potentials in a Group (a) first potentials of the lithium Group; (b) sum of the first three potentials of the boron Group*

As the atomic size increases within a Group, where the outer electronic configuration is the same, it becomes easier to remove the outermost electrons and this is shown by a decrease in the ionization potentials down a Group. Figure 7.7 shows this for the first ionization potential of the alkali metals and for the sum of the first three potentials of the boron Group. In the latter, the effect of the insertion of the first *d* shell is shown by the potentials of gallium, and that of the presence of the *f* series by the value for thallium. These effects are most marked in the boron Group, which follows immediately after the transition series, but they continue to show in the rest of the *p* block of elements, both in the ionization potentials and in small discontinuities in the chemistry of gallium, germanium, arsenic, selenium, and bromine. The complete graph of first ionization potential *versus* atomic number for all the elements is shown in Figure 7.8: ionization potentials increase from left to right across a Period and decrease with increasing atomic weight down a

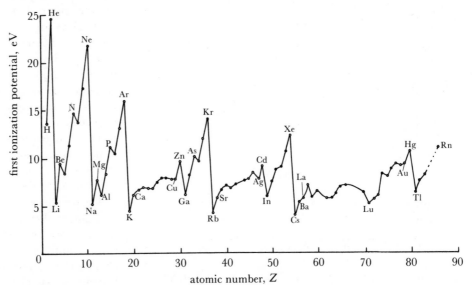

FIGURE 7.8 *Variation of the first ioniza-tion potential with Z*

Group. The variation for elements filling an *s* or *p* level is much greater than that for the *d* or *f* elements.

Since the electron affinities measure the energy of the process converse to ionization, the energy of gaining an electron, the stable configurations are similarly reflected in the electron affinity values. Less extensive data are available than for ionization potentials, but the values for the first electron affinities collected in Table 2.9 do indicate the stable configurations. Thus the addition of an electron to a halogen atom is exothermic, where the electron fills up to the rare gas configuration, while the addition of an extra electron to the rare gases is an endothermic process. Similarly, for the copper Group a large negative electron affinity marks the tendency to complete the $d^{10}s^2$ configuration. The values for the alkali metals, the chromium Group and the carbon Group show the tendency to attain, respectively, s^2, d^5 and p^3 configurations, while the values for the succeeding groups are markedly lower showing that the addition of a further electron to the filled or half-filled shell is a much less favoured process. However, less weight can be put on the trends in electron affinities as the values are more tentative than those for ionization potentials.

7.4 Atomic and ionic sizes

The definition and determination of the various sets of atomic and ionic radii is discussed in section 2.15. Here, their variation with position in the Periodic Table is treated. It will be clear from the discussion in Chapter 2 that the production of a set of atomic radii which applies equally to all the elements involves a number of assumptions and approximations. Atomic radii derived for metals, for covalent solids, and for gaseous molecular elements have to be reconciled. It is possible to measure covalent radii for many of the metallic elements in their covalent compounds, but the more reactive elements of the *s* block do not form covalent compounds and approximate figures have to be used. It is possible, however, to derive an approximately self-consistent set of atomic radii which apply to all the elements and these

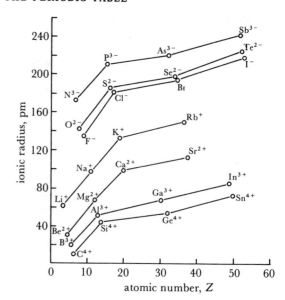

FIGURE 7.10 *Variation of ionic radii, corresponding to rare gas configurations, with Z, for Main Group elements*

are shown in Figure 7.9, plotted against the atomic number. A similar set for real or hypothetical cations and anions with the rare gas structures have also been calculated and these are shown for the Main Group elements in Figure 7.10. Figure 7.9 has many features in common with the ionization potential plot of Figure 7.8. The main discontinuities in size come between the rare gases and the alkali metals where the outermost electron has to enter a completely new quantum shell. There is thus a marked increase in size and a marked decrease in ionization potential at this point. Due to the imperfect shielding of the valency shell electrons by each other, the effective nuclear charge increases, on average, across a Period. The outer electrons become more and more tightly bound and the atomic radius decreases while the ionization potential increases. The discontinuities at the filled and half-filled shell configurations may also be discerned, that at the p^3 configuration being particularly clear. The slow changes across the *d* and *f* series contrast markedly with the sharp changes in the *s* and *p* blocks, and the general decrease in size across the lanthanide series has a noticeable effect in reducing the sizes of the following elements.

The variation in atomic size may be generalized as a decrease on going from left to right across a Period and an increase in going down a Group. These changes are the exact reverse of the ionization potential changes, as would be expected.

The changes in ionic radii generally reflect these changes in the atomic sizes but, as the valency electrons are removed in the formation of the ions, they do not show the intermediate discontinuities.

The parallelism between the changes in radius and ionization potential is not, of course, accidental but follows from the existence and arrangement of the atomic orbitals. Both the size and the ionization potential would be expected to change as the number of electrons in the atom increases, and this change would be discontinuous whenever a new orbital

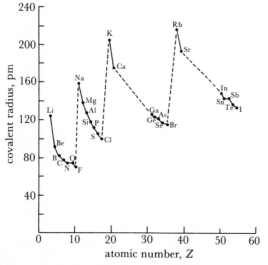

FIGURE 7.9 *Variation of 'covalent' atomic radii with Z*

was occupied. The inter-electronic forces, including both the electrostatic repulsions and the exchange forces, modify this pattern of change but leave the main outlines. The change embodied in the effective nuclear charge clearly affects both the extension of the electron cloud and the energy required to remove an outer electron. The property of electronegativity discussed in section 2.16 is a summarizing parameter which gives effect to the pattern of changes discussed above. It will be seen from the electronegativity values given in Table 2.13 that these increase towards the right of the Periods and decrease down the Groups. In addition, they reflect the other, smaller, variations which have been remarked; for example, the changes in the Main Groups are more pronounced than in the Transition Groups, and the discontinuity in properties of the elements from gallium to bromine when compared with the rest of their respective Groups is reflected in their electronegativity values.

7.5 Chemical behaviour and Periodic position

The detailed chemistry of the elements is discussed in the succeeding chapters. In this section, the skeleton of the Periodic properties is outlined to provide a framework for the more detailed account which follows.

Those elements where the outermost electrons are in a new quantum level, after a rare gas configuration, normally react by losing these loosely bound electrons and forming cations. This mode of behaviour is typical of the elements of the lithium, beryllium, and scandium Groups together with the lanthanide elements, which have the respective valency shell configurations, s^1, s^2, and d^1s^2. All these elements, with the exception of beryllium itself, lose these outer electrons completely with the formation of cations; M^+ in the lithium Group, M^{2+} in the beryllium Group, and M^{3+} for scandium, yttrium and the lanthanons.

The elements of the boron, carbon, nitrogen, oxygen, and fluorine Groups, where the outermost electrons are in p orbitals, show more complicated behaviour with varying oxidation states and in forming both covalent and ionic compounds. These p elements show a maximum oxidation state equal to the sum of the s and p electrons (the Group Oxidation State) and the other relatively stable oxidation states differ from the Group state by multiples of two. Thus the boron Group elements, with the configuration s^2p^1, show the Group oxidation state of III and the other stable state in this Group is I; the halogens, s^2p^5, show the Group state of VII and also the states V, III, I, and $-$I. The Group oxidation state is the most stable one for the lighter elements, especially of the earlier Groups, and a state two less than the Group state becomes the most stable for the heavier elements. Thus boron and carbon are stable in the III and IV states respectively, while their heaviest congeners are stable in the I and II states. In the Groups to the right of the Periodic Table, where a larger variety of oxidation states is possible, the picture is more complex and a further trend becomes apparent. This is the tendency of the lighter elements and those in the halogen Group to form anions—especially N^{3-}, O^{2-}, and X^- in the halogens.

The elements where the d orbitals are filling have a Group oxidation state equal to the sum of the s and d electrons. This state is shown in the earlier Groups but involves too many electrons in the later Groups of the transition series where only low states occur. The highest oxidation state shown in the Periodic Table is VIII, by ruthenium, osmium and xenon. The common oxidation states of the d elements vary in single steps, instead of the double ones shown by the p elements. Thus in the Group with the same number of valency electrons as the halogens, the manganese Group d^5s^2, the oxidation states found are VII, VI, V, IV, III, II, I, and 0. Another distinction between the behaviour of a d Group and a p Group is that the heavier elements of a d Group are more stable in the higher oxidation states, in contrast to the trend in the p Group. Thus, the most stable state of manganese is II while its heavier congener, rhenium, is stable in the IV and VII states. The lower oxidation states of the transition metals, particularly the II state, often occur as cations, while the higher states are commonly bound to oxygen or to the halogens in covalent molecules or anions.

The f elements of the lanthanide series show only one stable oxidation state, the III state. This corresponds to the configuration $4f^1$ for cerium(III) and to $4f^n$ for the other elements up to $4f^{14}$ in lutetium(III). As already noted, three electrons are lost, as if the configuration was $4f^n5d^16s^2$, despite the fact that most of the elements do not have the d electron in the ground state. One or two of these elements do show fairly stable oxidation states other than the III state, and most of these correspond to the f^0, f^7 or f^{14} configurations. Thus, cerium shows a IV state corresponding to f^0, europium has a II state, and terbium a IV state, corresponding to f^7; while ytterbium has a II state corresponding to f^{14}.

In the heaviest elements, where the $5f$ and $6d$ levels are filling, the pattern of oxidation states is less simple than with the lanthanides. As these two energy levels are very close, the earlier elements show a considerable variety of oxidation states with the maximum rising from III for actinium to VI at uranium and VII at neptunium and plutonium. The later actinide elements resemble the lanthanons more closely, and the III state becomes the most stable one at about curium.

These patterns of behaviour lead to the division of the Periodic Table into four major blocks: the s elements, the p elements, the d elements, and the f elements, together with a number of Groups which serve to bridge these divisions. The s elements are those of the lithium and beryllium Groups; the boron, carbon, nitrogen, oxygen and fluorine Groups make up the p block; the lanthanides and actinides form the f block; and the remaining transition elements, the d block. As the typical behaviour of d elements depends on the presence of both d electrons and available d orbitals, the scandium Group (which always loses its solitary d electron and forms the M^{3+} ion) and the zinc Group (which always preserves the filled d^{10} configuration) are not properly d elements and are best regarded as bridging Groups. The scandium Group links the s elements and the d block, while the zinc Group links the *d* block to the *p* elements, and both Groups show the appropriate intermediate properties. The chemistry of the scandium Group links strongly, also, with that of the *f* elements. Indeed, the general chemistry of the lanthanides in the III state is almost identical with that of yttrium and lanthanum. There are bigger differences between the chemistry of actinium and the actinide elements. It is convenient to treat scandium, yttrium, lanthanum, actinium, and the lanthanides all together, and to treat the actinides independently. The remaining Group in the Periodic Table is the helium Group. This Group forms the division between the *p* block and the *s* block, and the recently-discovered chemistry of xenon shows strong links with that of iodine. Finally, there is the lightest element, hydrogen, which falls into no Group so far discussed and is best regarded as a unique introductory element to the Periodic Table. The arrangement of the following chapters reflects this division of the Periodic Table. The chemistry of hydrogen and its compounds is treated first, giving a microcosm of the properties of the elements. Then follow chapters on the *s* elements, the scandium Group and the lanthanides, the actinides, the *d* elements, and the *p* elements (Figures 2.13 and 7.11).

7.6 Methods of showing the stabilities of oxidation states

In many cases, as implied in the last section, elements show a number of oxidation states and some method of determining and portraying the relative stabilities of these states is necessary. Similarly, it is useful to be able to compare relative stabilities and show the trends in stabilities among related elements. Of course, in a full and complete description of the chemistry of an element, these stabilities are clear from the range of compounds of a given oxidation state and their ease of formation and decomposition. However, it is impossible to give a complete description of the known chemistry of any element within the space available in a general textbook so that methods of summarizing and illustrating the general behaviour are required. Stabilities vary a good deal with the chemical environment—the temperature, solid, liquid, or gaseous state of the compound, solvent, presence of air or moisture, and so forth. However, there are two chemical states which are very common, as a solid and in solution in water.

The stability of an element in a particular oxidation state in the solid may be determined from the variety of ions or ligands with which it reacts to form solid compounds. Thus, a strongly oxidizing state will form compounds with non-oxidizable ligands only, and *vice versa* for a reducing state, while a stable state will give compounds with a wide variety

of ligands. For example, consider the relative stabilities of the II and III states of iron, cobalt, and nickel as shown by the existence of the solid compounds of the II state with oxychloride anions.

Fe(II) $Fe(ClO_4)_2 : ClO_3^-$ and ClO_2^- oxidize to Fe(III)
Co(II) $Co(ClO_4)_2$ and $Co(ClO_3)_2 : ClO_2^-$ oxidizes to Co(III)
Ni(II) $Ni(ClO_4)_2$, $Ni(ClO_3)_2$, and $Ni(ClO_2)_2$

The oxidizing power of the oxychloride ions increases in the order, perchlorate < chlorate < chlorite, so the existence of the compounds shown above illustrate that the order of stabilities of the II state is Fe < Co < Ni.

A convenient set of ligands, for the purpose of demonstrating stabilities, is provided by the halides which form compounds with almost all oxidation states. An element in a stable oxidation state will form all four halides, a strongly-reducing state will tend not to have a fluoride, while a strongly-oxidizing state will tend not to have an iodide or bromide. Similarly, an oxidizing state will show an oxide but no sulphide, while a reducing state will form a sulphide but no oxide. Thus, the existence and stabilities of the oxides, sulphides, and halides of the elements in their different states provides a useful general guide to the stabilities, in the solid state, of the various oxidation states. In the following chapters, Tables 12.3, 12.4, 12.5, of transition element halides and oxides (sulphides are omitted here as their stoichiometry is often in doubt) and Table 15.2, of *p* block element oxides, sulphides, and halides, are used to provide a general, overall view of the chemistry of these elements.

Stabilities in aqueous solution are expected to be broadly similar to stabilities in the solid state, but to differ in detail due to differences between lattice energies and hydration energies (compare Chapters 4 and 5). The relative stabilities of the various oxidation states of an element in solution are given by the free energy changes of the set of half-reactions connecting each pair of oxidation states. For example, the stabilities of the states of copper depend on the free energy changes of the half-reactions:

$$Cu^{2+} + e^- = Cu^+$$
$$Cu^+ + e^- = Cu$$
$$\text{and } Cu^{2+} + 2e^- = Cu$$

(These three free energies are not independent, of course, any one may be derived from the other two.)

Such free energies are related to the corresponding redox potentials, since $-\Delta G = nFE$, see p. 80. A full list of redox potentials for the half-reactions of all the elements is avail-

FIGURE 7.11 *Divisions of the Periodic Table*

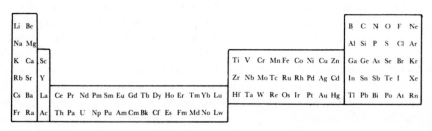

able, but a method is required for displaying these values to the best advantage. It has been suggested by Ebsworth (see references on page 282) that free energies may be usefully displayed graphically. In this, the oxidation states of the element are plotted against the free energy change, in one electron steps. The method is most readily discussed in terms of particular cases, for example uranium and americium whose potentials have the values shown below:

	E^0 (volts)	
	M = U	M = Am
$MO_2^{2+} + e^- = MO_2^+$	0·05	1·64
$MO_2^{2+} + 4H^+ + 2e^- = M^{4+} + 2H_2O$	0·33	
$MO_2^+ + 4H^+ + e^- = M^{4+} + 2H_2O$	0·62	1·26
$M^{4+} + e^- = M^{3+}$	−0·61	2·18
$M^{3+} + 3e^- = M$	−1·80	−2·32
$MO_2^{2+} + 4H^+ + 3e^- = M^{3+} + 2H_2O$		1·69

The diagrams, Figures 7.12 (a) and (b), are plotted by taking the value for the element itself as zero and plotting the free energy changes for the half-reactions against the oxidation states of the element. The free energies are given as $-\Delta G^0/F = nE^0$ where n is the number of electrons involved in the change. Using americium as our example the diagram is constructed as follows (see Figure 7.12).

1. Take the value for the element as zero.
2. For the change

$$Am^{3+} + 3e^- = Am$$

E^0 is −2·32 V, therefore nE^0 is −6·96 V and is so plotted.
3. For the change

$$Am^{4+} + e^- = Am^{3+}$$

E^0 is 2·18 V, therefore nE^0 is 2·18 V and this is added to −6·96 V to give −4·78 V, which is plotted for Am(IV).

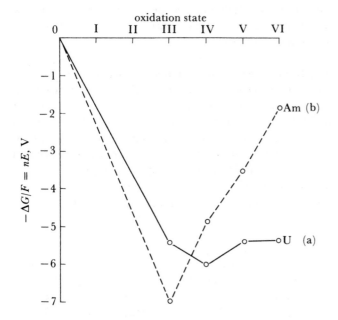

FIGURE 7.12 *Free energy versus oxidation state diagrams*, (a) *uranium*, (b) *americium*

4. For the change

$$Am(V) \rightarrow Am(IV) \qquad (AmO_2^+ \rightarrow Am^{4+})$$

E^0 is 1·26 V and nE^0 is 1·26 V. Adding this to −4·78 V gives −3·52 V to be plotted for Am(V).
5. For Am(VI), continuing in the same way one obtains −1·88 V.

Plots like these provide a kind of cross-section of the chemistry of the elements. The lower a state lies on the diagram, the more stable it is. Also, since a change from one state to another which involves moving down a slope involves a negative change of free energy, it is thermodynamically favourable. Similarly, a change which involves an 'uphill' move is unfavourable. These points about changes apply to those involving any two states, not only states which are nearest neighbours. Thus the changes from U(0) to any of the states, U(III), U(IV), U(V), or U(VI) are favourable, and the changes from U(IV) to any of the states U(0), U(III), U(V), and U(VI) are unfavourable. Clearly, the greater the slope, the greater the driving force, so that Am(0) goes to Am(III) more readily than to Am(IV).

A particular oxidation state may be represented on such a diagram by a point which is (a) a minimum, (b) a maximum, (c) a concave point (i.e. one lying below the line joining the two neighbouring points), (d) a convex point (i.e. one lying above the line joining its neighbours), or (e) a 'linear' point (i.e. one lying on the line joining its neighbours). Table 7.1 lists the stability properties of the oxidation states represented by these different types.

The properties of the states represented by concave, convex, and linear points are not restricted to being with respect to the two nearest neighbour states but may hold with respect to any two states. For example, in Figure 7.12, Am(V) is linear, not only with respect to Am(IV) and Am(VI), but also with respect to Am(III) and Am(VI) and the disproportionation products will include Am(III). In a similar way, in the case of phosphorus (Figure 15.25) the points representing the 0, I, and III states are all convex with respect to the −III and V states. That is, the element (0 state), hypophosphorous acid (I state) and phosphorous acid (III state) all disproportionate to phosphine, PH_3, (−III) and phosphoric acid (V state).

Applying these results to uranium and americium, Figure 7.12 shows, at a glance, that uranium(V) is unstable in aqueous solution, uranium(IV) is stable, and uranium(III) is relatively stable. Uranium(VI) is also stable, with the U(VI) → U(IV) change being mildly oxidizing. By contrast, americium(VI) is unstable and the Am(VI) → Am(III) change is strongly oxidizing. The most stable state of americium is Am(III) and americium(IV) and (V) both disproportionate, with the V state a little more stable than the IV state.

The elements lying between uranium and americium—neptunium and plutonium—also show III, IV, V, and VI states and the VI state becomes more oxidizing and less stable in order from uranium, through neptunium and

TABLE 7.1 Types of oxidation state in free energy diagrams

State represented by a point	Example from Figure 7.12	Further examples (Figure numbers in brackets)	Properties of the oxidation state
Minimum (a)	U(IV), Am(III)	Mn(II) (13.17) Cr(III) (13.13)	Stable relative to neighbouring states.
Maximum (b)	——	N(−I) (15.25)	Unstable relative to neighbouring states.
Concave (c)	U(III)	V(IV) (13.7) Re(IV) (14.18)	Relatively stable with respect to disproportionation into the neighbouring states.
Convex (d)	U(V)	Mn(III) (13.17) Re(III) (14.18)	Relatively unstable with respect to disproportionation.
Linear (e)	Am(III) and (IV)	All the intermediate states of Cl (15.47)	*Intermediate with respect to disproportionation.

*That is to say that, at equilibrium, there are approximately equal amounts of the original state and the states to which it has disproportionated.

plutonium, to americium. This is clear from Figure 11.1, in which the curves for all four elements are plotted. The VI state of uranium lies lowest and the others in the order U < Np < Pu < Am. In a similar way, the relative stabilities of the II and III states of iron, cobalt and nickel, referred to above, are clear from Figure 13.1.

These free energy diagrams will be extensively used in the following chapters but one or two reservations about them must be kept in mind. First, all the oxidation potential data are not of equal validity and some values may be in error so it is not wise to make too much of small effects—such as whether a point is slightly convex or slightly concave. It is unlikely, however, that there are any gross inaccuracies. The second point is that the properties listed in Table 7.1 are thermodynamic and there is no information about the rates of reactions. Thus a state may be thermodynamically unstable but persist in solution because its rate of reaction or disproportionation is slow. Similarly, the potential data apply only to systems in equilibrium, and the rate of attaining equilibrium may be slow. This applies particularly to systems involving solids—for example, an element. Many elements which are strongly reducing react only very slowly due to surface effects and the like. Throughout the later chapters, curves are plotted for potential data in acid solution, at a pH of 0. Similar data are available for alkaline solution, and a set of equivalent curves could be drawn for such media. These are less useful as many more states appear as solids (as hydroxides or hydrated oxides) in alkali.

The occurrence and stability of the solid halides, oxides, and sulphides of the elements, taken together with the free energy diagrams linking the different oxidation states, gives an adequate picture of the general chemistry of an element in its common compounds. It will thus not be necessary, in the following chapters, to describe in detail the chemistries of all the oxidation states of all the elements. In general, a stable state may be assumed to form compounds with all the common anions and ligands—all the oxyanions, pseudo-halides, organic acid anions, hydride, nitride, carbide, amide, sulphur anions and so on. Unstable states will form a much more limited set of compounds, down to those states represented by only one or two examples. A state which is unstable and oxidizing will form no compounds with oxidizable ligands such as nitrite or organic groups, and similarly, a reducing state will form no compounds with oxidizing ligands, as with the chlorite and chlorate compounds of iron mentioned above. By avoiding cataloguing such compounds of the elements, space is preserved for the mention of the more unusual compounds formed and as many of these as possible have been discussed in the later chapters, although, of course, some discussion of the simpler compounds is necessary in cases where their behaviour gives important information about the general chemistry of the elements. The later part of this book, therefore, may be regarded as an introduction and supplement to the systematic chemistry given in the major inorganic textbooks which are listed in the reading lists. For similar reasons, a detailed account of the extraction of individual elements is avoided in the following chapters and a general outline of extraction methods is given in the next section.

7.7 The occurrence and extraction of the elements

The abundance of elements in the Earth's crust varies very widely from about 50 per cent for oxygen to about three parts in 10^{-16} for actinium. The most abundant elements in the crust are given in Table 7.2. To these should be added the nitrogen of the atmosphere and the hydrogen and carbon in the seas and in plants to make up the complete list of very abundant elements.

TABLE 7.2 The most abundant crustal elements

	Mass, %		Mass, %
Oxygen	50	Sodium	2·6
Silicon	26	Potassium	2·4
Aluminium	7·5	Magnesium	1·9
Iron	4·7	Hydrogen	0·9
Calcium	3·4	Titanium	0·6

FIGURE 7.13 *Distribution of the elements*
The elements are divided into three classes, common, rare, and very rare, on the basis of their natural abundance combined with their accessibility. The boundaries between types are necessarily somewhat arbitrary but it has been taken that elements described as rare have had their chemistry less fully explored than their near neighbours in the Periodic Table, while the very rare elements have been subject only to preliminary studies.

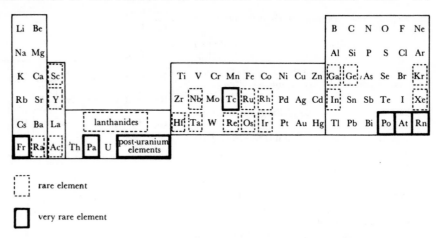

Such elements are, naturally, readily accessible and their chemistry is thoroughly studied. There are, in addition, many elements which, although rare in the overall composition, occur in concentrated deposits and are thus well known. One example is boron, which occurs to the extent of only three parts per million in the crust but is found in concentrated deposits as borax. A further class of accessible elements are those which are found native or are readily recovered from their ores, for example the precious metals silver and gold. On the other hand, there are a number of relatively rare elements which occur only in small proportions in the crust and are found only as trace constituents in the ores of more important minerals. These elements remain inaccessible to the chemist and their study has proceeded slowly, so there are many gaps in the Periodic Table where full experimental knowledge is not yet available. Figure 7.13 shows these 'rare' elements and this lack of knowledge must be borne in mind when assessing generalizations about Periodic behaviour. At present work is proceeding rapidly on many of these rare elements as they are found to have desirable properties. Among those in Figure 7.13, germanium, gallium, and indium have attracted attention for their semi-conductor applications, most of the heavier transition elements have been investigated for their effects and uses in atomic piles, and large-scale separation methods have been developed which have made pure samples of the individual lanthanide elements available.

The rarest elements of all are the artificial elements which have no naturally-occurring isotopes. These include all the post-uranium elements and a few lighter ones such as promethium and technetium. Supplies of many of these elements have become available recently from the fission products or synthesis products of the atomic piles, while the attempt to synthesize ever heavier elements continues, with half a dozen new elements produced in the last twelve years. Element 105 was announced recently.

The extraction processes for producing elements from their ores may be divided into three classes in order of increasing power:

(i) mechanical separation and simple heat treatment,
(ii) separations involving chemical reduction,
(iii) separations by electrolytic reduction.

Although the days of the gold rush are gone, mechanical separation on a huge scale is still the basic process of gold and diamond production; the final recovery of gold being by chemical reduction with zinc of a cyanide complex in solution. Also included in the first class of mild treatments are the recoveries, by distillation or thermal decomposition, of elements such as zinc or mercury. A further reaction of wide application in this class is the Van Arkel and De Boer process which is used to produce very pure metals on a small scale. In this process, elements which form volatile iodides are purified by a cyclic process in which the iodide is formed at a low temperature and decomposed on a heated wire, at a higher temperature, to the element and iodine. The iodine is recycled to form more iodide. For example, zirconium gives ZrI_4 when heated at 600 °C with iodine and this may be decomposed at 1800 °C, on a heated tungsten or zirconium filament, to zirconium and iodine.

Most commercial separations fall into the second class, commonly involving reduction by carbon, especially iron production. Reduction by other elements is also found on a small scale: examples include the preparation of pure molybdenum by hydrogen reduction and the reduction of titanium tetrachloride by magnesium in the Kroll process.

Electrolytic reduction represents the most powerful method available, but it is expensive compared to the chemical methods and is only used either for very reactive metals, such as magnesium or aluminium, or for the production of samples of high purity as in the electrolytic refining of copper (which has the additional advantage of allowing the recovery of valuable minor contaminants such as silver and gold). The main commercial application of electrolysis is, of course, in aluminium manufacture. Here, the ore bauxite, which is impure Al_2O_3, is purified by alkaline treatment, then dissolved in molten cryolite (Na_3AlF_6) and reduced electrolytically in this fused-salt system. The alkali metals and calcium and magnesium are also produced by electrolysis in a fused salt melt, while copper and zinc are among those elements recovered by electrolysis in an aqueous medium.

Although the type of process chosen for any one element is a complex function of the chemical properties, nature of the ore, and relative economics, and many elements are processed in different ways in different parts of the world,

FIGURE 7.14 *Methods used for the extraction of the elements*

(1) Reactive metals extracted electrolytically in a non-aqueous system. Main sources of alkaline earth minerals are insoluble sulphates and carbonates, while alkali metals often form deposits of soluble salts e.g. NaCl or KCl.

(2) Reactive metals of high charge with strong affinity for oxygen and occurring as oxyanions or double oxides. Separation is by electrolytic or chemical reduction, especially by active metal replacement.

(3) Elements occurring in sulphide ores or otherwise associated with sulphur. Extracted usually by roasting to oxide and then reducing or treating thermally.

(4) Elements occur native or in easily decomposed compounds yielding to thermal treatment.

(5) Non-metals which occur free, in the atmosphere, or as anions.

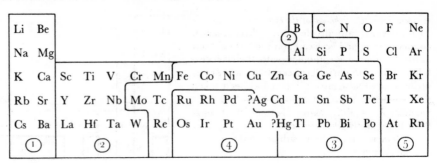

the very general picture given in Figure 7.14 of the mode of occurrence and method of recovery of the elements is broadly true. The extraction of the elements reflects, in some measure, their general chemistry and thus elements which are treated similarly tend to lie together in the Periodic Table. (This is also true of their occurrence. Elements of similar chemistry would have behaved similarly when the rocks crystallized out of the primitive magma, and later processes, such as leaching out of soluble salts, would further tend to concentrate similar elements in similar forms.)

The process of extraction by carbon reduction illustrates several interesting general points in the chemistry of metal recovery and is worth discussing in fuller detail. The extent of the reduction of one element from its oxide by a second element depends on the difference in free energy of the two oxidation reactions of the type:

$$M + \frac{x}{2}O_2 \rightarrow MO_x \quad \dots \dots \dots \dots (7.1)$$

It will be recalled that reactions which evolve free energy tend to occur spontaneously, so that the equilibrium between two elements and their oxides:

$$M'O + M'' \rightleftharpoons M''O + M' \quad \dots \dots \dots (7.2)$$

(or the corresponding equations for different oxide stoichiometries), will favour that oxide whose free energy of formation (equation 7.1) is most negative. The free energy change, ΔG, is separable into two components, the heat change ΔH, and the energy involved in the change of entropy, $T\Delta S$, where T is the absolute temperature:

$$\Delta G = \Delta H - T\Delta S$$

In the formation of a metal oxide according to equation (7.1), the heat change is usually favourable but, as the reaction uses up a gaseous component (the oxygen) which has a relatively large entropy, the entropy term is unfavourable and this energy increases with increasing T. As a result, the free energy change for metal oxide formation in equation (7.1) falls off with rising temperature in a broadly similar

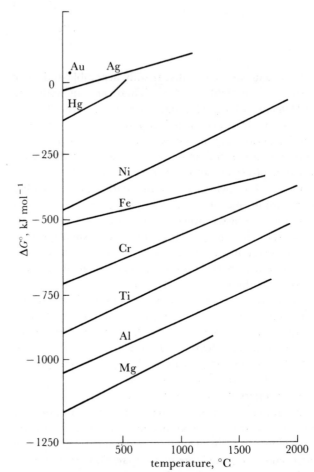

FIGURE 7.15 *The variation with temperature of the free energy of metal oxide formation according to equation* (7.1)

way for any metal, as in the examples shown in Figure 7.15. It will be seen that the metal oxides can be divided into two classes: first those which are intrinsically unstable at normal temperatures, such as gold oxide, or at accessible temperatures such as silver oxide or mercury oxide, and the second class which contains the elements with a favourable free energy of oxide formation at any accessible temperature. Gold, silver, mercury, fall into the class (i) (page 111) of elements which can be extracted by simple heat treatment alone, while the second class of elements contains those whose extraction process falls into class (ii) or (iii), requiring reduction. Any metal will reduce the oxide of any second metal which lies above it in Figure 7.15, according to equation (7.2), as the net change in the free energy will be negative (i.e. favourable) by an amount equal to the difference between the two curves at the appropriate temperature. For example, magnesium will reduce all the other oxides shown in Figure 7.15.

The formation of a metal oxide thus involves a free energy change which becomes less negative with increasing temperature, because the oxidation proceeds with the consumption of a gaseous component with a corresponding loss of entropy. The formation of carbon oxides is quite different. The main reaction between carbon and oxygen at higher temperatures is:

$$2C + O_2 \rightarrow 2CO$$

in which an excess of one mole of gas is produced for each mole of oxygen used. This reaction therefore involves a positive entropy change and its free energy becomes more negative as the temperature rises. At temperatures below 500 °C, the main reaction is:

$$C + O_2 = CO_2$$

where there is no overall change in the amount of gaseous reactant. The entropy change is thus small and the free energy of this reaction is almost independent of temperature. Figure 7.16 shows the overall change in free energy as carbon

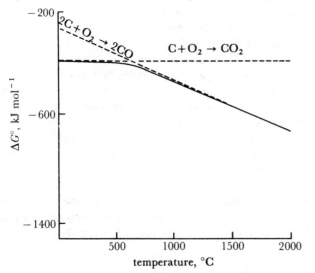

FIGURE 7.16 *The variation with temperature of the free energy of the reaction of carbon with oxygen*

is oxidized at increasing temperatures. The total free energy curve falls as T rises and will eventually cross every curve, of the type shown in Figure 7.15, for the free energy of metal oxide formation. It follows that, in principle, carbon may be used to reduce any metal oxide if a high enough temperature can be attained. In practice, of course, temperatures which would be high enough to allow carbon to reduce the more stable oxides such as TiO_2 or Al_2O_3 are not accessible

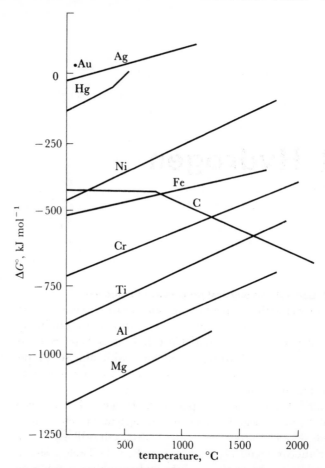

FIGURE 7.17 *Combined diagram of the free energy changes of the reactions of metals and carbon with oxygen*

economically on a large scale. The usefulness of carbon as a reducing agent may be seen in Figure 7.17, in which the curves of Figures 7.15 and 7.16 are superimposed. Of course, other chemical factors come into play in choosing reduction conditions, such as the formation by many elements of unwanted carbides, which limit the use of carbon in practice.

Once again there are important reservations to be kept in mind when applying such a thermodynamic analysis as that sketched above. Free energy changes are calculated on the assumption that the system is in equilibrium—which is far from the case in practice—and they give no indication of the kinetics of the reaction so that a particular reduction may have a favourable free energy change but be too slow. However, the thermodynamic analysis does distinguish reactions which will occur from those which will not, and it gives some indication of the conditions, for example of temperature, which are suitable. For a fuller account of this topic, and curves for sulphide and halide systems, the monograph by Ives (see page 281) should be consulted.

8 Hydrogen

8.1 General and physical properties of hydrogen

Hydrogen is the simplest of the elements with its one valency orbital and single electron. It can react only by gaining or sharing another electron and its behaviour is therefore fairly uncomplicated. Hydrogen combines with almost all the other elements, and, as its electronegativity value comes in the middle of the range, its bonds have a wide range of polarity from the strongly positive hydrogen of the hydrogen halides to the negative, anionic hydrogen in the active metal hydrides. The properties of the hydrides thus illustrate the variation in chemical properties with Periodic Table position. In addition, the small size of the hydrogen atom presents no steric barriers and the hydrides have the shapes expected from the electronic structure of the central atom. This small size does permit the close approach of other, non-bonded, atoms and special properties result, especially the *hydrogen bond* between atoms of high electronegativity values which is discussed later.

Hydrogen does not fit into any of the Groups of the Periodic Table and is best regarded as an introductory element to the Periodic classification. It does show significant analogies to three of the other Groups which are worth noting as a guide to its chemistry. In common with the halogens, it has a tendency to gain an electron and forms the hydride ion, H^-; in common with the alkali metals, it may lose an electron to form a cation, though this statement must be treated with considerable reserve as will be seen; in common with the carbon Group elements, hydrogen has a half-filled valency shell and forms covalent bonds with a wide range of polarities. This last comparison is especially apt, particularly if hydrogen is compared with carbon with one remaining free valency as in methyl, H_3C-. There is a very close relationship between the stability and properties of the hydrides and organometallic compounds of many elements, as the Table below illustrates, and it is often useful to regard the hydride as the parent compound of a homologous series of organometallic compounds, e.g. $HSnCl_3$, CH_3SnCl_3, $C_2H_5SnCl_3$, etc. As organometallics form a field of rapid growth which cuts across the old boundaries of organic and inorganic chemistry, this facet of hydrogen chemistry is of considerable current importance.

Hydrogen is the most abundant element in the universe and it is probable that the other elements are built up from hydrogen by nuclear fusion processes in the stars. On earth,

TABLE 8.1 Comparison of some metal-hydrogen and metal-alkyl compounds

Type	R = H	R = Alkyl, e.g. methyl
NaR	ionic, Na^+H^-	ionic, $Na^+CH_3^-$
AlR_3	polymeric solid linked by electron-deficient $Al\cdots H\cdots Al$ bridges	dimer linked by electron-deficient alkyl bridges, $Al\cdots CH_3\cdots Al$
SiR_4 to PbR_4	covalent gaseous molecules, $M-H$ decreases in stability $Si > Ge > Sn > Pb$	volatile covalent compounds, $M-R$ decreases in stability from Si to Pb though more stable than $M-H$
GeR_2	polymeric oxidizable solids of obscure structure	rings $(GeR_2)_n (n = 4, 5, 6)$ or polymeric solids
$RCo(CO)_5$	unstable complex hydride with σ $Co-H$ bond	unstable organometallic complexes with σ $Co-R$ bonds
$RPtX(PPh_3)_2$	hydride with $Pt-H$ bond stabilized by $Pt-$phosphine interaction	stable organometallic complex with σ $Pt-C$ bond stabilized by $Pt-PPh_3$ interaction

hydrogen occurs almost wholly in combination, especially as water and in organic compounds. There are five reported isotopes of hydrogen, shown below, of which the three lightest are of interest to the chemist:

H   normal or light hydrogen, mass = 1·008, natural abundance = 99·98%

H   deuterium or heavy hydrogen, mass = 2·015, natural abundance = 0·02%

H   tritium, mass = 3·017, natural abundance = 10^{-17}%, radioactive, with $t_{\frac{1}{2}}$ = 12·4 years, decay process $^3_1H = {}^3_2He + {}^{\ \ 0}_{-1}e$

H⎫
H⎬ these are recently reported artificial isotopes with short half-lives: the decay processes are
$^4_1H = {}^3_1H + {}^1_0n$ and $^5_1H = {}^5_2He + {}^{\ \ 0}_{-1}e$

Pure deuterium is normally separated from light hydrogen by the electrolysis of water. The lighter isotope is evolved preferentially and almost pure deuterium oxide remains by the time the bulk is reduced a millionfold. Tritium is most conveniently prepared by the irradiation of lithium with slow neutrons in a reactor:

$$^6_3Li + {}^1_0n = {}^3_1H + {}^4_2He$$

and the tritium is separated by oxidation to T_2O. Both deuterium and tritium are most readily produced as oxide and most deuterated or tritiated compounds are made directly from this isotopically substituted water. For example, deuterated acids may be made simply by solution:

$$P_2O_5 + 3D_2O = 2D_3PO_4$$
$$SO_3 + D_2O = D_2SO_4$$

or tritiated ammonia by the reaction of a nitride:

$$Mg_3N_2 + 3T_2O = 2NT_3 + 3MgO.$$

Since the chemical behaviour of all the isotopes of an element is identical (although rates of reaction may differ), deuterium or tritium substituted hydrides are widely used in studying the mechanisms of reactions involving hydrogen. (See section 2.3.)

Some of the important properties of hydrogen are listed in Table 8.2.

It has already been noted in section 2.15 that atomic and ionic radii show some variation with the chemical environment of the species. This variability is particularly marked in the ions and covalent molecules involving hydrogen, as Table 8.2 shows, since there is only a single nuclear charge on hydrogen and the $1s$ orbital is unshielded by any inner electron shells. The hydride ion, H^-, is especially sensitive to change of electric field intensity in its environment as it consists of two electrons in the field of the single nuclear charge and therefore has a very diffuse electron cloud. The free hydride ion has been calculated to have a radius of 208 pm, twice as large as that of helium, which has two electrons in twice the nuclear field. The measured values of the hydride ion radius in ionic lattices are much smaller than this and show considerable variation, as the values in the alkali metal hydrides illustrate:

MH	Li	Na	K	Rb	Cs
H^- radius (pm)	126	146	152	153	154

(Compare these values with 133 pm for F^- and 145 pm for O^{2-}, where the larger numbers of electrons are offset by the increase in the nuclear charges and consequent greater density of the electronic cloud.) The covalent radius of hydrogen also shows considerable variability, although the effect is less pronounced than with the anionic radius, apart from the high value for the hydrogen molecule itself.

The most striking change in dimension would come if the hydrogen atom were to lose its electron to become the positive ion, H^+. This is the bare proton with a radius of about 10^{-3} pm and is a hundred thousand times smaller than any other ion (compare Li^+, radius 60 pm). The charge density on the proton is thus enormously higher than on any other chemical species and it would have a powerful polarizing effect on any other molecule in its neighbourhood. As a result, the free proton has no independent existence in any chemical environment and always occurs in association. Thus, if an acid dissociates in water, H_3O^+ and not (except as a shorthand) H^+ is formed, and this 'hydrogen ion' or 'hydroxonium' ion is then further solvated just as any other cation. A similar situation holds for any other protonic solvent, as discussed in Chapter 5 (although it must be noted that in the Brönsted definition of an acid it is the proton which is transferred). In water, the proton may well exist as the solvated species $H_9O_4^+$, shown in Figure 8.1, as it has been shown that this species can be extracted from an aqueous acid solution into an immiscible organic base. This is comparable with the isolation of, say, the nickel ion from

TABLE 8.2 Properties of hydrogen

Heat of dissociation of H_2	$H_{2(gas)} = 2H_{(gas)}$, $\Delta H = 431\cdot7$ kJ mol^{-1} (439.2 kJ mol^{-1} for D_2)
Ionization potential	$H_{(gas)} = H^+_{(gas)} + e^-$, $I = 13\cdot595$ eV = 1309 kJ mol^{-1}
Electron affinity	$H_{(gas)} + e^- = H^-_{(gas)}$, $E = -71$ kJ mol^{-1} (exothermic)
Radius, anionic H^-	= 126 to 154 pm (measured in alkali hydrides)
	= 208 pm (calculated for free H^-)
cationic H^+	= 10^{-3} pm
covalent H	= 37·5 pm (from H_2 bond length)
	28 pm (from bond lengths of hydrogen halides)
	32 pm (from MH_4 distances in carbon Group hydrides)

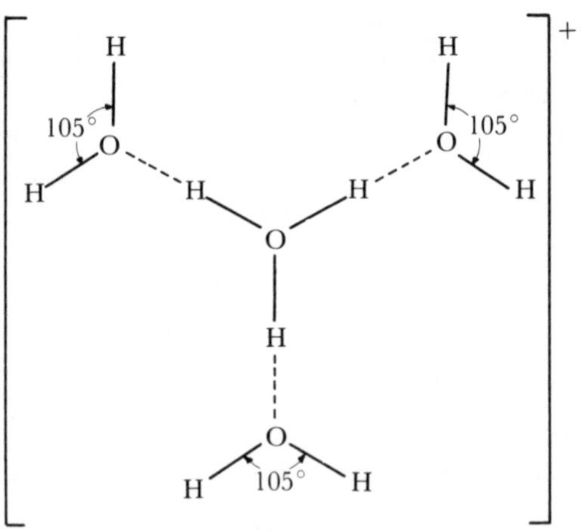

$$\left[\begin{array}{c} \end{array}\right]^{+}$$

FIGURE 8.1 *The species* $H_9O_4^+$

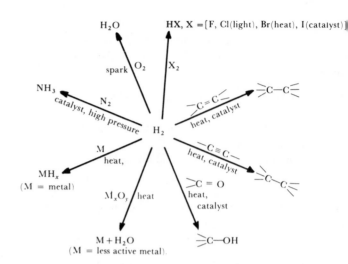

FIGURE 8.2 *Reactions of molecular hydrogen*

aqueous solution as the green hexahydrate, $Ni(H_2O)_6^{2+}$. Recent experiments suggest that a similar situation may exist in liquid ammonia, and it is well known in the strong acids such as hydrogen fluoride or sulphuric acid.

One further physical modification of the hydrogen molecule should be briefly mentioned: *ortho-* and *para-*hydrogen. These forms arise from the different ways in which the nuclear spins may be lined up. If the nuclear spins are parallel, the form is *ortho-*hydrogen, while if they are anti-parallel, the form is *para-*hydrogen. These two forms of molecular hydrogen have different physical properties, such as thermal conductivities, and the two co-exist at ordinary temperatures. (Other symmetrical molecules with nuclear spins, such as N_2 or Cl_2, have *ortho-* and *para-* forms but only H_2 and D_2 show significant differences in physical properties.) The more stable form at low temperatures is *para-*hydrogen and this makes up 100 per cent of hydrogen at absolute zero. At higher temperatures, the equilibrium proportion of *ortho-*hydrogen rises, and reaches its maximum of 75 per cent at room temperature. The equilibrium proportions at any given temperature may be calculated theoretically, and the interconversion may easily be followed experimentally. The interconversion is slow but subject to catalysis by a number of materials, especially by para-magnetic compounds, and this *ortho-para* conversion is frequently used in studies of catalysis.

8.2 Chemical properties of hydrogen

As the hydrogen molecule bond energy, of 431 kJ mol^{-1}, is high, molecular hydrogen is fairly unreactive at ordinary temperatures. At higher temperatures it combines, directly or with aid of a catalyst, with most elements. Some of the more important reactions are shown in Figure 8.2.

Atomic hydrogen may be produced in high-intensity electric arcs and is very short-lived and reactive. It finds use in welding where the recombination of the atoms takes place on the metal surface, yielding up the heat of dissociation and at the same time providing a protective atmosphere against oxidation.

Binary compounds of hydrogen (i.e., those containing hydrogen and one other element) are termed hydrides whether they contain the hydride ion, H^-, or are covalent, and this term is commonly extended to less simple hydrogen compounds, as in 'transition metal hydride complexes' and 'complex hydride ions'. It is convenient to discuss the chemistry of the hydrides in three groups, reflecting the three ways in which the hydrogen electron enters into bonding. These are (i) gain of an electron to form ionic compounds containing H^-, (ii) sharing the electron in covalent hydrides, (iii) forming metallic bonds with the electron delocalized in the so-called interstitial hydrides.

As the electron affinity of hydrogen is low compared with that of the halogens, the distribution of ionic hydrides in the Periodic Table is much more restricted than the distribution of ionic halides. Ionic hydrides are formed only by the elements of the alkali, alkaline earth and, possibly, the lanthanide Groups. Most other elements of the main Groups form covalent hydrides, while the transition elements give both covalent metal-hydrogen bonds in the complexes and also interstitial hydrides. Figure 8.3 shows the approximate distribution of the various hydride types within the Periodic Table. The boundaries between classes are by no means sharp, and a number of intermediate types are observed. In addition, there is some doubt whether the hydrides of a number of elements, especially of the heaviest elements, exist at all.

In passing across a Period, the type of hydride changes from ionic compounds at one extreme to volatile covalent molecules at the other. In the middle of the short Periods, the transition between the two types is marked by solid hydrides, say of magnesium and aluminium, of polymeric structure with bonding of intermediate character. Among the transition elements in the long Periods, the type changes through interstitial hydrides to hydrides of dubious existence at the right of the transition block, before coming to the covalent hydrides of the p elements. In any Main Group of the Periodic Table, the stability of the hydrides tends to fall with increasing atomic weight.

FIGURE 8.3 *Types of hydride in the Periodic Table*

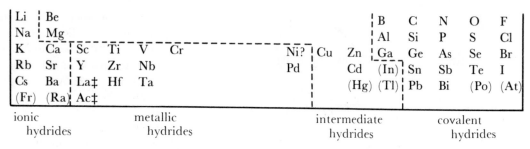

Li	Be											B	C	N	O	F
Na	Mg											Al	Si	P	S	Cl
K	Ca	Sc	Ti	V	Cr		Ni?	Cu	Zn	Ga	Ge	As	Se	Br
Rb	Sr	Y	Zr	Nb			Pd		Cd	(In)	Sn	Sb	Te	I
Cs	Ba	La‡	Hf	Ta				(Hg)	(Tl)	Pb	Bi	(Po)	(At)
(Fr)	(Ra)	Ac‡												

| ionic | metallic | intermediate | covalent |
| hydrides | hydrides | hydrides | hydrides |

‡ lanthanide or actinide elements

8.3 Ionic hydrides

The gain of an electron by a hydrogen atom gives the helium configuration, $1s^2$ and is analogous to halide ion formation. However, the formation of the hydride ion is much less favourable than the formation of a halide ion both because the hydrogen electron affinity is lower and because of the higher heat of formation of the bond in the hydrogen molecule, as Figure 8.4 shows. As a result only the most active

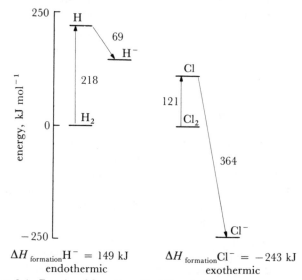

$\Delta H_{\text{formation}} H^- = 149 \text{ kJ}$
endothermic

$\Delta H_{\text{formation}} Cl^- = -243 \text{ kJ}$
exothermic

FIGURE 8.4 *Energies of formation of hydride and chloride ions*

elements, whose ionization potentials are low, form ionic hydrides. The alkali metal hydrides and CaH_2, SrH_2 and BaH_2 are the only compounds which are clearly ionic. Magnesium hydride, MgH_2, is intermediate between the ionic hydrides and the solid covalent hydrides like AlH_3. The dihydrides of the lanthanide elements are metallic and only approach the ionic type as the hydrogen content rises towards the MH_3 stoichiometry. However, the two lanthanide elements which show a relatively stable II state in their general chemistry (compare Chapter 10)—europium and ytterbium—do form ionic hydrides, EuH_2 and YbH_2. These two compounds are isomorphous with CaH_2 and differ in structure from the other lanthanide dihydrides which have the fluorite structure.

The ionic hydrides are formed by direct reaction between hydrogen and the heated metal. They are all reactive and reactivity increases with atomic weight in a Group. The alkali metal hydrides are more reactive than those of the

corresponding alkaline earth elements. The alkali metal hydrides have the sodium chloride structure (the radius of H^-, about 150 pm, is comparable with the halide radii, $F^- = 133 \text{ pm}$, $Cl^- = 181 \text{ pm}$). The alkaline earth hydrides, and EuH_2 and YbH_2, have the same structure as $CaCl_2$. The metal atoms are in approximately hexagonal close packing and each is surrounded by nine hydride ions in a slightly distorted lead dichloride structure. In the regular $PbCl_2$ structure, the metal atom is coordinated by six metal ions at the corners of a trigonal prism and three more beyond the rectangular faces. In the dihydrides, the metal atom has seven hydride ions at equal distances and two more distant ions (e.g. CaH_2 has seven Ca–H distances of 232 pm and two of 285 pm). In all these ionic hydrides, the metal-metal distances are less than they are in the elements so that the hydrides are denser than the metals.

Evidence for the ionic nature of these hydrides is based on two main counts:

(i) Molten LiH shows ionic conductance and the melt gives hydrogen at the *anode* on electrolysis. The other hydrides decompose before melting, but they may be dissolved in alkali halide melts without decomposition and, on electrolysis, give hydrogen at the anode.

(ii) It has been possible, by a combination of X-ray and neutron diffraction, to construct an electron density map for LiH. This shows that 0·8 to 1·0 of an electron has been transferred to hydrogen from each lithium atom (i.e. to give Li^+H^-). Thus lithium hydride is almost completely ionic. As polarization effects are greatest in lithium hydride (see Chapter 4) it follows that the other alkali hydrides are also ionic with full transfer of an electron from each metal atom.

(iii) The crystal structures of the hydrides show no indication of directional bonding (chains, sheets or discrete molecules) and are reasonable for ionic compounds with the radius ratios of the hydrides.

The ionic hydrides react readily, and often violently, with water or any other source of acidic hydrogen and with oxidizing agents. The reaction with acidic hydrogen may be represented as:

$$X^{\delta-} - H^{\delta+} + H^- = H_2 + X^-$$

Examples of this and other common reactions of the hydride ion are shown in Figure 8.5.

The ionic hydrides find use in the laboratory for drying solvents, and as reducing agents, though they have been

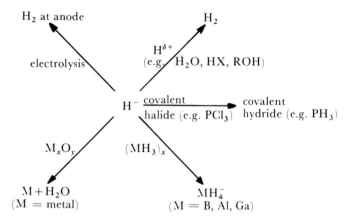

FIGURE 8.5 *Reactions of the hydride ion*

largely superseded in the latter application by the advent of the complex hydrides. On the industrial scale, NaH and CaH$_2$ which are relatively inexpensive and easily handled, find some application as condensing agents in organic syntheses and as reducing agents, e.g.:

$$CaH_2 + MO = CaO + M + H_2 \text{ (at 500–1 000 °C)}$$
$$CaH_2 + 2NaCl = 2Na\uparrow + CaCl_2 + H_2$$

Complex hydride anions. One important reaction of the alkali metal hydrides is in the preparation of the complex hydride anions. If lithium hydride is reacted in ether with aluminium trichloride, the tetrahydroaluminate, LiAlH$_4$, results (this compound is usually called by its non-systematic name of lithium aluminium hydride):

$$4LiH + AlCl_3 = LiAlH_4 + 3LiCl$$

Recent work has shown that the complex aluminium hydrides may also be made by direct reaction of the elements at high pressures, giving a much cheaper route as all the expensive lithium ends up in the product

$$Li + Al + 2H_2 \xrightarrow[\text{250 atm, 120–150 C}]{\text{ether solvent}} LiAlH_4$$

(also for Na or K).

NaBH$_4$, sodium tetrahydroborate (borohydride), LiGaH$_4$, and similar compounds may be prepared by one or other of these routes, as can the aluminium hexahydrides like Na$_3$AlH$_6$.

The alkali metal compounds MBH$_4$, MAlH$_4$, MGaH$_4$ and M$_3$AlH$_6$ contain M'H$_4^-$ or AlH$_6^{3-}$, anions, although there may be a weak interaction with the cation, especially in the case of lithium compounds, of the multicentred, electron deficient type discussed in section 8.6. The M'H$_4^-$ ions are tetrahedral and AlH$_6^{3-}$ is octahedral.

These complex hydrides, particularly LiAlH$_4$ and NaBH$_4$, are extremely useful reducing agents, especially the lithium compounds which are appreciably soluble in ether. Examples of their uses in the preparation of covalent hydrides are given in section 8.5, and organic applications include the reduction of aldehydes or ketones to alcohols and of nitriles to amines. The different complex hydrides vary in reactivity,

thus BH$_4^-$ is a milder reducing agent than AlH$_4^-$, and further modifications to the reactivity are possible by introducing organic substituents as in Na$^+$[HB(OCH$_3$)$_3$]$^-$. The introduction of such reagents in the last two decades has created a revolution in reductive preparations in both organic and inorganic chemistry. For example, yields in the preparation of the hydrides of the heavier elements of the carbon and nitrogen Groups have been raised to 80–90 per cent as compared with the 10–20 per cent which was common with the older methods using ionic hydrides.

Less electropositive elements also form borohydrides or aluminohydrides, but these are more closely related to the parent boron and aluminium hydrides, which are electron deficient covalent species and are discussed in section 8.6.

8.4 Metallic hydrides

Non-stoichiometric compounds between hydrogen and many transition metals have been known for a long time, but their nature and bonding have not been well understood. These compounds are brittle solids with a metallic appearance and with metallic conductivity and magnetic properties. Typical formulae are TiH$_{1.8}$, NbH$_{0.7}$ or PdH$_{0.6}$. These compounds were called 'interstitial' hydrides as it was originally conjectured that the hydrogen atoms were held in the interstices of the metal lattice. The problem was further complicated by the fact that most transition metals physically adsorb large amounts of hydrogen so that many spurious compounds were reported.

It is now known that all the lanthanide and actinide elements, the elements of the titanium and vanadium Groups, chromium and palladium combine exothermically and reversibly with hydrogen to form metallic hydrides in which the metal atoms are in a different structure from those of the element. These metallic hydrides have the idealized formulae shown in Table 8.3, but, as usually prepared, are commonly non-stoichiometric and hydrogen-deficient.

There is thus a 'hydride gap' in the middle of the Periodic Table where binary hydrides are not formed. The exothermic heats of formation decrease from left to right towards this gap, and it has been calculated that hydrides of the ferrous metals would have heats of formation of around zero.

Many of these hydrides have structures where the hydrogens occupy the tetrahedral sites in cubic, close-packed metal lattices. With all these sites occupied, this corresponds to the formula MH$_2$ and the fluorite structure. It is thought that VH, NbH and TaH have the hydrogens in tetrahedral sites and these compounds have slightly distorted body centred cubic metal lattices which are closely related to cubic, close-packed arrangements. In the lanthanide hydrides, and those of yttrium and the heavier actinides, the MH$_2$ phase can take up further hydrogen in octahedral sites. The lighter lanthanides form MH$_3$ phases in this way which remain cubic, but the heavier elements undergo a structural change around the composition MH$_{2.5}$ to give a hexagonal lattice. UH$_3$, PaH$_3$ and Th$_4$H$_{15}$ have more complex structures. In two of the hydrides, CrH and PdH, the hydrogens are in octahedral sites only. Palladium hydride (which has

TABLE 8.3 Metallic Hydrides (idealized formulae and structures)

ScH$_2$ (fluorite)	TiH$_2$ (fluorite)	VH (bcc) VH$_{1.6}$ (fluorite?)	CrH (anti-NiAs) CrH$_2$? (fluorite?)	—	—	—	(NiH)(c) (NaCl) CuH (wurtzite)
YH$_2$ (fluorite) YH$_3$ (hexagonal)	ZrH$_2$ (fluorite)	NbH (bcc?) NbH$_2$ (fluorite?)	—	—	—	PdH$_{0.7}$ (NaCl)	—
LaH$_2$ (fluorite) LaH$_3$ (cubic)	HfH$_2$ (fluorite)	TaH (bcc)	—	—	—	—	—

Lanthanide elements — MH$_2$ (fluorite) formed by Ce, Pr, Nd, Sm, (a), Gd, Tb, Dy, Ho, Er, Tm, (a), Lu
MH$_3$ (cubic) formed by Ce, Pr, Nd, Yb
MH$_3$ (hexagonal) formed by Sm, Gd, Tb, Dy, Ho, Er, Tm, Lu

AcH$_2$ (fluorite)

Actinide elements — MH$_2$ (fluorite) formed by Th, (b), Np, Pu, Am
MH$_3$ (hexagonal) formed by Np, Pu, Am
MH$_3$ (cubic, complex str.) formed by Pa, U

Notes: (a) EuH$_2$ and YbH$_2$ have orthorhombic, ionic hydrides discussed in section 8.3.
(b) Thorium dihydride has a distorted fluorite structure: a second hydride, Th$_4$H$_{15}$ also exists.
(c) NiH exists only under hydrogen pressures of 10^4 atm.

been prepared only up to the PdH$_{0.7}$ composition) forms a sodium chloride lattice but chromium hydride has a hexagonal, *anti*-nickel arsenide structure.

There has been much discussion of the bonding in these metallic hydrides (and in similar compounds such as some low-valency halides of the transition elements which also show metallic properties) but there is still no universally accepted picture. One theory which accounts for many of the facts regards the metallic hydrides as modified metals. Thus, a metal with n valency electrons is regarded as forming M^{n+} cations and having the n electrons per metal ion in completely delocalized orbitals. Then, as hydrogen atoms enter the lattice each acquires one of these delocalized electrons to form a hydride ion. Thus, a metal hydride, MH$_x$ contains M^{n+} ions, xH^- ions and $(n-x)$ delocalized electrons. The relative numbers and sizes of the metal cation and hydrogen anions govern the structure of the hydride, while the $(n-x)$ remaining metallic electrons give the hydride its metallic properties. For example, TiH$_2$ would be regarded as consisting of Ti^{4+}, two H$^-$ and two conduction electrons per formula unit.

The transition metal hydrides are usually prepared by direct combination between the metal and hydrogen at moderate temperatures and, often, high pressures. They may be decomposed by raising the temperature. This reversible hydrogenation is made use of in two ways. One is to provide a convenient source of very pure hydrogen as the metal may be hydrided and any impurities in the hydrogen removed. Then, by heating the hydride to a higher temperature, pure hydrogen is evolved. The second use is to provide the metal in a finely divided and highly reactive form. Many of the transition metal hydrides differ sufficiently from the metal in lattice parameters so that when the hydride is formed from the bulk metal, it is produced as a fine powder. The other hydrides are brittle and may be much more readily powdered than the metal. The powdered hydride is then heated to remove the hydrogen, leaving the metal in a suitable form and free of surface oxide for further reactions. In addition, the metal hydrides themselves often provide suitable starting materials for synthesizing other compounds of the metal.

A number of the hydrides find industrial application, particularly in powder metallurgy where the hydrogen evolved during fabrication gives a protective atmosphere. Another application is as a moderating material in atomic piles. A metal hydride, such as zirconium hydride, provides a higher density of hydrogen than conventional moderators, such as water, and they may be used to higher temperatures.

The one remaining transition metal hydride, copper hydride CuH, is quite different. It is formed endothermically by the reduction of copper salts with hypophosphorous acid and it decomposes irreversibly. It, and the relatively unknown hydrides of zinc and cadmium, probably resemble the solid covalent hydrides of aluminium and beryllium.

8.5 Covalent hydrides

The hydrogen atom may attain the inert gas structure by sharing an electron pair in a covalent bond. All the binary hydrides not already discussed fall into this group, and certain species are intermediate. Thus, MgH_2 has properties intermediate between ionic and covalent hydrides while CuH, ZnH_2 and CdH_2 are intermediate between metallic and covalent species. Covalent hydrides are formed by all the elements with an electronegativity down to about 1·5 (e.g. aluminium) which are just on the border for forming ionic hydrides. As hydrogen has an electronegativity of 2·1 it follows that bond polarities range from those, as in the hydrogen halides, where the hydrogen end is strongly positive, i.e. $H^{\delta +} - X^{\delta -}$, to those cases where the hydrogen end of the dipole is negative, as in $B^{\delta +} - H^{\delta -}$ or $Ga^{\delta +} - H^{\delta -}$. If the second element has an electronegativity less than about 1·2, the hydride becomes definitely ionic.

The covalent hydrides fall into two distinct classes. First there are the compounds of the carbon, nitrogen, oxygen and fluorine Groups which have normal electron pair bonds between the element and hydrogen. Secondly, there are compounds exemplified by the simplest boron hydride, B_2H_6, which do not have enough valency electrons to form electron pair bonds to all the hydrogens, and these are termed *electron-deficient*. Into this class fall the hydrides of boron, aluminium and gallium, together with beryllium hydride which has no ionic properties. MgH_2, ZnH_2 and perhaps CdH_2 and CuH have some affinities with this class. Other members are the borohydrides and aluminohydrides of elements which are not sufficiently electropositive to give the complex hydride ions: examples are $Be(BH_4)_2$ and $Al(BH_4)_3$. The electron deficient hydrides are treated in section 8.6.

Covalent electron pair bonds are also formed between hydrogen and the transition elements, in compounds where the metal is also bonded to ligands capable of forming π-bonds, such as CO, phosphines, arsines, sulphides, or NO. Such compounds as $(R_3P)_2PtH_2$—where R stands for a variety of aliphatic and aromatic substituents—or $(CO)_5MnH$, have metal to hydrogen covalent bonds of sufficient stability to allow for their isolation. Such covalent bonds to hydrogen seem to be most readily formed by those transition metals in the 'hydride gap' which do not form binary metallic hydrides. Some examples are given in Chapters 12 to 14.

There are three general methods available for the preparation of covalent hydrides, although many others are available in specific cases including the increasing use of direct combination of less reactive elements under high pressure and catalysis. The general methods are:

(i) simple direction combination, especially with the more reactive elements:

$$H_2 + Cl_2 \xrightarrow{\text{light}} 2HCl$$

(ii) acid hydrolysis of a binary compound of the element with an active metal:

$$Mg_3B_2 \rightarrow B_2H_6$$
$$Al_4C_3 \rightarrow CH_4$$
$$Ca_3P_2 \rightarrow PH_3$$

all by the action of dilute hydrochloric acid.

(iii) reduction of a halide or oxide by an ionic hydride or by a complex hydride:

$$GeO_2 + BH_4^- \rightarrow GeH_4$$
$$SiCl_4 + LiH \rightarrow SiH_4$$
$$AsCl_3 + LiAlH_4 \rightarrow AsH_3$$
$$R_2SbBr + BH_4^- \rightarrow R_2SbH.$$

The reactions of all these hydrides may be carried out in ether solution and BH_4^- may also be used in an aqueous system. A fourth method of synthesis, in increasing use, involves the interconversion of hydrides in a suitable discharge. This is particularly valuable for forming longer chains from simple hydrides or for forming long chain halides which may then be reduced to the hydride. The discharge may be of radio or microwave frequency or may be of the ozonizer type.

(iv) $GeH_4 \rightarrow Ge_2H_6 + Ge_3H_8 +$ hydrides up to Ge_9H_{20}
$$SiH_4 + GeH_4 \rightarrow GeH_3SiH_3 + Ge_2H_6 + Si_2H_6$$
$$SiCl_4 \rightarrow Si_2Cl_6 \xrightarrow{LiAlH_4} Si_2H_6$$
$$SiH_4 + PH_3 \rightarrow SiH_3PH_2 + (SiH_3)_2PH + Si_2H_5PH_2, \text{ etc.}$$

Of these four methods of preparation, (i) is of limited applicability but is being developed under extreme conditions of temperature and pressure, (ii) gives low yields but is the most direct way of obtaining the higher members of homologous series, (iii) is usually the best and most convenient method on a laboratory scale, while (iv) gives better yields than (ii) of the higher hydrides but requires a supply of the simple material. Higher hydrides, containing a chain of atoms of the central element, are known for many p block elements but, for most, chain lengths are short and stabilities are low. The exceptions are silicon and germanium, where hydrides M_nH_{2n+2} are characterized up to about $n = 10$. Both straight and branched chains are known, and hydrides containing mixed chains of silicon and germanium atoms are found. For example, pentagermane is found in all three isomeric forms

$$GeH_3GeH_2GeH_2GeH_2GeH_3 \quad (GeH_3)_2GeHGeH_2GeH_3$$
$$\text{and} \quad Ge(GeH_3)_4$$

and Si_2GeH_8 occurs as $SiH_3SiH_2GeH_3$ and $SiH_3GeH_2SiH_3$. Mixed hydrides containing silicon or germanium and certain other atoms are also fairly stable. Examples include the silicon-phosphorus hydrides SiH_3PH_2, both isomers $(SiH_3)_2PH$ and $SiH_3SiH_2PH_2$, and $(SiH_3)_3P$.

The p element hydrides are listed in Table 8.4.

The thermal stabilities of the hydrides decrease in each Group as the atomic weight of the central element increases. Except in the case of carbon compounds (and, to some extent, boron ones), the thermal stability decreases fairly rapidly with increasing molecular weight for the members of

TABLE 8.4 Hydrides of the p elements

B_2H_6 (and many higher hydrides: see Table 8.5)	C_nH_{2n+2}, etc. (no limit to n is known)	NH_3 N_2H_4	H_2O H_2O_2	HF
$(AlH_3)_x$ (solid polymer)	$Si_nH_{2n+2}^{(1),(3)}$ (characterized up to $n = 8$; straight- and branched-chain isomers occur)	PH_3 $[P_2H_4]$ 'PH'$^{(2)}$	H_2S H_2S_n (characterized up to $n = 6$)	HCl
$[(GaH_3)_x]$ (unstable oil)	$Ge_nH_{2n+2}^{(1),(3)}$ (characterized up to $n = 9$; straight- and branched-chain isomers occur)	AsH_3 $[As_2H_4]$ 'AsH' or 'AsH$_2$'$^{(2)}$	H_2Se	HBr
	$[SnH_4]$ $[Sn_2H_6]$ $[PbH_4]$?	$[SbH_3]$ $[BiH_3]$?	$[H_2Te]$ $[H_2Po]$?	HI $[HAt]$?

$[\ ]$ = unstable at room temperature
$[\ ]$? = existence unconfirmed or transient

1. The stability of the higher silanes and germanes apparently decreases with n. Mixed hydrides $Si_xGe_yH_{2(x+y)+2}$ are also known.
2. Solids of uncertain composition but containing P or As and H are widely reported but it is not clear if these are true compounds or element plus trihydride. There are reports of transient species P_3H_5 and As_3H_5.
3. A fairly well-established solid polymeric hydride of germanium, $(GeH)_x$, is known and a further compound $(GeH_2)_x$ may exist. Analogous Si compounds occur.

a homologous series. The variation of stability across a Period is irregular, with the carbon Group and the halogens forming the most stable hydrides, e.g. the hydrides of Ga < Ge > As < Se < Br in stability.

The structures of the hydrides are as predicted from the number of electron pairs on the central element, cf. Chapter 3. All the carbon Group hydrides are tetrahedral MH_4 molecules and the higher homologues are also based on tetrahedra. The nitrogen and oxygen Group molecules are based on tetrahedra with, respectively, one and two lone pairs. The bond angles decrease towards 90° with increasing atomic weight of the central element. The boron Group hydrides are discussed later. Hydrazine, N_2H_4, and hydrogen peroxide, H_2O_2, adopt the structures shown in Figure 8.6 which separate the lone pairs as much as possible.

The reactions of the hydrides are varied and many are familiar: for example, the reactions of hydrogen sulphide, hydrogen halides, water, and ammonia. In general terms, all the hydrides are reducing agents and react strongly with oxygen and halogens. Stability to oxygen varies from the relative stable germane, GeH_4, and hydrogen halides, to the cases such as silane, SiH_4, and phosphine, PH_3, which explode or inflame in air. With the halogens, reactions may also be violent, although iodine often reacts smoothly to cleave only one bond as in:

$$GeH_4 + I_2 = GeH_3I + HI$$

The hydrogen halides also react with many hydrides to give partial substitution:

$$SiH_4 + HX \xrightarrow{AlX_3} SiH_3X + H_2$$

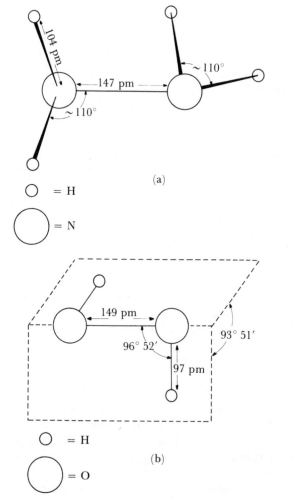

FIGURE 8.6 *The structures of* (a) *hydrazine*, (b) *hydrogen peroxide*

The hydrides of most elements are strong reducing agents which reduce many heavy metal salts to the element:

$$PH_3 + AgNO_3 \rightarrow Ag$$
$$SnH_4 + CuSO_4 \rightarrow Cu, \text{ etc.}$$

Exceptions are provided by the hydrides of the first short Period which are relatively unreactive. One reason for this probably lies in the absence of an easy reaction path. Thus silane, SiH_4, reacts readily as the silicon $3d$ orbitals can provide a means of coordinating an attacking reagent in a reaction intermediate; in methane, CH_4, there is no such pathway. Reaction depends on the breaking of a $C-H$ bond which requires much more energy and is correspondingly slower.

One important reaction of many covalent hydrides is ionization to give a positive hydrogen species. The hydrides of the most electronegative elements are already polarized, with the hydrogen positive. Although the formation of the free proton is impossible, as shown above, the hydrogen bonded to an electronegative element can ionize to give a proton associated with a neutral molecule (together with an anion). In the hydrides of the most electronegative elements, F, O and N, this is accomplished by self-ionization in the liquid phase when the proton is associated with a molecule of the hydride (compare section 5.5):

$$2NH_3 = NH_4^+ + NH_2^-$$
$$2H_2O = H_3O^+ + OH^-$$
$$2HF = H_2F^+ + F^-.$$

The hydrides of less electronegative elements do not self-dissociate in the liquid phase (largely because the liquids are less efficient ion-supporters), but they do dissociate when dissolved in an ionizing solvent such as water. Such hydrides are those of the other halogens, and of the other chalcogens, which dissociate in water or ammonia. The energies involved in such ionizations may be illustrated by the case of hydrogen chloride. The formation of the free proton in the gas phase, $H \rightarrow H^+$ requires 1340 kJ mol^{-1}, and the hydration energy of the gaseous proton is about -1090 kJ mol^{-1}. For hydrogen chloride:

$$HCl_{(gas)} \rightarrow H^+_{(gas)} + Cl^-_{(gas)} \quad \Delta H = \quad 1380 \text{ kJ mol}^{-1}$$
$$H^+_{(gas)} + Cl^-_{(gas)} \rightarrow H^+_{(aq)} + Cl^-_{(aq)} \quad \Delta H = -1460 \text{ kJ mol}^{-1}$$

so ionization of hydrogen chloride in aqueous solution is exothermic by about 80 kJ mol^{-1}. The hydrogen halides are fully dissociated in water but the other hydrides which dissociate in water, the chalcogen hydrides, do so only very weakly; the dissociation constant for

$$H_2S + H_2O \rightleftharpoons H_3O^+ + HS^-$$

is only about 10^{-7}. The hydrides of phosphorus and the other members of the nitrogen Group do not dissociate measurably in water. This rapidly decreasing tendency to dissociate in solution clearly parallels the fall in the polarity of the element-hydrogen bond.

The formation of H_3O^+ or NH_4^+ may be considered as the result of the donation of an electron pair to the proton by the oxygen or nitrogen atoms, $H_3N: \rightarrow H^+$ or $H_2O: \rightarrow H^+$. This is one case of the general donor-acceptor (Lewis base-acid) behaviour of the hydrides. The hydrides with one or more lone pairs of electrons may donate them to suitable acceptor molecules to form coordination complexes. This is, of course, the reaction involved in the solvation of cations by water or ammonia, but it is a general reaction possible for all the hydrides of the nitrogen, oxygen, or halogen Groups. In general, donor power falls as the number of lone pairs on the central atom increases and as the size of the central atom increases. Thus the hydrogen halides are weak donors as there are three lone pairs on the halogen atoms, and the heavier analogues of ammonia and water, such as SbH_3 or H_2Se, are also weak. Organic-substituted hydrides of the nitrogen and oxygen Group elements, especially the phosphines, R_3P, and the sulphides, R_2S (where R includes aliphatic and aromatic groups), do act as donors to a wide variety of species. Acceptor molecules with suitable empty orbitals available include the p acceptors of the boron Group (and beryllium), the nd acceptors such as the tetrafluorides of the carbon Group, and especially the $(n-1)d$ acceptors among the transition metals.

Acceptor molecules among the hydrides are largely confined to those of the boron Group, although there is some evidence of weak d orbital acceptor power in the carbon Group tetrahydrides. The complex hydride anions, BH_4^-, AlH_4^- and GaH_4^-, may be regarded as being formed by the acceptance of the electron pair on the hydride ion by the MH_3 species, $H^-: \rightarrow MH_3$. Donor-acceptor complexes are also formed between the boron Group hydrides and those of the nitrogen and oxygen Groups. Compounds such as H_3BNH_3 are formed at low temperatures but, on warming towards room temperatures, these compounds lose hydrogen and polymerize (in this case to the ring compound $B_3N_3H_6$ discussed in section 8.6). If organic derivatives are used, the complexes are more stable; both H_3BNMe_3 and Me_3BNH_3 are stable at room temperature. The much less-stable polymeric hydrides of aluminium and gallium also form such

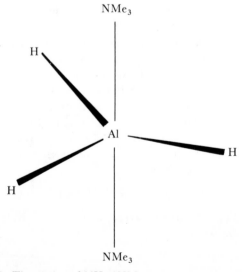

FIGURE 8.7 *The structure of* $AlH_3.2NMe_3$

complexes and considerable stabilization of the M — H bonds results. Complexes such as R_3NAlH_3 and R_3NGaH_3 are well characterized. These elements also make use of their d orbitals to accept more than one donor molecule. Thus, the compound $AlH_3 2NMe_3$ has the trigonal bipyramidal structure shown in Figure 8.7.

Many of these ions and hydrides form isostructural sets, all based on four electron pairs:

four bond pairs	tetrahedron	BH_4^-, CH_4, NH_4^+
three bond pairs and one lone pair	pyramid	NH_3, H_3O^+
two bond pairs and two lone pairs	V-shape	H_2O, H_2F^+

8.6 Electron-deficient hydrides

Boron and related elements form a range of hydrogen compounds which cannot be accounted for by classical ideas of electron-pair bonds between two atoms. It is found that the simple monomeric species, such as BeH_2 or BH_3, do not occur but that more complicated compounds are formed. Thus the simplest boron hydride is B_2H_6 while beryllium and aluminium form high molecular weight polymers, $(BeH_2)_x$ and $(AlH_3)_x$. The most fully studied compounds of this class are the hydrides of boron and these are treated first.

TABLE 8.5 The boron hydrides

Identified by Stock in his classical studies up to 1932	Modern discoveries
B_2H_6	B_6H_{12} (1963)
B_4H_{10}	B_8H_{12} (1964)
	B_8H_{14} (1968)
B_5H_9	B_8H_{18} (1965)
B_5H_{11}	B_9H_{15} (1957)
B_6H_{10}	iso-B_9H_{15} (1965)
B_6H_{12}? (Stock later withdrew his	$B_{10}H_{16}$ (1959)
identification of this	$B_{16}H_{20}$ (1968)
hydride)	$B_{18}H_{22}$ (1962)
$B_{10}H_{14}$	iso-$B_{18}H_{22}$ (1963)
Further involatile polymeric	$B_{20}H_{16}$ (1963)
materials, liquid and solid.	$B_{20}H_{26}$? (1962)

In addition a number of configurations are known as anions although no parent hydride is known; examples include $B_{12}H_{12}^{2-}$ and $B_{11}H_{14}^-$.

The work of Stock on the boron hydrides was one of the sources of the renaissance in inorganic chemistry in this century and it can be seen that it took another thirty years to add further compounds to his list. When it is realized that all these hydrides are inflammable, many are very unstable, and all appear as a mixture in the hydrolysis of magnesium boride, the full extent of Stock's experimental genius becomes apparent. As the dates in the second column show, his field underwent rapid development in the sixties.

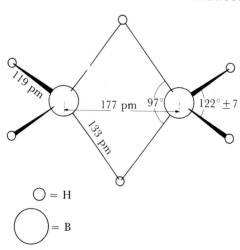

FIGURE 8.8 *The structure of diborane*

It is clear from Table 8.5 that the boron hydrides form an exceptional group of compounds. On simple valency grounds, the lowest hydride of boron is expected to be BH_3, but no such molecule has ever been discovered—despite considerable search. The simplest boron hydride is diborane, B_2H_6, which many studies have shown to have the bridge structure of Figure 8.8 (see section 6.5). There are only twelve valency electrons in B_2H_6 (3 per B and 1 per H), so that the molecule cannot be made up of electron pair bonds, which would require sixteen electrons for the structure of Figure 8.8. Such molecules, with insufficient electrons to form two-centre electron pair bonds between all the atoms, are termed *electron deficient*. The evidence of bond lengths and angles, and of the infra-red stretching frequencies, suggests that the terminal B — H bonds are normal single bonds. This leaves two hydrogen atoms and four electrons to be fitted into the picture. A number of bonding theories were put forward but the most satisfactory solution was that proposed by Longuet-Higgins. He suggested that, in place of the normal electron pair bond centred on two nuclei, a three-centre bond should be considered, made up of the hydrogen 1s orbital and appropriate hybrids on the two boron atoms. Such a bond may be indicated as B····H····B or as $\overset{H}{\underset{B \quad B}{\frown}}$. If the boron atoms form sp^3 hybrid orbitals, two of which form the bonds to the terminal hydrogens, then the others may overlap with the hydrogen 1s orbitals as shown in Figure 8.9. Figure 8.10 gives the energy level diagram for such bonds. The three atomic orbitals from B_1, H_A and B_2 combine to form a bonding, a nonbonding, and an antibonding, three-centred molecular orbital centred on these atoms. (Note that three atomic orbitals give three molecular orbitals.) The three atomic orbitals on B_1, H_B and B_2 give an identical set of three molecular orbitals. If two of the four electrons, left over after the terminal B — H bonds were formed, are placed in the $B_1H_AB_2$ bonding orbital and the other two in the $B_1H_BB_2$ bonding orbital, as in Figure 8.10b, the result is to use all the valency electrons and to fill only the three-centre molecular orbitals which are bonding. Diborane is thus described as having four two-centre B — H bonds and two three-centre B····H····B

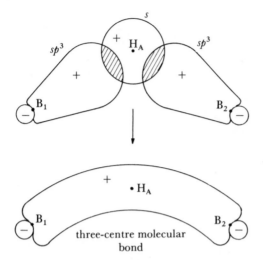

FIGURE 8.9 *Three-centred bonds in diborane*
A tetrahedral hybrid orbital on B_1, an s orbital on H_A, and a tetrahedral orbital on B_2 combine to form the bonding orbital shown, an antibonding orbital with nodes between each B and the H, and a non-bonding orbital which places the electron density mainly on the two B atoms. A similar set of three three-centre orbitals is formed by H_B with two other tetrahedral orbitals on B_1 and B_2.

bonds, all sigma, all holding two electrons, and accounting for the twelve valency electrons.

If the description of these three-centre bonds is compared with that of the many-centred bonds discussed in Chapter 3, no essential difference will be found, except that the earlier

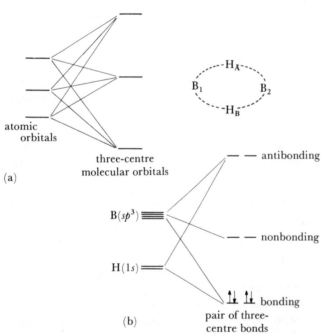

FIGURE 8.10 *Energy level diagram for three-centred bonds (diagrammatic)*, (a) *general form*, (b) *electronic structure for the pair of three-centred bonds in diborane*
Figure (a) shows the generalized energy level diagram for three atomic orbitals combining to form three three-centred molecular orbitals. The energy level diagram (b) corresponds to the two sets of three-centred orbitals formed by the s orbitals on H_A and H_B and two suitably directed orbitals on each of B_1 and B_2 in the diborane bridge. The six atomic orbitals give six three-centred molecular orbitals and, in diborane, the two bonding orbitals are filled.

discussion was confined to polycentred π bonds in molecules already held together by two-centred sigma bonds. The bonding in diborane amounts to polycentred sigma bonding.

Diborane will serve as a model compound for a wide range of other hydrogen species which are formulated with similar polycentred bonding. The higher boron hydrides are similarly formulated with three-centre $B\cdots H\cdots B$ bonds; one example is tetraborane, B_4H_{10}, shown in Figure 8.11(a). This has six two-electron $B-H$ bonds, one $B-B$ bond, and four two-electron three-centre $B\cdots H\cdots B$ bonds. The structures of higher boron hydrides may be built up similarly, but other types of polycentred bonds may also be present including three-centred $B\cdots B\cdots B$ bonds and even five-centred bonds involving five boron atoms, as found in B_5H_9, Figure 8.11.

In addition to the boron hydrides listed in Table 8.5, there is a wide range of boron hydride ions, of similar formulae and a variety of structures, and bonding in these species also involves multicentred bonding. Examples include $B_6H_6^{2-}$ and $B_{12}H_{12}^{2-}$ which have all the hydrogens present as terminal $B-H$ bonds. The B_6 skeleton is a regular octahedron and the B_{12} one a regular icosahedron held together by multicentred B_n bonds. Some of these ions contain atoms other than boron. Thus there is an ion $B_9C_2H_{11}^{2-}$ which can be regarded as derived from $B_{12}H_{12}^{2-}$ by removing one apex of the icosahedron (taking away one BH unit) and substituting two other B atoms by carbon atoms. A considerable number of similar *carboranes* has been reported recently.

Electron deficient bonding is by no means confined to boron compounds. Aluminium hydride is an insoluble polymer, $(AlH_3)_x$, and a recent structural determination by X-ray and neutron diffraction shows that it contains Al atoms surrounded octahedrally by six H atoms. The structure is a giant molecule of AlH_2Al bridges and is analogous to the structure of AlF_3. Thus this is an example of electron deficient bonding analogous to that in diborane. Another example of $Al\cdots H\cdots Al$ bridging is found in the molecule $Al_2H_5NMe_2$ which has the structure

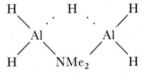

There is an exact boron analogue, $B_2H_5NMe_2$. In each molecule, the NMe_2 group may be regarded as replacing one of the hydrogen bridges in diborane or its aluminium analogue.

Beryllium hydride, $(BeH_2)_x$, is also polymeric and insoluble. The structure has not yet been determined but is reasonably postulated to be a chain polymer with bridging hydrogen

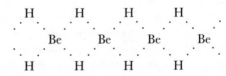

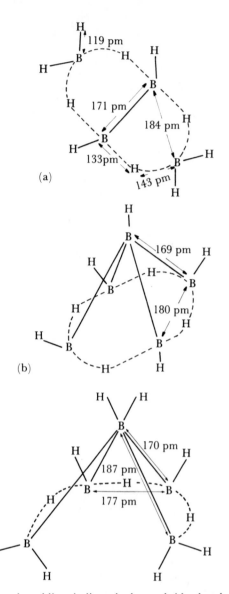

(a)

(b)

(c)

dotted lines indicate hydrogen bridge bonds

FIGURE 8.11 *Structures of* (a) *tetraborane,* (b) *pentaborane-9,* (c) *pentaborane*-11

(a) Tetraborane forms a boat-shaped molecule with four B····H····B three-centred bonds and a direct B−B single bond.

(b) Here the boron atoms form a square pyramid. The edges of the square base are bonded by three-centre B····H····B bonds while the apical B atom is bonded to the four base B atoms by five-centred bonds using boron orbitals directed into the body of the pyramid.

(c) Pentaborane-11 is an unsymmetrical square pyramid where, in addition to B····H····B bonds, there are three-centred bonds between three boron atoms in the triangular faces—the so-called 'central' B ⌄ B bond.
        B

Hydrogen bridging to beryllium is illustrated in a number of other species, for example, the ion $Et_2BeH_2BeEt_2^{2-}$ with the structure

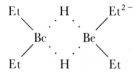

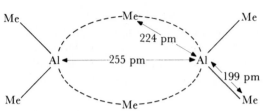

FIGURE 8.12 *The structure of* Al_2Me_6

This diagram gives the form of aluminium trimethyl dimer. The methyl group is bonded to two aluminium atoms in three-centred bonds formed by overlap of tetrahedral hybrid orbitals on each aluminium and on the carbon atom. This bonding resembles the B····H····B bond in diborane, and also the central B ⌄ B bonds
                   B
found in pentaborane-11.

It also seems reasonable to formulate gallane, GaH_3, which is an unstable, involatile oil, and possibly ZnH_2 and CdH_2, as hydrogen-bridged polymers. Electron deficient bonding is also found in some of the organic analogues of the hydrides. For example, trimethylaluminium exists as a dimer, $Al_2(CH_3)_6$, with a *methyl-bridged* structure very similar to that of diborane. This is based on three-centred bonds formed by an sp^3 hybrid on each aluminium together with an sp^3 hybrid on the carbon atom, as shown in Figure 8.12. BeH_2, ZnH_2, CdH_2 and CuH_2 are probably hydrides of this polymeric, electron-deficient type.

These electron-deficient bond systems are of reasonable strength. Thus, the boron hydrides resemble the silicon hydrides in thermal stability. The hydrides share the strong reducing properties of all hydrides and the boron hydrides react violently with oxygen. The electron-deficiency shows up mainly in the susceptibility of the hydrides to attack by electron pair donors, which provide the extra electrons to allow the formation of electron-pair two-centre bonds. Diborane, for example, readily reacts to give borine adducts, $H_3B \leftarrow D$, with suitable donors, D, such as the amines.

The boron Group of elements also makes use of the empty p orbital in the formation of the complex hydrides. The BH_4, AlH_4, and GaH_4 groups have been prepared and they decrease in stability in that order. Many elements form these complex hydrides and they range from the ionic compounds of the active metals, such as $Na^+BH_4^-$ or $Li^+AlH_4^-$, to the compounds of the less active metals which appear to be covalent and electron deficient. The structure of Figure 8.13, with three centre B····H····Be bonds has been reported for beryllium borohydride, $Be(BH_4)_2$, although details are not yet finally settled.

FIGURE 8.13 *The probable structure of beryllium borohydride,* $Be(BH_4)_2$.

The Be and two B atoms are arranged in a triangle held by B····H····B and B····H····Be three-centre bonds.

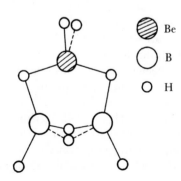

Be
B
H

For aluminium borohydride, $Al(BH_4)_3$, the Al atom is surrounded by six hydrogen atoms, forming three AlH_2B bridges while the low temperature form of $Zr(BH_4)_4$ probably contains zirconium surrounded tetrahedrally by four boron atoms linked by three-hydrogen bridges, of the form

Complex aluminium hydrides of the less electropositive metals probably have similar bridging structures. Even in $LiAlH_4$, the $Li-H$ distance is shorter than in LiH, suggesting some $Li\cdots H$ interaction which probably accounts for its solubility in ethers. The complex beryllium hydrides, Li_2BeH_4 and Na_2BeH_4 also behave more like electron deficient polymers than as species containing BeH_4^{2-} ions.

It has been seen that hydrogen-bridged, electron deficient species, where polycentred sigma bonding has to be postulated, occur quite widely. Although best known for boron, it is also established, or reasonably postulated, for compounds of beryllium, aluminium, zinc and gallium. As the number of structural determinations on hydrogen compounds of these elements increases, well-established cases should increase in number. It is also probable that a weaker interaction, of the form $Al-H\cdots Li$, occurs in complex beryllium, zinc, and aluminium hydrides in addition to electrostatic interactions. Thus electron deficient polycentred bonding is typical of hydrides of elements which have fewer valence shell electrons than valency orbitals. Its presence allows use of orbitals which would otherwise remain empty, and the interaction is sufficiently strong to give aluminium, and probably other elements, a coordination number greater than four. Similar conditions hold for the formation of methyl and similar bridges.

The boron group hydrides react readily with electron pair donors to revert to two-centre two electron bonding and adducts such as $H_3B.NMe_3$ are also readily formed. Figure 8.7. If the donor group is itself a hydride, further interaction may occur with elimination of H_2. For example, the product of the reaction between ammonia and diborane at low temperatures is the expected adduct, H_3BNH_3. On warming to room temperature, this compound loses hydrogen and gives a product of empirical formula, HBNH. The latter is actually the trimer, called borazole or borazine, whose structure is shown in Figure 8.14. The molecule, $B_3N_3H_6$, is isoelectronic with benzene, C_6H_6, and has a similar, planar structure with all the $B-N$ bonds the same and a delocalized π-bonding system. A wide variety of substituted borazines have been made, either by substitution reactions on borazine or, more commonly, by altering the composition of the starting adduct. Thus, $(CH_3)H_2BNH_3$ gives, on heating, $(CH_3)_3B_3N_3H_3$—called B-trimethylborazine (where the methyl groups are on the boron). The corresponding N-trimethylborazine is formed from the adduct $H_3BNH_2(CH_3)$. Many similar compounds can be made.

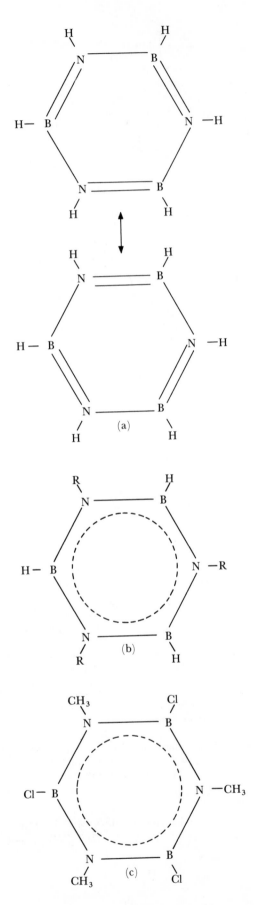

FIGURE 8.14 (a) *The structure of borazine* $(H_3B_3N_3H_3)$, (b) N-*trialkyl borazine*, (c) B-*trichloro-N-trimethylborazine*

If the ammonia is replaced by phosphine, PH_3, the adduct, H_3BPH_3, gives a different ring compound, $(H_2BPH_2)_3$. This is the analogue, not of benzene, but of cyclohexane, C_6H_{12}. The ring is no longer planar but chair-shaped and there is only a limited amount of π bonding involving the phosphorus d orbitals. Arsine, AsH_3, behaves similarly. Again, substituted compounds may be made by starting from substituted borines or phosphines.

8.7 The hydrogen bond

Compounds containing hydrogen bonded to a very electronegative element, especially F, O, or N, show properties which are consistent with an interaction between the hydrogen bonded to one electronegative atom and a second atom. This secondary bond is relatively weak and is termed the hydrogen bond. It may be written $X-H\cdots Y$, where X is the atom to which the hydrogen is bonded by a normal bond.

Evidence for hydrogen bonding is widespread. Some of the important facts are:

(i) *Evidence of molecular association* from melting points, boiling points, heats of evaporation, and, in some cases, molecular weight determinations. For example, the boiling points of the hydrides of the carbon, nitrogen, oxygen, and fluorine Group elements are shown in Figure 8.15. For the

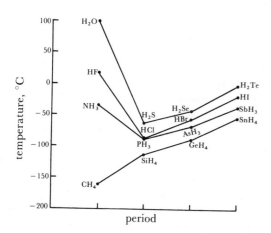

FIGURE 8.15 *Boiling points of Main Group hydrides*

carbon Group hydrides, the boiling point is a linear function of atomic weight. However, the positions of ammonia, water, and hydrogen fluoride are clearly anomalous. These unexpectedly high boiling points indicate an extra interaction between the molecules in the liquid, which is not found for the heavier hydrides. A similar anomaly appears in the latent heats of vaporization shown in Figure 8.16, and also in the melting points and the latent heats of fusion, showing that the interaction is also present in the solid. Figure 8.16 serves to give some idea of the size of the effect; the increases in the heats of evaporation, compared with those found by extrapolating from the heavier hydrides, is of the order of 20 ± 10 kJ mol^{-1}. In general, the hydrogen bond energies may be taken to lie in the range from 4 to 40 kJ mol^{-1}, about one tenth of normal bond energies but several times greater than normal weak interactions between molecules.

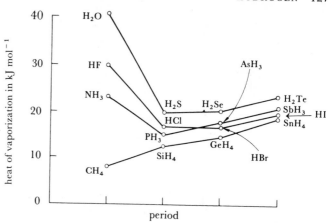

FIGURE 8.16 *Latent heats of vaporization of Main Group hydrides*

Molecular weight studies show clear evidence of association in many cases: perhaps the most striking case is the dimerization of the lower carboxylic acids. The heat of association of acetic acid:

$$CH_3COOH = \tfrac{1}{2}[CH_3COOH]_2$$

is found to be 28.9 kJ mol^{-1} (of monomer). Electron diffraction studies have shown that the structure of the acetic acid dimer is as given in Figure 8.17. The other carboxylic acids behave similarly.

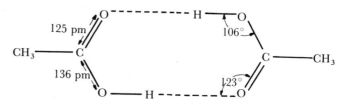

FIGURE 8.17 *The structure of the acetic acid dimer*

(ii) *X-ray studies.* The evidence from boiling points, etc., above gives no indication of the atomic arrangements in the hydrogen bond and this information might be expected from X-ray studies. However, as hydrogen is so light compared with other nuclei, it is not an efficient scatterer of X-rays and it is not usually possible to determine the positions of hydrogen atoms in this way. One exception is boric acid, H_3BO_3 which contains only light nuclei and whose structure consists of loosely bound sheets with the configuration shown in Figure 8.18.

In general, X-ray evidence gives only the $X-Y$ distance, and it is found that this is less than the sum of the van der Waals' radii, which would determine the distance in the absence of hydrogen bonding. See examples in Table 8.6.

(iii) *Neutron diffraction.* The scattering of neutrons does allow the location of hydrogen atom positions and a number of studies have appeared. Neutron diffraction, electron diffraction, and refined X-ray methods, now available for crystal studies, have contributed to the structures in Figure 8.19.

(iv) *Infrared studies.* When an $X-H$ grouping is involved in hydrogen bonding, the characteristic infrared stretching of the $X-H$ bond is shifted and it is found that this is always to lower frequencies. There is a general correlation between

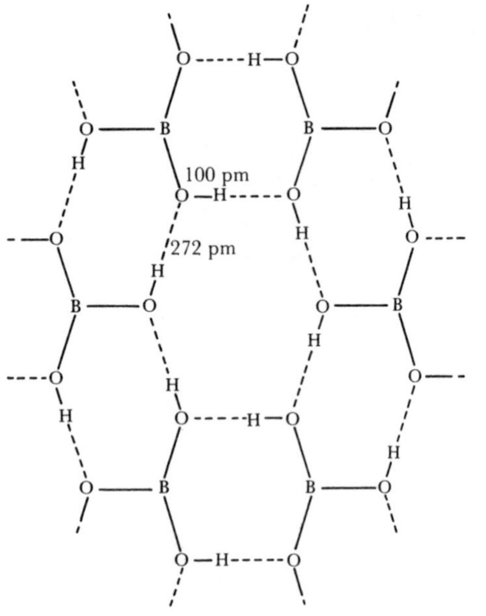

FIGURE 8.18 *Structure of boric acid,* H_3BO_3. *Hydrogen bonding holds the molecules in sheets*

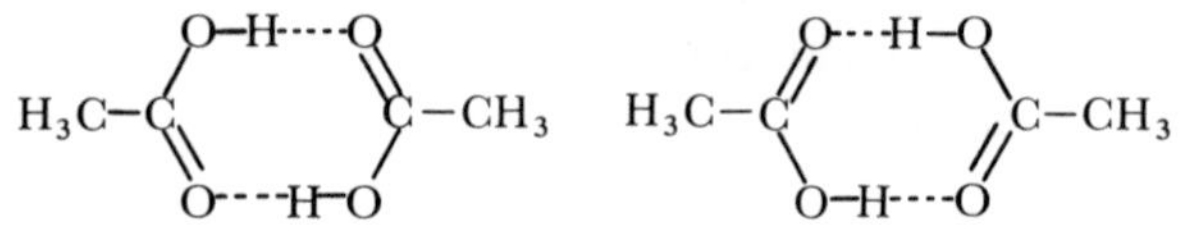

(a)

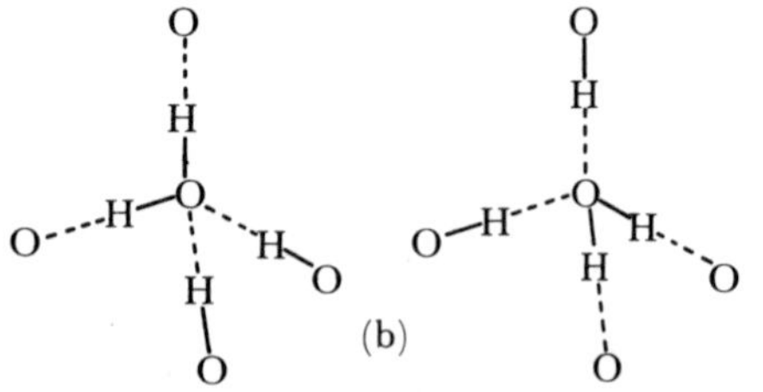

(b)

FIGURE 8.19 *Alternative configurations of hydrogen-bonded species leading to non-zero entropy at absolute zero:* (a) *acetic acid dimer,* (b) *ice*

the extent of this frequency shift and the hydrogen bond energy. Some examples are given in Table 8.6. As infrared spectra are readily measured, this frequency shift is usually the most convenient diagnostic test for hydrogen bonding.

(v) *Entropy data.* Evidence of hydrogen bonding may also be given by entropy data determined at low temperatures. As entropy is linked with disorder, a definition in terms of the number of possible arrangements of the system is possible. This is $S = -R\ln W$, where R is the gas constant and W is the probability of the state of the system. Now, a perfect crystal, when cooled to absolute zero (to eliminate thermal motions of the atoms), has only one possible arrangement of the atoms and the entropy at absolute zero will be $-R\ln 1 = 0$ (remember that certainty = a probability of one). However, if hydrogen bonding occurs in a crystal, more than one configuration may be possible at absolute zero. For example, in the acetic acid dimer, these are the two configurations of Figure 8.19a: similarly, ice has the oxygen

atoms surrounded by four hydrogen atoms and a number of configurations are possible, such as those shown in Figure 8.19b. As a result, such hydrogen-bonded molecules have non-zero values of the entropy at 0 K. In the case of ice, it has been calculated that the entropy should be $3 \cdot 39$ J K^{-1} mol^{-1} and it has been measured as $3 \cdot 39 \pm 0 \cdot 21$ J K^{-1} mol^{-1}.

The hydrogen bond can be shown to exist by the methods given above, and the next question is about its shape. The bond may be linear or bent and the hydrogen atom may lie symmetrically or unsymmetrically between X and Y. As far as present evidence goes, it appears that XHY are usually arranged linearly unless there is a good steric reason opposing this—as in the intramolecular hydrogen bonds in molecules like salicylic acid

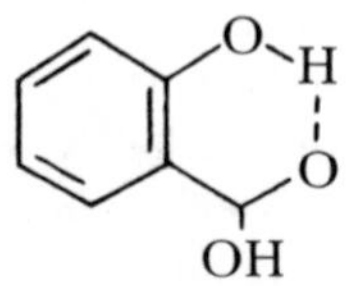

or nickel dimethylglyoxime (Figure 13.25).

The symmetry of the position of the H atom varies. In the bifluoride ion, FHF$^-$, the hydrogen is symmetrically placed between the two fluorine atoms, as is the hydrogen in the OHO links in the nickel dimethylglyoxime complex. In most other cases of known structure, however, the unsymmetrical position of the hydrogen is adopted, as in water or acetic acid. Apart from structure determinations, it should be noted that the entropy determination will distinguish symmetrical structures as these will have no residual entropy. It is in fact found that there is no residual entropy for HF$_2^-$ in potassium hydrogen fluoride. It appears that, in general, the stronger hydrogen bonds are the more symmetrical.

Table 8.6 summarizes the data available for some cases of hydrogen bonds involving F, O, and N. One or two cases are worth further mention.

The solvated proton. As discussed on p. 115, the free H$^+$ ion cannot exist in a chemical environment, and there are a number of structural studies which demonstrate its occurrence as H$_3$O$^+$. Thus the solid monohydrates of perchloric, sulphuric, and hydrochloric acids have been shown to exist as H$_3$O$^+$ClO$_4^-$, H$_3$O$^+$HSO$_4^-$ and H$_3$O$^+$Cl$^-$ respectively.

Recent studies of higher hydrates of protonic acids have shown that more complex forms of the hydrated proton exist, whose structures involve hydrogen bonding. For example, HCl.2H$_2$O is best formulated as H$_5$O$_2^+$Cl$^-$, while HBr.4H$_2$O is (H$_7$O$_3$)$^+$(H$_9$O$_4$)$^+$2Br$^-$.H$_2$O. The structures of these forms of the hydrated proton are summarized in Table 8.7. Most of the work was carried out by X-ray crystallography, so that the exact positions of the H atoms were not determinable, but the HAuCl$_4$ hydrate was also studied by neutron diffraction and the hydrogens were located.

The symmetric H$_5$O$_2^+$ ion has a short hydrogen bond length, similar to that in the nickel dimethylglyoxime molecule (Table 8.6). The unsymmetric H$_5$O$_2^+$ ion, H$_7$O$_3^+$ and H$_9$O$_4^+$ are structurally more like a H$_3$O$^+$ ion hydrogen bonded to one, two or three water molecules but with many of the hydrogen bond lengths distinctly short.

TABLE 8.6 Properties of hydrogen bonds

Bond	Compound	Bond length $X-Y$ distance, pm		Van der Waals' distance $X-Y$, pm	Depression of stretching frequency, cm^{-1}†	Bond energy, kJ mol^{-1}
$F-H-F$ (symmetrical)	$K^+HF_2^-$	226		270	2700	113
	$NH_4^+HF_2^-$	232		270		
$F-H\cdots F$ (unsymmetrical)	$(HF)_n$	255		270	700	28·0
$O-H-O$ (symmetrical, bent?)	$Ni(DMG)_2$*	240		280	1200	
$O-H\cdots O$ (unsymmetrical)	KH_2PO_4	248		280	900	
	$(HCOOH)_2$	267		280	600	29·7
	ice	276		280	400	18·8
$N-H\cdots N$	NH_4N_3	294	299	300		
	melamine	300		300	120	ca25
$N-H\cdots F$	NH_4F	263		285		
$N-H\cdots O$ (slightly bent)	$NH_4H_2PO_4$	291		290		
$N-H\cdots Cl$	$N_2H_4.HCl$	313		330	500	

*DMG = dimethylglyoxime
†The stretching frequency figures are rounded off to the nearest hundred wave numbers and represent averages of two frequencies in some cases.

TABLE 8.7 Hydrogen bonds in hydrated protons

Ion	Structure	$O\cdots O$ length (pm)	Compound	Comments
$H_5O_2^+$	$(H_2O-H-OH_2)^+$ probably symmetrical	242·4	$HClO_4.2H_2O$	$HCl.xH_2O$ $x = 2, 3)$ contain similar ions with short hydrogen bonds (242–250 pm)
	$(H_2O-H\cdots OH_2)^+$	257	$HAuCl_4.4H_2O$	Unsymmetric, $O-H = 99$ pm and $H\cdots O = 148$ pm. $H_5O_2^+$ units are linked through H_2O by further hydrogen bonds of 274 pm
$H_7O_3^+$	$(H_2O\cdots H-O-H\cdots OH_2)^+$ $\mid$ H	247 and 250	$HBr.4H_2O$	Central H_3O surrounded pyramidally by two outer H_2O and one Br
$H_9O_4^+$	similar to Figure 8.1	250 and two of 259	$HBr.4H_2O$	Central H_3O surrounded pyramidally by three outer H_2O

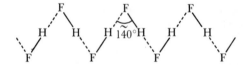

FIGURE 8.20   *The structure of hydrogen fluoride*

Hydrogen fluoride. This has been shown to have the zig-zag structure shown in Figure 8.20. There seems to be no convincing explanation of the wide $F-H-F$ angle. The hydrogen bonding in hydrogen fluoride persists into the gas phase, where small polymers are found.

Ammonium fluoride. NH_4F shows a different structure from the other ammonium halides. The latter have the CsCl or NaCl structure (the transition to NaCl taking place below 200 °C), but the fluoride has the wurtzite structure (Figure 4.4a), in which each N atom forms four $N-H-F$ bonds of length 263 pm to its four neighbouring N atoms which are arranged tetrahedrally around it. The structure resembles that around the O atoms in ice (Figure 8.19).

Apart from F, O, and N, there are few atoms which form hydrogen bonds. Chlorine occasionally enters into hydrogen bonding and the HCl_2^- ion is formed in the presence of large cations such as Cs$^+$ or NR$_4^+$. Indeed, recent work has shown that HBr_2^- and HI_2^- also exist under such conditions. There is no evidence as to their degree of symmetry. There is also some evidence for hydrogen bonding involving carbon, $C-H\cdots X$, if the carbon is bonded to electronegative groups as in HCN. However, all these other hydrogen bonds are much weaker and less well-substantiated than those involving F, O, and N.

The fact that hydrogen bonds occur only between strongly electronegative elements, and that small elements form the strongest bonds, suggests an electrostatic origin for the interaction. The relatively positive hydrogen in $O-H$, $N-H$ or $F-H$ bonds interacts with the dipole in neighbouring atoms to form the bond. This happens only with hydrogen, probably because of its small size and lack of inner shielding electron shells. Because most of the examples of hydrogen bonding are linear $X-H-Y$ arrangements, it follows that the attraction should have directional properties and Y is most strongly attracted when at 180° from X.

9 The 's' Elements

9.1 General and physical properties

Elements with their outermost electrons in an *s* level are those of the lithium Group (alkali metals) and of the beryllium Group (beryllium, magnesium and the alkaline earth metals). Many properties have been listed in the earlier chapters, including atomic weights and numbers (Table 2.5), atomic radii (Table 2.10), ionic radii (Table 2.12), ionization potentials (Table 2.8), electronegativities (Table 2.13), and structural details of halides, etc. (Table 4.1). Structures of *s* element compounds include NaCl (Figure 4.1a), CsCl (Figure 4.1b), and CaC_2 (Figure 4.13).

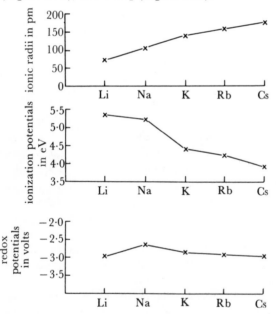

FIGURE 9.1 *The ionic radii, ionization potentials, and redox potentials of the alkali metals*
These three functions all show the relatively sharp changes between Li and Na, and—to a lesser extent—between Na and K, which is reflected in the general chemistry.

The energetics of formation of alkali halides and alkaline earth chalcogenides are discussed in section 4.3.

Other properties of the elements are given in Table 9.1 and important parameters in Figures 9.1 and 9.2.

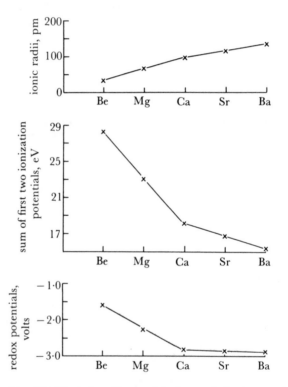

FIGURE 9.2 *The ionic radii, sum of the first two ionization potentials, and* M^{2+}/M *redox potentials of the beryllium Group of elements*
As in Figure 9.1, the major changes in properties come between the first and second members. The similarity between Ca, Sr, Ba is clear.

The elements are all prepared by electrolysis of the fused halides. The relatively volatile rubidium and cesium are also conveniently prepared in the laboratory by heating the chlorides with calcium metal and distilling out the alkali metal. The alkali metals have melting points ranging from 180 °C–29 °C and boiling points in the range 1340 °C–670 °C, in both cases falling with increased atomic weight so that cesium has the second-lowest melting point of any metal. The beryllium Group elements are much less volatile, with melting points ranging from 1300 °C–700 °C and boiling points six or seven hundred degrees higher. Again there is a general tendency for volatility to increase with atomic weight.

TABLE 9.1 The *s* elements

Element	Symbol	Electronic configuration	Abundance (ppm of the crust)	Accessibility*	Common coordination numbers
Lithium	Li	$[He]2s^1$	65	common	4, 6
Sodium	Na	$[Ne]3s^1$	28 300	common	6
Potassium	K	$[Ar]4s^1$	25 900	common	6
Rubidium	Rb	$[Kr]5s^1$	310	rare	6
Cesium	Cs	$[Xe]6s^1$	7	rare	6, 8
Francium	Fr	$[Rn]7s^1$	—	very rare	?
Beryllium	Be	$[He]2s^2$	6	becoming common	2, 4
Magnesium	Mg	$[Ne]3s^2$	20 900	common	6
Calcium	Ca	$[Ar]4s^2$	36 300	common	6
Strontium	Sr	$[Kr]5s^2$	300	common	6
Barium	Ba	$[Xe]6s^2$	250	common	6
Radium	Ra	$[Rn]7s^2$	trace	rare	

*See section 7.7 for a discussion of this point.

Francium and radium both occur only as radioactive isotopes. Francium is inaccessible, with the longest-lived isotope, ^{223}Fr, having a half-life of only 21 minutes. Its chemistry is little known, but, in the few reactions which have been studied, it resembles the other heavy alkali metals. For example, it has an insoluble perchlorate and hexachloroplatinate, like rubidium and cesium. The longest-lived isotope of radium, ^{226}Ra, $t_{\frac{1}{2}} = 1600$ years, is much more stable and radium chemistry is well-established.

The chemistry of the elements of the *s* block is dominated by their tendency to lose the *s* electrons and attain a stable, rare gas configuration. (Beryllium is an exception to this, see section 9.6). This tendency is shown by the low ionization potentials and strongly negative redox potentials. The metals are among the most powerful of chemical reducing agents and combine directly, and usually violently, with most non-metals to yield ionic compounds. The cations formed by these elements (M^+ by the lithium Group, M^{2+} by Mg and the alkaline earths) are very stable and form salts with the most strongly oxidizing or reducing anions. Cesium is particularly useful to stabilize large anions, as in the formation of the bihalide ions, HX_2^-, mentioned in section 8.7.

The reactivity increases down each Group and the lithium Group elements are more reactive than the corresponding members of the beryllium Group, so that metallic cesium is the most reactive and strongly reducing of all. The lightest elements are atypical, as their small size leads to a high charge density on the ions with resulting strong polarizing effects and high heats of solution. Lithium and beryllium show marked differences from their heavier congeners and sodium and magnesium also show distinct, though smaller, differences. There are many similarities between lithium and its diagonal neighbour in the beryllium Group, magnesium. Beryllium differs markedly from all the other elements. Its compounds are covalent and closely resemble those of

aluminium, its diagonal neighbour in the boron Group. It acts as an electron pair-acceptor and shows electron deficiency in its hydrogen and organic compounds. All the elements show only the Group oxidation states.

Salts of the *s* elements tend to be among the most soluble of their kind in water and other ionizing solvents, though solubilities vary widely with the anion. This is especially the case with the compounds of the alkaline earths where the double charge leads to both high lattice energies and high heats of solvation. AB_2 compounds tend to be soluble (e.g. the chlorides), while the insolubility of such AB compounds of the alkaline earths as the sulphates, oxalates, and carbonates, is well known in qualitative and quantitative analysis.

All the *s* elements dissolve in liquid ammonia to give intensely blue, conducting solutions which contain the metal ions and electrons which are free of the metal and appear to be associated with the solvent. (Beryllium and magnesium give only dilute solutions by electrolysis.) These 'solvated electrons' are very reactive and the metal-ammonia solutions are powerful reducing agents which can be used at low temperatures. (Compare section 5.7.) The heavier alkali metals may be recovered unchanged on evaporating off the ammonia, while lithium gives $Li(NH_3)_4$ and the alkaline earths yield hexammoniates, $M(NH_3)_6$. On standing, or in presence of catalysts such as transition metal oxides, the metals react with the ammonia to form the metal amide and hydrogen, e.g.:

$$M + NH_3 = MNH_2 + \tfrac{1}{2}H_2$$

(More accurately) $\quad e^-_{(solvated)} + NH_3 = NH_2^- + \tfrac{1}{2}H_2$

Similar, but more dilute, solutions are formed in amines and methyl ethers.

The reactions of the elements with a number of typical reagents are given in Table 9.2.

TABLE 9.2 Reactions of the s elements

M is used for one mole of an alkali metal or a half-mole of a beryllium Group element.

Reaction	Notes
$2M + X_2 = 2MX$	X = halogen: alkali metals also form polyhalides, e.g. CsI_3, $KICl_4$ or KIF_6, see Chapter 15.
$2M + Y = M_2Y$	Y = O, S, Se, Te: higher oxides are also formed (section 9.2) and also polysulphides M_2S_{2-6}.
$3M + Z = M_3Z$	Z = N, P: Li reacts even at room temperature, Mg to Ra at red heat. The alkali metals also react with As, Sb, and Bi. Metal solutions in ammonia give polyanions such as $(Bi_9)^{3-}$ or $(Sb_5)^{3-}$
$2M + 2C = M_2C_2$	Most rapidly with Li of the alkali metals. Ca, Sr, Ba and Ra at high temperatures. Be forms Be_2C.
$2M + H_2 = 2MH$	At high temperatures. Ionic hydrides. Not with Be, Mg.
$M + H_2O = MOH + \frac{1}{2}H_2$	By Be and Mg only slowly at room temperatures. $Be(OH)_2$ is amphoteric, all others are strong bases. $Mg(OH)_2$ is insoluble, others dissolve readily. All except beryllium hydroxide absorb CO_2 to give M_2CO_3, and alkali metals give $MHCO_3$ also.
$M + NH_3 = MNH_2 + \frac{1}{2}H_2$	With gaseous ammonia at high temperatures or liquid ammonia plus catalyst. Mg and Be only by reaction with amide of a more reactive metal, $Be + 2NaNH_2 \rightarrow Be(NH_2)_2 \rightarrow NaBe(NH_2)_3$.

9.2 Compounds with oxygen and ozone

The reaction between the s elements and oxygen may go further than to the simple oxide, and a number of higher oxides are formed when the metals are burned in air or are oxidized by O_2 in liquid ammonia. Peroxides, O_2^{2-}, are formed by all the elements except beryllium. In addition, sodium, potassium, rubidium, cesium, and calcium superoxides, O_2^-, have been prepared. The normal products of the combustion of the metals in an adequate supply of air are:

oxide	Li, Be, Mg, Ca, Sr
peroxide	Na, Ba (and Ra?)
superoxide	K, Rb, Cs.

The peroxides contain the ion $^-O-O^-$ and are salts of hydrogen peroxide (compare the relation of the hydroxides and oxides to water). The superoxides contain the ion, $O-O^-$. It will be recalled that oxygen, O_2, has its two outermost electrons unpaired in π^* orbitals. The superoxide ion and the peroxide ion have, respectively, one and two electrons more than in O_2. The superoxide ion has thus the configuration $(\pi^*)^3$ and has the paramagnetism corresponding to one unpaired electron and a bond order of one and a half. The peroxide ion has the configuration $(\pi^*)^4$, with no unpaired electrons and a bond order of one. It is isoelectronic with F_2. The MO_2 solids have the tetragonal lattice of calcium carbide. The increasing stability of the peroxides and superoxides with increasing cation size is noteworthy, and provides another example of the stabilization of large anions by large cations through lattice energy effects.

When the metals are treated with ozonized oxygen, or when ozone is passed into their solutions in liquid ammonia, the ozonides are formed. These are yellow or orange and contain the group $(O-O-O)^-$ which is paramagnetic with one unpaired electron.

9.3 Carbon compounds

If acetylene is passed through a solution of an alkali metal in liquid ammonia, or is reacted with the heated metal, the following reactions take place:

$$M + C_2H_2 = MHC_2 + \frac{1}{2}H_2$$
$$MHC_2 + M = M_2C_2 + \frac{1}{2}H_2$$

The carbon compounds M_2C_2 and MHC_2 are termed acetylides and contain the discrete anions, $(C\equiv C)^{2-}$ and $(C\equiv CH)^-$, arising from the displacement of both or one of the relatively acidic hydrogens in the acetylene molecule. Acetylides also result from the direct reaction between carbon and heated lithium, sodium, magnesium, and alkaline earth metals. All these compounds give acetylene on hydrolysis. The structure of calcium acetylide (commonly called calcium carbide) has been determined (Figure 4.13) and is related to that of sodium chloride (Figure 4.1a). These acetylides are the principal carbides (binary compounds of element and carbon) formed by the s elements, but two others of interest exist. Magnesium forms a carbide of formula Mg_2C_3, in addition to the acetylide, and this yields allylene, $HC\equiv C-CH_3$, as the main product on hydrolysis. This suggests the presence of a C_3 unit in this carbide. The product of direct combination between beryllium and carbon is the carbide Be_2C. This carbide probably contains single carbon atoms as the main hydrolysis product is methane. It has the anti-fluorite structure. All these carbides have many of the properties of ionic solids, with colourless crystals which are non-conducting at ordinary temperatures.

A second group of ionic compounds containing carbon is formed by the more reactive s elements. These are the metal alkyls and aryls such as ethyl sodium, $C_2H_5^- Na^+$, or phenyl-potassium, $C_6H_5^- K^+$. Such compounds are extremely reactive solids which inflame in air and react violently with almost all compounds apart from nitrogen and saturated hydrocarbons. They are involatile solids which decompose before melting and the evidence available indicates that they are ionized, R^-M^+. Those with aromatic ions are rather less unstable than those with aliphatic ones. The corresponding lithium and magnesium compounds are covalent and much less reactive. The alkyl-lithiums, for example, are liquids or low-melting solids which are soluble in ethers or hydrocarbons. They are relatively involatile and appear to exist as associated molecules with a highly polar $C-Li$ bond; butyl-lithium, for example, is hexameric in hydrocarbon solvents and dimeric in ether. These organolithium compounds find extensive uses in organometallic syntheses. They resemble the organomagnesium halides RMgX (Grignard reagents), in reactivity and these two types of reagent complement each other usefully. The more reactive ionic alkyls are much less useful as they are so difficult to handle, but they find some application where particularly vigorous conditions are required, for example:

$$PtX_4 \xrightarrow{\ MeLi\ } (PtMe_3X)_4 \xrightarrow{\ MeNa\ } (PtMe_4)_4$$

where it requires methyl sodium to complete the substitution.

9.4 Complexes of the heavier elements

The cations of the heavier s elements are very poor electron pair acceptors as their positive charge density is low. Solvates, such as hydrates or ammoniates, are not found for the heavier alkali metals although sodium gives a moderately stable tetrammoniate in the iodide, $Na(NH_3)_4I$. The alkaline earths are of higher charge density and more strongly hydrated. The best donor atom in complexing agents for these elements appears to be oxygen. Chelating agents with donor oxygen give a few complexes, of which the four-coordinated salicylaldehyde complex of the alkali metals shown in Figure 9.3 is typical. Potassium, rubidium, and cesium also give six-coordinated complexes $M(OC_6H_4CHO)(HOC_6H_4CHO)_2$.

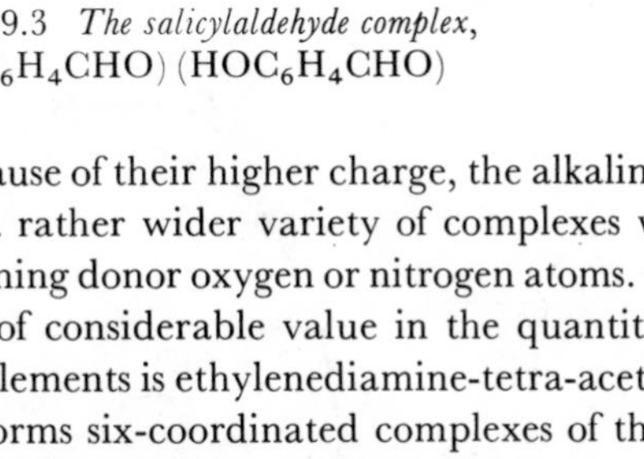

FIGURE 9.3 *The salicylaldehyde complex,* $Na(OC_6H_4CHO)(HOC_6H_4CHO)$

Because of their higher charge, the alkaline earth elements form a rather wider variety of complexes with compounds containing donor oxygen or nitrogen atoms. One complexing agent of considerable value in the quantitative analysis of these elements is ethylenediamine-tetra-acetic acid (EDTA). This forms six-coordinated complexes of the type shown in Figure 9.4, and is especially useful in the determination of calcium and magnesium.

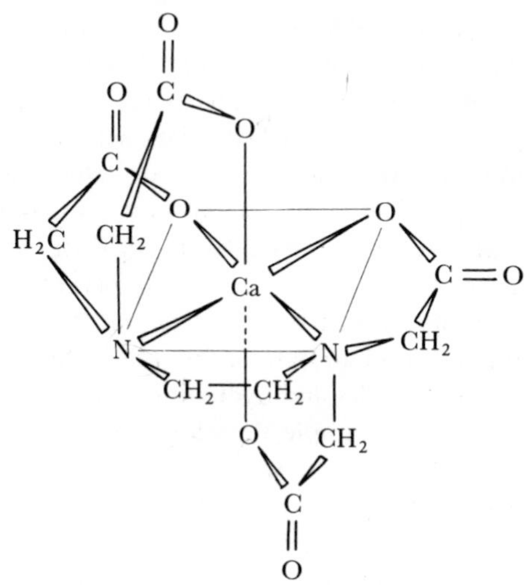

FIGURE 9.4 *EDTA complex of calcium*

Calcium is also complexed by polyphosphates and this reaction is the basis of methods of removing hardness from water.

9.5 Special features in the chemistry of lithium and magnesium

Lithium differs from the other alkali metals in a number of ways. These variations stem from the small size of the lithium cation (radius 68 pm, c.f. $Na^+ = 98$ pm) and its resulting higher polarizing power. This has already been seen in the more covalent nature of its alkyls. Magnesium resembles lithium in much of its chemistry, as the higher charge is offset by the greater size, and these elements are one example of the diagonal similarity which holds in parts of the first two short Periods.

Lithium and magnesium resemble each other in the direct formation of the nitride and carbide (Table 9.2), in their combustion in air to the normal oxide, and in the properties of their organic compounds, as already remarked in the previous sections. They are also similar to each other, and different from the heavier elements, in the thermal stability of their oxysalts and in the mode of decomposition of these. For example, lithium and magnesium nitrates decompose on heating to give the oxide and dinitrogen tetroxide, e.g.:

$$2LiNO_3 = Li_2O + N_2O_4 + \tfrac{1}{2}O_2$$

while the other alkali metal nitrates give the nitrite on heating ($NaNO_2$ decomposing further to a mixture of oxides):

$$MNO_3 = MNO_2 + \tfrac{1}{2}O_2$$

and the alkaline earth nitrates give nitrite and oxide mixtures.

In the case of the carbonates, lithium and magnesium (and also calcium and strontium) give the oxide on heating:

$$MgCO_3 = MgO + CO_2$$

while the carbonates of the other alkali metals and of barium and radium are stable to heat.

Lithium and magnesium are more strongly hydrated than their heavier congeners and their salts are similar in solubility. Most salts are soluble but the carbonate and phosphate are insoluble in each case. Magnesium fluoride and magnesium hydroxide are rather insoluble while lithium fluoride and hydroxide are both markedly less soluble than the sodium salts. Lithium and magnesium halides are also similar in being soluble in organic solvents such as alcohol. With larger anions, solubilities tend to fall with cation size in both Groups. Thus the alkali metal perchlorates are relatively insoluble for K, Rb, Cs and Fr and the solubilities of carbonates and sulphates are in the order Mg > Ca > Sr > Ba > Ra. The fluorides and hydroxides present an exception to this order—for example, the solubilities of the fluorides are in the order Mg < Ca < Sr < Ba—and this must result from the effect of the relatively small OH^- and F^- anions on the lattice energies.

Lithium and magnesium are also more strongly complexed by nitrogen donors, in ammonia and amines, than are the heavier elements. The ammoniated salts $Li(NH_3)_4I$ and $Mg(NH_3)_6Cl_2$, for example, can be precipitated from liquid ammonia solution by double decomposition reactions.

The high hydration energy of lithium more than compensates for its relatively high ionization potential in the case of reduction reactions which are carried out in water. The result is that the redox potential in water of lithium is as strongly reducing as that of cesium, although the latter is much more reactive under anhydrous conditions.

Magnesium has several important biological functions, particularly in catalyzing (sometimes in conjunction with calcium) the phosphate polymerization step in the chain by which energy obtained from oxidation of organic compounds is stored and utilized. Magnesium is also found in chlorophylls, the pigments which are responsible for nearly all the conversion of CO_2 and H_2O into organic molecules, using the energy of sunlight. The central coordination of magnesium in chlorophyll is to four nitrogen atoms in a macrocyclic ring, as shown in Figure 9.5. This environment is very

similar to that found for iron in haemoglobin (Figure 13.19) or for cobalt in vitamin B_{12}.

9.6 Beryllium chemistry

The size effects which produce the differences between the chemistry of lithium and the other alkali metals, have a much more marked effect on the chemistry of beryllium as compared with its heavier congeners. On passing from lithium to beryllium, the size of the atom grows less and the charge on the possible ion doubles. The result is to give such a high charge density on the hypothetical Be^{2+} ion that it is too polarizing to exist and all beryllium compounds are either covalent or contain solvated beryllium ions, such as $Be(H_2O)_4^{2+}$. Even the anhydrous halides are only feebly ionic; beryllium fluoride is one of the few metal fluorides which is not completely ionized in solution, while the conductivity of fused beryllium chloride is only one thousandth of that of sodium chloride under the same conditions.

The small size and strong hydration effects have a marked influence on the solubility of beryllium compounds as compared with those of the other elements of the Group. In water, the size effects increase the solvation energies more than the lattice energies, and the solubilities of compounds such as the sulphate, selenate, or oxalate are markedly greater than those of the corresponding calcium compounds. An extreme case is provided by beryllium fluoride which is a thousand times more soluble than magnesium fluoride.

In water, the hydrated beryllium ion differs from the other s element ions in being hydrolysed. $Be(H_2O)_4^{2+}$ exists in strongly acid solutions but in neutral or weakly acid solutions this is hydrolysed and polymerized, for example:

$$2Be(H_2O)_4^{2+} = (H_2O)_3Be-O-Be(H_2O)_3^{2+} + 2H_3O^+$$

and more complex species are formed. One stable form in dilute solutions is $[Be(OH)_3]_3^{3-}$ which is thought to have the ring structure of Figure 9.6. Beryllium hydroxide is eventually precipitated on going to more alkaline conditions, and this differs from the other s element hydroxides in being amphoteric and dissolving in excess base, probably with the formation of the species $Be(OH)_4^{2-}$.

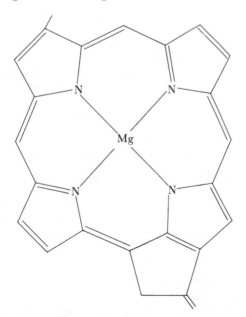

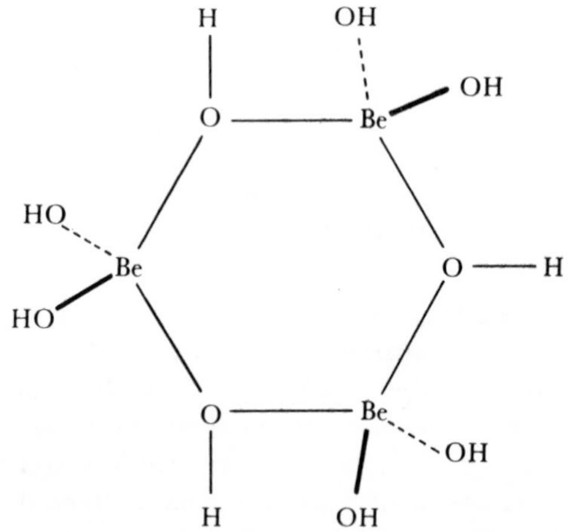

FIGURE 9.5 *The structure around magnesium in chlorophyll a*

FIGURE 9.6 *Structure of the* $[Be(OH)_3]_3^{3-}$ *ion*

FIGURE 9.7 *Beryllium dichloride forms:* (a) *the etherate* BeCl$_2$.2Et$_2$O, (b) BeCl$_2$ *polymer in the solid*

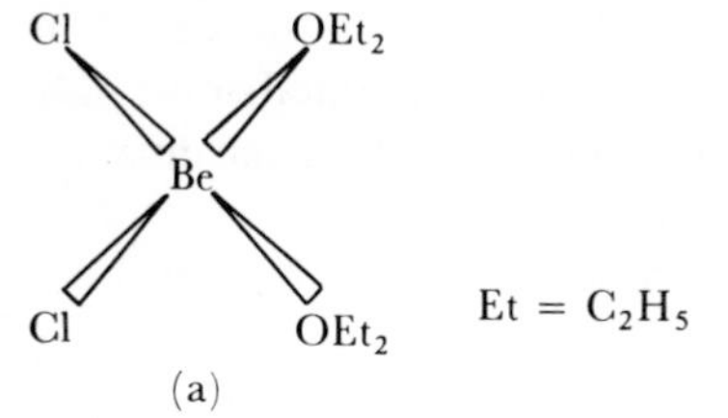

Et = C$_2$H$_5$

(a)

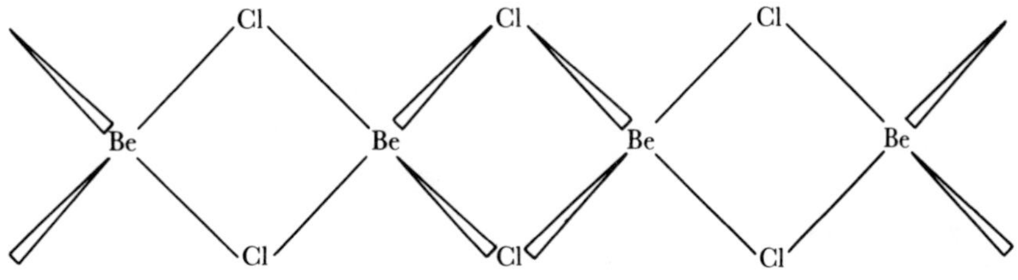

(b)

As would be expected from the above, beryllium forms more, and more stable, complexes than the other *s* elements. There is a strong tendency to assume a coordination number of four—in which use is made of all the valency orbitals. Thus the halides are monomeric and linear in the gas phase (section 3.5), but they dissolve in ether and a dietherate, BeX$_2$.2Et$_2$O, is recovered. This contains beryllium in an *sp*^{3} configuration with the ether oxygen atoms acting as lone pair donors (Figure 9.7a). A large number of similar four-coordinated complexes are formed with ligands such as ethers, aldehydes, and ketones. In the liquid and solid states, the beryllium halides polymerize in order to achieve four-coordination. In the solid, the unshared pair on a halogen atom of one molecule donates to the beryllium atom of the next and a long chain structure containing tetrahedral beryllium results (Figure 9.7b). Solid beryllium fluoride is a glassy material which also appears to contain chains but these are disordered and the material does not crystallize. The fluoride reacts with fluoride ion to form the tetrafluoro-beryllate ion, BeF$_4^{2-}$, which is tetrahedral and usually iso-structural with sulphate.

A more complex example, which illustrates the tendency of beryllium compounds to hydrolyse, and also to become four-coordinated, is provided by the compound called 'basic beryllium acetate'. This is a very stable compound formed by the partial hydrolysis of beryllium acetate:

$$4Be(OOCCH_3)_2 + H_2O$$
$$= Be_4O(OOCCH_3)_6 + 2CH_3COOH$$

It is a volatile solid which may be purified by sublimation. It is soluble in organic solvents such as chloroform, insoluble in water, and stable to heat and moderate oxidation. The structure is shown in Figure 9.8. The four beryllium atoms are placed tetrahedrally around the central oxygen and they are linked in pairs by the six acetate groups, each of which spans one edge of the tetrahedron. Similar compounds of other carboxylic acids have been prepared, and the existence of basic beryllium nitrate, Be$_4$O(NO$_3$)$_6$, has been reported

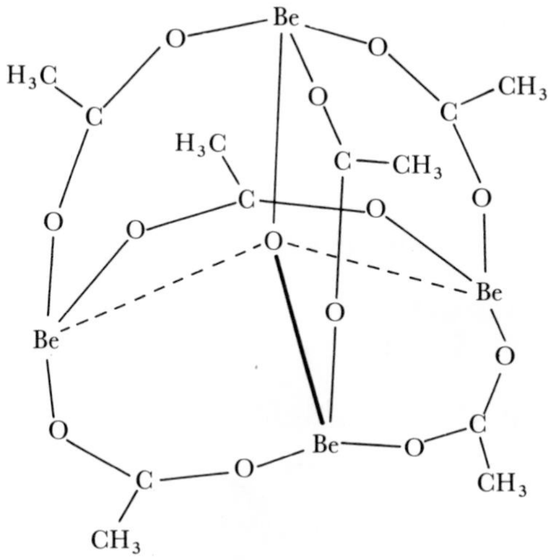

FIGURE 9.8 *Basic beryllium acetate,* Be$_4$O(OOCCH$_3$)$_6$

recently. In this molecule, the nitrate groups are linked to two beryllium atoms, Be$-$O$-$N(O)$-$O$-$Be, in the same way as the acetate groups in the basic acetate.

Complexes also exist in which nitrogen is the donor atom. For example, beryllium chloride readily takes up ammonia to give the ammine, Be(NH$_3$)$_4$Cl$_2$. This compound is very stable to thermal decomposition but the ammonia groups are readily displaced by water.

Beryllium also achieves four-coordination in the hydride and in the beryllium alkyls by polymerization through electron-deficient bridges. Thus beryllium dimethyl, Be(CH$_3$)$_2$, has a chain structure very similar to that of beryllium chloride, and the bridge bonding is the same as that in the aluminium trimethyl dimer of Figure 8.12.

Just as there is some resemblance in the properties of lithium and magnesium, so beryllium and its diagonal neigh-bour, aluminium, have similar properties. In this case, the resemblance in general chemistry is very close and beryllium is traditionally difficult to distinguish from aluminium in qualitative analysis. The basic resemblance is due to the high charge density on each element in the hypothetical cations, and to the existence of an empty p orbital or orbitals giving the elements acceptor properties. Among the detailed resemblances there may be noted:

(i) Similar redox potentials, $Be^{2+}/Be = -1.70$ V, $Al^{3+}/Al = -1.67$ V.

(ii) Both metals dissolve in alkali with the evolution of hydrogen.

(iii) The hydroxides are amphoteric, and the salts are readily hydrolysed.

(iv) The halides form polymeric solids, beryllium halides forming chains and aluminium halides, dimers. The halides are also similar in their solubilities in organic solvents, their electron pair acceptor properties (e.g. $AlCl_3.OEt_2$), and their catalytic effects.

(v) Both metals form carbides by direct combination and these yield mainly methane on hydrolysis.

(vi) Both elements form hydrides and alkyls which polymerize through electron-deficient bridges.

10 The Scandium Group and the Lanthanides

10.1 General and physical properties

The scandium Group of elements has the outer electronic configuration d^1s^2 and is formally part of the d block of the Periodic Table. However, as the chemistry is entirely that of the III oxidation state, which involves the loss of the d electron, this Group is best regarded as forming a transitional region between the s elements and the main d block. Following lanthanum, the third member of the scandium Group, the $4f$ shell fills for the next fourteen elements. As the general chemical behaviour in these *lanthanide elements* (or rare earths) is that of the III state and is very similar to that of the scandium Group proper, it is convenient to include these elements here. On the other hand, when the $5f$ shell fills following actinium, which is the fourth member of the scandium Group, the very small energy gap between the $5f$ level and the $6d$ level gives rise to a considerable variability in the chemistry of the actinide elements, and these are best treated separately. Properties of these elements are given in Table 10.1.

Of these elements, scandium and actinium are very rare and are not completely investigated. The lanthanide elements and lanthanum and yttrium all occur together and separation on a moderate scale has only been carried out recently. There has been recent interest in the lanthanide elements for use in lasers. One particularly important development has been the use of neodymium oxide dissolved in selenium oxychloride as a liquid laser. A liquid system has many technical advantages, but this was only the second

TABLE 10.1 Properties of the scandium elements and the lanthanides

Element	Symbol	Electronic configuration	Abundance ppm of crust	Radius of M^{3+} ion, pm	Oxidation states*
Scandium	Sc	$[Ar]3d^14s^2$	5	ca 70	III
Yttrium	Y	$[Kr]4d^15s^2$	28	90	III
Lanthanum	La	$[Xe]5d^16s^2$	18	106·1	III
Actinium	Ac	$[Rn]6d^17s^2$	trace	111	III
Cerium	Ce	$[Xe]4f^25d^06s^2$	46	103·4	III, IV
Praseodymium	Pr	$[Xe]4f^35d^06s^2$	5	101·3	III, (IV)
Neodymium	Nd	$[Xe]4f^45d^06s^2$	24	99·5	III, (IV?) (II?)
Promethium	Pm	$[Xe]4f^55d^06s^2$	unstable	97·9	III
Samarium	Sm	$[Xe]4f^65d^06s^2$	6	96·4	III, (II)
Europium	Eu	$[Xe]4f^75d^06s^2$	1	95·0	III, II
Gadolinium	Gd	$[Xe]4f^75d^16s^2$	6	93·8	III, (II?)
Terbium	Tb	$[Xe]4f^95d^06s^2$	1	92·3	III, (IV)
Dysprosium	Dy	$[Xe]4f^{10}5d^06s^2$	4	90·8	III, (IV?)
Holmium	Ho	$[Xe]4f^{11}5d^06s^2$	1	89·4	III
Erbium	Er	$[Xe]4f^{12}5d^06s^2$	2	88·1	III
Thulium	Tm	$[Xe]4f^{13}5d^06s^2$	0·2	86·9	III, (II?)
Ytterbium	Yb	$[Xe]4f^{14}5d^06s^2$	3	85·8	III, (II)
Lutetium	Lu	$[Xe]4f^{14}5d^16s^2$	0·8	84·8	III

*Bracketed states are unstable: states marked (?) are either unconfirmed or very unstable.

liquid system to be announced and it was by far the most powerful. The key aspect seems to be the use of a solvent which does not contain light atoms, such as hydrogen, so that most of the input energy is emitted in the laser beam and not transferred to heat the solution. Though the elements are becoming much more accessible, there are still gaps in their known chemistry, especially in the case of the later lanthanides. Properties listed earlier include atomic weights and numbers (Table 2.5), ionization potentials (Table 2.8), and electronegativities (Table 2.13).

The elements are electropositive and reactive, with the heavier elements resembling calcium in reactivity while scandium is similar to aluminium. Two of the elements, promethium and actinium, occur only as radioactive isotopes. Actinium is found associated with uranium and the most readily available isotope, ^{227}Ac, has a half-life of 22 years. It is, however, very difficult to handle as the decay products are intensely active and build up in the samples. Its chemistry fits in as that of the heaviest element of the scandium Group. The missing lanthanide, promethium, occurs only in radioactive forms with the longest-lived isotope, ^{145}Pm having $t_{\frac{1}{2}} = 30$ years. Its chemistry fits in with its place in the series.

Reactivity increases with increasing atomic weight in the scandium Group, just as in the s Groups. The elements are prepared by the reduction of the chlorides or fluorides with calcium metal. Some reactions of the metals are shown in Figure 10.1. These direct reactions with elements broadly

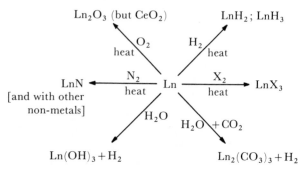

Ln = lanthanide element
X = halogen

FIGURE 10.1 *Reactions of the lanthanide elements*

parallel those of the s elements given in Table 9.2. The hydrides, formed by direct combination, illustrate the transitional character of this Group. They form stable MH_2 and MH_3 phases, which usually occur in non-stoichiometric form, and the MH_3 formula for the most highly hydrogenated species is never fully attained. The hydrides have some salt-like properties and appear to contain the H^- ion. There are also available extra delocalized electrons giving metallic properties, so that the overall properties of the hydrides are a mixture of the ionic character shown by the s element hydrides and the metallic character of the 'interstitial' hydrides of the d elements. The hydrides also resemble the ionic hydrides in their reactivity to oxygen and water.

10.2 Chemistry of the trivalent state

As Table 10.1 shows, the oxidation state of $+III$ is shown by all these elements and is the most stable state. Other states are found only where f electrons are present. The oxides, M_2O_3, and hydroxides, $M(OH)_3$, increase in basicity with increasing atomic weight. Scandium, because of its smaller size, is more easily hydrolysed in solution than the other ions, and its oxide is amphoteric. There does not appear to be a definite scandium hydroxide, although the species $ScO(OH)$ is well established [compare the existence of $AlO(OH)$]. The oxide, with water, forms the hydrous oxide, $Sc_2O_3.nH_2O$, which dissolves in excess alkali to form anionic species such as $Sc(OH)_6^{3-}$. The other elements form oxides and hydroxides which are basic only. These hydroxides are precipitated from solution by the addition of dilute alkalis and do not dissolve in excess alkali. Yttrium oxide and hydroxide are strong enough bases to absorb carbon dioxide from the atmosphere, while lanthanum oxide 'slakes', with evolution of much heat like calcium oxide, and rapidly absorbs water and carbon dioxide. The lanthanide element basicities decrease towards lutetium, which is similar to yttrium in the properties of its oxide. Actinium compounds are more basic than the lanthanum ones.

Among the trihalides, the fluorides are insoluble and their precipitation, even from strongly acidic solution, is a characteristic test for these elements. Scandium fluoride dissolves in excess fluoride with the formation of the complex anion, ScF_6^{3-}. The fluorides of the heavier lanthanide elements are also slightly soluble in hydrogen fluoride, probably because of complex formation. The other trihalides are very deliquescent and soluble (compare $CaCl_2$). $ScCl_3$ is much more volatile than the other trichlorides. It resembles $AlCl_3$ in this, but it is monomeric in the vapour (aluminium trichloride is dimeric) and it has no activity as a Friedel-Craft catalyst. The chlorides are recovered from solution as the hydrated salts and these, on heating, give the oxychlorides, $MOCl$ (with the exception of scandium which goes to the oxide). Actinium also forms oxyhalides but only by reaction with steam at 1000 °C, a treatment which produces oxide from the lower members of the Group. This is a further example of the increase in basicity from scandium to actinium. Bromides and iodides resemble the chlorides in general behaviour.

Among the oxysalts, most anions are to be found including strongly oxidizing ones. The carbonates, sulphates, nitrates, and perchlorates, for example, all resemble the calcium compounds. The carbonates, phosphates, and oxalates are insoluble while most of the others are rather more soluble than the calcium salts. Scandium carbonate differs from the others in dissolving in hot ammonium carbonate, with double salt formation, and this affords a method of separating scandium from yttrium and lanthanum. Double salts are very common and include double nitrates, $M(NO_3)_3.2NH_4NO_3.4H_2O$, and sulphates such as $M_2(SO_4)_3.3Na_2SO_4.12H_2O$. Such salts were used for separation of the lanthanides by fractional crystallization methods.

These salts are fully ionic and lanthanum is useful as

one of the few available stable ions with a charge higher than $2+$. The scandium ion is more readily hydrolysed than the others, and polymeric species of the type $[Sc-(OH)_2-Sc-(OH)_2-]_n$ have been identified with the chain length increasing as the acidity falls. The other ions are only slightly hydrolysed in the sense:

$$M(H_2O)_6^{3+} + H_2O \rightleftharpoons M(H_2O)_5(OH)^{2+} + H_3O^+$$

with the tendency to hydrolyse increasing as the size decreases.

Although these ions have a charge of $+3$, the tendency to complex-formation is relatively slight. When compared with transition metal ions, such as Fe^{3+} or Cr^{3+}, which readily form complexes, this reluctance to complex may be ascribed to the greater size of the scandium Group ions, and to their low electronegativity which decreases any possible covalent contribution to the bonding. The best donor atom is oxygen, and insoluble complexes are formed by β-diketones such as acetylacetone (Figure 10.2). Of some importance are the

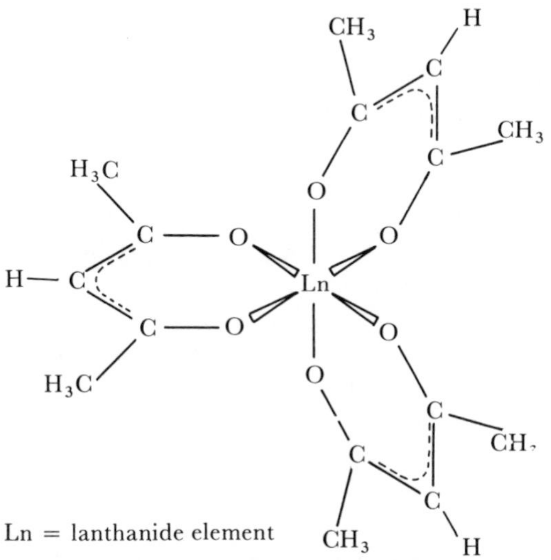

Ln = lanthanide element

FIGURE 10.2 *Acetylacetone complex of the lanthanide elements*

water-soluble complexes formed by chelate ligands such as EDTA, and especially the complexes formed by hydroxy-carboxylic acids such as citric acid, $HOOC.CH_2.C(OH)(COOH).CH_2COOH$, which are used in the separation of these elements by ion exchange methods (see section 10.3). The lanthanide elements are often six-coordinated in complexes, but higher coordinations are known. Eight-coordination is found in $La(acac)_3(H_2O)_2$, where acac = acetylacetone and in $Y(CF_3COCHCOCF_3)_4^-$. Shapes include antiprismatic (compare Figure 14.4b) and dodecahedral (compare Figure 12.15). Ten-coordination is found in $La(H_2O)_4EDTA.3H_2O$. The EDTA occupies six positions and there are four water molecules attached to the lanthanum. Three water molecules lie on one side of the lanthanum, the two nitrogen atoms of the EDTA lie opposite them, while the four EDTA oxygens and the final water molecule lie in a rather distorted medial plane. The complex nitrate, $Ce(NO_3)_5^{2-}$, contains ten-coordinate Ce. The nitrates are present at the apices of a trigonal bipyramid and

each nitrate is coordinated by two oxygens—

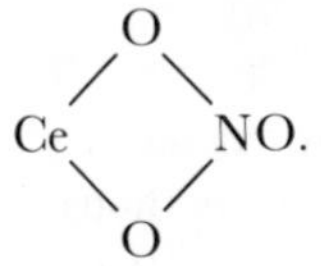

In $Ce(NO_3)_6^{3-}$, the coordination number is twelve. $Nd(BrO_3)_3.9H_2O$ contains the nine-coordinate $Nd(H_2O)_9^{3+}$ ion and similar ions exist for other lanthanides. Scandium, because of its small size, forms more stable complexes than the other elements. For example, the scandium acetylacetonate may be sublimed at about 200 °C, while all the others decompose on heating. Continuing this trend to the heaviest element, actinium is less ready to form complexes than the others. Thus, the lanthanides can be extracted into an organic solvent by means of tributylphosphate, $OP(OC_4H_9)_3$, which forms a complex, but actinium extracts much less readily under these conditions.

10.3 The separation of the elements

One separation of scandium is mentioned above, and actinium occurs separately, but yttrium, lanthanum, and the lanthanides are commonly found together in minerals. From the radii given in Table 10.1, it will be seen that the lanthanides are very close in size, with a small but regular decrease from lanthanum to lutetium, and that yttrium is close to dysprosium and holmium. A similar gradation is found in the redox potentials of the elements, M_{aq}^{3+}/M, which range from -2.52 V for La to -2.25 V for Lu, in steps of about $0.01–0.02$ V between each element and the next. Again yttrium, -2.37 V, fits in near dysprosium, -2.35 V. These resemblances in size and behaviour are much closer than those between elements in the same Group (compare Sr^{2+} and Ba^{2+} which differ by 19 pm) and mean that the chemistry of all the lanthanide elements is practically identical.

The slow decrease in size from La to Lu just about balances the normal increase in size between the elements in one period and the next. This is shown by the similarity between yttrium and the heavier lanthanides. The decrease is termed the *lanthanide contraction* and arises from the slow increase in effective nuclear charge as the f electrons are added. This accounts for the decrease in size and the increase in oxidation potential. As will be seen, the lanthanide contraction also affects the chemistry of the heavier transition metals.

As the elements and ions are so similar in size and properties, the separation of the individual lanthanide elements is extremely difficult. In the classical studies of the elements, fractional separations had to be adopted. These included fractional crystallization of double salts, such as the nitrates, fractional precipitation of the hydroxides and fractional decomposition of the oxalates. These processes were very slow, and as many as twenty thousand operations are reported in some cases before pure samples were obtained. The separation of the lighter elements, up to gadolinium, was relatively easy as cerium could be removed by oxidation to the relatively stable IV state, promethium was missing from the natural sources, and samarium and europium could be reduced to the II state. The heavier elements and yttrium were much more difficult to separate as such chemical aids were not available. Despite this, all the lanthanide elements had been separated and correctly characterized, before the advent of more powerful methods, in what was one of the most painstaking series of studies in chemistry (most of the work was completed before there was any theoretical guidance as to the total number of elements to be expected).

The separation problem is greatly simplified by the use of ion exchange or solvent extraction techniques (see Chapter 6). In the ion exchange method, a common technique uses the soluble citrate complexes. A cation exchange resin, which will be written resin-H, is used, a solution of the lanthanides is applied, and the acid formed washed out:

$$3Resin\text{-}H + Ln^{3+} \rightarrow Resin_3\text{-}Ln + 3H_3O^+$$

Then citric acid, buffered with ammonia to constant pH, is added, and the equilibrium:

$$Ln\text{-}Resin_3 + 3HCit \rightleftharpoons 3H\text{-}resin + LnCit_3$$

is set up (Ln is used as a general symbol for any lanthanide element). As the buffered citrate flows down the column, the concentration of lanthanide ions changes and the equilibrium reverses many times. As the heavier ions are smaller, they will be more strongly complexed by the citrate and so will tend to spend more time in solution and less on the resin. As a result, the heavier lanthanide elements are washed down the column first and will eventually be eluted. If the conditions are correct, the different elements will be separated into pure components. Figure 10.3 gives an example of such

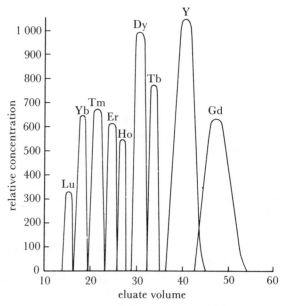

FIGURE 10.3 *Elution curve of the lanthanide elements from an ion exchange resin column*

an elution curve. The whole process is analogous to the classical fractionations but with the numerous operations taking place *in situ* on the column. The process leads to considerable dilution: in one example 0·4 g of mixed oxides per litre was used and collection of about fifty litres of eluate gave each element in about 80 per cent purity. Each fraction

would then be concentrated by precipitation of the oxalate and the exchange repeated to give pure samples.

A similar separation is possible by using solvent extraction methods in a counter-current technique. By extraction of lanthanides from a strong nitric acid solution **into tributyl-phosphate**, 95 per cent pure gadolinium has been prepared on a kilogramme scale. These methods were first developed in the atomic energy project during the Second World War as a means of separating and identifying rare earth isotopes produced during uranium fission. Similar methods have been most important in the characterization of the man-made post-plutonium elements.

10.4 Oxidation states other than III

A number of the lanthanide elements exist in oxidation states other than III and the most stable of these are Ce(IV) and Eu(II). The cerium(IV) state corresponds to the loss of the four outer electrons to give an $f^0 d^0 s^0$ rare gas configuration, while the europium(II) state corresponds to the loss of only the two *s* electrons to retain the half-filled f^7 shell.

Cerium(IV)

Cerium is the only lanthanide which exists in the IV state in solution. Ceric oxide, CeO_2, which is colourless when pure, is the product resulting from heating the metal, or decomposable cerium(III) oxysalts such as the oxalate, in air or oxygen. It is inert and insoluble in strong acids or alkalies. It does dissolve in acids in the presence of reducing agents to give cerium(III) solutions, and these, in turn, give cerium(IV) in solution on treatment with strong oxidizing agents such as persulphate. A yellow, hydrated, form of ceric oxide, $CeO_2.nH_2O$, is precipitated from such cerium(IV) solutions by the action of bases. The only other solid cerium(IV) compound known is the tetrafluoride, prepared by the action of fluorine on the trichloride or trifluoride. The aqueous chemistry of cerium(IV) resembles that of zirconium and hafnium, or of the four-valent actinides such as thorium.

The very high charge on the Ce^{4+} ion leads to its being strongly hydrated. The hydrated ion is acidic and hydrolyses to give polymeric species and hydrogen ions, except in strongly acidic solution. The solution of cerium(IV) in acid is widely used as an oxidizing agent, and its redox potential depends quite strongly on the acid used, ranging from 1·44 V in molar sulphuric acid to 1·70 V in perchloric acid. This variation probably arises from the formation of complex ions by association with the acid anion in nitric or sulphuric acids. As perchlorate shows no tendency to complex in this way, the potential in perchloric acid probably characterizes the plain hydrated ion, $Ce^{4+}.nH_2O$.

The high charge density means that cerium(IV) forms stronger complexes than the tripositive lanthanides. It is much more readily extracted by tributylphosphate, and the hexachloro-complex $CeCl_6^{2-}$ has been prepared as the pyridinium salt.

Europium(II)

Europium has the most stable divalent state of all the lanthanides. Europium(II) chloride is prepared as a solid by the action of hydrogen on the trichloride:

$$EuCl_3 + H_2 = EuCl_2 + HCl$$

Its dihydrate is very insoluble in concentrated hydrochloric acid (like $BaCl_2 \cdot 2H_2O$) and this is used as a means of purification. Europium in solution is readily reduced to the II state, for example by magnesium, zinc, or alkali metal amalgams, and it resembles calcium in this state. Thus the sulphate or carbonate may be precipitated from solution. The oxide does not exist, but EuS, EuSe or EuTe can all be prepared. EuH_2 is ionic and isomorphous with CaH_2.

The redox potential, $Eu^{3+} + e^- = Eu^{2+}$, is -0.43 V, so that europium(II) is a reducing agent of similar power to Cr^{2+}. A careful magnetic investigation has shown that the magnetic properties of Eu(II) are identical with those of Gd(III), over a wide range of temperature, confirming the $f^7 d^0 s^0$ arrangement in the ion. The dichloride, dibromide, and diiodide all have moments of 7·9 Bohr magnetons corresponding to the seven unpaired electrons.

Europium(II), although the most stable of the lower oxidation states, is a strong reducing agent as the potential shows, and its solutions are readily oxidized in air. The solids are rather more stable. Europium, together with ytterbium which also has a II state, dissolves in liquid ammonia to give a concentrated blue solution. The other lanthanides are either insoluble or give only weak solutions on electrolysis.

Other IV states

Other elements which form the IV state are praseodymium, neodymium, terbium, and dysprosium. Of these, only Tb(IV) can be accounted for by the tendency to equally-occupied f orbitals, in this case f^7. All the states are very unstable and have only been prepared as solid compounds.

Ignition of praseodymium compounds in air gives a complex oxidation product of approximate composition Pr_6O_{11}. Heating finely divided Pr_2O_3 in oxygen at 500 °C and 100 atmospheres gives the stoichiometric oxide, PrO_2. No binary fluoride, PrF_4, has been prepared but solid solutions of PrF_3 in CeF_3, containing less than 90 per cent PrF_3, do react completely with fluorine to the composition PrF_4/CeF_4.

There is no firm evidence for the existence of Nd(IV) in oxide systems, but the fluorination of NdF_3 in presence of CsF gives compounds containing 10–20 per cent Nd(IV) in the form of a double salt.

A higher oxide of terbium, of approximate composition Tb_4O_7, has long been known as a product of ignition. A careful study has yielded three oxide phases, in the range $TbO_{1.5 \text{ to } 1.8}$, in ignition products of oxalate or nitrate. The TbO_2 composition results from the reaction of atomic oxygen on Tb_2O_3. This, like PrO_2, has the fluorite structure. Terbium(IV) is also formed as fluoride by fluorination of the trifluoride. TbF_4 is isostructural with CeF_4. Terbium(IV) is probably the most stable of the IV states after cerium(IV), but it is an extremely powerful oxidizing agent and there is no question of its existence in an aqueous medium.

Dysprosium(IV) resembles neodymium(IV) in being found only in a fluorine system. Flourination of DyF_3, in presence of CsF, gives materials containing up to 50 per cent Dy(IV).

Other II states

Elements found in the II state are neodymium, samarium, gadolinium, thulium, and ytterbium. Ytterbium(II) corresponds to the completed f^{14} level. All these elements are much less stable in the II state than is europium. Yb(II) and Sm(II) may be prepared in water but are oxidized by water on standing; the others are found only in the solid state. The order of stability is Nb(II) $\approx$ Gd(II) < Tm(II) < Sm(II) < Yb(II).

Divalent neodymium and gadolinium are prepared, as the dichloride or di-iodide, by the reaction of the metals with the fused trihalides. $NdCl_2$ is isostructural with $EuCl_2$. Thullium di-iodide may be prepared in a similar manner by the action of Tm on TmI_3 at 600 °C. It is isostructural with YbI_2. These low valency halides tend to be non-stoichiometric and they have metallic conduction and other properties.

Samarium(II) occurs in a number of compounds, including the halides, sulphate, carbonate, phosphate, and hydroxide. It may be extracted from a lanthanide mixture, along with Eu(II), by reduction of the trichlorides with alcoholic sodium amalgam. The mixture of Eu(II) and Sm(II) chlorides is readily separated by controlled oxidation, which produces Sm(III) only.

A variety of ytterbium(II) compounds exist, including all those found for samarium, and also possibly the monoxide. The dihalides may be prepared by hydrogen reduction of the trihalides and, in the case of the di-iodide, by thermal decomposition. Yb(II) is more stable than Sm(II) and has been estimated to have an oxidation potential of -1.15 V with respect to the III state. YbH_2, like EuH_2, is ionic and isomorphous with CaH_2.

It will be noted that, although there are a fair number of examples of oxidation states other than the III state among the lanthanide elements, the III state is by far the most stable. Even the most stable of the other states, Ce(IV) and Eu(II), are very reactive and the majority are found only as a result of solid state reactions. This dominance of the III state in the lanthanides presents a marked contrast with the behaviour of the actinide elements and is probably a result of the larger energy gap between the $4f$ and $5d$ levels than that existing between the $5f$ and $6d$ levels.

10.5 Properties associated with the presence of f electrons

As is implied in the discussion in the earlier sections, the electrons in the $4f$ level are too strongly bound to be involved in the chemistry of the elements except under unusual conditions. In addition, the f orbitals appear to be too diffuse to enter into bonding generally, so that there are few chemical effects from the presence of f electrons or unfilled f orbitals. There are, however, electronic effects which show up in the spectra and magnetic properties of the lanthanides.

The lanthanide ions show absorptions in the visible or near ultraviolet regions of the spectrum, except La^{3+} with no

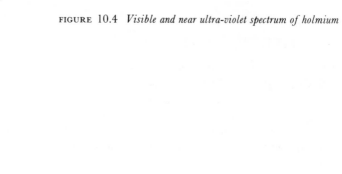

FIGURE 10.4 *Visible and near ultra-violet spectrum of holmium*

f electrons and Lu^{3+} with no empty f orbitals. These colours are due to transitions between f levels, $f-f$ transitions, and, as the f levels lie deep enough in the atom to be shielded from much perturbation by the environment, these transitions appear in the visible and near ultraviolet spectra as sharp bands. This is in contrast to the $d-d$ transitions found for the transition elements, which usually appear as broad bands due to environmental effects. Figure 10.4 illustrates a typical lanthanide ion spectrum. As these bands are so sharp, they are very useful for characterizing the lanthanides and for quantitative estimations. The positions of the absorptions shift with the f configuration, giving rise to the visible colours of the different ions as shown in Table 10.2.

TABLE 10.2 Typical colours of lanthanide compounds

f^1 or f^{13}	Ce(III), Yb(III)	uv absorption
f^2 or f^{12}	Pr(III), Tm(III)	green
f^3	Nd(III)	blue-violet
f^4 or f^{10}	Pm(III), Ho(III)	pink or yellow
f^5 or f^9	Sm(III), Dy(III)	cream
f^6, f^7 or f^8	Eu(III), Eu(II), Gd(III), Tb(III)	uv absorption
f^{11}	Er(III)	pink

Solutions of samarium(II) are red, and of ytterbium(II) are green. Notice that there is a general trend for f^n and f^{14-n} states to have the same or similar colours. Note also that the much more intense colours of cerium(IV) are not due to $f-f$ transitions, but to a different mechanism involving charge-transfer between ion and coordinated ligand.

All the f states, except f^0 and f^{14}, contain unpaired electrons and are therefore paramagnetic. These elements differ from the transition elements in that their magnetic moments do not obey the simple 'spin-only' formula (section 6.6). In the f elements, the magnetic effect arising from the motion of the electron in its orbital contributes to the paramagnetism, as well as that arising from the electron's spinning on its axis. (In the transition metals the orbital

contribution is usually quenched out by interaction with electric fields of the environment—at least to a first approximation—but the f levels lie too deep in the atom for such quenching to occur.) When the moments are calculated on the basis of spin and orbital contributions, there is excellent agreement between experimental and calculated values, as Figure 10.5 shows.

The one case in which contributions to the bonding from the f orbitals is possible is in complexes of the heavier elements in which the coordination number is high. Use of the s orbital together with all the p and d orbitals of one valency shell permits a maximum coordination number of nine in a covalent species. Thus, higher coordination numbers imply either bond orders less than unity or else use of the f orbitals. In addition, certain shapes (such as a regular cube) of lower coordination number also demand use of f orbitals on symmetry grounds. These higher coordination numbers have only become clearly established recently, but their occurrence in lanthanide or actinide element complexes suggest the possibility of f orbital participation. Examples include the ten-coordinate complexes mentioned above, $LaEDTA(H_2O)_4$ and $Ce(NO_3)_5^{2-}$ or 10-coordinate $La_2(CO_3)_3.8H_2O$; 11-coordinate $Th(NO_3)_4.5H_2O$ (coordination by four bidentate nitrate groups and three of the water molecules); and the 12-coordinate lanthanum atoms in $La_2(SO_4)_3.9H_2O$—with twelve sulphate O atoms around one type of La atom position.

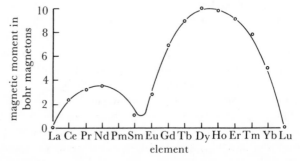

FIGURE 10.5 *Calculated (—) and experimental (°) values of the magnetic moments of the lanthanides*

11 The Actinide Elements

11.1 Sources and physical properties

All the elements lying beyond actinium in the Periodic Table are radioactive, and many of them do not occur naturally. Uranium and thorium are available as ores, and actinium, protactinium, neptunium, and plutonium are available in small amounts in these minerals. Their isolation is difficult and expensive and it is now more convenient to make them artificially. Plutonium is available on a large scale from fuel materials of uranium reactors.

There are two main nuclear reactions which are used in the synthesis of new elements of the actinide series. One is the capture of neutrons, followed by *beta* emission, which increases the atomic number by one unit. The second method is by the capture of the nuclei of light elements, ranging from helium to neon, which increases Z by several units in one step.

Atomic piles provide intense neutron sources and samples can be inserted into piles for irradiation, so the first method is readily carried out, but it is a process of diminishing returns as each element has to be made from the one before. For example, in the irradiation of plutonium 239, less than one per cent of the original sample appears as californium 252, after capture of thirteen neutrons. The main stages are

$$^{239}\text{Pu} \rightarrow {}^{241}\text{Pu} \rightarrow {}^{245}\text{Cm} \rightarrow {}^{247}\text{Cm} \rightarrow {}^{251}\text{Cf} \rightarrow {}^{252}\text{Cf}$$

% of					
original	100	30	10	1·5	0·7	0·3
sample					

Neutron capture by a nucleus increases its neutron/proton ratio until this becomes too high for stability. Then a neutron is converted to a proton, with emission of a β-particle, and an increase of one in the atomic number, for example:

$$^{242}_{94}\text{Pu} + {}^{1}_{0}\text{n} \rightarrow {}^{243}_{94}\text{Pu} \rightarrow {}^{0}_{-1}\text{e} + {}^{243}_{95}\text{Am}$$

As the synthesis of heavier elements by successive neutron capture depends on the build-up of intermediate elements, the yield of a heavier nucleus falls off sharply as the number of neutron addition steps from the starting material increases. The process is made even less favourable by the general decrease in nuclear stability with increasing atomic weight.

The heaviest elements may thus be obtained only by means which short-circuit the step-by-step addition of neutrons. One means was provided by the hydrogen bomb. In an atomic explosion, where there is a vast flux of fast neutrons, a number of neutrons are added to the target nucleus simultaneously, before the intermediate nuclei can decay. Thus, einsteinium and fermium were first discovered in the fall-out products of the first hydrogen bomb explosion. The use of controlled underground atomic explosions as a source of such elements has been suggested.

A second, and more amenable, method for jumping a number of places in one operation is to bombard the starting material with species containing several nuclear particles. Bombardment by *alpha*-particles is the easiest way of achieving this, and many of the actinides, such as ^{248}Cm, ^{249}Bk, ^{249}Cf, and ^{256}Md, were first made in this way, e.g.:

$$^{253}_{99}\text{Es} + {}^{4}_{2}\text{He} = {}^{256}_{101}\text{Md} + {}^{1}_{0}\text{n}$$

Alpha-particle bombardment requires the target to be the element with an atomic number of two less than the desired element, and such target elements will themselves be scarce and only obtained in small amounts in the case where the desired element is one of very high atomic weight. To make the very heaviest elements, therefore, bombarding nuclei heavier than the α-particle are required. Two of the last elements to be discovered were obtained by heavy nucleus bombardment in this way:

$$^{246}_{96}\text{Cm} + {}^{12}_{6}\text{C} = {}^{254}_{102}\text{No} + 4{}^{1}_{0}\text{n}$$

$$^{252}_{98}\text{Cf} + {}^{11}_{5}\text{B} = {}^{257}_{103}\text{Lw} + 6{}^{1}_{0}\text{n}$$

Table 11.1 indicates the current sources of each element.

The ease with which a radioactive element may be handled depends on the type and intensity of the radiation from the isotope and from its decay products which accumulate in the sample. This activity is roughly measured by the half-life, which is a measure of the rate of the decay process. The degree of activity of an isotope governs the extent to which its chemistry may be studied. In decay processes, the emitted particles are extremely energetic and break bonds and disrupt crystal structures. In addition, the energy given out appears largely as heat in the sample; for example, the heat produced in a millimolar solution of curium 242 salts in water would be sufficient to evaporate the solution to dryness in a short time. The breaking of bonds by emitted particles in a sample of a radioactive element is equivalent to a continuous process of self-reduction. As a result, the proof of the existence of an oxidizing state is impossible for very reactive isotopes. For example, the evidence for the oxidizing IV states of a number of the heavy elements had to await the production of isotopes which were more stable than the very short-lived ones originally available. The intense activity of the samples of these elements, and of their sources, demands special handling techniques which involve manipulating microgram amounts of the sample by remote control. Such work demands specialized facilities and training and is limited to a few laboratories in the world, the most famous being at the University of California where most of the transuranium elements were discovered.

In a number of cases, neutron bombardment, which is readily carried out, does not give rise to the isotope of longest half-life. An example is provided by berkelium where the most accessible isotope, Bk-249, has a much shorter half-life than Bk-247, which is only available in tiny amounts from ion bombardment.

The electronic configurations of these elements gave rise to considerable controversy in the early days of work in this field. The elements which were available before the advent of the atomic pile, especially thorium and uranium, strongly resemble the transition metals (hafnium and tungsten) in their chemistry, and the heaviest elements were accordingly placed in the d block. As new elements were studied, it became increasingly obvious that an f shell was being filled but it was not clear whether this started after actinium, paralleling the lanthanides, or later on in the middle of a d series. It became clear, however, that curium corresponded to gadolinium in properties and was thus the $f^7d^1s^2$ element, implying that the series are genuinely *actinides*. Later, it was

TABLE 11.1 Properties of the actinide elements

Z	Element	Symbol	Weight of most accessible isotope	Half-lives	Source	Electronic configuration
90	Thorium	Th	232·1 (natural mixture in ores)	10^{10} yr (Th-232)	natural	Rn $5f^06d^27s^2$
91	Protactinium	Pa	231	34×10^3 yr	natural and fuel elements	Rn $5f^26d^17s^2$
92	Uranium	U	238·1 (natural mixture)	$4\cdot5 \times 19^9$ yr (U-238) $7\cdot1 \times 10^8$ yr (U-235)	natural	Rn $5f^36d^17s^2$
93	Neptunium	Np	237	2×10^6 yr	fuel elements	Rn $5f^46d^17s^2$
94	Plutonium	Pu	242	410^5 yr, also Pu-244 $= 8 \times 10^7$ yr	fuel elements	Rn $5f^66d^07s^2$
95	Americium	Am	243	7 950 yr	fuel elements	Rn $5f^76d^07s^2$
96	Curium	Cm	248	5×10^5 yr, also Cm-247 $= 8 \times 10^7$ yr	neutron bombardment	Rn $5f^76d^17s^2$
97	Berkelium	Bk	249	314 d also Bk-247 $= 7\ 000$ yr	neutron bombardment	(Rn $f^8d^1s^2$ or $f^9d^0s^2$)
98	Californium	Cf	249	360 yr	neutron bombardment	(Rn $f^{10}d^0s^2$)
99	Einsteinium	Es	254	250 d	neutron bombardment	(Rn $f^{11}d^0s^2$)
100	Fermium	Fm	253	4·5 d	neutron bombardment	(Rn $f^{12}d^0s^2$)
101	Mendelevium	Md	256	1·5 h	ion bombardment	(Rn $f^{13}d^0s^2$)
102	(Nobelium)		254	3 s	ion bombardment	(Rn $f^{14}d^0s^2$)
103	Lawrencium	Lw	257	8 s	ion bombardment	(Rn $f^{14}d^1s^2$)

Notes: (i) The source given is the most recent for manageable quantities of the isotope in question, except that elements 100 onwards are not available in weighable amounts at present. (ii) Electronic configurations in parentheses are predicted ones. (iii) The isotopes ^{239}Pu, ^{241}Am, and ^{244}Cm, though of shorter half-lives ($2\cdot4 \times 10^5$ yr, 458 yr, and 163 d respectively) are currently available in larger quantities than the longer-lived species listed above. ^{258}Md of half-life = 2 months is reported.

possible to interpret the very complicated atomic spectra and determine the electronic configurations as given in the Table. Magnetic studies have also confirmed these. Notice that element 103, lawrencium, completes the actinide series. The recently synthesized elements 104 and 105 are presumably the heavier analogues of hafnium and tantalum in the 6*d* series.

11.2 General chemical behaviour of the actinides

Table 11.2 lists some important chemical properties of the actinides, with actinium included for comparison. The M^{4+}/M^{3+} redox potentials clearly show the increasing stability of the III state for the heaviest elements. Diagrams of the free energy changes per electron in oxidation-reduction reactions are shown in Figure 11.1, illustrating the stable oxidation states for the elements. The stabilization of the III state to the right of the actinide series is again shown here.

The most stable oxidation state of the elements up to uranium is the one involving all the valency electrons. Neptunium forms the VII state, using all its valency electrons, but this is oxidizing and the most stable state is Np (V). Plutonium also shows states up to VII and americium up to VI but the most stable states drop to Pu(IV) and Am(III) and later elements also tend to be most stable in the III state. This pattern of higher oxidation state stabilities has more in common with that across a *d* series than that of the lanthanides. In the *d* elements, all the electrons are used, in the Group oxidation state, until the middle of the series (Mn(VII) or Ru and Os(VIII)) and this state becomes more oxidizing across the series. Only for the later elements where the III state becomes dominant does the resemblance to the lanthanides appear. Notice that, although the IV state of berkelium is strongly oxidizing, it is more stable than the IV states of curium and americium. In this, it is showing a

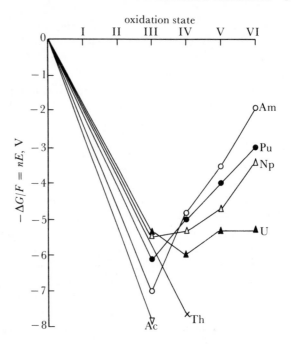

FIGURE 11.1 *Free energy changes per electron for actinide element oxidation-reduction couples in acid solution*
Here the free energy change (in volts) is plotted against the oxidation state for all the actinide elements, as far as data are available. The uranium and americium diagrams are discussed in section 7.6, Figure 7.12.

parallel to the properties of terbium, where the IV state, corresponding to the f^7 configuration, has some stability. Americium does not form the II state in aqueous media, but it has been reported in a chloride melt, so that there is some slight resemblance to europium which attains the f^7 con-

TABLE 11.2 Chemical properties of the actinides and actinium

Element	Oxidation states	Ionic radii pm M^{3+}	M^{4+}	Potential for $M^{4+}+e^- = M^{3+}$ (V)
Ac	III	111		
Th	(III), IV	(108)	99	
Pa	(III), IV, V	(105)	96	
U	III, IV, V, VI	103	93	-0.631
Np	III, IV, V, VI, VII	101	92	$+0.155$
Pu	III, IV, V, VI, VII	100	90	$+0.982$
Am	II, III, (IV), V, VI	99	89	$(+2.7)$
Cm	III, (IV?)	98	88	
Bk	III, IV	96		$(+1.6)$ (forming f^7 for Bk^{4+})
Cf	(II), III	95		
Es	(II), III			ca -1.1 for Es$^{3+}+e^- = $ Es^{2+}
Fm	(II), III			ca -0.1 for Md$^{3+}+e^- = $ Md^{2+}
Md	II, III			ca $+1.4$ for No$^{3+}+e^- = $ No^{2+}
No	II, III			(forming f^{14} for No^{2+})
Lw	III			

figuration in its moderately stable II state. The heavier actinides, though studied mainly by carrier methods, show some evidence for a II state in addition to the stable III state but attempts to oxidize Md^{3+}, No^{3+} or Lw^{3+} to the IV state were unsuccessful. Of the II states, it is interesting that Md(II) is moderately stable and that No(II) is markedly more stable than No(III), with No^{2+} requiring an oxidizing agent comparable with permanganate or ceric to form No^{3+}. The f^{14} configuration would probably be attained by No^{2+} but this state is relatively much more stable than the analogous Yb^{2+} in the lanthanides.

The regular trend in ionic radii resembles that shown by the lanthanide elements, and it is possible to talk of an *actinide contraction* similar to the lanthanide contraction and arising from a similar increase in effective nuclear charge due to poor screening by the f electrons. This actinide contraction means that the actinide elements should show similar ion-exchange behaviour to the lanthanides, and this has been made use of in a striking way in the identification of the newer heavy elements. The elements beyond curium have all very similar properties chemically, and the methods of synthesis mean that they are formed in only small amounts in the presence of excess target material. The identification of the heavier elements depends upon detecting their characteristic radiation (which can be predicted theoretically). The method which was successfully adopted was to dissolve the irradiated targets and pass the actinides in solution through an ion-exchange column and count the radiation from each fraction. Due to the tiny scale of the experiments, the 'column' was a few beads of resin and the fractions were single drops. The order of elution, and the elution positions, of the tripositive actinide ions and the tripositive lanthanide ions are the same on the same resin, and this was made use of in the first identification of the elements from americium up to mendelevium on a weighable scale (i.e. apart from the use of carrier methods). A composite elution diagram of these elements is shown in Figure 11.2 along with a similar diagram for the heavier lanthanides. The one-to-one correspondence in positions is clear. The scale of the operations is made evident by the fact that the first identification of element 101, mendelevium, was based on the count of five decompositions, i.e. of the fission of five individual atoms.

The actinide metals resemble the lanthanides, are of low electronegativity, and are very reactive. The metals are produced by electrolytic reduction of fused salts or by treating the halides with calcium at high temperatures. They are all extremely dense, with densities ranging from 12 to 20 g/cm^3. The direct reactions of the metals (for example, with oxygen, halogens and acids) are similar to those of the scandium Group elements. The metals also react directly with hydrogen with the formation of non-stoichiometric hydrides, such as Th_4H_{15}. Phases with idealized compositions MH_2 and MH_3 are most common. These hydrides are reactive and often form suitable starting materials for the preparation of other compounds. On heating, they decompose leaving the metal in a very finely divided and reactive form.

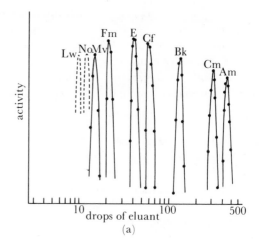

(a)

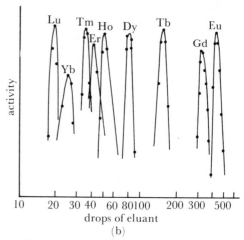

(b)

FIGURE 11.2 *Elution diagrams for* (a) *actinide and* (b) *lanthanide elements*

11.3 Thorium and protactinium

Thorium has been known since 1828. Its principal source is the mineral, monazite, which is a complex phosphate of thorium, uranium, cerium, and lanthanides. Thorium is extracted by precipitation as the hydroxide, along with cerium and uranium, and then separated by extraction with tributyl phosphate from acid solution. The metal is made by calcium reduction of the oxide or fluoride, and pure samples can be prepared in the *Van Arkel* process by decomposing the iodide, ThI_4, on a hot filament.

The only stable state, and the only one known until recently, is the IV state in which thorium resembles hafnium. This is very stable and, because of the large size, the Th^{4+} ion has a low enough charge density to be capable of existing without excessive polarization effects. This is the highest-charged ion known. The hydroxide is precipitated from thorium solutions and gives the oxide, ThO_2, on ignition. This is also formed directly from the metal and oxygen, and on ignition of oxy-salts. It is a stable and refractory material (m.p. 3050 °C) and is soluble only in hydrofluoric/nitric acid mixtures. The anhydrous halides, ThX_4, are prepared by dry reactions such as metal plus halogen or oxide plus hydrogen halide at 600 °C. The tetrafluoride is involatile but the others sublime in vacuum above 500 °C. Treatment of the halides with water vapour gives the oxyhalides, ThOX$_2$.

Dilute thorium solutions in acid contain the hydrated thorium ion, Th^{4+}nH$_2$O, but hydrolysis takes place on concentration or when the pH is raised. At a pH of about 6, the hydroxide, Th(OH)$_4$, is precipitated. This has a crystal structure containing chains with the thorium atoms being linked by oxy- and hydroxy-bridges.

The commonest salt is the nitrate, Th(NO$_3$)$_4$.5H$_2$O, and this is very soluble in water, alcohols, and similar solvents. The fluoride, oxalate and phosphate are very insoluble and may be precipitated even in strong acid solution (compare hafnium and cerium(IV). Thorium also gives a boro-hydride, Th(BH$_4$)$_4$ which sublimes in vacuum at about 40 °C.

The coordination of thorium(IV) is variable and tends to be high. ThCl$_4$ and ThBr$_4$ have the distorted eight-coordination of UCl$_4$ while seven-, eight-, and nine-coordinate Th atoms are all found in ThOCl$_2$ (and in the isomorphous Pa, U, and Np analogues). Eight-coordination to sulphur is found in the complex Th(S$_2$CNEt$_2$)$_4$. Nine-coordination, by sharing fluorines, is found in (NH$_4$)$_4$ThF$_8$ and in Na$_2$ThF$_6$ and the ThF$_9$ arrangement is similar to that shown in Figure 14.21.

No state other than IV exists for thorium in solution, but there is evidence for the tri-iodide, ThI$_3$, formed by heating the metal with the stoichiometric amount of iodine in vacuum at 555 °C. Using a deficiency of iodine gives the di-iodide. In a recent study, it has been shown that ThI$_3$, and two different crystal modifications of ThI$_2$, can be also prepared by heating ThI$_4$ with thorium metal. The tri-iodide converts to the di-iodide on further heating. The compounds all react with water with the evolution of hydrogen and formation of thorium(IV). These compounds are metallic with bonding which can be explained on the same model as that used for the metallic hydrides, section 8.4, page 118. One non-metallic thorium(III) compound is the white ThOF, prepared by reducing a ThF$_4$/ThO$_2$ mixture with thorium metal at 1500 K.

Protactinium was first identified in uranium in 1917. It is a product of uranium 235 decay and, in turn, gives actinium by alpha-particle emission:

$$^{235}_{92}\text{U} = {}^{4}_{2}\text{He} + {}^{231}_{90}\text{Th} = {}^{0}_{-1}\text{e} + {}^{231}_{91}\text{Pa} = {}^{4}_{2}\text{He} + {}^{227}_{89}\text{Ac}$$

A further isotope occurs in the decay of neptunium 227, but both these naturally-occurring isotopes have low concentrations in equilibrium and the element is most readily obtained by synthesis:

$$^{232}_{90}\text{Th} + {}^{1}_{0}\text{n} = {}^{233}_{90}\text{Th} = {}^{233}_{91}\text{Pa} + {}^{0}_{-1}\text{e}$$

This isotope has a half-life of 27·4 days but is a beta-emitter and more readily handled than the alpha-emitting Pa 231. The latter has, however, a much longer half-life, 34 000 years, is now available from the fuel elements of atomic piles and is currently used. Because of its relative scarcity until recently, and because of the strong tendency of its compounds

to hydrolyse and form polymeric colloidal particles which are adsorbed in reaction vessels, its chemistry is little investigated compared with that of its neighbours.

The oxide system is complex and compounds range in composition from PaO$_2$ to Pa$_2$O$_5$. The pentoxide is obtained on igniting protactinium compounds in air and is a white solid with weakly acidic properties, being attacked by fused alkali. On reduction with hydrogen at 1500 °C, the black dioxide PaO$_2$ is formed.

Among the halides, two fluorides are known. The penta-fluoride PaF$_5$ results from the reaction of bromine trifluoride on the pentoxide. It is a very reactive and volatile compound. The complex anion, PaF$_7^{2-}$, is known and was used in the classical isolation of the element. In this ion, the protactinium is nine-coordinate with Pa linked by two fluorine bridges to a neighbour on either side, giving a chain structure. The structure of the PaF$_9$ units is the same as that of the ReH$_9^{2-}$ ion shown in Figure 14.21. In the complex Na$_3$PaF$_8$ the PaF$_8^{3-}$ ion is a slightly distorted cube and the sodium ions are also eight-coordinated. The Na$_3$MF$_8$ compounds of uranium(V) and neptunium(V) are isostructural. The second fluoride, PaF$_4$, is a red, high-melting solid which results from the reaction of hydrogen and hydrogen fluoride on the oxide. Complexes MPaF$_5$ and M$_4$PaF$_8$ are known.

In recent studies, all the Pa^{IV} and Pa^V chlorides, bromides and iodides of the types PaX$_4$, PaX$_5$, PaOX$_2$, PaOX$_3$ and PaO$_2$X have been prepared, together with the complexes M$_2$PaX$_6$ and MPaX$_6$. PaCl$_5$ and PaF$_5$ are polymeric structures with pentagonal bipyramidal coordination, like β-UF$_5$. PaBr$_5$ consists of dimeric units with two bridging bromines giving six-coordinate Pa, while PaBr$_4$ is an infinite polymer with all the bromine atoms bridging pairs of Pa atoms giving PaBr$_8$ coordination.

The solution chemistry is obscure because of the formation of colloids, but anionic complexes like (PaOCl$_6$)$^{3-}$ have been claimed. Lower oxidation states may be obtained in solution by reduction with zinc amalgam. The tetravalent state is stable in absence of air but evidence for the III state is slight and based on polarographic results. The absorption spectrum of PaCl$_4$ in water shows three maxima and is similar to that of Ce^{3+}, providing some evidence for the presence of a single f electron in Pa (IV).

11.4 Uranium

Uranium is the longest-known of the actinide elements, having been discovered in 1789, but it attracted little interest until the discovery of uranium fission in 1939. It is now of vital importance as a fuel, and its chemistry has been very fully explored in the course of the atomic energy investigations.

Natural uranium contains two main isotopes, ^{238}U 99·3 per cent and ^{235}U 0·7 per cent, and also traces of a third, ^{233}U. The vital isotope from the nuclear energy point of view is ^{235}U because this reacts with a neutron, not by building up heavier elements as in the examples discussed in section 11.1, but by fission to form lighter nuclei. This fission process releases considerable energy and more neutrons, which, in turn, fission uranium 235 nuclei and allow the building-up of a chain of fissions. The energy of such nuclear processes is about a million times the energy released in chemical reactions, such as the burning of a fuel or the detonation of a high explosive, and this is the reason for the value of atomic fission as an energy source, and for the horror of fission as a source of explosive energy in a weapon. A typical fission process is:

$$^{235}_{92}U + ^{1}_{0}n = ^{92}_{36}Kr + ^{140}_{56}Ba + 3^{1}_{0}n + \text{about } 8 \times 10^9 \text{ kJ mol}^{-1}$$

The nuclei formed in fission fall into two main groups, a lighter set with masses from about 70 to 110 and a heavier one with masses from 125 to 160. Splitting into approximately equal nuclei is about a thousand times rarer than splitting to an unequal pair such as that shown in the equation. The neutrons evolved in the fission are either used in other fissions, absorbed by non-fissionable nuclei such as uranium 238, or escape through the surface of the uranium mass. The essence of running an atomic pile is to ensure that one neutron per fission is available to cause another fission. More than one leads to a rapidly increasing chain reaction and explosion, while less than one means that the process dies out. The absorption of neutrons by uranium 238 leads to the formation of heavier elements, of which the most important is plutonium which is itself a nuclear fuel. In appropriate conditions, more plutonium can be produced from the uranium 238 than the amount of uranium 235 consumed, and such an arrangement 'breeds' nuclear fuel in the 'breeder reactor' which is at present being studied in a number of countries.

In its chemistry, uranium resembles the three succeeding elements, neptunium, plutonium, and americium, in having four oxidation states, III, IV, V, and VI. The most stable state drops from VI for uranium through V for neptunium, IV for plutonium, to III for americium. This is illustrated by the oxides and halides found for these elements, shown in Table 11.3, and also by the free energy diagrams for the changes in oxidation state in acid solution. The VI state of uranium, in solution, is present as the uranyl ion, UO_2^{2+}, and the V state also occurs as an oxycation, UO_2^+. The IV and III states are present as simple cations, U^{4+} and U^{3+}. Since the change from U^{3+} to U^{4+}, and that from UO_2^+ to UO_2^{2+} (and the reverse changes) involve only transfer of an electron, these two pairs of redox reactions occur rapidly. On the other hand, oxidations such as U^{4+} to UO_2^{2+} involve oxygen transfer as well and are slow and often irreversible. In solution, the III, IV, and VI states all exist, but UO_2^+ has only a transitory existence. However, in a non-oxide medium, such as anhydrous hydrogen fluoride or a chloride melt, uranium(V), although still unstable, is well represented.

Uranium metal, produced by reduction of the tetra-fluoride with calcium or magnesium:

$$UF_4 + 2Ca = U + 2CaF_2$$

is reactive and combines directly with most elements. It dissolves in acids but not in alkalis. Direct reaction with hydrogen at about 250 °C gives the hydride, UH_3, usually in a form with a small deficiency of hydrogen. This is a very reactive compound which is a useful starting material for the preparation of uranium compounds of the III and IV states. The normal product is the IV compound, for example:

$$UH_3 + H_2Y = UY_2 \ (Y = O, S \text{ at about } 400 °C)$$
$$UH_3 + HF = UF_4$$
$$\text{but } UH_3 + HCl = UCl_3 \ (Cl_2 \text{ gives } UCl_4)$$

The uranium-oxygen system is very complex as the oxidation states are of comparable stability and non-stoichiometric phases are common. The main uranium ore, *pitchblende*, is an oxide approximating to UO_2 in composition. The other stoichiometric oxides are U_3O_8, which is the ultimate product of ignition, and UO_3 which is obtained by the decomposition of uranyl nitrate, $UO_2(NO_3)_2$, at about 350 °C. The trioxide can be reduced to the dioxide by the action of carbon monoxide at 350 °C, and it goes to U_3O_8 on heating to 700 °C. All three oxides dissolve in nitric acid to give uranyl, UO_2^{2+}, salts.

The known halides of uranium are listed in Table 11.3 and the interconversions of the fluorides and chlorides are shown in Figures 11.3 and 11.4, respectively. The hexahalides are

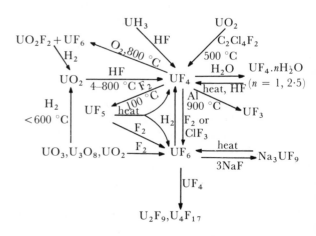

FIGURE 11.3 *Reactions and interconversions of uranium fluorides*

octahedral. UF_6 is volatile, sublimes at 56 °C, and has been used in the separation of uranium isotopes by gaseous diffusion. The compound is a powerful fluorinating agent and is rapidly hydrolysed. Uranium pentafluoride and pentachloride both readily disproportionate to give the hexahalide and the tetrahalide. Uranium(V) also occurs in fluoride complexes, UF_8^{3-} and UF_6^-. The latter is formed in HF solution, and may be the reason for the relative stability of the V state in this medium. The most stable halides of uranium are the tetrahalides, to which the higher halides are readily reduced and the trihalides oxidized. The structures of the tri-, tetra-, and penta-halides of uranium are all polymeric with high coordination numbers. UCl_3 (like many

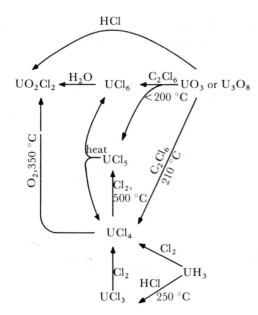

FIGURE 11.4 *Reactions and interconversions of uranium chlorides*

other lanthanide and actinide trihalides), crystallizes with nine-coordinated U^{3+} ions, in a structure where the U atom has three chloride ions coplanar with it, three above this plane, and three below. UI_3, and the tribromides and tri-iodides of Np, Pu, and Am, have an eight-coordinated layer lattice. The tetrachloride has a distorted eight-coordinated structure, while the pentafluoride has two forms. In α-UF_5, an octahedron of fluorines around the uranium is completed by the sharing of fluorine atoms and the formation of long chains of linked octahedra. In the β-form of UF_5, the uranium atom is seven-coordinated with three fluorines attached to only one uranium and the other four shared with four different uraniums in a polymeric structure. In U_2F_9, all the uranium atoms have nine fluorines at equal distances and this (black) compound probably contains crystallographically equivalent uranium(IV) and uranium(V) atoms.

Oxyhalides of uranium(VI), UO_2X_2, are also known. These are made from the oxides or halides by partial substitution:

$$UO_3 + 2HF \xrightarrow{400\ ^\circ C} UO_2F_2 + H_2O$$

In solution, hydrolysis occurs for all oxidation states, being least for the III state. Uranium(III) and (IV) exist as ions in strong acid, and uranium(IV) hydrolyses in more dilute solution in a similar manner to Th^{4+}. The U^{4+} ion also gives insoluble precipitates with similar reagents—F^-, PO_4^{3-}, etc. —as does Th^{4+}. Uranium(V) has a strong tendency to disproportionate to U^{4+} and UO_2^{2+}. It is most stable in fairly acid solution at a pH of about 3. Uranium(VI) also hydrolyses in solution, this time giving double hydroxy bridges so that polymers of the type . . . $UO_2(OH)_2UO_2(OH)_2$ . . . are formed. Uranium(VI), as uranyl, forms the only common uranium salts, and the most usual starting material is uranyl nitrate, $UO_2(NO_3)_2.nH_2O$, ($n = 2$, 3, or 6). This is soluble in water and in a variety of organic solvents.

The stereochemistry of uranium, and of the other actinides, shows a tendency to high coordination numbers, as illustrated by the halide structures above. A tendency to eight-coordination for M^{4+} actinide ions seems to be general. For example, uranium and thorium form an acetylacetonate, $M(C_5H_7O_2)_4$, which has the eight coordinating oxygen atoms at the corners of a square antiprism, as shown in Figure 11.5. (The square antiprism is most readily visualized as a cube with the top face twisted 45° relative to the bottom

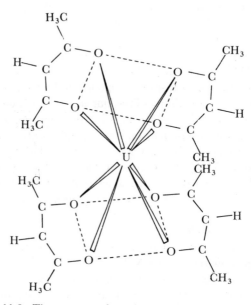

FIGURE 11.5 *The structure of uranium (or thorium) acetylacetonate,* U(acac)$_4$
The uranium (or thorium) atom is coordinated to eight oxygen atoms which are arranged in a square antiprism around it.

one.) In uranyl compounds, the UO_2 group is linear and complexes exist in which four, five, or six donor atoms lie coplanar with the uranium atom, giving six-, seven- or eight-coordination overall. An example is shown in Figure 11.6,

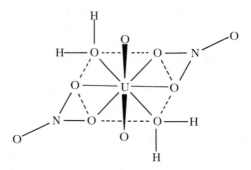

FIGURE 11.6 *The structure of uranyl nitrate hydrate,* $UO_2(NO_3)_2.2H_2O$
The UO_2 group is linear and two water molecules and the two nitrate groups are coordinated in the central plane to the uranium. As the nitrate groups here are bidentate, the uranium has a co-ordination number of eight.

where one form of hydrated uranyl nitrate has two water molecules and the two nitrate groups coordinated to uranium through both oxygens, all in the central plane, to give the eight-coordinated structure shown.

11.5 Neptunium, plutonium, and americium

The three elements, neptunium, plutonium, and americium, which follow uranium, also show the four oxidation states, III, IV, V, and VI. In addition, strong oxidation in alkaline solution by reagents like ozone, XeO_3, or periodate, produces plutonium and neptunium in the VII state. Attempts to prepare plutonium(VIII) have so far been unsuccessful. The VII states exist in strong alkaline solution and have been isolated as Li_5MO_6 (M = Np, Pu). On acidification, reduction to the VI state occurs rapidly. The redox potential of the Np(VII)/Np(VI) couple at pH = 14 has been put at 0·61 volt at 25 °C.

The III, IV, V, and VI states are present in solution as the M^{3+}, M^{4+}, MO_2^+, and MO_2^{2+} ions, as in the case of uranium, and they have a similar tendency to hydrolysis. However, the relative stability of the four states varies among the elements. The VI state becomes increasingly unstable and oxidizing from uranium to americium, and AmO_2^{2+} solutions are as strongly oxidizing as permanganate. The V state in solution, as MO_2^+, is the most stable state for neptunium, while PuO_2^+ and AmO_2^+, while less stable than NpO_2^+, are more stable than UO_2^+. Thus, UO_2^+ disproportionates in aqueous solution to uranium(VI) and uranium(IV) (see Figure 11.1) while NpO_2^+ is stable. In contrast, in hydrofluoric acid solution, $CsUF_6$ only dissolves while $CsNpF_6$ disproportionates to give $NpF_4 + NpF_6$. This example is a useful reminder that stabilities differ markedly in different media, and our generalizations apply only to stabilities in aqueous solutions in air. The IV state is the most stable one for plutonium, while americium resembles the following elements in being most stable in the III state.

This behaviour in solution is shown by the free energy diagrams and is paralleled in the solid state as the Table of halides and oxides shows. The most marked omissions in this list of compounds are the lack of NpF_5 and of tetrahalides of plutonium. A number of attempts have been made to prepare NpF_5 and $PuCl_4$ but these have all failed, despite the stability of neptunium(V) and plutonium(IV) in solution. In the case of neptunium, this suggests that the stability of the V state depends on the possibility of forming NpO_2^+. This is supported by the disproportionation of the neptunium(V) complex fluoride, NpF_6^- mentioned above. The lack of plutonium tetrahalides while the IV state is stable in solution probably reflects the high solvation energy of the Pu^{4+} ion.

These three elements are similar to uranium and are reactive metals which combine with most elements. They all three form hydrides, but these resemble the lanthanide hydrides more than uranium hydride. Stoichiometric hydrides, MH_2, are formed and also non-stoichiometric systems up to a composition, $MH_{2.7}$. The properties of the metals show increasing similarities to those of the lanthanide elements, and americium resembles these quite closely. **Americium is the analogue of europium and americium(II) does exist, at least in a halide melt.**

These three elements provide the most striking evidence, in their chemical properties, that the heaviest elements are,

TABLE 11.3 Halides and oxides of uranium, neptunium, plutonium and americium

Uranium	UF_6		UF_5	U_2F_9	UF_4	UF_3
				U_4F_{17}		
	UCl_6		UCl_5		UCl_4	UCl_3
					UBr_4	UBr_3
					UI_4	UI_3
	UO_3	U_3O_8			UO_2	
Neptunium	NpF_6				NpF_4	NpF_3
					$NpCl_4$	$NpCl_3$
					$NpBr_4$	$NpBr_3$
						NpI_3
		Np_3O_8			NpO_2	
Plutonium	PuF_6			Pu_4F_{17}	PuF_4	PuF_3
						$PuCl_3$
						$PuBr_3$
						PuI_3
					PuO_2	Pu_2O_3
Americium					AmF_4	AmF_3
						$AmCl_3$
						$AmBr_3$
						AmI_3
					AmO_2	Am_2O_3

in fact, filling an f shell and are not a fourth d series. While **thorium and uranium show similarities to hafnium and tungsten, neptunium and plutonium are clearly not the** analogues of rhenium and osmium. Neptunium(VII) is strongly oxidizing while rhenium(VII) is stable and there is no indication as yet of an VIII state for plutonium analogous to osmium(VIII). Even the VI state of these two elements is relatively unstable and americium, with its stable III state, conforms to the expected pattern and links up with the chemistry of the succeeding elements.

11.6 The heavier actinide elements

Less information is available for the heavier actinides. Curium, berkelium and californium are now available in microgramme amounts, so that their macroscopic chemistry may be studied, and sufficient einsteinium was available for the absorption spectrum of Es(III) in solution to be recorded. The later elements have only been available in much smaller amounts and have been studied by carrier methods. In this way the behaviour of these elements has been elucidated by finding which one of a mixture of metal ions is accompanied by their characteristic radiation in the course of a chemical reaction. For example, Md accompanies Eu but not La when solutions of the M^{3+} ions are reacted with sodium amalgam and extracted. As this reaction forms Eu^{2+} but leaves La^{3+} unreduced, the formation of Md^{2+} is indicated.

The III state, the most stable for americium, is also the most stable state for the remaining actinides with the exception of nobelium. The only higher states are the IV state of curium and berkelium. Curium(IV) does not exist in solution and, for example, attempts to oxidize Cm^{3+} with Na_4XeO_6 (which produces Pu(VII)) were unsuccessful. Curium(IV) is represented by CmF_4 and Rb_2CmF_6 in the solid state. Thus curium, the f^7 actinide, shows a resemblance to gadolinium, the f^7 lanthanide, in its solution behaviour. Berkelium(IV), although strongly oxidizing, is rather more stable and is found in solution and in the solid state. In solution, Bk^{3+} may be oxidized by bromate, and the oxidizing power of Bk(IV) is similar to that of cerium(IV).

The availability of microgramme amounts of berkelium and californium has allowed the preparation of a number of solid compounds. Berkelium(IV) is represented by BkO_2 and Cs_2BkCl_6 while the III states are found as BkOCl, Bk_2O_3, $Cs_2NaBkCl_6$, Cf_2O_3, CfOCl and $CfCl_3$.

From californium to nobelium, there is evidence for the existence of a II state, but this is strongly reducing except for mendelevium, where it is mildly reducing, and for nobelium. No(II) is stable and No(III) is a strong oxidizing agent, and this can be rationalized as showing a strong tendency to form the $f^{14}d^0s^0$ state which is the expected one for No^{2+}. This very stable No(II) state, and the formation of the II state by earlier elements, shows a much stronger tendency by the heavier actinides to go to the lower oxidation state than was shown by the lanthanides.

Lawrencium behaves as expected for the $f^{14}d^1s^2$ configuration by forming only Lw(III)—$f^{14}d^0s^0$—and resisting oxidation or reduction.

Recent work, particularly on the heavier elements, has thus filled in the picture of the actinide elements as a whole. There are some quite close resemblances to the lanthanides, especially in the formation of a stable III state by Am, Cm, Bk, Cf, Es, Fm, Md, and Lw, but there are distinct differences. These arise both in the range of higher oxidation states shown by the earlier elements and in the newer evidence for the formation of the II state from californium to nobelium. The whole body of results indicates that, while there are helpful analogies with both the lanthanides and with the transition metals of the 5d series, the chemistry of the actinides presents an individual pattern reflecting the relatively small difference between the 5f and 6d energies for all these elements.

11.7 Possible new elements

In the elements up to about $Z = 100$, the mode of decay is principally by alpha- or beta-particle emission and it is the likelihood of these events which governs the half-lives. It will be seen from Table 11.1 that the half-lives tend to decrease as Z increases. However, at about the atomic number of the heaviest elements, another mode of decomposition, spontaneous fission, becomes the dominant one. In this mode half-lives again decrease with increasing atomic number, and it has been estimated that, by about element 110, the half-

life will have dropped to about 10^{-4} second. Such calculations are approximate, and there is the possibility of certain nuclear configurations having unexpected stability, but the probability at present is that only a few more elements can be prepared and identified by present techniques.

The actinide series is complete at lawrencium, element 103. Elements 104 and 105 have recently been reported. Element 105 was prepared by bombarding californium with ^{15}N nuclei, and it decays with a half-life of 1·6 s

$$^{260}_{105}Ha \rightarrow {}^{256}_{103}Lw + {}^4_2He$$

The name commemorates Otto Hahn. Controversy still surrounds element 104, first reported by a Russian group and named khurchatovium, as later American work did not reproduce the original observations and produced isotopes of 104 with different properties. The element has certainly been synthesized but it is not clear whether the Russian or the American group was the first to produce it.

These two elements are expected to be the 6d congeners of hafnium and tantalum, respectively. The 6d shell would be full at element 112, and then the 7p shell should be occupied until element 118 which would be a new rare gas.

If it should be possible to go further, there is the interesting possibility that a g shell, the 5g level, would start to be occupied somewhere about element 123.

Two nuclear configurations which should represent 'islands of stability' are represented by the elements with 114 protons or 126 protons, i.e. by isotopes of elements 114 or 126. There are a number of very recent proposals by various American research groups to attempt the syntheses of these elements by bombarding accessible heavy nuclei with ions of nuclei of much higher atomic number than the boron or carbon nuclei used to synthesize the heavier actinides. For example, the bombardment of uranium with selenium is proposed as a route to element 126 or of gadolinium with tin as a route to element 114. Such preparations obviate the 'process of diminishing returns' mentioned in section 11.1, but at the price of building very large and expensive accelerators to produce sufficiently energetic bombarding ions.

If such experiments succeed, it is now thought that the elements produced will be much more stable than indicated by a simple extrapolation of the half-lives observed for the heaviest known elements: a half-life of ten years is predicted. It would thus be possible to study the chemistry of these two elements in detail, even though most of their near neighbours would be very unstable. Element 114 would probably be 'ekalead', the highest member of the carbon Group and would be expected to have a stable oxidation state of II and an unstable, possibly non-existent, IV state. Element 126 presents a variety of possibilities, including the chance that it might contain 5g electrons.

12 The Transition Metals: General Properties and Complexes

12.1 Introduction to the transition elements

The elements of the transition block are those with d electrons and incompletely filled d orbitals. The zinc Group, with a filled d^{10} configuration in all its compounds, is transitional between the d block and the p elements and is discussed later.

The Groups of the d block contain only three elements and correspond to the filling of the $3d$, $4d$ and $5d$ shells respectively. In between the $4d$ and $5d$ levels is interposed the first f level, the $4f$ shell, which fills after lanthanum. It has already been seen (Chapter 10) that the occupation of this level is accompanied by a gradual decrease in atomic and ionic radius from La to Lu and the total lanthanide contraction is approximately equal to the normal increase in size between one Period and the next. The result is that in the transition Groups there is the normal increase of about 20 pm in radius between the first and second members (filling the $3d$ and $4d$ shells), but the expected increase between the second and third members is just balanced by the lanthanide contraction so that these two elements are almost identical in size. This effect is illustrated by the radii given in Table 12.1, where the normal increases in the alkaline earths and in the scandium Group contrast sharply with the figures for the succeeding Groups.

As the pair of heavy elements have almost identical radii, and therefore very similar characteristics in other ways (e.g. ionization potentials, solvation energies, redox potentials, lattice energies), their chemistry is very similar. Thus each transition Group typically divides into two parts—the lightest element with its individual chemistry, and the pair of heavy elements with almost identical chemistries. The three elements within each Group have a number of properties in common, of course. They show the same range of oxidation states in general, though not beyond the manganese Group, but these differ in relative stabilities. All the d and s electrons are involved in the chemistry of the earlier elements, so that the Group oxidation state is the maximum state shown. Once the d^5 configuration is exceeded, there is less tendency for all the d electrons to react, and the Group

oxidation state is not shown by iron (though Os(VIII) and Ru(VIII) exist), nor by any elements of the cobalt, nickel, or copper Groups. Since, in the Group oxidation state, all the valency electrons are involved and since the properties of the elements then depend on valency and size only, there are similarities between the properties of Main Groups and Transition Groups of the same Group oxidation state. Thus sulphates and chromates, both MO_4^{2-}, are iso-structural, while molybdenum and tungsten show higher coordination numbers with oxygen (especially six), just as does tellurium. The principal differences between the first and the heavier elements in a transition Group are those of size, and stability of oxidation states. The larger elements commonly show higher coordination numbers, and the higher oxidation states are more stable for the heaviest elements. Thus, chromium(VI) is strongly oxidizing while molybdenum and tungsten are stable in the VI state.

The effects of the lanthanide contraction die out towards the right of the d block. In the titanium and vanadium Groups, which immediately follow the lanthanides, the heavier elements are practically identical and their separation is more difficult than the separation of a pair of lanthanides. The next two Groups show clear differences between the two last elements, though these are still slight. In the platinum metals, the differences are increasing until, in the copper Group, there are few points of resemblance between silver and gold. Finally, in the zinc Group, the pattern approaches that in a p Group and zinc and cadmium resemble each other with mercury as the singular member.

Table 12.2 summarizes this dicussion in terms of the oxidation states shown by the d elements, and the stabilities of these. The general behaviour is also illustrated by Tables 12.3 to 12.5 which give the oxides, fluorides, and other halides of the transition elements. A stable state will show all these compounds while a strongly oxidizing state will be more likely to have a fluoride than an iodide; similarly, a reducing state will be more likely to show a heavier halide than a fluoride.

TABLE 12.1 Radii showing the effect of the lanthanide contraction (Pauling)

M^{2+}, pm	M^{3+}, pm	M^{4+} (calc), pm	atomic radii, pm		
Ca = 99	Sc = 70	Ti = 68	Ti = 132	V = 122	Cr = 117
Sr = 113	Y = 90	Zr = 74	Zr = 145	Nb = 134	Mo = 129
Ba = 135	La = 106	Hf = 75	Hf = 144	Ta = 134	W = 130

TABLE 12.2 Oxidation states of the transition elements

Group O.S.	Ti	V	Cr	Mn	Fe	Co	Ni	Cu	
	<u>IV</u>	V	VI (ox)	VII (ox)					d^0
	III (red)	<u>IV</u>	(V) (d)	(VI) (d)					d^1
	(II) (red)	III (red)	(IV) (d)	(V) (d)	(VI) (ox)				d^2
		(II) (red)	<u>III</u>	IV (ox)	(V) (ox)				d^3
	(0)	(I)	II (red)	(III) (ox)	(IV) (ox)				d^4
	(−I)	(0)	(I)	<u>II</u>	III	(IV) (ox)			d^5
		(−I)	0	(I)	<u>II</u>	III	(IV?)		d^6
			(−I)	0		<u>II</u>	(III) (ox)		d^7
			(−II)	(−I)	0	(I)	<u>II</u>	(III) (ox)	d^8
						0	(I)	<u>II</u>	d^9
							0	I	d^{10}

Group O.S.	Zr	Nb	Mo	Tc	Ru	Rh	Pd	Ag	
	<u>IV</u>	<u>V</u>	<u>VI</u>	<u>VII</u>	VIII (ox)				d^0
	(III) (red)	(IV) (d)	V	VI (d)	(VII) (d)				d^1
	(II) (red)	III (red)	IV	(V) (d)	VI				d^2
		(II) (red)	III	<u>IV</u>	(V)	(VI) (ox)			d^3
			(II)	?	IV				d^4
				?	<u>III</u>	IV			d^5
			0		II	<u>III</u>	IV		d^6
				0		II	III?		d^7
					0	(I)	<u>II</u>	(III) (ox)	d^8
						0	(I?)	II (ox)	d^9
							0	<u>I</u>	d^{10}

Group O.S.	Hf	Ta	W	Re	Os	Ir	Pt	Au	
	<u>IV</u>	<u>V</u>	<u>VI</u>	<u>VII</u>	VIII (ox)				d^0
	(III) (red)	(IV) (d)	V	VI	(VII)				d^1
	(II) (red)	III (red)	IV	(V)	VI				d^2
		(II)	(III)	<u>IV</u>	(V)	(VI) (ox)			d^3
			(II)	<u>III</u>	<u>IV</u>	(V) (ox)	(VI) (ox)		d^4
				(II)	III	<u>IV</u>	(V) (ox)		d^5
			0	(I)	(II)	<u>III</u>	<u>IV</u>		d^6
				0	(I?)	(II?)	?		d^7
					0	(I)	<u>II</u>	<u>III</u>	d^8
						0			d^9
						(−I)	0	I	d^{10}

Notes: ox = oxidizing, red = reducing, unstable states bracketed, d = disproportionates, most stable state(s) for any given element underlined. State 0 usually in carbonyls and related complexes: the element itself is not counted as a 0 state here.

TABLE 12.3 Transition element oxides

			Oxidation state				Other compounds
+II	+III	+IV	+V	+VI	+VII	+VIII	
TiO	Ti_2O_3	$\underline{TiO_2}$ $\underline{ZrO_2}$ $\underline{HfO_2}$					
VO	V_2O_3	VO_2 NbO_2 $(TaO_2?)$	$\underline{V_2O_5}$ $\underline{Nb_2O_5}$ $\underline{Ta_2O_5}$				
CrO	Cr_2O_3	CrO_2 MoO_2 WO_2	Mo_2O_5 $(W_2O_5?)$	CrO_3 $\underline{MoO_3}$ $\underline{WO_3}$			
$\underline{MnO}$	Mn_2O_3 Re_2O_3*	MnO_2 TcO_2 ReO_2	(Re_2O_5)	TcO_3 ReO_3	Mn_2O_7 $\underline{Tc_2O_7}$ $\underline{Re_2O_7}$		Mn_3O_4 Also Tc_2S_7 Re_2S_7
FeO	Fe_2O_3 Ru_2O_3*	$\underline{RuO_2}$ $\underline{OsO_2}$		$(RuO_3)*$ $(OsO_3)*$		RuO_4 $\underline{OsO_4}$	$\underline{Fe_3O_4}$
$\underline{CoO}$ RhO	$(Co_2O_3)*$ $\underline{Rh_2O_3}$ Ir_2O_3	$(CoO_2)*$ RhO_2 $\underline{IrO_2}$		(IrO_3) (peroxide?)			Co_3O_4
$\underline{NiO}$ $\underline{PdO}$ $(PtO)*$	$(Ni_2O_3)*$ $(Pd_2O_3)*$ $(Pt_2O_3)*$	$(NiO_2)*$ $(PdO_2)*$ $\underline{PtO_2}$		$(PtO_3)*$			
$\underline{CuO}$ AgO	$(Ag_2O_3?)$ Au_2O_3						Cu_2O $\underline{Ag_2O}$ Au_2O

Most stable compounds underlined.
*Hydrous oxides of these states are reported.

TABLE 12.4 Transition element fluorides

| | Oxidation state | | | | | Notes and other compounds |
+II	+III	+IV	+V	+VI	+VII	
	TiF$_3$ (ZrF$_3$)	TiF$_4$ ZrF$_4$ HfF$_4$				
VF$_2$	VF$_3$ NbF$_3$ TaF$_3$?	VF$_4$ (NbF$_4$)	VF$_5$ NbF$_5$ TaF$_5$			Nb$_6$F$_{15}$
CrF$_2$	CrF$_3$ MoF$_3$	CrF$_4$ MoF$_4$ WF$_4$	CrF$_5$ MoF$_5$ WF$_5$(d)	(CrF$_6$) MoF$_6$ WF$_6$		(CrF)
MnF$_2$ (ReF$_2$)	MnF$_3$ (ReF$_3$)	MnF$_4$ ReF$_4$	TcF$_5$ ReF$_5$	TcF$_6$ ReF$_6$	ReF$_7$	
FeF$_2$	FeF$_3$ RuF$_3$	RuF$_4$ OsF$_4$	RuF$_5$ OsF$_5$	RuF$_6$ OsF$_6$		*Note* No OsF$_8$
CoF$_2$	CoF$_3$ RhF$_3$ IrF$_3$	RhF$_4$ IrF$_4$	(RhF$_5$) (IrF$_5$)	RhF$_6$ IrF$_6$		
NiF$_2$ PdF$_2$	PdF$_3$	PdF$_4$ PtF$_4$	PtF$_5$	PtF$_6$		PdF$_3$ = Pd^{2+}(PdF$_6$)$^{2-}$
CuF$_2$ AgF$_2$	AuF$_3$					Ag$_2$F, AgF

Most stable compounds underlined. d = disproportionates.

TABLE 12.5 Transition element halides*

	Oxidation state				
$+II$	$+III$	$+IV$	$+V$	$+VI$	Notes
TiX_2	TiX_3	TiX_4			
(ZrX_2)	ZrX_3	$\underline{ZrX_4}$			
$HfCl_2, Br_2$	$HfCl_3, Br_3$	$\underline{HfX_4}$			also HfCl?
					VBr_4 very unstable
VX_2	$\underline{VX_3}$	VCl_4, Br_4			Nb_6X_{14}, Ta_6X_{14}
$(NbBr_2)$	NbX_3	NbX_4	$\underline{NbX_5}$		all $= (M_6X_{12})^{2+}(X^-)_2$ (a)
$TaCl_2$	$TaCl_3, Br_3$	TaX_4	$\underline{TaX_5}$		$Nb_3Br_8, Nb_3I_8, Nb_6I_{11}$
CrX_2	$\underline{CrX_3}$				
MoX_2	MoX_3	MoX_4	$MoCl_5$		MoX_2 and WX_2
WX_2	WX_3	WX_4	WCl_5, Br_5	WCl_6, Br_6	$= (M_6X_8)^{4+}X_4^-$
$\underline{MnX_2}$					
		$TcCl_4$		$TcCl_6$	
(ReX_2)	ReX_3	$\underline{ReX_4}$	$ReCl_5, Br_5$	$ReCl_6$	$ReCl_3, ReCl_4$ are trimers
					ReX_2 in complexes only
$\underline{FeX_2}$	$FeCl_3, Br_3$				
	$\underline{RuX_3}$	$RuCl_4$			
$OsCl_2, I_2$	$\underline{OsX_3}$	OsX_4			OsI
$\underline{CoX_2}$					
	$\underline{RhX_3}$				
$(IrCl_2)?$	$\underline{IrX_3}$	$(IrCl_4)$			
$\underline{NiX_2}$					Platinum trihalides
$\underline{PdX_2}$					may be mixtures of
$\underline{PtX_2}$	$PtX_3?$	$\underline{PtX_4}$			$Pt(II) + Pt(IV)$.
					$PtCl_2$ structure consists
					of Pt_6Cl_{12} units
$CuCl_2, Br_2$					Also $\underline{CuX}$
					$\underline{AgX}$
	$AuCl_3, Br_3$				$AuCl, I$

Most stable compounds underlined.

*The symbol X is used when the chloride, bromide and iodide all occur.

(a) All species $M_6X_{12}^{n+}(X^-)_n$ for $n = 2, 3$ and 4 occur for M = Nb, X = Cl; M = Ta, X = Cl, Br. $Nb_6I_{11} = (Nb_6I_8)^{3+}(I^-$

This division between the lighter transition elements and the two heavier Periods is quite marked and is reinforced in practice by the relative inaccessibility of most of the heavier elements. The latter have therefore been less fully studied, although a considerable amount of work is currently in progress, and often the less available member of the pair (for example Hf, Nb, Tc) is not well-known. In addition, there are strong horizontal resemblances, especially among ions of the same charge, and horizontal trends in properties, with increasing number of d electrons in a given oxidation state, that make it convenient to divide the discussion of the transition block into two sections, one on the first row elements (Chapter 13) and one (Chapter 14) on the heavier elements of the second and third rows.

The pattern of oxidation state stabilities outlined above is complex and there are exceptions to most of the generalizations which can be made about it. The picture is complicated by the use of the term 'stability' in a number of different senses. In the most general sense, it is used to mean that a compound exists in air at around room temperatures: that is, that it is thermally stable at room temperature, that it is not oxidized by air, and that it is not hydrolysed, oxidized or reduced by water vapour. In turn, terms such as thermal stability, may cover a number of processes. Thus a higher oxide such as MO_2 might decompose thermally to $M + O_2$ or to $MO + \frac{1}{2}O_2$ and the free energy change of each process would have to be evaluated before conclusions could be drawn about the stability. Similarly, a compound may exist for a long time at room temperature, not because it is thermodynamically stable with respect to decomposition, but because the decomposition process occurred at a negligible rate. Thus, whether a compound can be kept 'in a bottle' depends on a wide variety of thermodynamic and kinetic factors.

Despite these difficulties, some attempts are being made to examine, predict, and rationalize stabilities, although most treatments to date are either limited in scope or are empirical. One example of a general approach which may be quoted is that of Sheldon who proposes that the pre-ferred oxidation state of a transition element (defined as that which, in simple binary compounds such as the oxides or halides, is the most stable under normal labora-tory conditions) is related to the quantity $rH/40$. Here, r is half the interatomic distance and H the heat of atomiza-tion of the metal—both well-known quantities. This ex-pression leads to the following predictions for the most stable oxidation states of the transition elements:

VI for W, Re, Os, Ir
V   for Nb, Mo, Tc, Ru, Ta
IV for Ti, V, Zr, Rh, Hf, Pt
III for Cr, Fe, Co, Ni, Pd, Au
II for Mn, Cu, Ag.

If these predictions are compared with Tables 12.2′ to 12.5, it will be seen that they are surprisingly accurate for such a simple formula. The only really poor predictions are those for nickel, palladium and silver where the states predicted to be stable are non-existent or very unstable.

A much more fundamental and searching analysis is that, discussed in the next Chapter, on the stability of trihalides of the first row of transition elements, but this is limited to one particular decomposition reaction. Further work on rigorous thermodynamic analysis is to be expected and should lead to a greatly increased understanding of stabilities and Periodic trends.

12.2 The transition ion and its environment: ligand field theory

In the discussion of the energy levels of an atom given in the earlier chapters, the levels of a given p, d or f set were treated as of equal energy. This is true of isolated atoms, or of those in an electric field which is spherically symmetrical around the atom, but is not true when the atom lies in an unsym-metrical field. This may be readily seen by considering an atom which is strongly coordinated to two other groups in a linear configuration. If these groups lie in the $\pm z$ directions, the orbitals which point along the z axis will lie in the field of these ligands and be perturbed by them more than orbitals lying in other directions. As ligands are regions of negative charge (they coordinate through lone electron pairs and also have negative charges or the negative end of a dipole directed towards the central atom or ion) the z-directed orbitals on the central atom will be in a region of higher negative field than the non-z orbitals and electrons will avoid entering them as far as possible. This means that such orbitals as p_z, d_{z^2} and, to a lesser extent, d_{xz} and d_{yz}, are of higher energy than the remaining ones and the degeneracy of the p and d set is split in such a z-directed field in the man-ner shown in Figure 12.1. The size of such a splitting will depend on the size of the ligand field and this, in turn, depends on the distance to the ligand and thus on the intensity of the attraction between the central atom and the

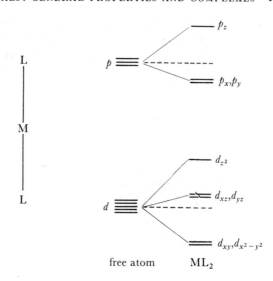

FIGURE 12.1 *Energy level diagram for a linear field*
The z axis is taken as the direction of the coordinated groups. The ligands are regions of high negative charge density, so orbitals on the central atom with components in the $\pm z$ direction are less stable than those with no such component. Among the d orbitals, the d_{z^2} orbital is most strongly destabilized as it has the greatest density in the z direction.

ligand. Such ligand fields occur in all chemical environ-ments. Their effects are generally negligible, except for d orbitals, either because the fields are small (as in the case of f orbitals) or because the orbitals are equally populated as is usually the case for p orbitals. Thus our discussion of ligand fields is confined to d element chemistry.

The strength of the ligand field effect is marked in the case of the transition elements as their ions are small, and the M^{2+} and M^{3+} ions are thus centres of high charge density and are strongly coordinated by lone pair donors such as water or ammonia. The case of the first row of the transition block is particularly interesting as the energy differences introduced by ligand field splittings are of the same size as the exchange energy losses involved in electron pairing. The effects of the fields of different ligands on ions of differing numbers of d electrons, and in different environments, show up in the numbers of unpaired d electrons. These are readily determined by magnetic measurements. Ligand field effects are also seen in a number of other properties such as ionic radii, lattice energies, reaction mechanisms, and electronic spectra, but it was the magnetic effects which first attracted attention and gave rise to the current interest in *ligand field theory*. The application of the theory may be examined in more detail in the case of octahedral complexes of the first row transition elements. This is the commonest coordination number shown by these elements, in solution, in solvated or coordinated individual ions, and also in solids such as the oxides or fluorides. In later sections, the extension of the theory to other coordination numbers and to the heavier elements will be discussed.

12.3 Ligand field theory and octahedral complexes

Regular six-coordination is most readily pictured by placing the ligands at the plus and minus ends of the three coordinate axes. In the xy plane, the positions of such ligands relative to the d orbitals is shown in Figure 12.2a, while the corresponding diagrams for the xz and yz planes are shown in Figures 12.2b and c. In the xy plane, the orbital d_{xy} lies between the

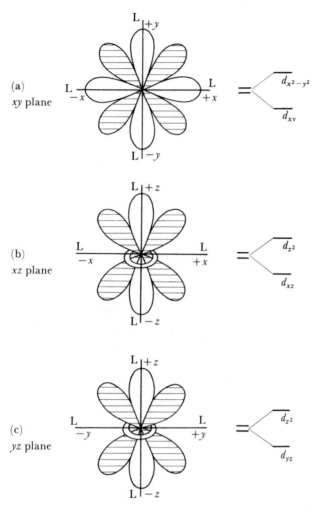

FIGURE 12.2 *Positions of ligands and d orbitals in an octahedral complex:*
(a) *the xy plane*, (b) *the xz plane*, (c) *the yz plane*
The d_{xy}, d_{yz}, and d_{zx} orbitals are indicated by shading.

ligands while $d_{x^2-y^2}$ points directly at the ligands. The latter orbital is therefore most affected by the ligand field and is raised in energy relative to the former. In a similar manner, in the xz plane, d_{z^2} is raised relative to d_{xz} and it is also raised relative to d_{yz} as the Figure for the yz plane shows. If the alignments of the three orbitals, d_{xy}, d_{xz} and d_{yz}, relative to the ligands are compared, it will be seen that these are identical. It follows that in the full three-dimensional case, these three orbitals are identical in energy and are stabilized relative to the other two. The remaining orbitals, d_{z^2} and $d_{x^2-y^2}$, are also identical in energy and are destabilized. (It is easier to accept this if it is recalled that d_{z^2} is compounded of two orbitals similar to $d_{x^2-y^2}$.) The combined energy level diagram is therefore composed of two upper orbitals, of equal energy, and three lower orbitals, which are also

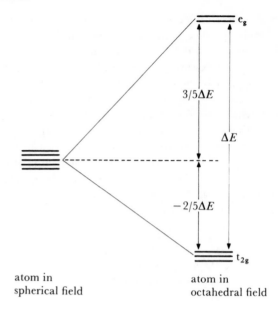

FIGURE 12.3 *Energy level diagram for the d orbitals in an octahedral field*

degenerate (Figure 12.3). The energy zero is conveniently taken as the weighted mean of the energies of these two sets of orbitals; the lower trio are thus stabilized by $-2/5\Delta E$ while the upper pair are destabilized by $3/5\Delta E$, where ΔE is the total energy separation. A number of ways of labelling the lower and upper sets of orbitals will be found in different texts and some are collected in Table 12.6. The e_g and t_{2g} symbols are symmetry labels arising from Group Theory and are now the most commonly used symbols. It may help in remembering them that e signifies a doubly-degenerate state and t, a triply-degenerate one.

TABLE 12.6 Some common symbols for the orbital sets in an octahedral field

| d_{z^2} and $d_{x^2-y^2}$ | e_g | d_γ | γ_3 | Γ_3 |
| d_{xy}, d_{yz}, and d_{xz} | t_{2g} | d_ε | γ_5 | Γ_5 |

The energy gap ΔE is often labelled 10Dq.

Consider the case of a d^1 system in an octahedral field, for example the hydrated titanium(III) ion, $Ti(H_2O)_6^{3+}$. The orbitals are split as in Figure 12.3, and the single d electron naturally enters the lowest available one, here one of the t_{2g} set. In doing so it gains energy, equal to $-2/5\Delta E$, relative to the energy it would have had if the octahedral splitting had not occurred. In this case the energy gain is equal to about 10·5 kJ mol^{-1}. This energy gain, relative to the case of five equal d orbitals, is termed the *ligand field stabilization energy*, or, since it was first remarked in crystals, the crystal field stabilization energy or CFSE. This was first observed when it was found that calculations of the lattice energy of simple transition element oxides and fluorides, by the electrostatic method which was so successful for s element salts, gave answers which did not agree with the experimental values. Including the effect of the crystal field on the *d* orbital energies (section 12.7) led to full agreement. The CFSE is an additional energy increment to the system which has to be added to the other attractive and repulsive energies, both in solids and in calculation of solvation energies and the like.

The size of ΔE is most readily measured spectroscopically by observing the energy of the electronic transitions between the t_{2g} and e_g orbitals. The energy usually lies in the visible or near ultra-violet region of the spectrum and it is such $d-d$ transitions which are responsible for the colours of most transition metal compounds. The magnitude of ΔE depends on the ligand and on the nature of the transition metal ion. One of the simplest examples is that of titanium(III) complexes, where the configuration is d^1. The transition from the t_{2g} to e_g level of the single electron, gives rise to a single absorption band in the visible region (Figure 12.4). The

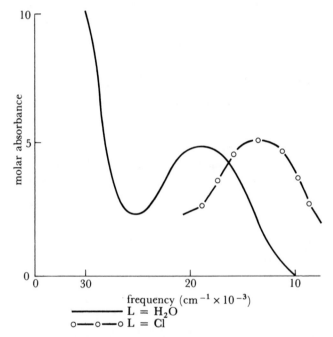

FIGURE 12.4 *Representation of the electronic spectra of* Ti(III) *complexes* The full line (———) is a (slightly simplified) representation of the spectrum of $Ti(H_2O)_6^{3+}$ with a maximum at 20 300 cm^{-1}. The broken line ((-o-o-o-) represents the spectrum of $TiCl_6^{3-}$, with a maximum at 13 000 cm^{-1}. The steeply-rising portion of the curve for the hydrate is the edge of the strong, allowed, charge-transfer transition in the ultraviolet.

position of this band gives the size of ΔE, compare the discussion in section 6.5. When the ligand in $Ti^{III}L_6$ changes from L = H_2O to L = Cl$^-$, the position of the absorption band moves to lower energy, thus ΔE for Cl$^-$ is smaller than for H_2O. By examining a whole series of complexes with different ligands, L, the size of ΔE for each ligand may be determined, both for titanium(III) and for other metals in various oxidation states. It is found that the order of increasing ligand effects is approximately constant from one transition ion to the next and increases in the order:

$$I^- < Br^- < Cl^- < F^- < H_2O < NH_3 < en < NO_2^-$$
$$< CN^- \quad (en = ethylenediamine).$$

The ligand field increases by a factor of approximately two from halide to cyanide. A large number of other ligands are to be fitted into this series, of course, but these are representative ligands, and cyanide has the strongest ligand field of all common ligands. The series is termed the *spectrochemical series*. The main effects of the transition metal ion are those due to charge and to Period. The splitting increases by about 30 per cent between members of the same Group in successive Periods, and there is an increase of roughly 50 per cent on going from the divalent to the trivalent ion of any element. These trends are illustrated by the values shown in Table 12.7. The $d-d$ transitions for configurations other than d^1 are rather more complicated and are discussed in section 12.11.

TABLE 12.7 Values of the ligand field splitting ΔE in octahedral complexes

Ion		6Cl$^-$	6H$_2$O	6NH$_3$	6CN$^-$
			(ΔE, kJ mol^{-1})		
Ti^{3+}	$3d^1$		243		
V^{3+}	$3d^2$		226		
Cr^{3+}	$3d^3$	163	213	259	314
Mn^{3+}	$3d^4$		(250)		
Fe^{3+}	$3d^5$		162		
Co^{3+}	$3d^6$		222	296	406
Mo(III)	$4d^3$	230			
Rh(III)	$4d^6$	243	322	406	
Ir(III)	$5d^6$	297			
Pt(IV)	$5d^6$	347			
V^{2+}	$3d^3$		151		
Cr^{2+}	$3d^4$		(170)		
Mn^{2+}	$3d^5$		92	approx.100	
Fe^{2+}	$3d^6$		126		393
Co^{2+}	$3d^7$		113	121	
Ni^{2+}	$3d^8$	88	100	130	
Cu^{2+}	$3d^9$		(150)	(180)	

Values for d^4 and d^9 configurations are approximate because of distortion in these octahedral complexes.

In the case of titanium(III), there is no ambiguity as to the location of the *d* electron, and this is so for the d^2 and d^3 configurations also. The *d* electrons enter the t_{2g} set of orbitals with parallel spins to give CFSE values of $-4/5\Delta E$ for d^2 and $-6/5\Delta E$ for d^3. However, in the case of d^4, two alternative configurations are possible. The first three electrons enter the three t_{2g} orbitals while the fourth may either remain parallel to the first three, thus producing maximum exchange energy, and enter the higher-energy e_g level, or it may pair up with one of the electrons already present in the t_{2g} level and produce maximum crystal field stabilization energy. The first configuration is termed the high-spin or weak field configuration, while the arrangement with the paired electrons is the low-spin or strong field case. In the case of d^4, the CFSE for the $t_{2g}^3 e_g^1$ configuration is $-6/5\Delta E$ $+3/5\Delta E = -3/5\Delta E$, while the CFSE for the low-spin t_{2g}^4 configuration is $-8/5\Delta E$, so that the adoption of the low-

TABLE 12.8 Crystal field stabilization and exchange energies in octahedral configuration

Number of d electrons	Electron configuration t_{2g}			e_g		CFSE ΔE	Exchange energy' K
1	↑					$-2/5$	0
2	↑	↑				$-4/5$	1
3	↑	↑	↑			$-6/5$	3
4 high-spin	↑	↑	↑	↑		$-6/5+3/5$	6
4 low-spin	↑↓	↑	↑			$-8/5$	3
5 high-spin	↑	↑	↑	↑	↑	$-6/5+6/5$	10
5 low-spin	↑↓	↑↓	↑			$-10/5$	$3+1$
6 high-spin	↑↓	↑	↑	↑	↑	$-8/5+6/5$	10
6 low-spin	↑↓	↑↓	↑↓			$-12/5$	$3+3$
7 high-spin	↑↓	↑↓	↑	↑	↑	$-10/5+6/5$	$10+1$
7 low-spin	↑↓	↑↓	↑↓	↑		$-12/5+3/5$	$6+3$
8	↑↓	↑↓	↑↓	↑	↑	$-12/5+6/5$	$10+3$
9	↑↓	↑↓	↑↓	↑↓	↑	$-12/5+9/5$	$10+6$
10	↑↓	↑↓	↑↓	↑↓	↑↓	$-12/5+12/5$	$10+10$

Exchange energies shown separately for the parallel and antiparallel sets.

spin configuration means the gain of ΔE in excess of the CFSE in the high-spin configuration. On the other hand, the exchange energy of four parallel electrons is $6K$ (see section 7.2) while that of the three parallel electrons in the low-spin configuration is only $3K$. Which configuration is actually adopted therefore depends on the relative sizes of ΔE and K. The K values are difficult to determine but remain approximately constant for atoms of the same quantum shell. In the case of the first transition series the loss of exchange energy usually lies within the range of values found for ΔE. Thus, for any configuration where alternative electronic arrangements are possible, there will be a particular value of ΔE where the change from high-spin to low-spin values takes place. A large value of ΔE obviously favours the low-spin arrangement, hence the alternative name of strong field configuration. Alternative electronic configurations are possible for d^4, d^5, d^6 and d^7 ions in octahedral complexes. Table 12.8 lists the values of the CFSE and exchange energy for all the d configurations, while Table 12.9 shows the differences for high- and low-spin configurations in the states d^4 to d^7. Notice that there are no examples known of intermediate configurations such as $t_{2g}^4 e_g^1$. In all cases the electrons are either paired as far as possible or parallel as far as possible. Table 12.10 gives the approximate magnetic moments, based on the 'spin-only' formula for each configuration. The 'spin-only' formula is usually a good approximation for transition metal ions although orbital coupling occurs to a small extent in most cases and is marked in the cases of Co^{2+} and Co^{3+}. Apart from such exceptions, experimental magnetic moments usually agree with those calculated by the 'spin-only' formula (section 6.6) to within ten per cent, quite close enough to distinguish high-spin from low-spin configurations.

The crystal field stabilization energy and the possibility of alternative electronic configurations are the main phenomena which the theory of bonding in compounds of d

TABLE 12.9 Balance of exchange and crystal fields energies for states of alternative configurations in the octahedral field

Number of d electrons	Gain in CFSE of low-spin relative to high-spin configuration	Loss in exchange energy of low-spin relative to high-spin configuration
4	ΔE	$3K$
5	$2\Delta E$	$6K$
6	$2\Delta E$	$4K$
7	ΔE	$2K$

TABLE 12.10 'Spin only' magnetic moments for octahedral arrangements

Number of d electrons	Magnetic moment, Bohr magnetons High-spin	Low-spin
1	1·73	
2	2·83	
3	3·87	
4	4·90	2·83
5	5·92	1·73
6	4·90	0·00
7	3·87	1·73
8	2·83	
9	1·73	
10	0·00	

'Spin only' moment equals $2\sqrt{[S(S+1)]}$, where $S = \frac{1}{2}n$ = number of unpaired spins × spin quantum number.

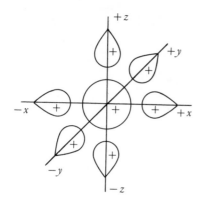

$$\psi_{a_{1g}} = \phi_s + 1/\sqrt{6}\,(l_x + l_{-x} + l_y + l_{-y} + l_z + l_{-z})$$

(a)

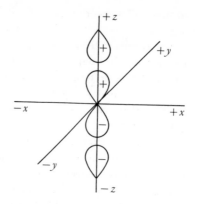

$$\psi_{t_{1u}} = \phi_{p_z} + 1/\sqrt{2}\,(l_z - l_{-z})$$

(d)

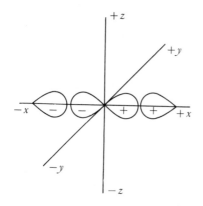

$$\psi_{t_{1u}} = \phi_{p_x} + 1/\sqrt{2}\,(l_x - l_{-x})$$

(b)

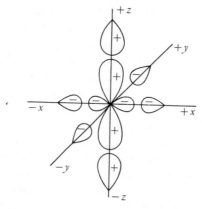

$$\psi_{e_g} = \phi_{d_{z^2}} + 1/(2\sqrt{3})\,(2l_z + 2l_{-z} - l_x - l_{-x} - l_y - l_{-y})$$

(e)

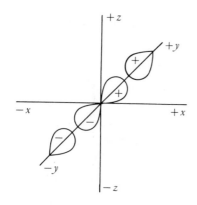

$$\psi_{t_{1u}} = \phi_{p_y} + 1/\sqrt{2}\,(l_y - l_{-y})$$

(c)

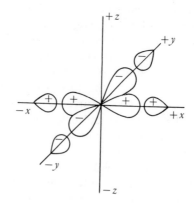

$$\psi_{e_g} = \phi_{d_{x^2-y^2}} + \tfrac{1}{2}(l_x + l_{-x} - l_y - l_{-y})$$

(f)

FIGURE 12.5 *The six bonding molecular orbitals formed by the six ligand orbitals and the s, p, and d_{z^2} and $d_{x^2-y^2}$ orbitals on the central atom in an octahedral complex*

The symbols a_{1g}, t_{1u} etc. are symmetry-indicating labels. Here they are used as a convenient way of distinguishing the orbitals. Note that there are three levels in the set labelled t and two in the set labelled e (see page 158)

elements has to treat. This theory may be formulated in two independent ways, as an electrostatic theory or as a molecular orbital theory. The results of these two approaches are very similar so that either may be used as seems most appropriate and the combined theory is termed the *ligand field theory*. The term crystal field theory is sometimes reserved for specific application to the electrostatic version, but usages vary among different authors.

These approaches will be examined briefly in turn. In the electrostatic approach, the bond energy is held to arise purely from the electrostatic attractions between the central ion and the charges, dipoles and induced dipoles on the ligands, with the repulsions between dipoles, induced dipoles, and charges, on different ligands taken into account. To these forces is to be added the CFSE arising from the *d* electron arrangement in *d* orbitals split by the ligand field. The situation is similar to the electrostatic treatment of ionic crystals with the addition of the crystal field stabilization energy. As no covalent bonds enter into this treatment, the energy level diagram is simply that of the atomic energy levels in the transition element, of which the vital part is the *d* electron diagram as shown in Figure 12.3. This approach has the major advantage that the electrostatic calculations can actually be carried out, without drastic approximations, so that energies of formation and reaction mechanisms or stabilities can be predicted. The disadvantage of the theory is that it neglects the clear evidence that some covalent bonding does occur in transition metal compounds and it can throw no light on those cases, such as nickel carbonyl, where the central element is in a low, zero, or even negative oxidation state when the electrostatic forces would be weak or non-existent.

Covalent bonding is incorporated in the molecular orbital theory. This assumes pure covalent bonding and constructs seven-centred (in an octahedral complex) molecular orbitals from the six ligand orbitals holding the lone pairs which are to be donated to the central atom, together with six orbitals of suitable energy and symmetry, on the central atom. In a transition metal of the first series, these six atomic orbitals are the $3d_{x^2-y^2}$, $3d_{z^2}$, $4s$, and the three $4p$ orbitals. These are the metal orbitals directed towards the ligands and of the right energy. The bonding molecular orbitals from these combinations of atomic orbitals with ligand orbitals are shown in Figure 12.5; the numerical constants are chosen to weight the contributions of the ligand orbitals so that each adds up to unity. The anti-bonding orbitals corresponding to these six are those with the sign reversed between the central orbitals and the ligand combinations. Some of these are shown in Figure 12.6, which also illustrates that orbitals such as d_{xy} cannot form sigma bonds with any ligand combination. It will be noticed that these are delocalized polycentric sigma orbitals, similar in general type to those used in diborane. The six atomic orbitals combine with the six ligand orbitals to form six bonding molecular orbitals and six antibonding molecular orbitals, while the remaining three $3d$ orbitals are nonbonding. The energy level diagram of the molecular orbitals in the octahedral complex is shown in Figure 12.7.

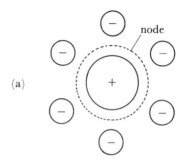

(a)

$$\psi_{a_{1g}}^* = \phi_s - 1/\sqrt{(6)}\,(l_x + l_{-x} + l_y + l_{-y} + l_z + l_{-z})$$

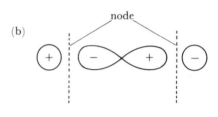

(b)

$$\psi_{t_{1u}}^* = \phi_{p_x} - 1/\sqrt{(2)}\,(l_x - l_{-x})$$

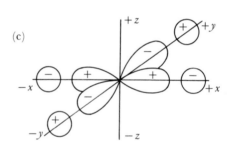

(c)

$$\psi_{e_g}^* = \phi_{d_{x^2-y^2}} - \tfrac{1}{2}(l_x + l_{-x} - l_y - l_{-y})$$

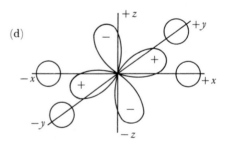

(d)

$\phi_{d_{xy}}$ — no σ-bonding combination possible

FIGURE 12.6 *Some non-bonding and antibonding combinations in an octahedral complex*

Figures (a) to (c) are antibonding combinations of the *s*, one *p*, and one *d* orbital, with ligand orbitals, corresponding to some of the bonding orbitals in Figure 12.5. Figure (d) shows how the d_{xy} orbital is wrongly aligned to form any sigma bond; the other two t_{2g} orbitals are similarly non-bonding. The labels a_{1g}^* etc. are used to distinguish the antibonding orbitals corresponding to the bonding orbitals of Figure 12.5

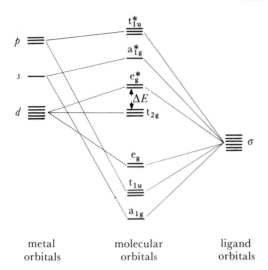

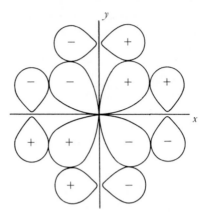

π bonding involving t_{2g} orbitals

metal
orbitals

molecular
orbitals

ligand
orbitals

FIGURE 12.7 *Energy level diagram for the bonding, non-bonding, and anti-bonding molecular and atomic orbitals in a sigma-bonded octahedral complex* The levels are labelled as in Figure 12.5 and 12.6.

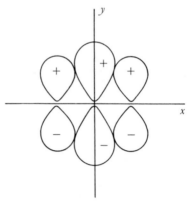

π bonding involving p orbitals

FIGURE 12.8 *Pi-bonding in an octahedral complex*
The t_{2g} and p orbitals on the central atom may combine with orbitals of π symmetry on the ligands, in an octahedral complex.

The twelve electrons from the ligand lone pairs enter the bonding orbitals. Then the nonbonding t_{2g} set of $3d$ orbitals and the antibonding e_g^* pair are the next five orbitals available to accommodate the d electrons on the metal. Thus the descriptions in the electrostatic and molecular orbital theories are very similar. Both theories produce five orbitals in a lower set of three and an upper set of two, separated by ΔE, to accommodate the d electrons. In the electrostatic theory these sets are, respectively, the atomic d_{xy}, d_{yz}, and d_{zx} orbitals and the d_{z^2} and $d_{x^2-y^2}$ atomic orbitals, while, in the molecular orbital theory, the lower set are the same atomic orbitals and the upper set are antibonding molecular orbitals composed of the d_{z^2} and $d_{x^2-y^2}$ atomic orbitals with ligand orbital contributions.

Thus in their essential description of the varying magnetic properties and $d-d$ transitions, these two theories are identical. The molecular orbital theory has the advantage that it is easily extended to include π-bonding and it gives a prediction of the alteration of energy levels in such a case. It is also more useful for the interpretation of spectra as it provides information, not only about the d levels, but also about the higher energy antibonding orbitals to which excitations occur when higher energy quanta are absorbed. On the other hand, the molecular orbital theory suffers from the disadvantage of all wave mechanics, that it is impossible to calculate bond energies, heats of formation and the like directly.

In practice, these two theories may be used interchangeably, as most convenient. Both rely on experimental data to fix the energy levels; for example, ΔE is usually determined spectroscopically.

The extension of the molecular orbital theory to include π-bonding is indicated in Figure 12.8. The t_{2g} and p orbitals are correctly positioned to give π overlaps with the ligands. The energy level diagram of Figure 12.9 shows how π-bonding, by involving the t_{2g} set of orbitals, increases the energy

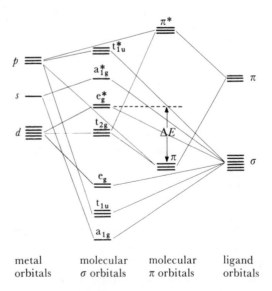

metal
orbitals

molecular
σ orbitals

molecular
π orbitals

ligand
orbitals

FIGURE 12.9 *Energy level diagram for the molecular orbitals in an octahedral complex where pi-bonding occurs*
The important effect of π-bonding is to lower the t_{2g} orbitals and thus increases ΔE.

gap ΔE and hence increases the splitting and the CFSE. The situation shown in Figure 12.9 is the simplest with the total number of ligand orbitals equal to nine, the same as the total number of metal orbitals. It is more likely that the ligands will contribute six, or even twelve π orbitals, and this will produce a number of sets of intermediate π^0 or π^{0*} molecular π orbitals (compare the discussion in section 3.9). However, such more complex behaviour does not affect the principal argument, that the formation of π bonds increases ΔE by lowering the level of the t_{2g} orbitals. Any π^0 levels are likely to be near, or above, the e_g^* levels in energy.

Ligands to the right of the spectrochemical series, especially cyanide and phosphines (which lie between NO_2^- and CN^-) have the capacity to π-bond using either π^* orbitals or d orbitals on the ligand, as indicated in Figure 12.10.

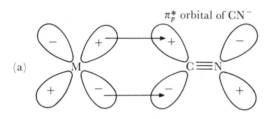

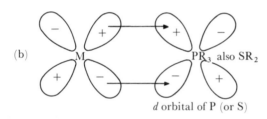

FIGURE 12.10 *Suitable pi-bonding ligand orbitals*
Ligands commonly form π-bonds, either with antibonding pi orbitals, as with cyanide (a), or by using d orbitals, as with phosphines or sulphides (b).

12.4 Coordination number 4

Two configurations other than octahedral are common among the elements of the first transition series. Both involve four-coordination and are the tetrahedral and square planar configurations. The square planar configuration may be considered as derived from the octahedral one by removing the ligands on the z-axis. An elongation of the M—L distances on the z-axis leads to a decrease of the interaction between the ligand field and those metal orbitals with components in the z direction. The energy levels therefore split as indicated in Figure 12.11a with the d_{z^2} level and the d_{xz} and d_{yz} levels falling below the others. The $d_{x^2-y^2}$ and d_{xy} levels rise slightly as the metal-ligand distances in the xy plane shorten a little because of the decrease in repulsion from the z ligands. Such an intermediate case corresponds to elongation of the metal-ligand distances on the z-axis, to unsymmetric substitution on the z-axis in a complex such as

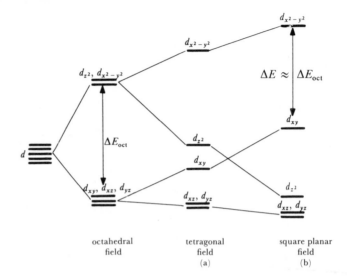

FIGURE 12.11 *Effect on orbital energies of removing ligands on the z axis, starting from an octahedron:* (a) *the tetragonal field with non-equivalent z ligands,* (b) *the square planar field with no z ligands*

MX_4Z_2, (both of which are tetragonal distortions) or to the case of the five-coordinated square pyramidal configuration. If the z ligands are removed completely to give the square planar configuration, the energy level diagram of Figure 12.11b results. Here, the d_{z^2} and d_{xy} levels have crossed over. Notice that, as the configuration in the xy plane is similar to that in the octahedron, the energy separation between $d_{x^2-y^2}$ and d_{xy} remains practically the same as the octahedral ΔE.

In the tetrahedral arrangement, the d orbitals are split into a lower set of two (called e in this symmetry) and an upper set of three (t_2). No orbital points directly at a ligand in the tetrahedral case, but the d_{xy} type lies closer to the ligand than $d_{x^2-y^2}$ or d_{z^2}. Figure 12.12 shows that the distances are as half the side of a cube compared with half the face diagonal. This lack of direct interaction between orbitals and ligands in the tetrahedral configuration reduces the magnitude of the crystal field splitting by a geometrical factor equal to $2/3$, and the fact that there are only four ligands instead of six reduces the ligand field by another $2/3$. The total splitting in the tetrahedral case is thus approximately $2/3 \times 2/3 = 4/9$ of the octahedral splitting. In theory, alternative electron configurations are possible for the cases d^3, d^4, d^5 and d^6 in the tetrahedral field but the gain in CFSE is reduced by the smaller size of the splitting, and the loss of exchange energy is never counterbalanced. Thus all

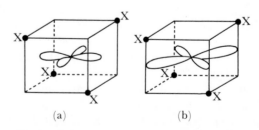

FIGURE 12.12 *Alignment of the d orbitals and the ligands in a tetrahedral complex:* (a) *the* $d_{x^2-y^2}$ *orbital,* (b) *the* d_{xy} *orbital*

TABLE 12.11 CFSE values in a tetrahedral field

Number of d electrons	Electronic configuration e_g	t_{2g}	CFSE (as ΔE_{tetr}) (assuming $\Delta E_{tetr} = 4/9\ \Delta E_{oct}$)	(as ΔE_0)
1	↑		$-3/5$	-0.27
2	↑ ↑		$-6/5$	-0.53
3	↑ ↑	↑	$-6/5+2/5$	-0.36
4	↑ ↑	↑ ↑	$-6/5+4/5$	-0.18
5	↑ ↑	↑ ↑ ↑	$-6/5+6/5$	0.00
6	↑↓ ↑	↑ ↑ ↑	$-9/5+6/5$	-0.27
7	↑↓ ↑↓	↑ ↑ ↑	$-12/5+6/5$	-0.53
8	↑↓ ↑↓	↑↓ ↑ ↑	$-12/5+8/5$	-0.36
9	↑↓ ↑↓	↑↓ ↑↓ ↑	$-12/5+10/5$	-0.18
10	↑↓ ↑↓	↑↓ ↑↓ ↑↓	$-12/5+12/5$	0.00

tetrahedral complexes are high-spin. The CFSE values for these high-spin configurations are given in Table 12.11.

If Table 12.11 is compared with Table 12.8, it will be seen that the CFSE in an octahedral configuration is always greater than that in the corresponding tetrahedral configuration, except in the cases of d^0, d^5 and d^{10} where both values are zero. The configurations with the next smallest CFSE loss in the tetrahedral field compared to the octahedral field, are d^1 and d^6 if the octahedral state is high-spin. The adoption of four-coordination rather than six-co-ordination is expected, in general, to be accompanied by loss of energy of formation as only four interactions occur instead of six. In addition, there is commonly a loss of CFSE as well.

These considerations, and the data for the CFSE values, allow a prediction of the most probable cases in which four-coordination will be found for transition elements of the first series. The tetrahedral configuration is expected to be un-favourable compared with the octahedral one, except in the case of large ligands with low positions in the spectro-chemical series, or in the ions with 0, (1), 5, (6), or 10 d electrons. The larger ligands will experience steric hindrance to formation of six-coordinated complexes and the inter-actions will also be reduced due to the increased metal-ligand distances. The low position in the spectrochemical series ensures that any loss of CFSE is not too serious due to the low intrinsic value of ΔE (similarly, a low charge on the transition metal ion affects the ΔE value so divalent ions should form tetrahedral complexes more readily than tri-valent ones, and *a fortiori* for lower charges). Finally, the d configurations listed are those with lowest CFSE differences. In practice, tetrahedral complexes are typically formed by the halides (except fluoride) and related ligands.

The case of the square-planar configuration is rather different. Because the bond distances in the xy plane are essentially the same in octahedral and square planar con-figurations, steric effects are negligible and the increase in attractions due to forming six bonds rather than four will normally overwhelmingly favour the regular octahedral complex. Changes in CFSE for low numbers of d electrons either favour the octahedral case or are small. Consider, however, the case of d^8. In an octahedral complex, there are two electrons in the e_g level, while the square planar con-figuration allows these to be paired in the d_{xy} orbital (Figure 12.13) with a gain in CFSE of about $2\Delta E_{oct}$. (This is only

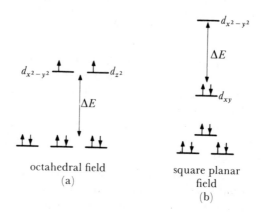

FIGURE 12.13 *The configuration of electrons in a d^8 complex:* (a) *octa-hedral, and* (b) *square planar*

approximate as the lower levels in the square planar case do not match the octahedral t_{2g} levels, but they are quite close.) This is offset by a loss in pairing energy from the $13K$ of the octahedral case to $12K$ for two sets of four electrons. However, if ΔE is sufficiently large, it is possible for the gain in CFSE to overbalance the loss in bonding interactions and the small loss in exchange energy. It is found that square planar complexes are indeed formed by d^8 ions with ligands to the right of the spectrochemical series: for example Ni^{2+} forms a square planar cyanide complex, $Ni(CN)_4^{2-}$, while its hydrate or ammine are octahedral, e.g. $Ni(NH_3)_6^{2+}$. The larger ΔE values of heavier elements or of more highly-charged elements extend the scope of formation of square complexes.

Thus all complexes, even halides, of platinum(II) and gold(III) (both d^8) are square planar.

This extra CFSE found in square planar complexes of d^8 elements also occurs for the d^7 and d^9 configurations. However, the CFSE gain is only ΔE. Table 12.12 lists the species which typically form square planar complexes. While such square complexes of the first row elements are comparatively rare, those formed by heavy transition elements are the majority of the representatives of these oxidation states, especially for the d^8 configurations.

TABLE 12.12 Species forming square planar complexes

d electron configuration	Species	Approx CFSE	Unpaired electrons
d^8	Ni(II), Pd(II), Pt(II) Rh(I), Ir(I) Au(III)	$2\Delta E_{oct}$	0
d^9	Cu(II), Ag(II)	ΔE_{oct}	1 (in $d_{x^2-y^2}$)
d^7	Co(II)	ΔE_{oct}	1 (in d_{xy})
d^6	Fe(II)	$\frac{1}{2}\Delta E_{oct}$	2
d^4	Cr(II)	$\frac{1}{2}\Delta E_{oct}$	4

The Cr(II) spin corresponds to one electron in each of the four stable orbitals, and the Fe(II) value indicates two filled and two half-filled orbitals. The CFSE are with respect to the octahedral configuration, those for d^4 and d^6 being for the weak field configuration.

In configurations other than d^8, the CFSE gain relative to octahedral is small, and most of these configurations are distorted in the octahedral case, so it is often difficult to decide what has happened. For example, copper(II) compounds often show four short bonds in a square plane with two longer ones, or even three sets of pairs of bonds with different lengths.

The distortion mentioned in the last paragraph arises whenever the d_{z^2} and $d_{x^2-y^2}$ orbitals are unequally occupied. If, for example, there is one electron in the d_{z^2} orbital, ligands on the z axis are more shielded from the nuclear field than are ligands on the x and y axes. The ligand-metal distances in the z direction are therefore shorter than those in the xy plane. If the electron is, instead, in the $d_{x^2-y^2}$ orbital the four distances in the xy plane are shorter. Such distortions, which are less simple than described here, are one manifestation of the Jahn-Teller Theorem, which states that if a system has unequally-occupied, degenerate energy levels it will distort so as to raise the degeneracy. Cases where distortions are expected are d^4 high-spin, d^7 low-spin, and d^9. Though distortions can theoretically involve the t_{2g} level as well, these are too small to be detected.

12.5 Stable configurations

With all the factors discussed in the last three sections in mind, it is possible to make some general remarks about the stabilities of various d configurations. These apply particularly to the first row elements which only occasionally show coordination numbers other than four or six (see section 12.6). The extent to which these apply in practice will become clear when the chemistry of the individual elements is discussed.

The traditionally stable configurations of the empty, half-filled, and filled shells should still be stable for d elements; the stability in the last two cases stemming from the high exchange energy as well as from the general symmetry of the electron clouds. In octahedral environments, d^5 will be unusually stable only in the high-spin arrangement. Hence, it should be more stable in the divalent ion Mn^{2+} than in the trivalent d^5 element, Fe^{3+}, and also more stable with ligands of relatively low field in each case. The configurations d^0 and d^{10} may well be found in stable tetrahedral complexes.

In the octahedral configuration, the states d^3 and low-spin d^6 are also expected to have special stability as they correspond to filled and half-filled t_{2g} orbitals. In this case, stability increases with increasing ΔE and should be most marked for trivalent ions, here Cr^{3+} (d^3) and Co^{3+} (d^6).

It will be noted that the d^4 configuration in an octahedral environment lies between two others that are especially stable, so that d^4 species are expected to be unstable, both to oxidation to d^3 and to reduction to d^5. In particular, the $M^{2+}d^4$ state should readily oxidize to the $M^{3+}d^3$ ion where the increase in charge increases ΔE and the CFSE, while the $M^{3+}d^4$ ion should be readily reduced to the $M^{2+}d^5$ ion which is favoured by the reduced ΔE. Such behaviour is indeed found, the ions in question being respectively Cr^{2+} and Mn^{3+}.

At the d^8 configuration, there is competition between the square planar and the octahedral arrangements. In nickel(II) the latter is more common but the former is dominant for Pd(II), Pt(II), and Au(III). In both configurations, d^8 tends to be more stable than either d^7 or d^9 which have neither the CFSE of square planar d^8 nor the equally-occupied e_g arrangement of octahedral d^8. There is some tendency for d^9 to go d^{10} but a number of factors come into play here as the oxidation states of the copper Group elements show. It might be noted that d^{10} species have some tendency to linear configurations as Cu(I) and Au(I), as well as Hg(II), show. This is in addition to their ready adoption of tetrahedral structures, and the two-coordinate species are found for relatively low charge densities on the ions, either in large ions or in M^+ species.

The discussion is summarized in Table 12.13

12.6 Coordination numbers other than four or six

While the majority of compounds of the first row elements exhibit octahedral, tetrahedral or square planar coordination, there are now well-established examples of a range of other coordinations. For the heavier transition elements, octahedral and square planar configurations are common, but higher coordination numbers are well known. An alternative form of six-coordination, the trigonal prism of Figure 12.14a is also found but is rare. It is the configuration

TABLE 12.13 General stabilities of d configurations

Number of d electrons	Comments
0	Stable in earlier Groups as the Group oxidation state.
1, 2	Tend to reduce to d^0, e.g. Ti^{3+}, Ti^{2+}, V^{3+}.
3	Stable, especially for trivalent ion, e.g. Cr^{3+}.
4	Unstable, e.g. Cr^{2+} oxidizes to Cr^{3+}; Mn^{3+} reduces to Mn^{2+}.
5	Stable, especially Mn^{2+}; Fe^{3+} is also relatively stable.
6	Stable in spin-paired state, e.g. Co^{3+} (except in hydrate); Fe^{2+} is also relatively stable. Pt(IV) very stable.
7	Relatively unstable, Co^{2+} oxidizes except as hydrate or halide complex; Ni^{3+} is scarcely known.
8	Stable, Ni^{2+} as octahedral and square planar complexes; Pd(II), Pt(II), Au(III) very stable square planar species.
9	Relatively unstable except in case of Cu(II).
10	Stable, e.g. Ag^+; zinc Group never show states above II.

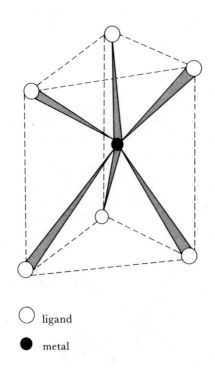

(a)   ◯ ligand   ● metal

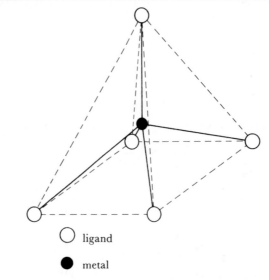

(b)   ◯ ligand   ● metal

FIGURE 12.14 *Lower-symmetry coordinations (a) trigonal prism form of six-coordination (b) square pyramid form of five-coordination*
Trigonal prismatic coordination is often found when the donor atoms are sulphurs, as in MoS_2 or in the dithiolene complexes $M(S_2C_2R_2)_3$ where M = V, Mo, Re. In the square pyramid, note that the metal atom lies above the base plane making the base-L/metal/apex-L angle greater than $90°$.

of arsenic in NiAs (Chapter 4) and is found for molybdenum in MoS_2.

Coordination numbers greater than six are most usually found among the compounds of the second or third row transition elements, and also for the lanthanides and actinides as illustrated in Chapters 10 and 11. For these higher co-ordination numbers, it is usual to find several alternative shapes, and the energy differences between them are commonly very small. There is at present no generally accepted, unambiguous, method available for predicting which shape will be adopted, either for the higher coordination numbers or for five-coordination.

Coordination Number Two. This is found in d^{10} configurations

such as copper(I), silver(I), gold(I) and mercury(II). The arrangement is always linear, as in $Ag(NH_3)_2^+$ or $AuCl_2^-$, and is simply that derived from the electron pair repulsion theory.

Coordination Number Five. Five-coordination is known for most of the first row elements, but is uncommon for the heavier transition metals. As well as the trigonal bipyramid (Figure 3.23), square-pyramidal five-coordination is found for transition metal compounds. As the discussion of five electron pairs in Chapter 3 showed, there is little energy difference between the trigonal bipyramid and square pyramid, and the balance of stability is readily tipped one way or the other by additional stabilizations arising from the d electron configuration, the shape and properties of the ligands, or from interactions in the solids. The square pyramid configuration has the central metal atom raised *above* the plane of the four base atoms (Figure 12.14b). Notice that this increases the bond angles at the central atom, and is the opposite distortion from that found in AB_5L species, like BrF_5, derived from the octahedron, where the repulsion of the lone pair decreases BAB angles so that the central atom lies *below* the base plane. (Compare BrF_5 with Figure 13.10.)

The small differences in energy between the two shapes is strikingly illustrated by the compound

$$(Cr\ en_3)[Ni(CN)_5],\tfrac{3}{2}H_2O,$$

whose crystal structure contains both square pyramidal and distorted trigonal bipyramidal $Ni(CN)_5^{3-}$ ions. Table 12.14 gives some illustrative examples of five-coordinate complexes, chosen either for their relative simplicity or because a family of compounds of one type exist. In some of the structures, distortions from the ideal shapes occur, while many other cases are known where the structure is intermediate between the two geometries.

TABLE 12.14 Examples of five coordination

	Trigonal bipyramid	Square pyramid
titanium(IV)	$TiCl_5^-$, $TiOCl_2(NMe_3)_2$	Y_2TiO_5
(III)	A	
vanadium(IV)	$VOCl_2(NMe_3)_2$	$VO(acac)_2$ (Figure 13.10)
(III)	A	
niobium(V)	$NbCl_5$ (in gas phase only)	
tantalum(V)	$TaCl_5$ (in gas phase only)	
chromium(III)	A	
(II)	B	
molybdenum(V)	$MoCl_5$ (in gas phase only)	$Mo_3O_{10}^{5-}$ includes Mo^VO_5 units
manganese(II)	B	C
rhenium(V)	$ReOX_4^-$ (X = Cl, Br, I)	
iron(0)	$Fe(CO)_5$	
(II)	B	C
(III)		$Fe(S_2CNEt_2)_2Cl$
cobalt(II)	B	C
nickel(II)	$Ni(CN)_5^{3-}$ (a), B	$Ni(CN)_5^{3-}$ (a)
platinum(II)	$Pt(SnCl_3)_3^{3-}$	
copper(II)	B, $Cu(bipy)_2X$	$[Cu(NO_3)_2(C_5H_5NO)_2]_2$, $Cu(acac)_2$ quinoline
zinc(II)	B	C, see also Figure 12.14

A = $M^{III}X_3L_2$ (X = halogen, L = ligand like NMe_3).
B = $[M^{II}$ (tetra N)Br]$^+$ where tetra N = $(Me_2NCH_2CH_2)_3N$, compare Appendix B.
C = $(terpy)M^{II}Cl_2$ (terpy = terpyridyl, compare Appendix B).
(a) Pentacyanonickel(II) is found in both shapes in the $Cr(en)_3^{3+}$ compound.

Coordination Number Seven. Seven-coordinate complexes are commonly found in one of three shapes. The most regular is the regular pentagonal bipyramid adapted by ReF_7, OsF_7 and by the ions MF_7^{3-} (M = Zr, Hf) in their sodium salts. As the in-plane angle is only 72°, some distortion of the five equatorial ligands is likely to reduce steric interactions, either by buckling out of the plane or by variation of the M−F bond lengths in the equatorial plane. Some uranyl compounds, such as $UO_2F_5^{3-}$, also show this shape as the

UO_2 unit has to be linear and the other five substituents lie in a plane at right angles to this, around the uranium atom (compare Figure 11.6 for the corresponding UO_2L_6 situation).

A second seven-coordinate arrangement found, is that formed by inserting an extra substituent into the triangular face of an octahedron, and spreading out the three ligands forming this face. One example is $NbOF_6^{3-}$ illustrated in Figure 14.4a (which emphasises that the seventh ligand is on the three-fold axis).

The third shape found for seven-coordination is that obtained by inserting a substituent above one of the rectangular faces of a trigonal prism. This structure is shown in the ammonium salt of ZrF_7^{3-} (Figure 14.1) or by MF_7^{2-} (M = Nb, Ta). These structures are all very similar in energy and change from one to another is readily induced, for example by the change of counterion in the hepta-fluorozirconium complexes.

Coordination Number Eight. While the eight-coordinate shape of highest symmetry is the cube, this is common only in extended structures like the ionic CsCl or CaF_2 structures or in metallic body-centred cube forms. A smaller repulsion between ligands results when one face of the cube is twisted by 45° relative to the opposite one to form the square anti-prism. This structure is shown by TaF_8^{3-} (Figure 14.4b) and ReF_8^{2-}, or by $Zr(acac)_4$ and $U(acac)_4$ (Figure 11.5).

The other common eight-coordinate structure, the dodecahedron or bis-bisphenoid (bisdisphenoid), may also be derived from the cube. As every second corner of a cube defines a tetrahedron (as can be seen in Figure 12.12), the cube may be regarded as two interpenetrating tetrahedra. If one of these tetrahedra is flattened and the other elongated, the eight-coordinate structure of Figures 14.2 or 14.16 results. A tetrahedron distorted in this way is called a bisphenoid, hence the name bisbisphenoid. If the distortions are equal

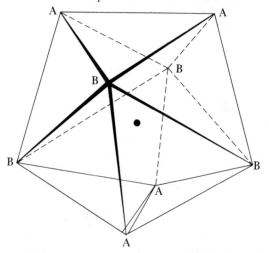

ligand atoms at A and B   ● metal atom

FIGURE 12.15 *The dodecahedron*
This eight-coordinate shape is related to that of the cube by elongating one of the interpenetrating tetrahedra (A vertices) and flattening the other (B vertices). For the regular figure, all the faces are equilateral triangles.

and produce equilateral triangular faces, the structure is a regular dodecahedron, shown in Figure 12.15 where the vertices labelled A define one bisphenoid, and those labelled B, the other. A number of structures approximate to a regular dodecahedron, for example the MF_8^{4-} species of Figure 14.2, but quite distorted forms also occur. These result particularly from the presence of bidentate ligands (see section 12.7) where the two donor atoms are close together. Thus $Co(NO_3)_4^{2-}$, with two oxygens bridging Co to each nitrate, forms a distorted dodecahedron with the Co—O bond lengths equal to 204 pm in the elongated bisphenoid and averaging 255 pm in the flattened one. A similar distorted structure is found in the chromium(V) peroxy compound $K_3Cr(OO)_4$ where both oxygens of the O—O group are bonded to Cr.

The eight-coordination of $UO_2(NO_3)_2.2H_2O$, Figure 11.6, shows a fourth shape—the hexagonal bipyramid. This is only found for such MO_2 species where the O—M—O group has to be linear.

Coordination Number Nine. The one configuration found for nine-coordination is that formed by placing one ligand above each of the three rectangular faces of a trigonal prism. This structure is illustrated in Figure 14.21 for ReH_9^{2-} and is also shown by the lanthanide hydrates like $Nd(H_2O)_9^{3+}$ mentioned in Chapter 10.

Higher Coordination Numbers. As a transition element of the d block has only nine valence shell orbitals (five d plus one s plus three p) the maximum coordination number is limited to nine unless metallic bonding (compare close-packed structures with coordination number twelve, Chapter 4), purely electrostatic bonding, or electron deficient bonding (compare section 8.6) occurs. One example is $Zr(BH_4)_4$ which forms two hydrogen bridges to each BH_4 group at room temperature, but transforms to a three-bridge structure at $-160°C$, $Zr(H)_3BH$. Thus the Zr is twelve-coordinate in the low temperature form.

If f orbitals are available, this restriction no longer holds as illustrated by the ten-coordinate $La(H_2O)_4$ EDTA compound mentioned in Chapter 10.

12.7 Effect of ligand on stability of complexes

The type of ligand has a distinct influence on the stabilities of complexes and this must be discussed in more detail. The spectrochemical series gives one classification of the ligands and some points about this should be noticed. In an extended form of the series, it is seen that position depends largely on the donor atom in the order:

Halogen or S < O < N < P < C

Thus, nitrate which donates through O is much weaker than nitrite which donates through N. An even more striking example is the thiocyanate group, SCN, which is sometimes found coordinated through S, when it has a weak ligand field, and sometimes coordinates through N, when it shows a much stronger field. A similar example is shown by the isomeric form of nitrite, nitrito, which coordinates as ONO,

donating through O, instead of the NO_2 form of nitrite which donates through N. The nitrito form has a weaker ligand field than the nitrite form. These features govern the effect of the ligands by way of their contribution to the crystal field stabilization energy, but this term may be only a minor contribution to the overall stabilization, and it is necessary to take care not to equate strong crystal field unthinkingly with stability.

A number of attempts have been made to find general relationships which indicate overall stability. One of the most extensive is Pearson's classification into *hard and soft* ligands and metal ions (or, using the Lewis nomenclature where a lone pair donor is a base and an acceptor is an acid, the metal ions are classified as hard or soft acids, and the ligands as hard or soft bases). In this, a ligand is a hard base if it is non-polarizable as for most ligands with a first row donor atom, and the ligand is a soft base if it is polarizable, as with sulphur or phosphorus ligands. Similarly, a metal ion is a soft acid if it has easily polarizable electrons, or is large, or has a low charge, while a hard acid is a metal ion of high charge, small size and with valence electrons which are not polarizable. The classification extends beyond transition elements and some examples of hard and soft ligands and metal ions are given in Table 12.15.

TABLE 12.15 Classification of ligands and ions under hard/soft formalism

Ligands or bases
Hard H_2O, ROH, R_2O, OH^-, OR^-, NO_3^-, RCO_2^-, SO_4^{2-}, CO_3^{2-}, $C_2O_4^{4-}$, PO_4^{3-} (donors through O) NH_3, NR_3, NHR_2, NH_2R, Cl^-
Soft R_2S, RSH, RS^-, SCN^- (S donors) R_3P, R_3As, I^-, CN^-, H^-, R^-, $S_2O_3^{2-}$
Borderline py, Br^-, N_3^-, SO_3^{2-}, NO_2^-
Metal ions or acids (charges are formal only)
Hard Mn^{2+}, Cr^{3+}, Fe^{3+}, Co^{3+}, Ti^{4+}, VO_2^+, VO^{2+}, Zr^{4+}, MoO^{3+} H^+, all s element ions, M^{3+} for M = Al, Ga, In, Sc, Y, Ln
Soft Cu^+, Ag^+, Au^+, Hg_2^{2+}, Hg^{2+}, Pd^{2+}, Pt^{2+}, Pt^{4+} Tl^+, Tl^{3+}
Borderline Fe^{2+}, Co^{2+}, Ni^{2+}, Cu^{2+}, Zn^{2+}, Ru^{2+}, Os^{2+}, Rh^{3+}, Ir^{3+} Sn^{2+}, Pb^{2+}, Sb^{3+}, Bi^{3+}

The most important generalization about stabilities is then that soft ligands form stable complexes with soft metal ions, and hard ligand-hard ion complexes are also stable. Mixtures of hard ion-soft ligand or soft ion-hard ligand are less stable. A second general trend is, that the substitution of a ligand of a particular type tends to make the metal ion behave more as that type of acid. Thus a borderline ion like Fe^{2+}, becomes harder when a number of hard ligands are coordinated so that the intermediate $Fe(H_2O)_3^{2+}$ (say) is more likely to add further water molecules than would a ferrous ion coordinated by soft ligands.

This hard-soft classification has been greatly extended and ramified, and some features are still subject to controversy. However, the basic predictions derived from the simple version are most useful. One example is provided by thiocyanate, SCN^-, which can bond through S or through N (in isothiocyanates). The S-bonded form is a soft ligand, and is preferred in complexes of soft ions such as Hg^{2+}. The N-bonded form is harder and is most usually found in first row complexes. A neat example is $HgCo(NCS)_4$ which is an extended lattice formed of $Hg-SCN-Co$ links.

The next effect is the *chelate effect*. If a ligand contains more than one donor atom it may coordinate to more than one position on the cation giving ring formations. The existence of such chelation in a complex is accompanied by increased stability and this is shown by increased heats of formation and resistance to substitution, and also by the higher position in the spectrochemical series of the chelating ligand as compared to an analogous non-chelating ligand. An example is provided by the case of ethylenediamine, $H_2NCH_2CH_2NH_2$, where both nitrogen atoms can donate, forming a five-membered ring. Such a reagent with two donor atoms is termed *bidentate*. It is observed that the treatment of ammonia complexes with ethylenediamine results in the displacement of the ammonia:

$$Co(NH_3)_6^{3+} + 3 \text{ en} = Co(en)_3^{3+} + 6NH_3$$
$$(en = ethylenediamine)$$

The driving force of such a replacement is probably the entropy change. The equation above has four particles on the left hand side and seven on the right, so that reaction proceeds with increase of entropy. For example, the change in free energy in the copper complexes are:

$$Cu(en)_2^{2+} - Cu(NH_3)_4^{2+}$$
difference in $\Delta G = 18\cdot0 \text{ kJ mol}^{-1}$
difference in $\Delta H = 10\cdot9 \text{ kJ mol}^{-1}$
difference in $T\Delta S = -7\cdot1 \text{ kJ mol}^{-1}$
(Recall that $\Delta G = \Delta H - T\Delta S$)

This entropy term thus provides about 40 per cent of the free energy change, whereas the substitution of one monodentate ligand by another, for example the exchange of ammonia for water, usually has only a small entropy effect which is commonly neglected in calculations. The optimum ring size for elements of the transition series appears to be a five-membered ring. Thus the substitution of 1,3-propanediamine for ethylenediamine, increasing the ring size to six atoms (Figure 12.16b), results in a loss of free energy of formation of 16.3 kJ mol^{-1} in the case of the copper complex above and 35.6 kJ mol^{-1} in the case of the nickel complex $Ni(diamine)_3^{2+}$.

The chelating effect is increased still further if the chelating molecule has more donor atoms. The triamine (trien) $H_2NCH_2CH_2NHCH_2CH_2NH_2$ forms an even more stable complex than ethylenediamine. Chelating agents with up to

six donor atoms are available, an example of the latter being ethylenediaminetetra-acetic acid, EDTA, shown in Figure 9.4. The common complexing and precipitating agents of analysis are nearly all chelating agents which form five- or six-membered rings with the metal atom which is being determined. Some examples are shown in Figure 12.16. Polydentate chelating agents may give rise to complexes of unusual coordination, for example, the less common five-coordination in the zinc complex shown in Figure 12.17.

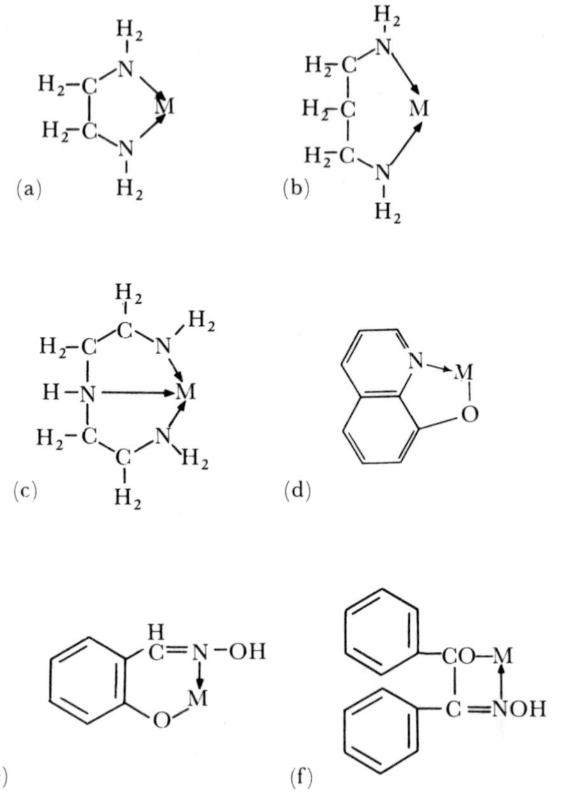

FIGURE 12.16 *Some chelating ligands:* (a) *ethylenediamine* (en), (b) *1,3-propanediamine*, (c) *diethylenetriamine* (trien), (d) *8-hydroxyquinoline* (oxine), (e) *salicylaldoxime*, (f) *α-benzoinoxime* (cupron)

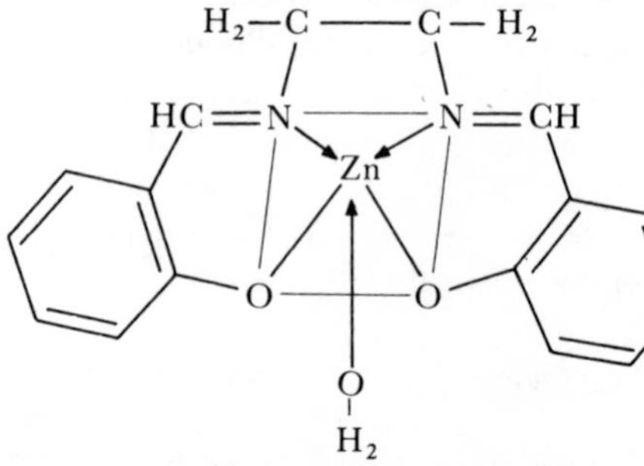

FIGURE 12.17 *A five-coordinated zinc complex, NN'-disalicylidene-ethylenediamine zinc hydrate*

The quadridentate ligand coordinates through its two nitrogen atoms and two oxygen atoms to the zinc, giving a structure of a shallow square pyramid, with the zinc atom 30 pm above the plane of the four donor atoms. The water molecule is coordinated to the zinc with the $Zn-OH_2$ bond in the direction of the axis of the pyramid.

Quantitative indication of the process of forming a complex comes from the evaluation of the stability constants which characterize the equilibria corresponding to the successive addition of ligands. That is, we can consider the steps

$$M + L \rightleftharpoons ML$$
$$ML + L \rightleftharpoons ML_2$$

and so on down to

$$ML_{n-1} + L \rightleftharpoons ML_n$$

These are characterized by equilibrium constants K_1, K_2, ..., K_n such that

$$K_1 = [ML]/[M][L]$$
$$K_2 = [ML_2]/[ML][L]$$

and,

$$K_n = [ML_n]/[ML_{n-1}][L]$$

These constants K, are termed *stepwise formation constants*. An alternative formulation is to consider the overall formation reaction

$$M + nL \rightleftharpoons ML_n$$

characterized by the nth *overall formation constant* β_n.

$$\beta_n = [ML_n]/[M][L]^n = K_1 K_2 \ldots K_n$$

In most preparations, the complex is formed in aqueous solution, and the stability constants refer to steps where L replaces coordinated water. The evaluation of formation constants usually calls for a good deal of experimental ingenuity and will not be discussed here. We may simply note some values and their interpretation. For example, the logarithms of the successive formation constants of various nitrogen complexes formed in aqueous solution are shown in Table 12.16.

TABLE 12.16 Stepwise formation constants of some nitrogen complexes

Ligand		Co^{2+}	Ni^{2+}	Cu^{2+}	Zn^{2+}
NH_3	$\log K_1$	2·1	2·8	4·2	2·4
	$\log K_2$	1·6	2·2	3·5	2·4
	$\log K_3$	1·1	1·7	2·9	2·5
	$\log K_4$	0·8	1·2	2·1	2·2
	$\log K_5$	0·2	0·8	−0·5	
	$\log K_6$	−0·6	0·03		
en	$\log K_1$	6·0	7·5	10·6	5·7
	$\log K_2$	4·8	6·3	9·1	4·7
	$\log K_3$	3·1	4·3	−1·0	1·7
trien	$\log K_1$	8·1	10·7	16·0	8·9
6-en	$\log K_1$	15·8	19·3	22·4	16·2

en = ethylenediamine (Figure 12.16a): trien = diethylene-triamine (Figure 12.16c).

6-en = pentaethylenehexamine, the analogous 6-nitrogen molecule (compare Appendix B)'

The values in Table 12.16 illustrate a number of features which give a quantitative indication of the stability properties discussed above. There is first a general tendency for K values to fall as the number of ligands increases. This is probably due, at least in part, to the statistical effect that the number of sites for substitution is reduced as substitution proceeds.

The values for nickel show a steady progression to $Ni(NH_3)_6^{2+}$ and zinc, similarly, goes steadily to $Zn(NH_3)_4^{2+}$. The cobalt values show that the sixth ligand is unstable, probably reflecting the effect of the e_g electron, and this is even more marked for copper where the fifth NH_3 is only added in presence of a large excess of ammonia and the sixth is not taken up at all. This reflects the general tendency of the d^9 ions to form four strong bonds and two, much longer, weaker ones.

It will also be noted, comparing K_1 values for example, that the stability order for M^{2+} ions increases to copper. This is part of a more extensive sequence, the Irving-Williams order, which shows that stability constants vary

$$Mn(II) < Fe(II) < Co(II) < Ni(II) < Cu(II) > Zn(II)$$

towards a particular ligand.

The values also illustrate the effect of chelation on the stabilities. If we compare $\log \beta_2$ for NH_3 with $\log K_1$ for $H_2NCH_2CH_2NH_2$, we are comparing values where two N atoms are coordinated in each case. As $\beta_2 = K_1 . K_2$, $\log \beta_2 = \log K_1 + \log K_2$. It will be seen that $\log K_1$ for the ethylenediamine complexes is uniformly greater than $\log \beta_2$ for ammonia, and likewise $\log K_2$ and $\log K_3$ for en are greater than the corresponding $\log \beta_4$ and $\log \beta_6$ values for NH_3. Similarly, the tridentate chelate, trien, gives a value of $\log K_1$ which exceeds $\log \beta_3$ for the analogous NH_3 species. Finally, the hexadentate ligand pentaethylenehexamine has a value of $\log K_1$ (coordination of all six nitrogens) which is greater than $\log \beta_3$ for the $M(en)_3^{2+}$ species and than $\log \beta_6$ for $M(NH_3)_6^{2+}$. It is also noteworthy that while copper and zinc fail to become six-coordinated to ammonia, they do form six-coordination to nitrogen when the ligand is a chelating amine,

The effect of chelate ring size is illustrated by the values for 1,3-propanediamine (Figure 12.16b) which has $\log K_1 = 10·0$ and $\log K_2 = 7·2$ for the copper compounds.

Anomalous changes in stability constants may often indicate electronic changes. One example is provided by the Cr(II) dipyridyl values where $\log K_1 = 4·5$, $\log K_2$ is 6·0, instead of the expected decrease, and $\log K_3$ is 3·5. In this case, the d^4 Cr^{2+} species is high-spin in its hydrate, and low-spin in $Cr(dipy)_3^{2+}$. The anomalous value for K_2 suggests that spin pairing occurs as the second dipyridyl is added.

A full set of stability constants is to be found in the Chemical Society special publication no. 25, of that title, which is given on page 283.

The properties of the ligands affect the replacement reactions which they undergo. It has already been seen that a chelating ligand will replace a monodentate ligand with the same donor atom. Such chelating ligands also commonly come higher in the spectrochemical series than the corresponding simple one: en lies above ammonia for example. However, care must be taken, in general, to avoid equating high ligand field strength with high energy of formation or high stability of a complex. Obviously, in all configurations where there is some ligand field stabilization energy, the size of the field of a given ligand is an important factor in its reactions, but the CFSE varies with the metal atom, as has been seen. It is safe to expect a general correlation between stability and ligand field strength, for example, cyanide is expected to replace water in general, but the detailed behaviour in any one case is a balance of different effects. A most marked example of this is the case of the transition metal complexes containing metal-hydrogen bonds, such as $(R_3P)_2PtH_2$. Hydrogen lies high in the spectrochemical series with ΔE values approaching those of cyanide, yet its compounds with the transition metals are notoriously unstable and isolable compounds containing σ-M—H bonds (as opposed to metallic bonds in non-stoichiometric hydrides) are found only in the presence of a few stabilizing ligands. This instability may arise from the fact that, when metal-hydrogen bonds dissociate, hydrogen atoms are produced. These are very reactive and combine readily with each other. By contrast, when the normal metal-ligand bond dissociates, stable lone pair entities like water, ammonia, or halide ions are produced and the metal-ligand bond can reform in equilibrium with small quantities of ligand.

This is an extreme case, but it is common to find an unthinking equation of high ligand field with stability of complexes. This does not follow particularly when comparing ligands of similar strength. Thus $Co(NH_3)_6^{3+}$ is prepared in water and is stable indefinitely to exchange of the coordinated ammonia for water, while $Cr(NH_3)_6^{3+}$ can only be prepared in liquid ammonia and rapidly exchanges the ammonia groups for water when dissolved in water.

12.8 Isomerism

The use of polydentate chelating agents aids the study of *stereoisomerism* in complexes. In octahedral complexes, both optical and geometrical isomerism is possible, as illustrated by the ethylenediamine complexes of cobalt in Figure 12.18. The existence of optical isomers may be shown by rotatory dispersion measurements and by the resolution of isomers in favourable cases. The scheme for the resolution of the tris-ethylenediamine complex of cobalt(III) is shown in Figure 12.19, making use of naturally occurring *d*-tartaric acid to separate the isomers. The existence of geometrical isomers is generally noticed when two compounds of different properties are found to have identical molecular formulae. In favourable cases the identity of the isomers may be determined by experiment, as in the separation of the optically active forms of the *cis*-$Co(en)_2X_2^+$ compound shown in Figure 12.18. It is also possible to distinguish geometrical isomers in some cases by substitution reactions with a bidentate ligand. Such a ligand can replace two monodentate ligands lying *cis* but not two lying *trans*, as they are too far apart. Figure 12.20 shows an example of such a

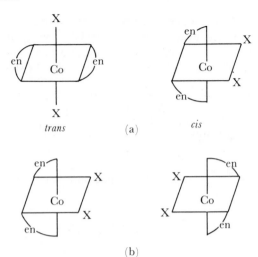

FIGURE 12.18   *Optical and geometrical isomerism in octahedral complexes: the example of* $Co(en)_2X_2^+$

Figure (a) shows the geometrical isomers, *cis* and *trans* forms, while Figure (b) shows the non-superimposable optical isomers of the *cis* form. The *trans* form has a plane of symmetry and is inactive.

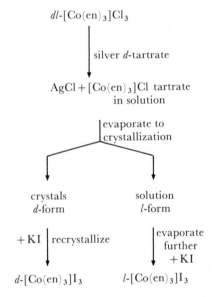

FIGURE 12.19   *Resolution into optical isomers of* $Co(en)_3I_3$

reaction scheme. It must be noted that such proofs of configuration by reaction are not fully definite, as change of configuration may occur during reaction. The final proofs depend on structural evidence, usually from crystallography, although other techniques such as nmr may be useful on occasion.

Stereoisomerism is not confined to octahedral coordination, of course. Figure 12.21 illustrates a classic case of isomerism among square planar platinum complexes.

Other types of isomerism occur among complexes. The case of linkage isomerism in which a ligand can coordinate through one or other of alternative atoms has been exemplified above by the cases of nitro-nitrito and thiocyanate-isothiocyanate. The nitro-nitrito isomerization occurs in the same compound, for example, both forms of the cobalt-

cis—violet form of $[Cr(en)_2Cl_2]^+$

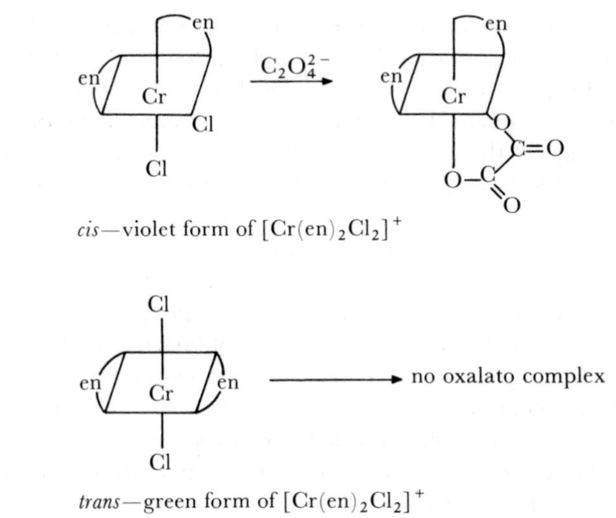

trans—green form of $[Cr(en)_2Cl_2]^+$

FIGURE 12.20 *Differentiating between geometrical isomers by a replacement reaction with a bidentate ligand*

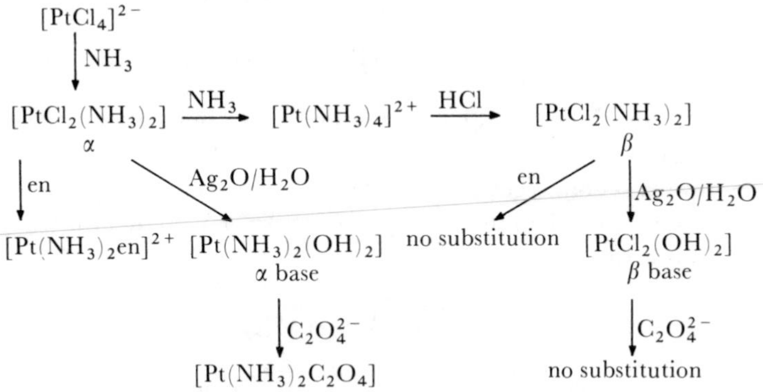

A number of other types of isomer have been defined but the names are of little importance. Some examples are given in Table 12.17.

TABLE 12.17 Further types of isomerism in complexes

(a) $[Co(NH_3)_4Cl_2]NO_2$ and $[Co(NH_3)_4Cl(NO_2)]Cl$

    $[Pt(NH_3)_4Cl_2]Br_2$ and $[Pt(NH_3)_4Br_2]Cl_2$

(b) $[Co(NH_3)_6][Cr(CN)_6]$ and $[Cr(NH_3)_6][Co(CN)_6]$

    $[Pt^{II}(NH_3)_4][Pt^{IV}Cl_6]$ and $[Pt^{IV}(NH_3)_4Cl_2][Pt^{II}Cl_4]$

(c) $[Pt(NH_3)_2Cl_2]$ and $[Pt(NH_3)_4][PtCl_4]$

(a) is the case of isomers with one group either as a ligand or as a counter-ion; (b) and (c) show different cases of polymerization or coordination isomerism. Here, isomers of the same analytical formula contain species of different coordination, oxidation state, or degree of polymerization.

Hence the α form of $[PtCl_2(NH_3)_2]$ is cis and the β form is trans.

FIGURE 12.21 *Stereoisomerism in square planar complexes of platinum*

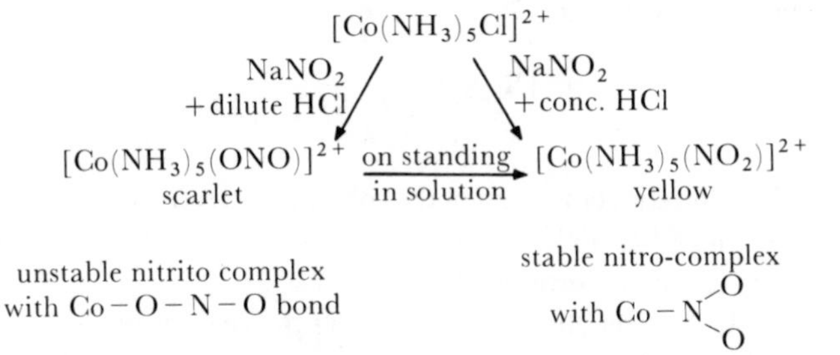

FIGURE 12.22 *Linkage isomerism: nitro- and nitrito-pentamminecobalt(III)*

pentammine are known, Figure 12.22. Another example is provided by the NCS group, which is found bonded through either N or S in certain compounds of palladium, platinum and copper. A good example is provided by the series $LCu(NCS)_2$ which is brown, $LCu(NCS)(SCN)$ which is yellow-green, and $LCu(SCN)_2$ which is deep green. Here, L may be one of a number of bidentate nitrogen ligands.

12.9 Mechanisms of transition metal reactions

A number of different classes of reaction of transition element complexes exists and we review some of the better-established cases.

The simplest type of reaction is that of oxidation-reduction by *electron-transfer*. For example, if a mixture of ferricyanide, $Fe(CN)_6^{3-}$, and ferrocyanide, $Fe(CN)_6^{4-}$, is made, and if an electron is lost from the $Fe(CN)_6^{4-}$ ion and gained by the $Fe(CN)_6^{3-}$ ion, an oxidation-reduction reaction has taken place, even though there is no change in the overall composition of the mixture. The presence of such reactions is most readily demonstrated by making one of the components with a radioactive isotope. If the radioactivity is found, at a later time, to be spread between both components the reaction is indicated (star indicates the radioactive isotope)

$$*Fe(CN)_6^{4-} + Fe(CN)_6^{3-} \rightleftharpoons *Fe(CN)_6^{3-} + Fe(CN)_6^{4-}$$

As there is no nett change in the reaction mixture, there is no heat change in such a reaction. There is, however, a requirement for activation energy. This is because the $Fe-CN$ bond length in ferricyanide is shorter than that in ferrocyanide. Thus the simple electron transfer between ions in their equilibrium configurations would produce ferrocyanide ions with compressed bonds, and ferricyanide ions with extended ones, that is the product ions would be vibrationally excited. The electron transfer takes place between matched ions, thus there is an activation energy required to stretch the bonds in ferricyanide and compress those in ferrocyanide to the intermediate matching configuration, by vibrational excitation. The matched anions must approach closely, but they do not have to be in actual contact for the electron transfer to occur.

Similar electron transfer reactions occur for a variety of matched ions, such as MnO_4^-/MnO_4^{2-} (permanganate/manganate), $IrCl_6^{2-}/IrCl_6^{3-}$ or $Mo(CN)_8^{3-}/Mo(CN)_8^{4-}$. The indications of the electron transfer mechanism are, first, that the reaction is second order (that is, the rate is a function of the concentration of each ion) and, second, the reactions are much faster than ones involving ligand exchange which are discussed below. Electron transfer reactions are favoured by the presence of ligands like cyanide or phenanthroline which allow electron delocalization from the metal through a conjugated system.

A slower group of electron transfer reactions is that exemplified by the $Co(H_2O)_6^{3+}/Co(H_2O)_6^{2+}$ system. Here, not only are the bond lengths different in the two oxidation states, but cobalt(III) is spin-paired with configuration t_{2g}^6 while cobalt(II) is high spin in the hydrate, configuration $t_{2g}^5 e_g^2$. Thus electron transfer between the ground state configurations would yield excited electronic states in the products as shown by the first row of configurations below the equation.

$$*Co(H_2O)_6^{3+} + Co(H_2O)_6^{2+} \rightleftharpoons *Co(H_2O)_6^{2+} + Co(H_2O)_6^{3+}$$

ground state initial configurations

$$t_{2g}^6 \qquad t_{2g}^5 e_g^2 \qquad t_{2g}^6 e_g^1 \qquad t_{2g}^5 e_g^1$$

pre-excitation needed

$$t_{2g}^5 e_g^1 \qquad t_{2g}^6 e_g^1 \qquad t_{2g}^5 e_g^2 \qquad t_{2g}^6$$

To overcome this, a further excitation energy is needed to produce an excited electronic configuration, such as that shown in the second line below the equation, in addition to the vibrational excitation needed to adjust the bond lengths. Thus, electron transfer reactions which involve changes in spin pairing need a higher excitation energy, and are slower than those which require only the matching of bond lengths.

A related class of reactions is that where oxidation-reduction takes place by *electron-transfer accompanied by transfer of a ligand*. In the transition state, the transferred ligand (sometimes in conjunction with other species) forms a link between the two metals, e.g. $L_5Cr-Cl-CoL'_5$. This type of mechanism is often termed *inner sphere* or *bridged*. Such reaction mechanisms are most readily established when the complex to which the ligand is transferred is *substitution-inert*. The rates of substitution processes vary enormously. For most d configurations substitutions take place very rapidly, but one or two configurations are substitution-inert, that is to say, reaction times are measurable in hours or days rather than in fractions of a second. Note that this inertness has nothing to do with the thermodynamic stability of the reactants or products but is a question of the reaction rate. The most common inert configurations are d^3 and d^6, for example Cr(III), Co(III), Mo(III), W(III), Re(IV), Rh(III), Ir(III), Ru(II), Os(II), Pd(IV), and Pt(IV). One striking example, used by Taube, is the study of the oxidation mechanism by using chromium(II) oxidized to chromium(III). As the latter is substitution-inert, any transfer of an atom or group during the oxidation process will be detected by its appearance in the chromium(III) complex. For example, in the reaction

$$Cr(H_2O)_6^{2+} + Co(NH_3)_5X^{2+}$$
$$Cr(II) \qquad\qquad Co(III)$$
$$= Cr(H_2O)_5X^{2+} + Co(NH_3)_5(H_2O)^{2+}$$
$$Cr(III) \qquad\qquad Co(II)$$

the transfer, in the oxidation process, of the group X has been demonstrated for $X^- = $ halide$^-$, NCS$^-$, N$_3^-$, SO$_4^{2-}$ and PO$_4^{3-}$. Although the equation has been balanced by giving the cobalt(II) product as the pentammine hydrate, exchange in the labile cobalt(II) complex means that the hexahydrate would be recovered if the reaction was carried out in water.

The electron transfer reactions make up one group of transition metal reactions, the other major class are *ligand substitution reactions*. The mechanisms of ligand replacement reactions may be discussed in the light of ligand field theory. Two limiting modes of reaction for an octahedral complex are conceivable: either a ligand may be removed, leaving a five-coordinated intermediate which then picks up the substituting ligand, or else the incoming ligand may become coordinated to the original complex, giving a seven-coordinated intermediate, which subsequently expels one of the original ligands. That is, either:

$$MX_6 \xrightarrow{slow} MX_5 \xrightarrow{Y} MX_5Y \qquad (12.1)$$

or,

$$MX_6 + Y \rightarrow MX_6Y \rightarrow MX_5Y \qquad (12.2)$$

The first labelled S_N1 (substitution, nucleophilic, first order) and the second S_N2, as it is second order. That is, the rate of S_N1 reactions is governed by the first, dissociation, step which is far slower than the subsequent uptake of Y, so that the rate law is simply

$$\frac{d[MX_5Y]}{dt} = k[MX_6] \qquad (12.3)$$

It should be noted that reaction (12.1) is almost indistinguishable from that in which the solvent, S, enters the sixth position and is then displaced by the incoming ligand

$$MX_6 + S \xrightarrow{slow} MX_5S \xrightarrow{Y, fast} MX_5Y \qquad (12.4)$$

The rate law for (12.4) would be

$$\frac{d[MX_5Y]}{dt} = k'[MX_6][S] \qquad (12.5)$$

and, as the concentration of the solvent is effectively constant, $k'[S]$ may be replaced by a constant k'', giving equation (12.5) the same form as (12.3).

In the S_N2 reaction, again making the reasonable assumption that the first step is slow and the expulsion of a ligand from the seven-coordinate intermediate is fast, the rate law would be

$$\frac{d[MX_5Y]}{dt} = k[MX_6][Y] \qquad (12.6)$$

Whether a given Y will replace any particular ligand is governed by the usual balance of energies with the additional factor of the change in CFSE. If the latter is an important factor, it will allow not only a prediction of whether the substitution will occur but also a prediction of the path. The CFSE of the five- and seven-coordinated intermediates may be calculated and the most likely path determined. Consider the case of substitution in a low-spin cobalt(III) complex (which has six d electrons and a strong CFSE contribution). It may be shown that the possible intermediates have the CFSE values shown (all given in terms of the octahedral splitting ΔE_{oct}):

Shape:	octahedral	pentagonal bipyramid (7)	trigonal bipyramid (5)	square pyramid (5)
CFSE:	2·4	1·55	1·25	2·0

The square pyramid therefore corresponds to the lowest loss of CFSE of all the possible intermediates in this case, the loss being equal to about 92 kJ mol^{-1} if the ammine complex is the case in point. It can be calculated that the total activation energy for substitution in the cobalt(III) hexammine complex ion ranges from 521 kJ mol^{-1} for the trigonal bipyramidal intermediate, through 431 kJ mol^{-1} for the pentagonal bipyramid, to a range between 25 and 395 kJ mol^{-1} for the square pyramid depending on whether the empty site in the last case is occupied by a solvent molecule or not. Thus the CFSE variation accounts for about 50 per cent of the total activation energy in each case. It would be predicted that the mechanism involving the five-coordinated square pyramidal configuration for the intermediate is the most likely and the activation energy found by experiment, about 140 kJ mol^{-1}, appears to bear this out.

It must be borne in mind that, while the kinetic study of reactions is a valuable guide to mechanisms, the order of the reaction is not necessarily the same as the molecularity of the critical step in the reaction path. One example is provided by systems where a rapid equilibrium is first set up

$$MX_6 + Y \rightleftharpoons MX_6.Y \xrightarrow{slow} MX_5Y + X \quad \ldots (12.7)$$

and the slow step is the elimination of X from the intermediate, associated, species $MX_6.Y$. The actual rate-determining step is the slow one, which is unimolecular but the rate law will involve the constants for the forward and reverse steps of the equilibrium and the concentration of both MX_6 and Y: that is, the rate law will have the overall form of a second order reaction. In this case, the $MX_6.Y$ does not involve coordination of Y to M but denotes some association with a definite lifetime, such as the formation of an ion pair. Such species become especially important when reactions are carried out in less polar solvents than water, such as acetone or methanol.

A further important class of reactions exists which are apparently S_N2, but where the basic step is unimolecular, and the overall kinetics result from the combination of this step with a pre-equilibrium. This is the class of reactions where the attacking ligand is OH$^-$ and there is a replaceable hydrogen present in the ligands on the metal. Thus the reaction

$$Co(NH_3)_5Cl^{2+} + OH^- \rightleftharpoons Co(NH_3)_5OH^{2+} + Cl^- \quad (12.8)$$

obeys the rate law

$$\frac{d[Co(NH_3)_5Cl^{2+}]}{dt} = k[Co(NH_3)_5Cl^{2+}][OH^-] \quad (12.9)$$

(note that in the rate equation the square brackets indicate concentrations). Such reactions are about 10^6 more rapid than most ligand replacements. It is believed that this reaction is not S_N2, but instead involves the initial abstraction of a proton from one of the ammonia ligands in a fast pre-equilibrium,

$$Co(NH_3)_5Cl^{2+} + OH^- \rightleftharpoons Co(NH_3)_4(NH_2)Cl^+ + H_2O \quad (12.10)$$

and this is followed by a slower, rate-determining step involving the expulsion of chloride from this amide complex

$$Co(NH_3)_4(NH_2)Cl^+ \xrightarrow{slow} Co(NH_3)_4(NH_2)^{2+} + Cl^- \quad (12.11)$$

and the five-coordinate amido-intermediate rapidly reacts with water to give the hydroxy complex

$$Co(NH_3)_5(NH_2)^{2+} + H_2O \xrightarrow{fast} Co(NH_3)_5(OH)^{2+} \quad (12.12)$$

Thus the rate-determining step is unimolecular, but the rate law is second order, with a constant which is a combination of the forward and backward constants of reaction (12.10) and the constant for the slow step (12.11). Thus, the overall reaction (12.8) is the sum of (12.10), (12.11) and (12.12) obeying the overall rate law (12.9) but the critical step is unimolecular. As the amido-complex $Co(NH_3)_4(NH_2)Cl^+$ is the conjugate base (see Chapter 5) of the starting complex $Co(NH_3)_5Cl^{2+}$, this mechanism is usually known as the S_N1 CB (substitution, nucleophilic, unimolecular, conjugate base) mechanism. Such reactions are found whenever OH$^-$ attacks a complex ion which contains ionizable hydrogen atoms.

These remarks about mechanisms apply, of course, to coordinations other than the octahedral one. In square planar complexes, an additional effect, the *trans effect*, becomes important. In reactions of square planar complexes of the general form:

$$(MLX_3) + Y = (MLX_2Y) + X$$

the group X which is displaced may be *cis* or *trans* to the ligand L and two isomeric products are possible, one or both of which may be observed. The common ligands may be arranged in an order of increasing ability to direct incoming substituents to a position *trans* to themselves:

$$H_2O < NH_3 < Cl^- < Br^- < NO_2^- < CN^-$$

The knowledge of *trans* directing powers opens up methods of synthesizing desired compounds. Consider the method of synthesizing the *cis* and *trans* isomers of $[Pt(NH_3)_2(Cl)_2]$. If the *cis* isomer is required, the starting material must be the tetrachloroplatinate ion:

$$\begin{bmatrix} Cl & \diagdown \diagup & Cl \\ & Pt & \\ Cl & \diagup \diagdown & Cl \end{bmatrix}^{2-} \xrightarrow{NH_3} \begin{bmatrix} Cl & \diagdown \diagup & NH_3 \\ & Pt & \\ Cl & \diagup \diagdown & Cl \end{bmatrix}^{-} \xrightarrow{NH_3} \begin{bmatrix} Cl & \diagdown \diagup & NH_3 \\ & Pt & \\ Cl & \diagup \diagdown & NH_3 \end{bmatrix}$$

The first ammonia enters to give $Pt(NH_3)Cl_3^-$ and then, as the *trans* effect of the chloride ion is greater than that of ammonia, the second ammonia molecule enters *trans* to chlorine giving the *cis* isomer. To prepare the *trans* isomer, the starting material must be the tetrammine. In the intermediate $Pt(NH_3)_3Cl^+$, the *trans* effect of the chloride is greater than that of the ammonia, and the second chloride enters *trans* to the first to give the *trans* isomer:

$$\begin{bmatrix} H_3N & \diagdown \diagup & NH_3 \\ & Pt & \\ H_3N & \diagup \diagdown & NH_3 \end{bmatrix}^{2+} \xrightarrow{Cl^-} \begin{bmatrix} H_3N & \diagdown \diagup & Cl \\ & Pt & \\ H_3N & \diagup \diagdown & NH_3 \end{bmatrix}^{+} \xrightarrow{Cl^-} \begin{bmatrix} H_3N & \diagdown \diagup & Cl \\ & Pt & \\ Cl & \diagup \diagdown & NH_3 \end{bmatrix}$$

The examples given serve only to illustrate the type of work which is being carried out in this field and are by no means even a complete summary of the known results. The field is obviously of importance for the understanding of transition metal chemistry as all reactions of transition elements in solution are reactions between complexes. Thus, when simple reactions involving transition metal ions in solution are discussed, the species actually involved are the complexes formed between ion and solvent molecule.

For octahedral complexes, S_N1 mechanisms and electron transfers with or without ligand transfer are well established, together with those involving ion-pairing or conjugate base pre-equilibria which are unimolecular in their rate-determining step, although following a second order rate law, such as equation (12.6) for the overall sequence of reactions. There is, however, little firm evidence of true S_N2 mechanisms, going through a seven-coordinate intermediate, for ligand substitution in octahedral complexes. In square planar complexes, and also in five-coordinate species, there is little steric hindrance to an increase of coordination number, and both S_N1 and S_N2 mechanisms are available for these lower coordination numbers.

One mechanism which is related to this question is that of *oxidative addition* in which an increase in oxidation state is accompanied by an increase in coordination number. This occurs most readily where a low coordination number is stable in a low oxidation state of an element while a higher one is the preferred one for a higher oxidation state. One striking example is provided by platinum where the $+II$ state is d^8 and found in square planar configurations while the $+IV$ state is d^6 and most stable in the octahedral configuration (Table 12.13). Thus a change

$$Pt^{II}L_4 + X_2 = Pt^{IV}L_4X_2$$

is particularly favoured, compare section 14.8. Such a

reaction may lead to oxidations by relatively unreactive species under very mild conditions, as in the reaction at room temperature and normal pressure

$$Ir(CO)Cl(PPh_3)_2 + H_2 = Ir(CO)(H)_2Cl(PPh_3)_2.$$

12.10 Structural aspects of ligand field effects

If the radii of divalent ions in a given Period are considered, it is found that the change of radius across a transition series follows a curve with two minima, such as that shown for the first transition series in Figure 12.23. A similar effect is

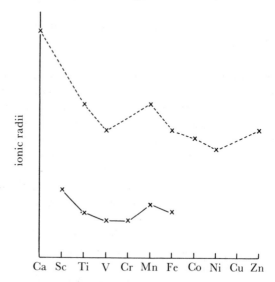

FIGURE 12.23 *Radii of the divalent* (. . . .) *and trivalent* (———) *ions of the First Transition Series*

observed for trivalent ions but the minima come one element later. A smooth curve may be drawn through the values for the radii of the d^0, d^5, and d^{10} ions representing the fall in radius with increasing number of d electrons as the increase in nuclear charge is imperfectly balanced by the shielding of a d electron (similar to the lanthanide contraction discussed in Chapter 10). The actual radii for the ions with empty, half-filled, and filled shells lie on this curve, as the d electron clouds are spherically symmetrical. The radii of ions corresponding to other d configurations lie below this curve, and the largest differences are found for those ions with the largest excess of t_{2g} over e_g electrons. (The discussion and Figure refer here to high-spin configurations in the octahedral field—extensions to other cases are obvious, though less well documented.) Consider first the case of Ti^{2+} (divalent scandium is unknown). In this ion, there are two d electrons and these lie in the t_{2g} level, that is, in orbitals which are directed away from the ligands. Electrons in such orbitals have only a slight effect in screening the nuclear charge of the transition ion from the ligands—a much smaller screening effect than if they were disposed in a spherically symmetric shell. It follows that the metal-ligand distance is shorter than if these two electrons were spherically distributed, thus the octahedral radius is shorter, when measured, than that found by interpolation between the spherical d^0 and d^5 configurations. (It will be recalled from Chapter 2 that ionic radii are derived from measured ion-ion or ion-molecule distances.) The difference between measured and interpolated radii is even greater for V^{2+} where there are three electrons in the t_{2g} orbitals, and then the difference decreases for Cr^{2+} with one electron in the e_g set which is directed at the ligands. In the d^5 ion, Mn^{2+}, the configuration is $t_{2g}^3 e_g^2$ which is symmetrical, and the radius lies on the d^0—d^{10} curve. A similar pattern is repeated from d^6 to d^{10}.

Such a variation in radius will contribute a corresponding variation to the lattice energy of isostructural compounds of a given formula involving transition metal ions. Such a lattice

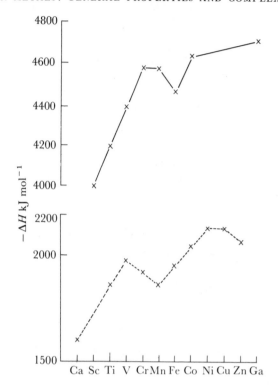

FIGURE 12.25 *The hydration heats of the divalent* (----) *and trivalent* (——) *ions of the First Transition Series*

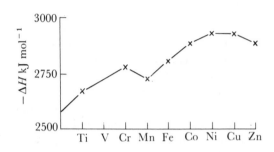

FIGURE 12.24 *The lattice energies of the difluorides of the elements of the First Transition Series*

energy curve is shown in Figure 12.24. The lattice energy rises as the ionic radii fall, as the lattice energy is inversely proportional to the interionic distance. Thus, the points for the d^0, d^5, and d^{10} ions lie on a smoothly rising curve, and those for intermediate configurations lie above this curve. If the crystal field stabilization energies are subtracted from the experimental values, the resulting points lie on the curve. Notice that the CFSE values and the effects on radii both result from the ligand field splitting and the resulting electron configuration. Whether the lattice energy values are regarded entirely as an effect of CFSE or the radial effects are included depends on whether calculated or experimental values are used (calculated values already take account of the variation in radius).

A similar curve is found if other thermodynamic values are considered. Figure 12.25 shows the hydration energies of the divalent and trivalent ions of the first row, that is the energy of:

$$M^{n+}_{(gas)} + 6H_2O_{(gas)} = M(H_2O)_6^{n+}_{(gas)}$$

In each case, a curve with two maxima and a minimum at the d^5 configuration is observed, and the values fall on a smooth curve rising from d^0 to d^{10} when the CFSE values are subtracted from the experimental values. Notice that such a curve may be used in reverse to find the crystal field stabilization energies for different configurations, and hence the value of ΔE. Such thermodynamically determined values of the ligand field splitting agree closely with those derived from spectroscopic data.

12.11 Spectra of transition element complexes

When only one d electron is present, a simple spectrum consisting of a single band is observed (Figure 12.4). The band is relatively broad both because the excited and ground states interact to some extent with their environment (usually with molecules of the solvent) and because the equilibrium bond lengths in the excited state would be greater than those in the ground state so that the upper state is produced with vibrational modes excited. These vibrational levels are too close to be resolved into individual bands but they give a general broadening of the d—d absorption.

The effect of an octahedral ligand field on the d^1 ion may be shown as in Figure 12.26a. The t_{2g} and e_g levels are separated more widely as the ligand field increases. The symbol ^2D at the left is a symmetry label which describes the d^1 configuration in the absence of the ligand field splitting. We shall use such symbols here simply as labels and readers who study this subject further will find out how they are derived and their full significance.

Consider now the d^9 configuration in the octahedral field. We can describe this in a way which is related to the description of d^1 and use this as a method of determining the d^9 splitting diagram. In the ground state, d^9 can be regarded as derived from the filled d^{10} shell by forming a single hole in the e_g levels. Thus d^1 has a single electron in the t_{2g} level, outside a filled shell configuration, and d^9 has a single hole in the e_g level. In d^9, when a t_{2g} electron is excited to the e_g level, the latter level becomes filled and the hole appears in the t_{2g} level. Thus, the excitation may be described as the transition of a hole, that is, an electron vacancy, from the e_g

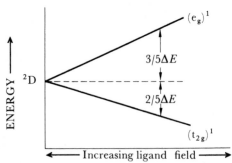

FIGURE 12.26a *Energy level diagram for the d^1 configuration in an octahedral field*

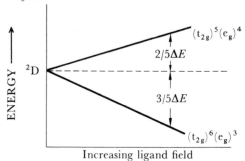

FIGURE 12.26b *Energy level diagram for the d^9 configuration in an octahedral field*

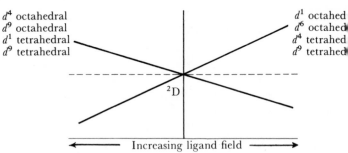

FIGURE 12.27 *Combined energy level diagram for single-electron or single-hole configurations in octahedral and tetrahedral fields.*
The zero crystal field level is shown at the origin and the left portion is the reciprocal of the right hand one. Such diagrams are termed *Orgel diagrams* after their originator.

to the t_{2g} level, while the d^1 transition is described as the transition of an electron from t_{2g} to e_g. Thus the transition in d^9 is shown on an energy level diagram which is the *reciprocal* of that for d^1, as shown in Figure 12.26b.

Consider now a tetrahedral field. Here, the lowest level is the doublet e level and the upper state is the triplet t_2 level. Thus the energy level diagram for d^1 configuration in a tetrahedral field is the reverse of d^1 octahedral, and therefore qualitatively the same (though the ΔE value is only about 4/9) as that for d^9 octahedral. Similarly, d^9 tetrahedral is qualitatively the same as d^1 octahedral.

Yet another related configuration is that of d^6 high spin. Here, each d orbital is singly occupied and the sixth electron, in the octahedral field, is in one of the t_{2g} orbitals with an opposed spin. Now, transitions which require spin reversal are formally forbidden, and give rise to very weak bands if they can be observed at all, so that we may neglect any transition of one of the set of five electrons with parallel spins. Thus the only band we are likely to observe is that due to the excitation of the single antiparallel electron from the t_{2g} level to the e_g level. Thus the energy level diagram for d^6 high spin in the octahedral field is the same as that for d^1 in an octahedral field.

In a similar manner, the high spin d^4 configuration in the octahedral field may be described, by the hole formalism, in exactly the same way as d^9 octahedral. Furthermore, d^6 tetrahedral and d^4 tetrahedral are qualitatively described by the same diagrams as, respectively, d^4 octahedral and d^6 octahedral.

Thus, to summarize, by combining Figures 12.26a and 12.26b into one, as shown in Figure 12.27, we can describe all those configurations with one electron in excess of an empty or half-filled d shell and, by the hole formalism, all the configurations with one electron less than a half-filled or filled d shell. This description applies both to the octahedral field, and to the tetrahedral field which causes the reciprocal splitting. Thus Figure 12.27 describes qualitatively the effect of the ligand field on d^1 octahedral, d^6 octahedral, d^4 tetrahedral and d^9 tetrahedral, by the portion to the right of the origin, and on d^4 octahedral, d^9 octahedral, d^1 tetrahedral and d^6 tetrahedral by the portion to the left of the origin. For all these configurations, there is only one state above the ground state, separated by ΔE, so that there is only one transition and it occurs at a position in the spectrum corresponding to ΔE.

If the d^2 configuration is now considered, the energy level diagram is found to be more complicated. This is because, while the spin and orbital quantum numbers of a single d electron give rise to only one state, the ^2D state above, the combinations of these quantum numbers for two d electrons give rise to four different states. Some of these have the two electron spins antiparallel, so that transition to them would involve spin reversal. These states may therefore be neglected and we are concerned with two states for d^2, which are labelled ^3F (the ground state) and ^3P and are shown in the centre of Figure 12.28. In an octahedral field, the ^3F state splits up, in a similar manner to the ^2D state in Figure 12.26, but into three components while the ^3P state remains unsplit. Furthermore, interaction between the levels derived from ^3F and ^3P causes some of these levels to curve away from each other, to give the dependence of energy level on ligand field which is shown in Figure 12.28. This interaction depends on a single parameter, B—the Racah parameter—and this may be evaluated in the course of the assignment. Thus, the d^2 diagram, for an octahedral field, has three states above the ground state to which an electron may be excited without changing its spin. This means that the spectrum of a d^2 complex contains three bands, in place of the single band of d^1, and the positions of these three bands together give the value of the ligand field for the complex. By the same arguments that applied to d^1, the energy level diagram is qualitatively the same for d^2 octahedral and for d^7 octahedral (high spin) and also for d^8 tetrahedral and d^3 tetrahedral, while the reciprocal diagram applies to d^8 and d^3 octahedral and also to d^2 and d^7 tetrahedral. All these levels are shown in Figure 12.28. A d^8 spectrum is illustrated in Figure 12.29.

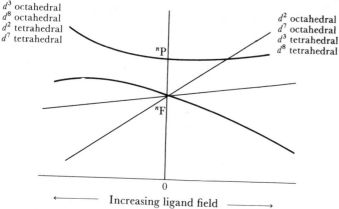

d^3 octahedral
d^8 octahedral
d^2 tetrahedral
d^7 tetrahedral

d^2 octahedral
d^7 octahedral
d^3 tetrahedral
d^8 tetrahedral

^nP

^nF

0

⟵ —— Increasing ligand field —— ⟶

FIGURE 12.28 *Combined Orgel diagram for two-electron and two-hole configurations in octahedral and tetrahedral fields.*
The multiplicities are, for d^2 and d^8, $n = 3$ and for d^3 and d^7, $n = 4$. States of lower multiplicity are omitted. On the d^3 octahedral side of the diagram, the upper state which is derived from the ^4P interacts with the highest level derived from the ^4F and these levels curve away from each other. The other two levels derived from ^4F do not interact with any other and vary linearly with the ligand field.
On the d^2 octahedral side of the diagram, the same two states also interact (derived now from ^3P and ^3F) but, as the ^3F state is now the lowest one, the interaction is less marked.

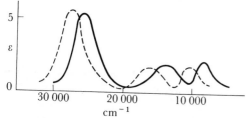

5

ε

0

30 000   20 000   10 000
cm^{-1}

FIGURE 12.29 *The electronic spectra of* $Ni(H_2O)_6^{2+}$ (———) *and* $Ni(NH_3)_6^{2+}$ (– – – –).
Each shows the three bands expected for a d^8 system, corresponding to $\Delta E = 8500$ cm^{-1} for the hydrate and $\Delta E = 11\,300$ cm^{-1} for the ammine. The curve is idealized: the experimental spectrum shows a splitting for the middle band.

Finally, the d^5 high spin configuration has all d orbitals half-filled with parallel spins in the ^6S state. Any electron transition must involve spin reversal so that the $d-d$ bands in a d^5 complex, such as a manganese(II) compound, are very weak. There are four excited states which involve only one spin reversal, and these split up into a total of ten states in the ligand field. Thus, a d^5 complex might show anything up to ten very weak absorptions in the visible and ultraviolet, but since these transitions are formally forbidden, their intensities are about 100 times weaker than normal $d-d$ transitions in an octahedral field. Hence most d^5 compounds look almost colourless, witness the very pale pink of the majority of manganese(II) species.

A further point about intensities is the difference between octahedral and tetrahedral complexes. A further selection rule forbids transitions between states with the same symmetry with respect to a centre of inversion (see Appendix C) and this applies to octahedral complexes but not to tetrahedral ones as the tetrahedron has no inversion centre. Thus tetrahedral complexes are commonly more strongly coloured than octahedral ones, as in the case of the cobalt(II) com-

plexes where the blue of the $CoCl_4^{2-}$ species is much more intense than the pink of the octahedral aquo complex, $Co(H_2O)_6^{2+}$. All these intensities are much less than those of fully allowed transitions, see section 6.5, so that $d-d$ transitions are only observed where allowed bands of the ligands or of the total complex are sufficiently far into the ultra-violet that they do not swamp the weaker $d-d$ bands.

Low spin complexes may be analyzed in a similar way to the above, although the detailed arguments are more difficult and will not be treated here. Further effects found in the spectra must also be considered in a full discussion. For example, Jahn-Teller effects may cause absorption bands to split, as also will interaction between orbital and spin angular momentum. However, all these effects may be taken into account and it is found that experimental spectra agree well with those predicted by application of ligand theory.

The major effects in the spectra of transition complexes may be summarized as follows. For configurations d^n where $n = 1, 4, 6$ and 9, one band is expected and its position gives ΔE directly. Configurations with $n = 2, 3, 7$ or 8 give rise to three bands whose positions depend on two parameters, ΔE and B, the latter measuring the interaction between energy levels derived from F and P states. As there are three bands depending on only two parameters, both ΔE and B may be evaluated unambiguously. Finally, when $n = 5$, a much larger number of transitions may be observed which may be analyzed to give the value of ΔE. Low spin configurations for $n = 4, 5, 6$ and 7 may be similarly analyzed to give ΔE values, though the treatment is more complicated. Apart from allowed transitions, the most intense absorptions are found in tetrahedral complexes, those in octahedral species are weak, while transitions in any d^5 high spin complex are extremely weak indeed.

12.12 Complexes with π bonding
Although the bonding between transition metal ion and ligand may involve π bonding in many cases, one particular type of π complex deserves special attention. This is the case, of which nickel carbonyl is the classic example, where the complex is formed by the metal in a low or zero oxidation state. The stability of such compounds depends on an unusual energy effect as there can be no electrostatic stabilization.

Carbonyl complexes are formed by the first row elements from vanadium to nickel in their zero valent states, and also by many of their heavier congeners. Table 12.18 lists the reported metal carbonyls, including some examples of those containing more than one metal.

Apart from vanadium carbonyl, which is paramagnetic with one unpaired electron, all these compounds are diamagnetic. They have a rare gas configuration of eighteen electrons around the central atom, if each carbonyl group is regarded as providing two electrons, and the binuclear complexes are regarded as having a metal-metal bond with the two bond electrons shared between the two atoms. The monomeric carbonyls have symmetrical structures (octahedron, trigonal bipyramid, and tetrahedron respectively). Figure 12.30 shows the structures of two of the binuclear

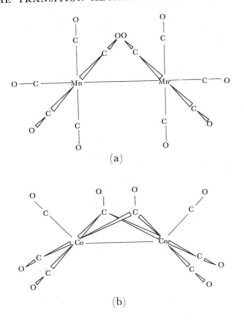

FIGURE 12.30 *Binuclear carbonyls:* (a) $Mn_2(CO)_{10}$, (b) $Co_2(CO)_8$

compounds. In $Mn_2(CO)_{10}$ only the Mn–Mn bond holds the two halves of the molecule together. The coordination around each Mn is approximately octahedral and the two $Mn(CO)_5$ pyramids are staggered with respect to each other. In $Co_2(CO)_8$, there are two bridging CO groups as well as the Co–Co bond. The iron compound, $Fe_2(CO)_9$ has a structure related to that of $Co_2(CO)_8$ with the ninth CO also occupying a bridging position so that there is three-fold symmetry around the Fe–Fe axis.

The higher carbonyls have more complex structures which we describe only briefly. Of those with three metal atoms, the simplest structure is that of $FeMn_2(CO)_{14}$. This has a linear metal atom skeleton, $(CO)_5Mn–Fe(CO)_4–Mn(CO)_5$, with the planar $Fe(CO)_4$ group inserted into the Mn—Mn bond of manganese decacarbonyl. Each group has its CO ligands in staggered positions relative to the next. The $Os_3(CO)_{12}$ structure is also simple, with an equilateral triangle of Os atoms held together only by Os–Os bonding, and four terminal carbonyls on each Os making an approximately octahedral configuration around osmium. The $Ru_3(CO)_{12}$ molecule is isostructural with the osmium compound but the tri-iron species has a less symmetrical structure with one $Fe(CO)_4$ group bonded by Fe—Fe bonds to an $Fe_2(CO)_8$ moiety with both an Fe–Fe bond and $Fe(CO)_2Fe$ bridging.

The higher species contain a tetrahedron or octahedron of metal atoms held together both by metal-metal bonding and by bridging carbonyls. The $Rh_6(CO)_{16}$ structure contains four CO groups which bridge three metal atoms.

A range of species which may be regarded as derivatives of the carbonyls are also known. These include cations such as $Mn(CO)_6^+$, anions like $Re(CO)_5^-$ or $Co(CO)_4^-$, the carbonyl hydrides like $HMn(CO)_5$ or $HCo(CO)_4$, the similar alkyl derivatives with a sigma C—metal bond like

$MeMn(CO)_5$, and finally the carbonyl halides like $Mn(CO)_5Cl$. In addition to alkyl derivatives, a range of carbonyls substituted by isoelectronic silicon, germanium or tin species have been developed over the last decade. Examples include $H_3MMn(CO)_5$, $H_3MCo(CO)_4$ and $(H_3M)_2Fe(CO)_4$ (for M = Si or Ge) and the analogous R_3M derivatives where R = alkyl or aryl and M = Si, Ge, Sn or (in some cases) Pb. In the above compounds there is a sigma two-electron bond between the substituent and the metal carbonyl moiety. In addition, there is a wide variety of ligands which may replace the CO groups in compounds such as $Ni(PF_3)_2(CO)_2$ or $Co(CO)_3(NO)$. The trifluorophosphine group, PF_3, is very similar in ligand properties to CO, and PF_3 analogues of many of the carbonyls are known, also a wide range of mixed species. For example, all members of the series $Ni(CO)_x(PF_3)_{4-x}$ are reported.

The case of nitric oxide is interesting. In most compounds, it is best regarded as a source of three electrons and donates as the cation in $M^-(NO)^+$; that is, one electron transfers to the metal (taking it to a negative oxidation state) and then the NO^+ ion (which is isoelectronic with carbonyl) coordinates to the metal ion. Thus two nitric oxide groups replace three carbonyls, as in $Fe(NO)_2(CO)_2$, or odd numbers of nitric oxide molecules occur in the molecule allowing the formation of mononuclear complexes of those elements with an odd number of electrons, as in the cobalt compound above.

The carbonyls are made by the action of carbon monoxide, usually under pressure, with the finely divided metal (for iron or nickel) or with the metal halide, in presence of a reducing agent. The more complex carbonyls are usually made from the simpler ones. For example, $Fe_2(CO)_9$ is formed by photolysing a solution of $Fe(CO)_5$ in a hydrocarbon. The mixed metal species are often made by reaction of a carbonyl halide with the alkali metal salt of the anion in organic solvent.

The bonding in the carbonyls and related compounds is illustrated schematically in Figure 12.31. The essential feature is that the ligand donates a lone pair to an empty orbital on the metal to form a sigma bond, and the metal donates electrons from a filled orbital of π symmetry back to an empty orbital on the ligand. The basic requirements are thus a lone pair on the ligand together with an empty orbital of pi symmetry with respect to the metal-ligand bond, and an empty sigma orbital on the metal together with enough electrons to fill a π orbital for back donation. These requirements explain the limited formation of carbonyl compounds as shown in Table 12.18. Elements to the left of the d block have insufficient electrons to form the number of pi bonds required, while elements to the right have no empty orbitals to accept the sigma electrons.

The carbon monoxide molecule, and the nitrosonium ion, have a lone pair on the donor atom (C and N respectively, the lone pair on the O atom in each molecule is too tightly bound to be donated) and the acceptor orbitals are the antibonding π orbitals from their internal π system (such orbitals are only antibonding as far as the C–O or N–O bonding is con-

FIGURE 12.31 *The sigma and pi-bonds between a metal and* CO

(a) the sigma bond formed when the lone pair on the CO donates into a suitable metal orbital; (b) the pi-bond formed when electrons from a filled metal orbital donate back into the empty π^* antibonding orbital on the CO. Cross-hatching indicates the orbitals which originally held the donated electrons.

TABLE 12.18 Properties of the carbonyls

$V(CO)_6$	$Cr(CO)_6$	$Mn_2(CO)_{10}$	$Fe(CO)_5$	$Co_2(CO)_8$	$Ni(CO)_4$
green-black d. 70 °C octahedron paramagnetic	white d. 130 °C. s octahedron	yellow m 154 °C. s Fig. 12.30	yellow m −20°. b 103 °C trig. bipyramid	orange m 51 °C. s Fig. 12.30	colourless m −19·3 °C, b 42 °C tetrahedron
			$Fe_2(CO)_9$ orange d. 100 °C	$Co_4(CO)_{12}$ black . d. 60 °C	
			$Fe_3(CO)_{12}$ green d. 140 °C. s	$Co_6(CO)_{16}$	
$Mo(CO)_6$ white d. 180 °C, s octahedron	$Tc_2(CO)_{10}$ white m 160 °C, s	$Ru(CO)_5$ colourless m −22 °C $Ru_3(CO)_{12}$ orange	$Rh_2(CO)_8$? orange d. 76 $Rh_4(CO)_{12}$ red		
				$Rh_6(CO)_{16}$ black	
$W(CO)_6$ white d. 180 °C, s octahedron	$Re_2(CO)_{10}$ white m 177 °C, s	$Os(CO)_5$ colourless m 15 °C	$Ir_2(CO)_8$? green s	$Pt(CO)_4$ transient existence
			$Os_3(CO)_{12}$ yellow	$Ir_4(CO)_{12}$ yellow $Ir_6(CO)_{16}$	

Mixed metal carbonyls are also known including
$ReMn(CO)_{10}$ $MCo(CO)_9$ (M = Mn, Re)
$M_2Fe(CO)_{14}$ (M = Mn, Re) $MM'_2(CO)_{12}$ (M = Fe, Ru, Os: M' = Ru, Os)
$Co_2Rh_2(CO)_{12}$ and $Co_2Rh_4(CO)_{16}$

m = melting point, b = boiling point, d. = decomposes, s = sublimes in vacuum. $Rh_2(CO)_8$ and $Ir_2(CO)_8$ have not been reported since their initial discovery. $Ru(CO)_5$ and $Os(CO)_5$ are less studied than the $M_3(CO)_{12}$ species.

cerned, they are perfectly good orbitals for general bond formation). Where the donor atom is phosphorus, as in PF_3 or organic phosphines, the acceptor orbital is the empty $3d$ orbital on the phosphorus atom. Similar use of d orbitals is made by other common donor atoms such as As or S.

The importance of the double-donation theory is that it gives a mechanism which prevents the build-up of charge in the molecule. If only sigma donation occurs, there is a large build-up of negative charge at the central atom, with no intrinsic charge to offset this as in complexes of charged ions. However, the pi donation, in the opposite direction, removes this excess of negative charge and strengthens the sigma bond. This in turn removes charge from the ligand and strengthens the pi effect—such a mechanism has been called *synergic*. The relative importance of the sigma and pi bonds varies with the metal and ligand: some authorities regard the back-donation in nickel carbonyl as providing the major part of the bond energy.

A further large class of pi-bonded transition metal complexes, which must be dismissed in only a few words here, is the group containing aromatic molecules as ligands. The two classical examples are ferrocene, with the cyclopentadienyl ion $C_5H_5^-$, as the aromatic ligand (with the charge, this species has six pi electrons in aromatic delocalized orbitals), and dibenzene chromium where the neutral benzene molecule is the aromatic system. These molecules contain the metal atom sandwiched between the organic parts, $C_5H_5 - Fe - C_5H_5$ and $C_6H_6 - Cr - C_6H_6$, and the planes of the organic rings are parallel, Figure 12.32.

Cyclopentadiene compounds are formed by most transition metals. Types include dicyclopentadienides like ferrocene, and molecules with one or three rings which may con-

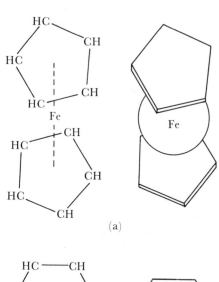

(a)

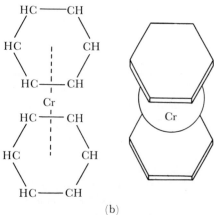

(b)

FIGURE 12.32 *The structure of metal sandwich compounds:* (a) *ferrocene,* $(C_5H_5)_2Fe$, (b) *dibenzene chromium,* $(C_6H_6)_2Cr$

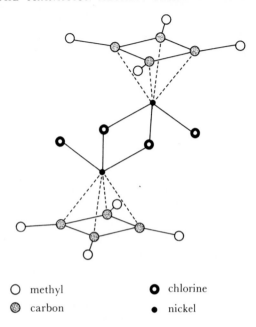

O methyl   ⊙ chlorine
⊚ carbon   • nickel

FIGURE 12.33 *Di(tetramethylcyclobutadienylnickeldichloride)*, $(C_4Me_4NiCl_2)_2$

tain other ligands as well, such as $C_5H_5Mo(CO)_3Cl$. Dibenzene compounds are less widespread. Other aromatic systems appear in similar compounds including the cation of cycloheptatriene $(C_7H_7)^+$ and the dianion of cyclobutadiene $(C_4H_4)^{2-}$. The latter system has long appeared as a hypothetical aromatic system with no evidence of its actual occurrence. The parent hydride is still unknown, but the substituted molecule with four phenyl or methyl groups in place of the hydrogens has been attached to a nickel atom as in the compound shown in Figure 12.33. The four-membered ring is planar and the aromatic electrons appear to be fully delocalized.

The bonding in these aromatic sandwich compounds is too complicated to be described in detail here. The basic bond in ferrocene is a single bond of π symmetry between the iron atom and each ring. This bond is formed by overlap of the d_{xz} and d_{yz} orbitals on the iron (the z axis is the molecular axis, and these two d orbitals are of equal energy) with that aromatic orbital on each ring which has one node passing across the ring. These two metal d orbitals and the two ring orbitals combine to give two bonding π orbitals, of equal energy, and two antibonding orbitals. Each of these orbitals is 'three-centred' on each ring and the iron atom. There are four electrons in ferrocene which fill the two bonding orbitals. In the corresponding cobalt and nickel compounds, the extra electrons enter the antibonding orbitals, making these compounds less stable and giving cobaltocene one unpaired electron and nickelocene two unpaired electrons (as there are two degenerate π^* orbitals). One component of the π bond is shown in Figure 12.34. Other overlaps add smaller contributions to the bonding, but this is the main interaction. The basic feature is that there is a single metal-ring bond and the rings are aromatic, undergoing aromatic substitution and similar reactions.

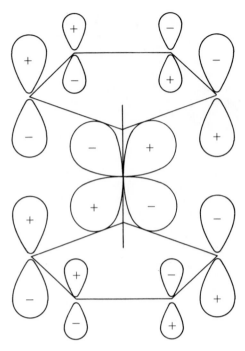

FIGURE 12.34 *One component of the principal ring-metal-ring bond in ferrocene*
This figure shows the interaction between the ring orbitals and the d_{xz} orbital of the iron (where the z axis is the ring-metal-ring axis). The ring-d_{yz}-ring interaction is similar.

Many compounds are known which contain both carbonyl and pi-bonded organic ligands. The wide variety of types is illustrated by the examples $\pi\text{--}C_5H_5M(CO)_x$ (M = V, Nb, Ta, x = 4: M = Mn, Tc, Re, x = 3: M = Co, Rh, Ir, x = 2) and the dimers $[\pi\text{--}C_5H_5M(CO)_x]_2$: (M = Cr, Mo, W, x = 3: M = Fe, Ru, Os, x = 2: and M = Ni, Pt, x = 1). Mixed metal species include

$$\pi\text{--}C_5H_5(CO)_2FeMo(CO)_3(\pi\text{--}C_5H_5)$$

or

$$\pi\text{--}C_5H_5(CO)_3MoMn(CO)_5,$$

and other pi ligands include, in addition to cyclopentadienyl C_5H_5, the groups C_4H_4, C_6H_6 or C_7H_7 or derivatives of these. Just as C_5H_5 may be regarded as contributing five electrons to the central atom, these latter groups contribute respectively four, six or seven electrons. Also well-known are allyl derivatives like $\pi\text{--}C_3H_5Mn(CO)_4$ where the C_3H_5 group contributes three electrons and diene derivatives where the substituent contributes four electrons, as in $\pi\text{--}C_6H_8Re(CO)_3H$.

This extremely brief survey must suffice as an introduction to the topic of π-bonded complexes which is rapidly growing into a major area of transition metal chemistry. It includes the major portion of the chemistry of the transition elements in low oxidation states. It also makes up a considerable part of the field of organometallic chemistry which is of fundamental importance, not only in the development of chemistry as a science, but in its applications especially in catalysis, and in its relation with the biological activity of metallic compounds.

13 The Transition Elements of the First Series

13.1 General properties

The transition elements of the first series, dealt with here, are those elements, from titanium to copper, where the $3d$ level is filling. Table 12.2, in the last chapter, shows their oxidation states, while Tables 12.3, 4, and 5 give their oxides, fluorides, and other halides. Other properties which are listed earlier include the electronic configurations, Atomic Weights and Numbers (Table 2.5) and the Ionization Potentials (Table 2.8). Some values of the Redox Potentials, especially of species used in quantative analysis, are given in Table 5.3. The free energy diagrams of all the elements are shown on a small scale in Figure 13.1 so that a general comparison may be made. The individual diagrams for important systems are given on a larger scale in the later sections.

The lists of the oxides and halides and the free energy diagrams give largely complementary pictures of the stabilities of the various oxidation states, in the solid state and in aqueous solution respectively. At titanium, the Group oxidation state of IV is the most stable and the lower states become increasingly reducing. Moving along the series, the Group oxidation state becomes more unstable and more oxidizing so that, at manganese, only a few compounds of the VII state are known and all are strongly oxidizing. Beyond manganese, the Group state disappears and only a few, unstable, strongly oxidizing states greater than III exist. Among the lower states, either the II or the III state is the most stable state from chromium onwards, and the relative stability of these two states varies with the number of d electrons as was indicated in the last chapter, with the II state finally becoming the most stable at nickel and copper. This variation in stability is shown up in the free energy diagrams by the increasing height above zero of the Group state up to manganese, and by the increasing instability of the III state at the right of the Figure. (See also section 13.10.)

Oxidation states which lie between the Group state and the II or III states have a tendency to disproportionate. Some of them, such as Cr(IV) and (V), and Mn(V) and (VI), are very rare and poorly represented. The one state of moderate stability among these intermediate ones is manganese(IV), which owes part of its importance in chemistry to the insolubility of MnO_2 in neutral or basic solution.

All the elements of the first transition series are common and all are important commercially, titanium and copper in their own right, and the others largely as constituents or

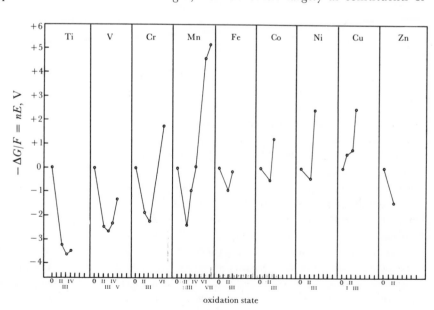

FIGURE 13.1 *Oxidation state free energy diagrams of the First Transition Series elements*

It is clear that the Group Oxidation state becomes increasingly unstable towards manganese. Thereafter, all states above III are unstable, and the variation in the relative stability of the II and III states (and I for copper) is of most importance. (Compare section 7.6.)

coatings of iron and steel. Most are produced in the blast furnace, and very pure nickel is produced in the Mond process by formation of the carbonyl, its volatilization, and finally its thermal decomposition. High grade copper for electrical applications is prepared by electrolysis. Titanium is prepared by magnesium or calcium reduction of the tetrachloride. As it readily forms a brittle nitride, carbide, or hydride, the final reduction is carried out in an atmosphere of argon, the first large-scale industrial use of a rare gas for such a purpose.

The metals are reactive and electropositive with both properties decreasing towards the right of the series. However, a number form a thin layer of coherent and impermeable oxide on the surface and so resist attack. In particular, this is the action of chromium, hence its use in chromium plating to protect iron.

The transition elements form non-stoichiometric compounds in which a number of small elements, hydrogen, boron, carbon, or nitrogen are incorporated into metallic lattices. Such compounds are termed interstitial, reflecting the original theory that the small atoms simply fitted into spaces in the metallic lattice. It is now known that, although the metal atom arrangements in such compounds are commonly in one of the forms characteristic of metals (cubic or hexagonal close packed or body-centred cube), the arrangement of the metal atoms in the interstitial compound is rarely the same as that in the metal itself. The best theory of these compounds regards the included atoms as 'metallic'. That is, they show larger coordination numbers than normal —e.g. six for carbon in TiC—and contribute some, at least, of their valency electrons to the common pool of delocalized electrons in the metallic bonding. The interstitial compounds show many metallic properties—hardness, metallic appearance, high conductivity with negative temperature coefficient, high melting point—which support this view. The interstitial compounds are among the hardest known and show extremely high melting points, e.g.:

	m.p., °C	hardness, mohs
TiC	3140	8–9
HfC	4160	9
W_2C	3130	9–10

The binary compound $4TiC + ZrC$ has the highest recorded melting point of 4215 °C. (The hardnesses given above are on Moh's scale on which diamond = 10). The earlier metals, of the titanium, vanadium, and chromium Groups form interstitial compounds with small metal-nonmetal ratios, such as MC, M_2C, MN, or MH_2 (all the formulae are idealized, nonstoichiometry marks most of these phases), and these have regular structures. Thus the MN and MC compounds are sodium chloride structure and the M_2C ones have the zinc blende structure. These compounds are very unreactive.

The elements of the later Groups form less-well defined compounds with complex metal-nonmetal ratios such as Fe_3C or Cr_7C_3. Such compounds are similar in general physical properties but are much more reactive chemically.

They are attacked by water and dilute acids to give hydrogen and mixtures of hydrocarbons.

Borides, and also silicides, are similar to the carbides and nitrides in structures and properties. These provide a link between the interstitial compounds and metal alloys.

13.2 Titanium, $3d^24s^2$

The free energy diagram for titanium is shown in Figure 13.2. The titanium(IV) state is the most stable one and is well-characterized with a wide variety of compounds. The III state is reducing but reasonably stable; in water, it is a reducing agent somewhat stronger than tin(II). The II state is very strongly reducing. The only solid compounds are unstable polymerized solids and it rapidly decomposes water.

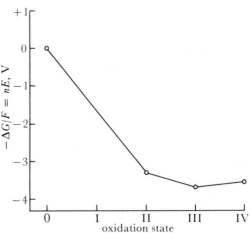

FIGURE 13.2 *The oxidation state free energy diagram of titanium*
The IV state is stable with the lower states reducing. The titanium redox potentials are subject to some uncertainty, and Ti(II) may lie higher than shown, corresponding to its marked reducing power.

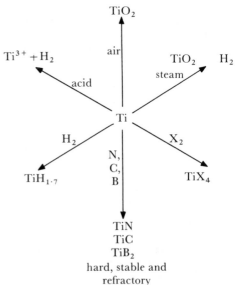

FIGURE 13.3 *Reactions of titanium*

The metal is inert at low temperatures but combines with a variety of reagents at elevated temperatures, as shown in Figure 13.3. Titanium dioxide (which is used as a white

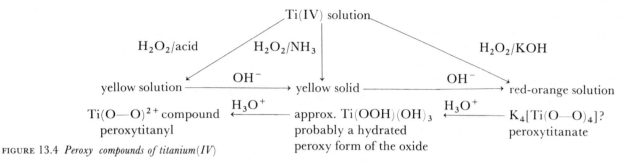

FIGURE 13.4 *Peroxy compounds of titanium(IV)*

pigment) is prepared by hydrolysis of the tetrachloride (produced in the course of the metal extraction) or by purification of the naturally-occurring form, called *rutile*. The oxide is weakly acidic and weakly basic, dissolving in concentrated base or acid but easily hydrolysing on dilution:

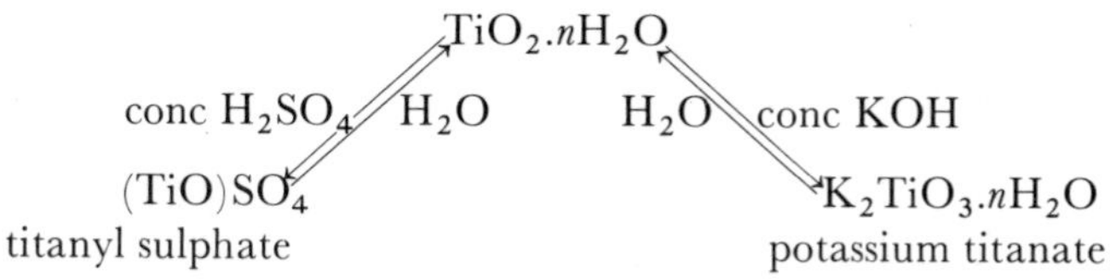

$$\text{(TiO)SO}_4 \overset{\text{conc H}_2\text{SO}_4}{\underset{\text{H}_2\text{O}}{\nwarrow}} \text{TiO}_2.n\text{H}_2\text{O} \overset{\text{H}_2\text{O}}{\underset{\text{conc KOH}}{\searrow}} \text{K}_2\text{TiO}_3.n\text{H}_2\text{O}$$

titanyl sulphate   potassium titanate

No definite hydroxide, such as Ti(OH)_4, exists, the compound produced on hydrolysis being a hydrated form of the oxide. The titanyl group, TiO^{2+}, does not appear to exist as the monomeric cation. The structure of titanyl sulphate, for example, consists of $(\text{TiO})_n^{2n+}$ chains held together in the crystal by the sulphate groups. The alkali metal titanates are of unknown structure and do not contain discrete anions.

Titanium(IV) forms peroxy compounds which are the formula analogues of the oxy-compounds as shown in Figure 13.4. The yellow colour in acid is extremely intense and is the basis of the colorimetric determination of titanium.

The titanium halides may all be made by direct reaction of the metal and halogen, and TiCl_4 may be converted to the others by the reaction of the hydrogen halide. All the tetrahalides are readily hydrolysed, although TiF_4 is more stable than the others, and the intermediate oxyhalides may be isolated under careful conditions:

$$\text{TiX}_4 \overset{\text{H}_2\text{O}}{\longrightarrow} \text{TiOX}_2 \overset{\text{H}_2\text{O}}{\longrightarrow} \text{TiO}_2.n\text{H}_2\text{O}$$

Titanium tetrachloride forms the hexachloride in concentrated hydrochloric acid, and salts, M_2TiCl_6, are known but are very unstable. The hexafluoride TiF_6^{2-}, is readily formed and is very stable. The shape is regular octahedral.

The tetrahalides act as acceptors to a wide variety of donor ligands such as R_3P, R_2O or py, to give complexes TiX_4L_2. These are usually in the *cis* configuration unless the ligands are bulky. Among five-coordinate complexes are the unusual hydride derivative $\text{TiCl}_4.\text{AsH}_3$, where the arsine molecule is thought to occupy an equatorial position in a trigonal bipyramid, and the compound $\text{Et}_4\text{N}^+(\text{TiCl}_5)^-$ where the TiCl_5^- ion may be trigonal bipyramidal like the isoelectronic SnCl_5^- ion. The oxychloride also forms a five-coordinate species $\text{TiOCl}_2.\text{NMe}_3$.

Chloro-compounds of titanium(IV), including TiCl_4 and

$(\text{C}_5\text{H}_5)_2\text{TiCl}_2$, when reacted with an aluminium alkyl such as AlEt_3, act as catalysts for the polymerization of alkenes in the *Ziegler-Natta* process. This is discussed in more detail under aluminium in section 15.4.

One further simple titanium(IV) compound which is of interest is the anhydrous nitrate, $\text{Ti(NO}_3)_4$. The nitrate groups form a tetrahedon around the titanium, but are each bonded through two oxygen atoms so that the titanium is eight-coordinate. The coordination is nearly regular dodecahedral.

Titanium(III) compounds are readily formed by reduction and, as they contain a d electron, are coloured. Some preparations are shown in Figure 13.5. The existence of

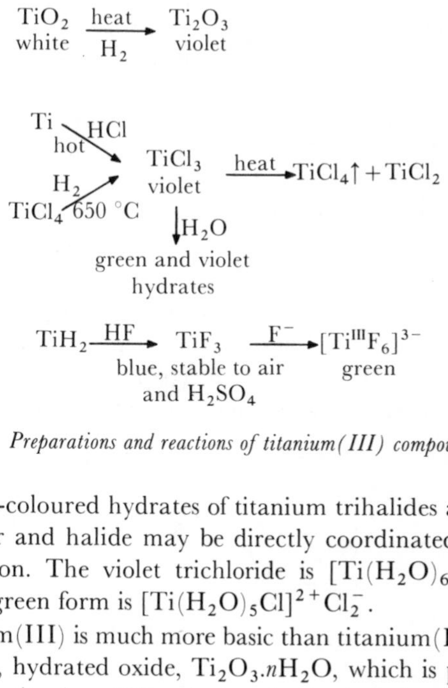

$$\underset{\text{white}}{\text{TiO}_2} \overset{\text{heat}}{\underset{\text{H}_2}{\longrightarrow}} \underset{\text{violet}}{\text{Ti}_2\text{O}_3}$$

$$\underset{\substack{\text{hot} \\ \text{H}_2 \\ \text{TiCl}_4\ 650\ ^\circ\text{C}}}{\text{Ti}} \searrow \underset{\text{violet}}{\text{TiCl}_3} \overset{\text{heat}}{\longrightarrow} \text{TiCl}_4\uparrow + \text{TiCl}_2$$

$$\Big\downarrow \text{H}_2\text{O}$$

green and violet
hydrates

$$\text{TiH}_2 \overset{\text{HF}}{\longrightarrow} \underset{\substack{\text{blue, stable to air} \\ \text{and H}_2\text{SO}_4}}{\text{TiF}_3} \overset{\text{F}^-}{\longrightarrow} \underset{\text{green}}{[\text{Ti}^{\text{III}}\text{F}_6]^{3-}}$$

FIGURE 13.5 *Preparations and reactions of titanium(III) compounds*

differently-coloured hydrates of titanium trihalides arises as both water and halide may be directly coordinated to the titanium ion. The violet trichloride is $[\text{Ti(H}_2\text{O})_6]^{3+}\text{Cl}_3^-$ while the green form is $[\text{Ti(H}_2\text{O})_5\text{Cl}]^{2+}\text{Cl}_2^-$.

Titanium(III) is much more basic than titanium(IV) and the purple, hydrated oxide, $\text{Ti}_2\text{O}_3.n\text{H}_2\text{O}$, which is precipitated from titanium(III) solutions by base, is insoluble in excess alkali.

The titanium(III) ion is a d^1 system with the electron in a t_{2g} orbital. Only one $d-d$ transition is possible, Figure 12.26a, and a simple spectrum with one band in the visible region is observed, Figure 12.4. Magnetic measurements on titanium(III) compounds give values close to the spin-only value of 1.73 bohr magnetons for one unpaired electron.

Titanium(II) is a very unstable state represented only by solid compounds. The dihalides may be made by heating the trihalides, as indicated in Figure 13.5 above, but the dihalide disproportionates to metal plus tetrahalide at a temperature below that of its formation so that it is never obtained un-contaminated by metal. The oxide, TiO, is made by heating the dioxide with titanium. It has the sodium chloride structure but is seldom obtained in the stoichiometric form.

Titanium(II) is thought to be the active intermediate in one of the few reversible systems of *nitrogen fixation* so far announced. A number of halides (including $TiCl_4$) yield a species which fixes N_2 gas when reduced by organometallic compounds but these produce only NH_3 on hydrolysis with the concomitant destruction of the active intermediate (see under nitrogen, section 15.6). In a recent study, it was shown that nitrogen could be fixed and converted to ammonia by a cycle, involving titanium alkoxides, which also regenerated the starting reagent so that an overall catalytic system is possible. The main steps of the reaction are

$$Ti(OR)_4 + 2Na = Ti(OR)_2 + 2RONa \qquad (13.1)$$

$$Ti(OR)_2 + N_2 = [Ti(OR)_2N_2]_n \qquad (13.2)$$

$$[Ti(OR)_2N_2]_n + 4Na = [\text{intermediate}] + 6ROH =$$
$$2NH_3 + Ti(OR)_4 + 4RONa \qquad (13.3)$$

Thus, the overall reaction is

$$N_2 + 6e^- + 6ROH = 2NH_3 + 6RO^-$$

The sodium in equations (13.1) and (13.2) is used as its naphthalide complex, $Na^+C_{10}H_8^-$. In this, the valency electron of the sodium is transferred to the lowest antibonding π level of the aromatic hydrocarbon, where it is readily available for reductions but is in a more manageable form than if the metal was used alone (compare section 9.3). The critical step is the uptake of N_2, from the gas phase at normal temperature and pressure, by the titanium(II) alkoxide. The formulation of the reduced intermediate is still uncertain.

This cycle does not consume titanium and is thought to model the biological nitrogen-fixation process which also probably acts through the reversible formation of a low valency transition metal intermediate to which the nitrogen

becomes coordinated. A model for the coordinated nitrogen may be provided by the nitrogen complexes such as those formed by ruthenium discussed in section 14.6.

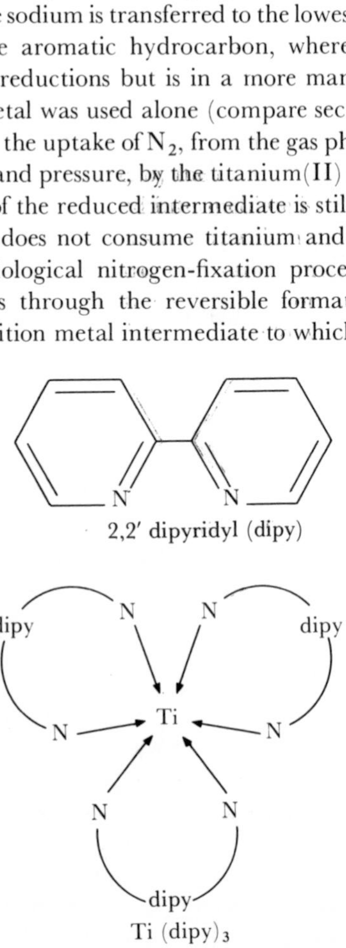

One compound of titanium in the oxidation state 0 has been reported. This is the compound Ti(dipy)$_3$, where dipy = 2,2'-dipyridyl (Figure 13.6). The oxidation states $-$I and $-$II are reported in the same system, in the compounds Li(Ti dipy$_3$)3.5THF and Li$_2$(Ti dipy$_3$).5THF respectively. These compounds result when $TiCl_4$ is reduced in the presence of dipyridyl by lithium in tetrahydrofuran (THF).

13.3 Vanadium, $3d^34s^2$

The free energy diagram for vanadium is shown in Figure 13.7. There are five valency electrons and oxidation states

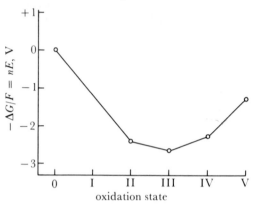

FIGURE 13.7 *The oxidation state free energy diagram of vanadium* (Compare section 7.6.)

from V to $-$I are known, with the ones from II to V of importance. Vanadium(V) and (IV) are both stable, with the former mildly oxidising and represented mainly by oxy-species. Vanadium(IV) is stable, but shows dispro-portionation reactions in the solid when volatile species, like VF$_5$, can be driven off. Vanadium(III) is reducing but less so than Ti(III); it is stable to water and is only slowly oxidized in air. Vanadium(II) is strongly reducing; it attacks water and is rapidly oxidized in air. Again, in the solid state, its compounds tend to disproportionate, yielding vanadium(III) and the element. The oxidation state free energy diagram shows that none of the intermediate states disproportionates in solution.

The metal resembles titanium in readily forming carbides and nitrides, but as it is seldom used alone this is less of a disadvantage. Its main use is in steels and it is usually pro-duced as an iron alloy, ferrovanadium, which is suitable for direct incorporation in the steel preparation. The metal it-self may be isolated by direct reduction with aluminium of the pentoxide, V_2O_5. Pure samples of the metal are best prepared by the van Arkel process in which the iodide is decomposed on a hot filament under vacuum. The pure metal resembles titanium in being relatively unreactive at low temperatures. When heated, it combines directly, as shown in Figure 13.8. Notice that the halogens, other than fluorine, produce IV or III state compounds only.

The V state is represented among the simple compounds by V_2O_5, VF$_5$, and the oxyhalides, VOX$_3$. The pentoxide is

2,2' dipyridyl (dipy)

Ti (dipy)$_3$

FIGURE 13.6 *2,2'dipyridyl, and its titanium*(0) *compound,* Ti(dipy)$_3$

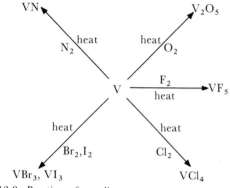

FIGURE 13.8 *Reactions of vanadium*

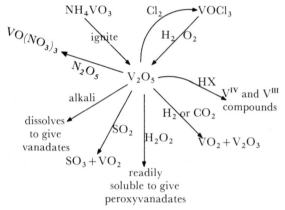

FIGURE 13.9 *Preparation and reactions of vanadium pentoxide*

behave similarly, e.g. in the chromate-dichromate equilibrium, but many fewer species are known. The other elements with a strong polyanion-forming tendency are molybdenum and tungsten.

Vanadium forms a red peroxy cation in acid peroxide solution and this is probably the peroxy-analogue of the vanadyl cation, $V(O-O)^{3+}$. In alkaline solution, a yellow pervanadate is formed with two peroxy groups per vanadium; the simplest formulation is $[V(O-O)_2O_2]^{3-}$. These peroxy compounds are less stable and well-known than the titanium or chromium peroxides.

The only pentahalide of vanadium is VF_5. This results from fluorine plus the metal at 300 °C or by disproportionation:

$$VF_4 \xrightarrow[N_2]{600\,°C} VF_5\uparrow + VF_3$$

The pentafluoride is a volatile white solid melting at 20 °C. The crystal structure is composed of chains of VF_6 octahedra with two bridging fluorines in *cis* positions to each other. Such a structure is indicated by the useful nomenclature: $VF_4(F_{2/2})$ to show the four F atoms bonded only to one vanadium and two F atoms shared between two vanadiums, and also the overall six-coordination. A related structure is shown by the oxyfluoride ion, $VO_2F_3^{2-}$, which also contains six-coordinated vanadium with *cis* fluorine bridges in the structure $VO_2F_2(F_{2/2})$. The O atoms are *trans* to the bridging fluorines. The hexafluoro- complex ion also exists and is made by reacting VCl_3 and an alkali halide in liquid BrF_3 which acts as a fluorinating and oxidizing agent:

$$Cl^- + VCl_3 + BrF_3 = VF_6^- + ClF + BrF$$

Vanadium pentafluoride reacts with air to form the oxyfluoride, which also may be made by oxidizing or fluorinating suitable compounds:

$$VF_3 + O_2 \qquad VF_5 \qquad VOCl_3 + HF$$
$$\searrow \qquad \downarrow{\scriptstyle air} \qquad \swarrow$$
$$VOF_3$$

m.p. 300 °C, very reactive

prepared, and reacts, as shown in Figure 13.9. It is amphoteric, dissolving in acid and base to give a variety of species. In strong base, the mononuclear vanadate ion, VO_4^{3-}, is present and, as the pH is reduced, these units link up into binuclear species, and then into polynuclear ions until, at pH about 6, $V_2O_5.nH_2O$ is precipitated. In more acidic solutions, this dissolves to form cationic vanadyl species. The approximate species present at different pH values are shown in Table 13.1, together with the colour changes. The full structures of the different anions are not known and all species are probably hydrated.

This example of polymerism in solution is an extreme instance of a general tendency to form polymeric oxyanions in this region of the Periodic Table. Titanium and chromium

TABLE 13.1 Vanadium(V) species in basic and acidic solutions

pH	above 12	12-9	9-7	7-6·5
number of V atoms	1	2	3 → 4	5 → ∞
approximate formula	$(VO_4)^{3-}$	$[V_2O_6(OH)]^{3-}$	$(V_3O_9)^{3-}$	$V_5O_{14}^{3-}*$, $V_5O_{16}^{5-}*$
colour	← colourless →			red → red-brown

pH	6·5-2·2	below 2·2		
number of V atoms	∞	10 ⇌ 1 (V₁ being favoured as the pH is lowered)		
approximate formula	$V_2O_5.nH_2O*$	$V_{10}O_{28}^{6-}$ ⇌ VO_2^+ (or VO^{3+})		
colour	brown	yellow		

*These species are representative of solids precipitated at the appropriate pH; they do not necessarily correlate with the formulae of species in equilibrium with them in solution. The other formulae apply in solution; the V_{10} species is rather well established and, for example, V_9 or V_{11} give a much poorer fit to the data.

The second oxyfluoride, VO_2F, has been reported more recently. The oxychloride is made by the reaction of chlorine on any of the vanadium oxides, and the oxybromide similarly. The oxyhalides are relatively volatile and are therefore probably simple monomeric molecules, especially in the gas phase. This has been confirmed in the case of $VOCl_3$, where an electron diffraction study shows a tetrahedral molecule.

$VOCl_3$ is violently hydrolysed by water and is inert to metals, even to sodium. It shows no reaction with salts and dissolves many non-metals which suggests possible uses as a non-aqueous solvent.

Vanadium(IV) is readily prepared by mild reduction of vanadium(V), for example by ferrous salts. The oxide, VO_2, is formed from V_2O_5 in this way and is dark blue. It reverts to the pentoxide on heating in air. It is also mildly amphoteric, but is much more basic than acidic. It dissolves in acid to give blue solutions of the vanadyl(IV) ion VO^{2+}, and a variety of salts containing this cation are known. It dissolves in alkali to form vanadites which are readily hydrolysed and relatively unstable. A number of vanadate(IV) anions are known, such as VO_3^{2-}, VO_4^{4-}, and $V_4O_9^{2-}$. These are generally prepared by reaction of VO_2 with alkali, or alkaline earth oxides in the fused state. The best known vanadyl(IV) salts are the sulphate, $VOSO_4$, and the halides, VOX_2. A number of complexes of vanadium(IV) are known, all derived from the vanadyl ion. Examples include the halides, VOX_4^{2-}, and corresponding oxalate and sulphate compounds. Neutral complexes are formed by the enol forms of β-diketones as in vanadylacetylacetonate, $VO(acac)_2$, which has the square pyramidal coordination shown in Figure 13.10. Lone pair donors can coordinate weakly in the sixth position to complete the octahedron as in $VO(acac)_2C_5H_5N$.

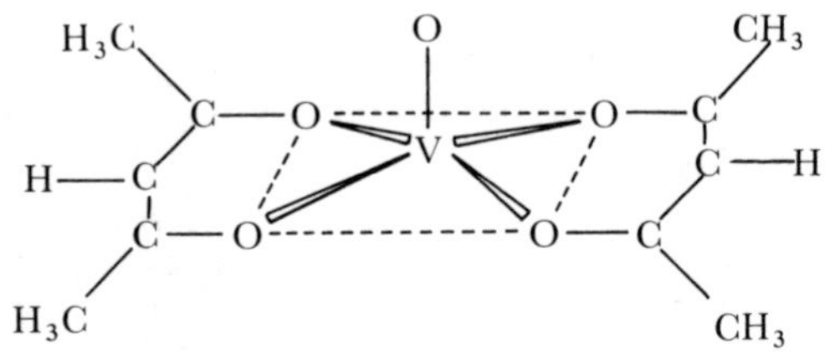

FIGURE 13.10 *Structure of vanadylacetylacetonate, $VO(acac)_2$*

The tetrachloride is formed by direct reaction and gives the tetrafluoride with HF. Hydrolysis leads to the VOX_2 molecules. Figure 13.11. shows some of these inter-relationships. The tetrabromide is very unstable and VI_4 is known only in the vapour above VI_3. VCl_4 contains one d electron, which might be expected to lead to some distortion in the shape, but the molecule is tetrahedral in the gas. The liquid is dimeric. Notice, in Figure 13.11, the tendency for these halogen compounds of the IV state to disproportionate to give vanadium(III).

Vanadium(III) is produced by fairly strong reduction by hydrogen or red-hot carbon. It resembles other trivalent ions such as chromium or ferric in size and general properties, apart from reduction behaviour. The solid oxide and trihalides are known and the state exists in aqueous solution as the green $V(H_2O)_6^{3+}$ ion.

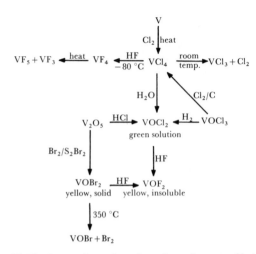

FIGURE 13.11 *Preparation and reactions of vanadium tetrachloride*

This state has no acidic properties: the action of alkali on the solution of V^{3+} precipitates hydrated $V(OH)_3$ which has no tendency to redissolve. The green hydroxide oxidizes rapidly in air, and the solutions are also oxidized in air.

All four trihalides exist. VBr_3 and VI_3 are the products of direct combination with the metal while VF_3 and VCl_3 result from the disproportionation of the tetrahalides, as shown in Figure 13.11. The trichloride itself disproportionates on heating and some of the vanadium chloride relationships are given in Figure 13.12. It will be seen that heating

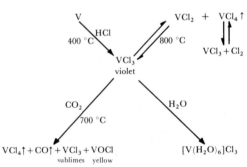

FIGURE 13.12 *The vanadium-chlorine system*

with removal of chlorine gives VCl_2, while heating in excess of chlorine gives the tetrachloride. VCl_3 forms a 2:1 adduct with trimethylamine, $VCl_3.2NMe_3$, which is a trigonal bipyramid with axial NMe_3 groups. The triiodide also disproportionates

$$VI_3 \rightleftharpoons VI_2 + VI_4 \quad \text{(decomposes)}$$

The structure of VI_3 is a layer lattice like BiI_3 (Figure 4.10).

Vanadium(III) also forms a triacetate which exists as a dimer. Its structure is a sandwich one, with four bridging acetates and one terminal one on each vanadium, $AcOV(OAc)_4VOAc$. This is similar to the chromium(II) acetate (Figure 13.16) or copper acetate structure (Figure 13.27) with terminal acetates in place of the two water molecules.

Oxyhalides, VOX, are reported for $X = F$, Cl and Br. VOF has the rutile structure with both F and O in the oxygen positions of TiO_2. The corresponding TiOF is isostructural.

The V^{3+} ion forms a variety of complexes, typical of transition ions. The commonest shape is octahedral and most complexes contain vanadium coordinated to oxygen, nitrogen or halogen. The hexafluoride and related complexes are stable, e.g. VF_6^{3-} and $VF_4(H_2O)_2^-$, but other halogen complexes are more readily oxidized. A cyanide, $V(CN)_6^{3-}$, can be made in alcohol but precipitates $V(CN)_3$ on addition of water. All the complexes are labile as expected of a d^2 ion where the empty t_{2g} orbital presents a low-energy path for attack.

The vanadium(II) state is strongly reducing and evolves hydrogen from water, although it is more stable in acid solution. The oxide, VO, is made by reaction of the pentoxide with vanadium and is commonly non-stoichiometric. All four halides, VX_2, exist. The chloride and iodide result from disproportionation of the trihalides, the dibromide from reduction of VBr_3 with hydrogen. VF_2 is made by the action of HF on VCl_2 or by the reduction of VF_3 by H_2/HF at 1200 °C. VF_2 forms blue crystals with the rutile structure while VI_2 exists in the CdI_2 layer lattice (see Figures 4.3 and 4.10). The dihalides dissolve in water to give violet solutions, from which $V(OH)_2$ is precipitated by alkali. The solutions soon turn green due to formation of $V(H_2O)_6^{3+}$. All these compounds are unstable and readily oxidized. The divalent ion in solution is probably octahedral and a few complexes are isolable, including the cyanide, $K_4V(CN)_6.7H_2O$.

Low oxidation states are represented by a few compounds including dipyridyl complexes of vanadium I, 0, and $-I$, $V(dipy)_3^+$, $V(dipy)_3$, and $V(dipy)_3^-$, the latter being isolated as the etherate of the lithium salt, as in the case of titanium. The carbonyl, $V(CO)_6$, is known and differs from the sequence of first row carbonyls in being monomeric in the gas and therefore in not having 18 electrons on the vanadium. The anion, $V(CO)_6^-$, is also known and is a further case of $V(-I)$. In all the low oxidation state compounds, the geometry is octahedral.

13.4 Chromium, $3d^54s^1$

The oxidation state energy diagram for chromium, Figure 13.13, illustrates the increased oxidizing power of the Group oxidation state of VI, as compared with the cases of titanium

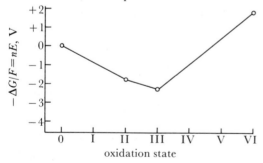

FIGURE 13.13 *The oxidation state free energy diagram for chromium*
It is clear that Cr(III) is stable, Cr(VI) is strongly oxidizing, and Cr(II) is reducing. The other states disproportionate, but data are not very reliable.

and vanadium, and the stability of the III state. This is further shown in the solid state by the existence of CrO_3 and

the unstable CrF_6 only (Tables 12.3–5). The III state corresponds to the d^3 configuration and is very stable. Other states shown range to chromium $(-II)$ in carbonyl anions but only the II state is well-represented, the IV and V states disproportionating very readily.

Chromium metal is formed by aluminium reduction of Cr_2O_3. Like titanium and vanadium it is unreactive at low temperatures, due to the formation of a passive surface coating of oxide. It dissolves in hot hydrochloric or sulphuric acids and reacts with oxygen or the halogens on heating to give chromium(III) compounds, and with hydrogen chloride to give $CrCl_2$ and hydrogen.

The compounds of chromium(VI) are limited in number. They include CrO_3, CrO_2Cl_2, CrO_2F_2, CrF_6, and the chromates, CrO_4^{2-} and dichromates, $Cr_2O_7^{2-}$. The oxide is acidic and dissolves in alkali to give solutions of the chromate ion. On acidification, these give dichromate solution, which exists in strongly acid solutions down to a pH of 0. In very concentrated acid the trioxide is precipitated and there are no cationic forms of chromium in solution. This behaviour is similar to that of vanadium(V) but the chromium(VI) is more acidic and polymerization does not go so far. (There are some indications of trichromates and tetrachromates, however.) The equilibrium between chromate and dichromate is rapid and the two forms coexist over a wide range of pH. Many chromates of heavy metals, such as Pb^{2+}, Ag^+ or Ba^{2+}, are insoluble and these may be precipitated from chromium(VI) solutions, even at a pH where the major part exists as dichromate, as the equilibrium is so rapidly established. The CrO_4^{2-} ion is tetrahedral while dichromate has two tetrahedral CrO_4 groups joined through an oxygen with the Cr—O—Cr angle about 115°.

If a dichromate solution is heated in the presence of chloride ion and concentrated sulphuric acid, the red, hydrolysable chromyl chloride, CrO_2Cl_2, distills out (b.p. 117 °C). An intermediate chlorochromate ion, CrO_3Cl^-, is also known. The chromyl bromide and iodide do not exist and the fluoride is made only by the action of fluorine on chromyl chloride, so the latter compound distinguishes chlorine from the rest of the halogens and is used in this way in qualitative analysis. The chromyl halides are covalent compounds which hydrolyse in water to chromate, and there is no evidence for any CrO_2^{2+} cation.

A number of peroxy compounds of chromium are known. The best-known are the deep blue CrO_5, the blue ion $Cr_2O_{12}^{2-}$, and the red ion CrO_8^{3-}. When hydrogen peroxide is added to acidified chromate solution, a deep blue, transient colour appears. This is unstable in water but extracts into ether (the colour test for chromium) and a stable solid pyridine adduct, $CrO_5.C_5H_5N$, may be prepared. This is a peroxy-analogue of the trioxide and the structure of the CrO_5L species (where $L = H_2O$, ether, or pyridine in the sequence above) is a tetrahedral arrangement around the chromium of an oxygen, L, and two peroxy (O—O) groups coordinated edgeways on (Figure 13.14a). The peroxydichromo-ion is a peroxy analogue of dichromate and has the structure of Figure 13.14b, as suggested by its containing two

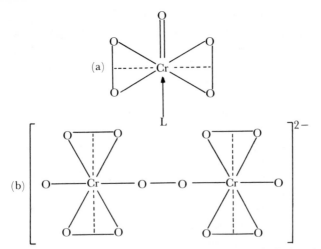

FIGURE 13.14 *Peroxy compounds of chromium:* (a) $CrO_5.L$, (b) $Cr_2O_{12}^{2-}$

Bonding between the peroxy group and metals is not fully understood but there is evidence for overlap between the π orbitals of the peroxy group and a metal orbital. If the $O-O$ group is regarded as a single ligand, then the configurations at Cr are tetrahedral.

and a half peroxy groups per chromium. The red compounds, $M_3^ICrO_8$, are prepared in alkali, are paramagnetic with one unpaired electron, and form mixed crystals with the pentavalent niobium and tantalum compounds $M_3^IM^VO_8$. The red peroxide is therefore formulated as a compound of chromium(V), $Cr(O-O)_4^{3-}$.

Chromium(V) compounds are rare and unstable. They include CrF_5, the peroxy compound above, some oxyhalide ions such as $CrOF_4^-$ and $CrOCl_5^{2-}$, and the oxyanion, CrO_4^{3-}, which results when potassium chromate is heated in molten KOH (compare $K_2Mn^{VI}O_4$). A recent structural study indicates that the CrO_4^{3-} ion is of distorted tetrahedral structure as might be expected for a d^1 species.

Chromium(IV) compounds are also rare. CrO_2 and CrF_4 exist and there are reports that $CrCl_4$ and $CrBr_4$ exist in the vapour phase at high temperatures in mixtures of the trihalides and halogen. The complex ion, CrF_6^{2-}, has been isolated and mixed oxides apparently containing chromium(IV) are known. The structure of one of these, $Ba_2(CrO_4)$, has been partially determined and appears to contain discrete CrO_4^{4-} ions. Chromium(IV) is also found in some mixed oxidation state species. The sulphide Cr_5S_8 is formulated as $Cr_4^{III}Cr^{IV}S_8$, and the selenium analogue is similar. In a structural study on the oxygen species $M_2^+Cr_3O_9$, linked chains of $Cr^{VI}O_6$ octahedra and $Cr^{IV}O_4$ tetrahedra were indicated.

By far the most stable and important oxidation state of chromium is chromium(III). It is the most stable of all the trivalent transition metal cations in water and a wide variety of compounds and complexes are known. The complexes are octahedral, inert to substitution, and have a half-filled t_{2g} set of orbitals.

The oxide is green Cr_2O_3, with the corundum structure, and is used as a pigment. It, and the hydrated form precipitated from Cr^{3+} solution by OH^- ions, dissolve readily in acid to give $Cr(H_2O)_6^{3+}$ ions and also in concentrated alkali

to give the chromites. The species in the latter solutions are not identified but may be $Cr(OH)_6^{3-}$ and $Cr(OH)_5(H_2O)^{2-}$. They are readily hydrolysed and precipitate hydrated oxide on dilution. Cr_2S_3 is a black stable solid made by direct combination and is inert to non-oxidizing acids.

All four anhydrous trihalides are known and may be prepared by the standard methods. The trichloride gives the dichloride and chlorine when heated to 600 °C, and sublimes in the presence of chlorine at this temperature. It is a green flakey solid which is rather insoluble in water, except in the presence of Cr^{2+} ions. It is thought that these assist the solution process by attaching through a Cl bridge to Cr^{3+} in the crystal—$Cr_{solid}^{3+}-Cl-Cr_{soln}^{2+}$—which then transfers an electron to give divalent chromium in the solid. This Cr^{2+} does not fit the crystal lattice and dissolves to repeat the process, $Cr_{solid}^{2+}-Cl-Cr_{soln}^{3+} \rightarrow Cr_{soln}^{2+}$. A similar effect is found in the case of chromium(III) complexes in solution. These are inert to substitution except in the presence of chromium(II) which may behave similarly in abstracting a ligand *via* a bridging and oxidation process.

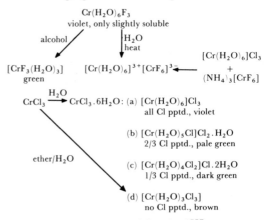

FIGURE 13.15 *Aquo-halogeno ions of chromium(III)*

The trihalides in aqueous solution give rise to violet $Cr(H_2O)_6^{3+}$ ions and to a number of aquo-halogeno ions, some of which are indicated in Figure 13.15.

A wide variety of chromium(III) complexes exist, all octahedral. The hexammine, $Cr(NH_3)_6^{3+}$, and similar complexes with variety of substituted amines and related molecules are found, and all possible aquo-ammine mixed complexes have been prepared. There are also a wide variety of aquo-X and ammino-X mixed complexes (X = acid radical like halide, thiocyanate, oxalate, etc.) and even aquo-ammino-X species. A commonly used example is Reinecke's salt, $NH_4[Cr(NH_3)_2(NCS)_4].H_2O$, where the large monovalent anion is widely used as a precipitant for large cations. Apart from the simple hydrated cation, hydrated chromium also occurs in an extensive series of alums, $M^ICr(SO_4)_2.12H_2O$, where Cr^{3+} replaces Al^{3+} which is of similar size.

The unusual three-coordination appears to occur in the chromium(III) species $Cr(NR_2)_3$ which have been prepared for R = isopropyl and R = $SiMe_3$. The configuration may be planar and its adoption is probably aided by the presence of the bulky ligands.

A fair number of chromium(II) compounds are known,

including all the dihalides and CrO. A considerable variety of complexes are also found, although all are unstable to oxidation. In water, the sky-blue $Cr(H_2O)_6^{2+}$ ion is formed. This is a strong reducing agent which is just too weak to reduce water. It has a number of uses as a reducing agent including the removal of traces of oxygen from nitrogen. The solutions are less stable when not neutral, and hydrogen is evolved. They react rapidly with the air. Chromium(II) complexes include the hexammine, $Cr(NH_3)_6^{2+}$, and related ions. The dipyridyl complex is found to disproportionate to give a chromium(I) species:

$$2Cr(dipy)_3^{2+} = Cr(dipy)_3^+ + Cr(dipy)_3^{3+}$$

The halides can be prepared by reaction of the appropriate HX on the metal at 600 °C or by the reduction of the trihalides with hydrogen at a similar temperature. The iodide, CrI_2, and CrS, may also be made by direct combination.

Among the chromium(II) salts, the hydrated acetate, $Cr_2(CH_3COO)_4.2H_2O$, is the commonest. It is readily prepared by adding chromium(II) solution to sodium acetate when it is precipitated as a red crystalline material. It has the unusual bridged structure shown in Figure 13.16, with a

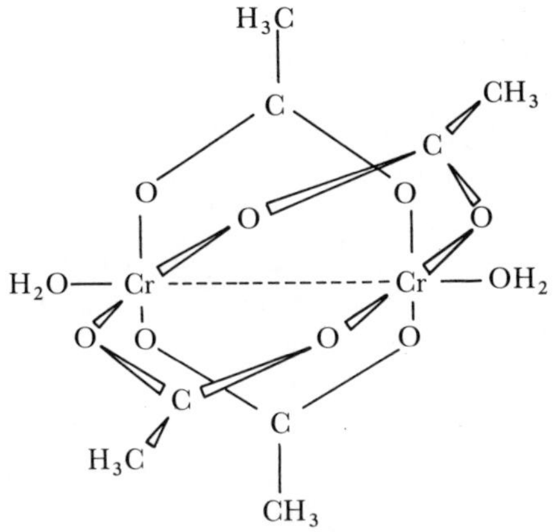

FIGURE 13.16 *Structure of chromium(II) acetate,* $Cr_2(CH_3COO)_4.2H_2O$
The oxygen atoms of the acetate groups lie in a square plane around each Cr and the acetates link the edges together. The Cr−Cr distance is short, implying direct metal-metal bonding. Compare the copper acetate, Figure 13.27.

very short Cr—Cr distance suggesting metal-metal interaction.

Lower oxidation states are represented by a few compounds including the chromium(I) dipyridyl complex above. The 0 state is shown in the stable, octahedral carbonyl, $Cr(CO)_6$, in carbonyl ions, $Cr(CO)_5X^-$, and in dibenzene chromium (Figure 12.32). Carbonyl anions of chromium $(-I)$ and $(-II)$ are the compounds $Na_2[Cr_2(CO)_{10}]$ and $Na_2[Cr(CO)_6]$ respectively, prepared by sodium reduction of the hexacarbonyl. Chromium(I) is also found in the cyanide, $K_3Cr(CN)_4$, and the zero state in $K_6Cr(CN)_6$ which may be used to prepare $Cr(diphos)_3$ and $Cr(triphos)_2$ (for di- and tri-phosphine ligands see Appendix B).

13.5 Manganese, $3d^5 4s^2$

Manganese, with seven valency electrons, shows the widest variety of oxidation states in the first transition series. The oxidation state energy diagram, Figure 13.17, shows that

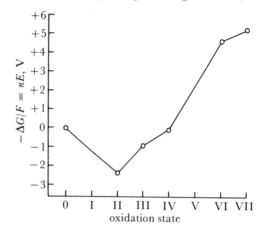

FIGURE 13.17 *Oxidation state free energy diagram for manganese*
The stability of Mn(II) is striking and it is clear that the III and VI states disproportionate. Compare Figure 13.1 and section 7.6.

Mn(II) is the most stable state, and comparison with the diagrams for previous elements shows that manganese(VII) is the most strongly oxidizing of all the high oxidation states known in aqueous solution. Many of the intermediate states are rare, with a strong tendency to disproportionate.

Manganese is abundant in the earth's crust and its principle ore, pyrolusite, is a crude form of the dioxide, MnO_2. The metal is obtained by reduction with aluminium, or in the blast furnace. The metal resembles iron in being moderately reactive and dissolving in cold, dilute non-oxidizing acids. It combines directly with most non-metals at higher temperatures, sometimes quite vigorously. Thus it burns in N_2 at 1200 °C to give Mn_3N_2 and in Cl_2 to give $MnCl_2$. The product of high temperature combination with oxygen is Mn_3O_4.

The VII state is strongly oxidizing and strongly acidic. The red heptoxide, Mn_2O_7, is a volatile unstable molecule produced by heating permanganate in sulphuric acid. Manganese heptoxide can decompose or react with explosive violence. The structure is two tetrahedra linked through oxygen. On treatment with chlorosulphonic acid, the explosively unstable oxychlorides result

$$Mn_2O_7 + HSO_3Cl \rightarrow MnO_3Cl + MnO_2Cl_2 + MnOCl_3$$
$$\qquad\qquad\qquad\quad \text{black} \qquad \text{brown} \qquad \text{green}$$

corresponding to the VII, VI, and V states. They are more stable when handled in solution in carbon tetrachloride. The oxyfluoride, MnO_3F, has been reported, but the only stable representative of the VII state is the permanganate ion, MnO_4^-, which is most common as the potassium salt. This is a strong oxidizing agent in acid solution:

$$MnO_4^- + 8H_3O^+ + 5e^- = Mn^{2+} + 12H_2O, \ 1.51 \text{ V}$$

and is also strong in basic media:

$$MnO_4^- + 2H_2O + 3e^- = MnO_{2(solid)} + 4OH^-, 1.23 \text{ V}$$

when manganese dioxide is precipitated. However, in concentrated alkali the anion of the dibasic acid, manganate, is formed in preference, and this reverts to permanganate plus dioxide on acidification:

$$MnO_4^- + OH^- = MnO_4^{2-} \xrightleftharpoons[OH^-]{H^+} MnO_4^- + MnO_2$$

The permanganates are, of course, purple, while the manganate ion is an intense green, and is the only stable representative of the manganese(VI) state. Permanganate is about the strongest accessible oxidizing agent which is compatible with water and even it undergoes slow decomposition in acid solution to give MnO_2 and oxygen.

The manganese(V) state is also represented by only one type of compound, the blue MnO_4^{3-} or MnO_3^- ion produced by the action of alkaline formate on MnO_4^{2-}.

Manganese(IV) does not have a very extensive chemistry although black MnO_2 is well-known and is a common precipitate from manganese compounds in an oxidizing medium. It is very insoluble and usually only dissolves with reduction, as in its reaction with HCl:

$$MnO_2 + 4HCl = MnCl_2 + Cl_2 + 2H_2O$$

A few complex salts are known including the hexachloride and fluoride, MnX_6^{2-}. The only simple halide reported is the fluoride, MnF_4.

The III state is represented by the oxide, Mn_2O_3, and the fluoride, MnF_3. The oxide stable at high temperatures, Mn_3O_4, is correctly formulated as a mixed oxide of the II and III states, $Mn^{II}Mn_2^{III}O_4$. No other simple trihalides exist but the red $MnCl_5^{2-}$ complex ion is known, as is the corresponding fluoride. $(NH_4)_2MnF_5$ contains chains of MnF_6 octahedra linked by *trans* bridging fluorides, in the structure $MnF_4(F_{2/2})$. Complex ions of manganese(III) include $MnCl_3L$ and $MnCl_3L_3$ types, where L is a nitrogen ligand, and the acetylacetonates $MnCl_2acac$, and $MnClacac_2$ are also reported.

In solution manganese(III) is unstable both to reduction and to disproportionation:

$$2Mn^{3+} + 6H_2O = Mn^{2+} + MnO_{2(solid)} + 4H_3O^+$$

If manganese(II) in sulphuric acid solution is oxidized with permanganate in the stoichiometric ratio, an intensely red solution results which contains all the manganese as manganese(III), presumably as a sulphate complex. This solution is as strongly oxidizing as permanganate and was once used as an alternative oxidizing agent for sulphate media. A rather similar oxidation of the acetate gives $Mn(OAc)_3.2H_2O$ as a solid. This manganese(III) acetate is readily prepared and used as a starting material for most manganese(III) studies.

In contrast to all the above unstable or poorly represented states, the manganese(II) state is very stable and widely represented. It is the d^5 state and all the compounds contain five unpaired electrons, except for the cyanide and related complexes, for example $Mn(CN)_6^{4-}$ and $Mn(CN)_5(NO)^{3-}$, which are low-spin with only one unpaired electron. All the high-spin manganese(II) compounds are very stable and resist attack by all but the most powerful oxidizing or reducing agents. They have all very pale colours, for example the hydrate $Mn(H_2O)_6^{2+}$ is pale pink, as the absorptions due to the d electron transitions are very weak. This is because a $d-d$ transition which involves the reversal of an electron spin (as it must be in a high-spin d^5 system) is an event of low probability, compared with one which is 'spin-allowed'. The absorptions in manganese(II) compounds are therefore about a hundred times weaker than the general run of transition compound absorptions.

By contrast to the high-spin compounds, the low-spin complexes are much more reactive and oxidize readily, for example:

$$Mn(CN)_6^{4-} \xrightarrow{air} Mn(CN)_6^{3-}$$

Manganese(II) is unstable in these low-spin compounds probably because the crystal field stabilization energy—though large enough to cause spin-pairing—is only a little greater than the loss of exchange energy due to spin pairing. In the trivalent state, the CFSE is increased because of the greater charge. A somewhat similar situation arises in the case of cobalt(III) in water.

The stability of the high-spin manganese(II) state is illustrated by the wide variety of stable compounds formed. Some of these are listed in Table 13.2.

A variety of complex ions exist, including $Mn(NH_3)_6^{2+}$ and octahedral complex ions with EDTA, oxalate, ethylenediamine, and thiocyanate but the hexahalide ions are unknown. The equilibrium constants for the formation of such ions in solution are low, as the Mn^{2+} ion is relatively large (Figure 12.23) and there is no CFSE. Thus one important source of the energy required to displace the coordinated water molecules is lacking.

A few examples exist of manganese in a square-planar environment. In $Mn(acac)_2.2H_2O$, the bidentate acetylacetonate groups form a square plane around the manganese and the water molecules complete a distorted octahedron: when the compound is dehydrated, it is probable that the $Mn(acac)_2$ remaining is truly square planar. The sulphate, $MnSO_4.5H_2O$, is isostructural with $CuSO_4.5H_2O$ and therefore contains square planar $Mn(H_2O)_4^{2+}$ units.

Some tetrahedral manganese units are found, especially the halogen anions, MnX_4^{2-}, which may be prepared as the salts of large cations. These tetrahedral complexes are unstable in water, or other donor solvents, and go to octahedral complexes of the solvent.

In its low oxidation states, manganese forms the cyanide, $K_5Mn(CN)_6$, and the carbonyl halides, like $Mn(CO)_5Cl$, in the I state, the carbonyl, $Mn_2(CO)_{10}$, in the 0 state, and the anion, $Mn(CO)_5^-$, in the $-I$ state. The latter is a trigonal bipyramid, as is the related $Mn(NO)(CO)_4$ with NO in an equatorial position.

13.6 Iron, $3d^64s^2$

When iron is reached, in the first transition series, the elements cease to use all the valency electrons in bonding, and the Group oxidation state of VIII is not found. The highest

TABLE 13.2 Examples of high-spin manganese(II) compounds

MnX_2	X = F, Cl, Br, I	Isomorphous with the Mg halides.	Stable at red heat.
MnY	Y = O, S, Se, Te	Sodium chloride structure.	Very stable when dry but the hydrated forms slowly oxidize to MnO_2 in air.
$Mn(OH)_2$		This is a true hydroxide, not a hydrated oxide, isomorphous with $Mg(OH)_2$.	
$MnSO_4$		Very stable, even at red heat. The hydrate is isomorphous with copper sulphate.	
$Mn(ClO_4)_2$		Very soluble. Stable to 150 °C, then the perchlorate oxidizes the Mn^{2+} to the dioxide.	
$MnCO_3$		Insoluble. Very stable for a transition metal carbonate. It goes to MnO and CO_2 at about 100 °C.	
$Mn(OOCCH_3)_2$ and other organic acid salts		Stable: prepared by heating $Mn(NO_3)_2$ with the acid anhydride.	

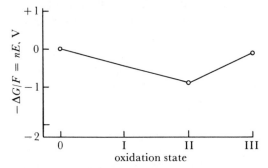

FIGURE 13.18 *Oxidation state free energy diagram for iron*
No potential values for higher states than III are known: the few Fe(IV) compounds known are strongly oxidizing. The II state is the more stable but, compare Figure 13.22, the III state is also relatively stable.

state of iron is VI and the main ones are II and III. The oxidation energy diagram, Figure 13.18, shows that the III state is only slightly oxidizing while the II state lies at a minimum and is stable in water. By comparison with the II and III states of other transition elements, the iron(II) and (III) states lie much closer together in stability, and this accords with the well-known properties of ferrous and ferric solutions which are readily interconverted by the use of only mild oxidizing or reducing agents.

Iron is the most abundant of the fairly heavy elements in the earth's crust and is used on the largest scale of any metal. Its production in the blast furnace is well-known. The element is reactive and quickly forms a coat of hydrated oxide on the surface in moist air. This is non-coherent and flakes away revealing fresh surfaces for attack. Iron combines at moderate temperatures with most non-metals and it readily dissolves in dilute acids to give iron(II) in solution, except with oxidizing acids which yield iron(III) solutions. Very strongly oxidizing agents, such as dichromate or concentrated nitric acid, produce a passive form of the metal, probably by forming a coherent surface film of oxide.

Iron forms three oxides, FeO, Fe_2O_3, and Fe_3O_4, which all commonly occur in non-stoichiometric forms. Indeed, FeO is thermodynamically unstable with respect to the structures with a deficiency of iron. The three oxides show

TABLE 13.3 The iron oxides

FeO (black)	Prepared by thermal decomposition of ferrous oxalate at a high temperature, followed by rapid quenching to prevent disproportionation to $Fe + Fe_3O_4$.	Structure is sodium chloride—i.e. O^{2-} ions ccp and Fe^{2+} ions in all the octahedral sites.
Fe_2O_3 (brown)	Occurs naturally. Otherwise by ignition of hydrated ferric oxide, precipitated from a ferric solution by ammonia.	The structure has the oxide ions ccp with Fe^{3+} ions randomly distributed over the octahedral and tetrahedral sites. A second form is hexagonal.
Fe_3O_4 (black)	Occurs naturally as *magnetite*. It is the ultimate product of strong ignition in air of the other two oxides.	The structure is inverse spinel*: that is the O^{2-} ions are ccp, the Fe^{2+} ions are in octahedral sites, and the Fe^{3+} ions are half in octahedral sites and half in tetrahedral sites.

*It will be recalled from Chapter 4 that a normal spinel is an oxide AB_2O_4, with A a divalent metal ion and B a trivalent one. The oxide ions are cubic close packed and the A atoms are in tetrahedral sites while the B atoms are in octahedral sites. The inverse structure of magnetite may be ascribed to the much greater CFSE of d^6 (low spin) ferrous ions in octahedral sites instead of tetrahedral ones. The d^5 ferric ions have no CFSE in either type of coordination.

a number of structures, some based on a cubic close packed array of oxide ions. Properties of the oxides are summarized in Table 13.3.

It will be seen that the structures of the cubic forms of all these oxides are related. If the cubic array of oxide ions is taken, then all the structures result from different dispositions of ferrous and ferric ions in the octahedral and tetrahedral sites. This explains the tendency to non-stoichiometry and the interconversion of oxides. If the oxide lattice has its octahedral sites filled up with Fe^{II} ions, the FeO structure builds up as the Fe/O ratio approaches one. If a small portion of the Fe^{II} is missing or replaced by Fe^{III} in the ratio of two Fe^{III} for every three Fe^{II}, defect FeO forms result, with a stability maximum at $Fe_{0.95}O$. If the process is continued until there are two Fe^{III} atoms for every Fe^{II} and if half these Fe^{III} ions enter tetrahedral sites, the structure becomes that of Fe_3O_4. Conversion of the remaining Fe^{II} to Fe^{III} gives the cubic form of Fe_2O_3.

If hydrated ferric oxide in alkali is treated with Cl_2, a red-purple solution of iron(VI) is obtained containing the ferrate ion, FeO_4^{2-}. The sodium and potassium salts are very soluble but the barium compound may be precipitated. The ferrate ion is stable only in a strongly alkaline medium, in water or acid it evolves oxygen:

$$2FeO_4^{2-} + 10H_3O^+ = 2Fe^{3+} + 3/2O_2 + 15H_2O$$

Ferrate is a stronger oxidizing agent than permanganate. It is a tetrahedral ion and the potassium salt is isomorphous with K_2CrO_4 and with K_2SO_4.

A pentavalent oxyanion, FeO_4^{3-}, is also reported.

Iron(IV) may occur in the complex $Fe(diars)_2Cl_2^{2+}$, where diars = the *ortho*-diarsine derivative of benzene, $o\text{-}C_6H_4(AsMe_2)_2$, which acts as a bidentate ligand through the lone pair on each arsenic atom. The complex is produced as the salt with a large anion, $FeCl_4^-$ or ReO_4^-, by oxidation of the iron(III) complex with 15 M nitric acid.

Apart from the oxides, solid compounds of the II and III states are represented by all the halides except FeI_3. Salts of Fe^{II} are known with nearly all stable anions and Fe^{III} also forms a wide range of salts, but reducing anions are oxidized, as in the case of I^-.

In solution, the relative stabilities of the III and II states vary widely with the nature of the ligand. As Fe^{II} is d^6 and Fe^{III} is d^5, changes in CFSE have an important effect on these relative stabilities. The effect is illustrated by the potentials below:

$$\begin{array}{ll} Fe(H_2O)_6^{3+} + e^- = Fe(H_2O)_6^{2+} & 0.77 \text{ V} \\ Fe(CN)_6^{3-} + e^- = Fe(CN)_6^{4-} & 0.36 \text{ V} \\ Fe(phen)_3^{3+} + e^- = Fe(phen)_3^{2+} & 1.12 \text{ V} \end{array}$$

(phen = *o*-phenanthroline, $C_{12}H_8N_2$, a bidentate aromatic nitrogen ligand)

$$Fe(C_2O_4)_3^{3-} + e^- = Fe(C_2O_4)_2^{2-} + C_2O_4^{2-} \quad 0.02 \text{ V}$$

The cyanide and phenanthroline complexes are low-spin while the other two are high-spin, in all cases both in the II and III states. Since the ΔE value for the trivalent d^5 Fe^{III} ion is larger than that of the divalent d^5 ion, Mn^{II}, the

cyanide of ferric iron is less unstable, from the CFSE versus exchange energy point of view, than the manganese(II) hexacyanide. There is a gain in CFSE on going from ferricyanide to the d^6 low-spin ferrocyanide, due to the additional t_{2g} electron, but this is relatively small. Ferricyanide acts as a mild oxidizing agent while ferrocyanide is stable. In addition, the d^5 ferricyanide is relatively labile to substitution and the cyanide may be replaced by water and other ligands, as in $Fe(CN)_5(H_2O)^{2-}$. Thus ferricyanide in solution is much more poisonous than is ferrocyanide.

In aqueous solution, ferric iron shows a strong tendency to hydrolyse. The hydrated ion, $Fe(H_2O)_6^{3+}$, which is pale purple, exists only in strongly acid solutions at a pH of about 0. In less acidic media, hydroxy complexes are formed:

$$Fe(H_2O)_6^{3+} + H_2O = Fe(H_2O)_5(OH)^{2+} + H_3O^+$$
$$Fe(H_2O)_5(OH)^{2+} + H_2O = Fe(H_2O)_4(OH)_2^+ + H_3O^+$$

These occur up to pH values of 2 to 3 and are yellow in colour, the typical colour of ferric salts in solution in acid. At lower acidities, above a pH of 3, bridged species are formed and the solutions soon form colloidal gels. As the pH is raised hydrated ferric oxide is precipitated as a reddish-brown gelatinous solid. This precipitate probably does not contain any of the hydroxide, $Fe(OH)_3$, and part of it is probably in the form FeO(OH) and part as the hydrated oxide. The hydrated oxide readily dissolves in acid and is also slightly soluble in strong bases, so the ferric state is weakly acidic as well as moderately basic. The basic solutions in strong alkali probably contain the $Fe(OH)_6^{3-}$ ion which has been isolated as the strontium and barium salts.

Ferric iron forms many complexes with ligands which coordinate through oxygen, especially phosphate anions and polyhydroxy-organic compounds such as sugars. It also forms intensely red thiocyanate complexes, used to detect and estimate trace quantities of iron. These colours are destroyed by fluoride due to the formation of FeF_6^{3-}. In contrast, Fe^{III} forms no ammonia complex (ammonia precipitates the oxide from aqueous solutions) and is only weakly coordinated by other amine ligands. If the ligand field is sufficiently strong to produce spin pairing, much more stable complexes are formed and this occurs with dipyridyl and phenanthroline.

Iron(III) forms the three-coordinate compound $Fe[N(SiMe_3)_2]_3$ like the chromium analogue. Structural studies show the presence of planar three-coordinate FeN_3 and $FeNSi_2$ groups indicating extended π bonding.

Ferrous iron also forms a variety of complexes. In aqueous solution it exists as the $Fe(H_2O)_6^{2+}$ ion, which is pale sea-green in colour. This is slowly oxidized by air in acid, and is very readily oxidized when the hydrated oxide is precipitated in alkali. The anhydrous halides combine with ammonia gas to give the hexammine, $Fe(NH_3)_6^{2+}$, but this is unstable and loses ammonia when brought into contact with water. Stable complexes are formed, however, by chelating amines such as ethylenediamine. All these examples are octahedral.

Tetrahedral complexes are rare but the anions of the heavier halogens, FeX_4^{2-}, can be precipitated by large cations.

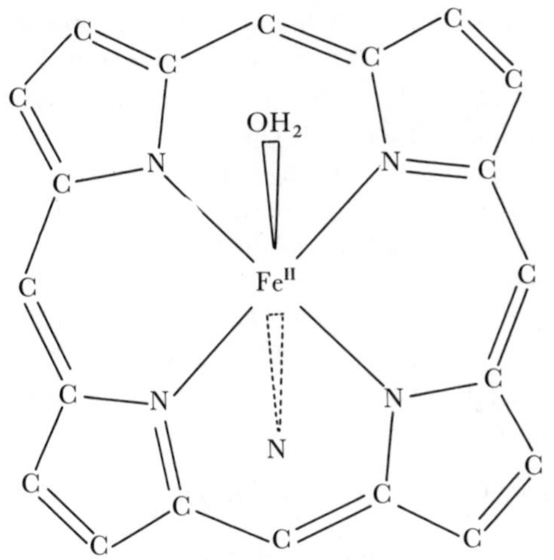

FIGURE 13.19 *The environment of the iron atom in haemoglobin*
The four N atoms of the porphin ring are coplanar with the iron. The fifth position on the iron is occupied by a nitrogen atom from a long side-chain on one of the rings, leaving the sixth site in the octahedron around the iron to hold an oxygen molecule, a water molecule, or some other group.

The most famous complex of ferrous iron is the complex, haem, which exists in haemoglobin. The central porphin ring system is shown in Figure 13.19. Side chains are attached to the porphin skeleton and an imidazole ring on one of these is coordinated to a fifth position on the iron atom while a water molecule occupies the sixth position, as shown in Figure 13.20a. This water molecule may be replaced by an O_2 molecule and this process is reversible, 13.20b, providing the mechanism for the transport of oxygen by the red blood cells in the body. The iron remains in the II state throughout this equilibrium and oxidation to Fe^{III} removes its power to coordinate O_2. The oxygen site on the iron may also be occupied by CN^-, CO, or PF_3 and the coordination in these cases is strong and irreversible. This is one reason for the poisonous nature of these substances (although cyanide, in particular, has other reactions on the body as well). Iron occurs in a number of other important biological compounds

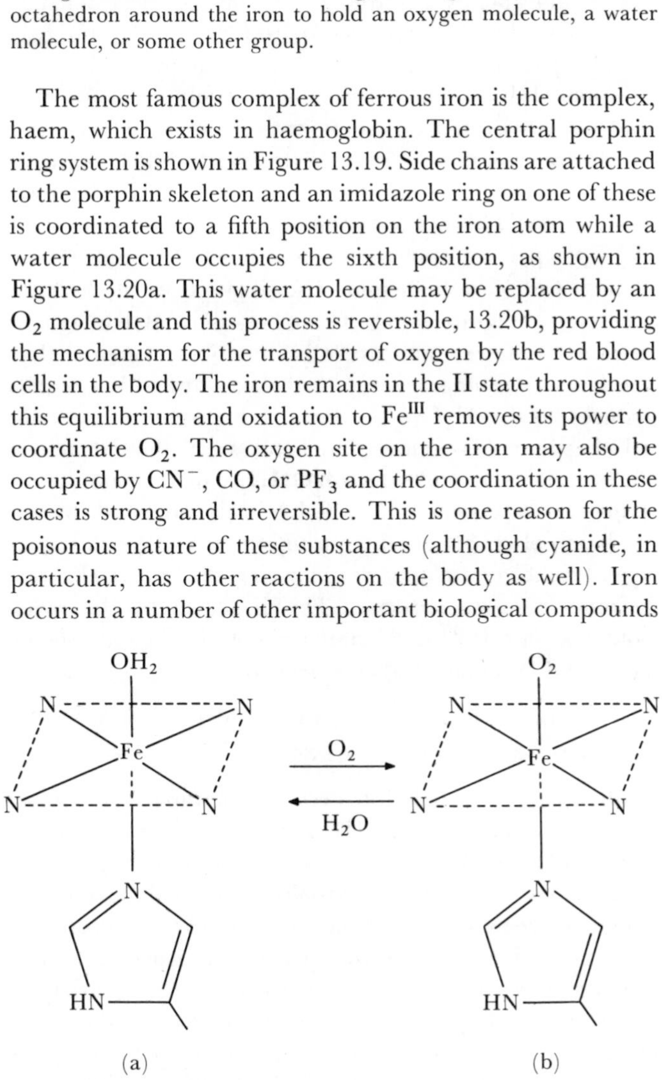

FIGURE 13.20 *The mode of oxygen carriage by haemoglobin: (a) with a water molecule which is reversibly replaced by an oxygen molecule (b)*
Recent studies indicate that the O_2 molecule may be coordinated 'sideways on' to the Fe. The iron remains in the II state in both (a) and (b).

including myoglobin, which is used to store oxygen in the muscles, and in various cytochrome pigments. In the latter, a chain of transfers occurs which links dehydrogenation of alcohols or fatty acids with the conversion of O_2 to H_2O via a series of oxidation-reduction steps involving Fe(II)/Fe(III) (or Cu(I)/Cu(II)) conversions in various cytochromes.

A well-known reaction among iron complexes is the formation of *Prussian blue* by the reaction of ferrocyanide with ferric solution, and of *Turnbull's blue* by the reaction of ferricyanide with ferrous. In addition, ferrous plus ferrocyanide gives a white potassium salt of ferrousferrocyanide, and ferric plus ferricyanide gives brown-green ferricferricyanide. It now appears that all these compounds are related

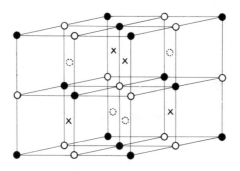

FIGURE 13.21 *The structure of Prussian blue and related compounds*
If none of the cube centre sites are occupied, the structure is that of ferric ferricyanide (both black and white Fe positions occupied by Fe^{III}): if every second cube centre site (marked with a dotted circle) is occupied by K^+, the structure is that of soluble Prussian blue (black = Fe^{II}, white = Fe^{III}): if all the centre sites are occupied by K^+ (crosses as well as dotted circles) the structure is that of dipotassium ferrousferrocyanide (black and white sites = Fe^{II}).

structurally and that Prussian and Turnbull's blues are identical. The basic structure is that shown in Figure 13.21. Ferric ferricyanide, $Fe[Fe(CN)_6]$, contains a cubic array of iron(III) atoms of unit cell length equal to 510 pm. This is the structure in the Figure when the atoms are identical and the Figure contains eight unit cells. If every second atom becomes Fe^{II} and one potassium ion is placed in the centre of every alternate small cube, a structure of unit cell length equal to 1020 pm results. This corresponds to the so-called soluble Prussian blue, $K[Fe(Fe(CN)_6)]$ where one iron is Fe^{II} and one is Fe^{III}, the alternative cases being indistinguishable from the X-ray data. Recent Mossbauer results show the presence of ferrocyanide ions and free ferric ions in Prussian blue: any interaction between Fe^{3+} and the cyanide N atoms is weak. If all the irons are Fe^{II}, the unit cell reverts to 510 pm in length. A potassium ion placed at the centre of each small cube gives the formula, $K_2FeFe(CN)_6$, of the ferrousferrocyanide ion. The insoluble blues correspond to the formation of alkali-free complexes by replacing the potassium by ferrous or ferric ions as follows. If the cyanides are regarded as lying in the cube edges and coordinated through carbon to one iron atom and through nitrogen to the next iron atom, the framework of Figure 13.21 corresponds to a superlattice of formula $Fe^{II}Fe^{III}(CN)_6^-$ in the case of the blue compounds. The compounds with ferric iron and ferrous iron then become $Fe^{2+}[FeFe(CN)_6]_2^-$ or $Fe^{3+}[FeFe(CN)_6]_3^-$. Other complex ferricyanides and ferrocyanides are related to these structures. Thus the cupriferricyanide ion, $CuFe(CN)_6^-$, is the same structure with the Fe^{II} ions replaced by Cu^{II} ions. Such structures probably hold for all ferrocyanide or ferricyanide complex ions of heavy metals apart from the alkali and alkaline earth metals.

Ferrocene, $(C_5H_5)_2Fe$, is discussed in section 12.12. It contains Fe(II) and is oxidizable to the ferrocinium ion, $(C_5H_5)_2Fe^+$, which contains Fe(III). It is interesting that a boron hydride analogue of ferrocene exists. In this, the $B_9C_2H_{11}^{2-}$ ion, which presents an open face consisting of a pentagon of three boron and two carbon atoms, replaces the $C_5H_5^-$ ions. Compounds $(B_9C_2H_{11})Fe(C_5H_5)^-$ and $(B_9C_2H_{11})_2Fe^{2-}$ are formed which are ferrocene analogues, and these are oxidizable to Fe(III) compounds, such as $(B_9C_2H_{11})_2Fe^-$, which are analogues of the ferrocinium ion. This *carborane* ion, $B_9C_2H_{11}^{2-}$, has been shown to replace the cyclopentadienyl ion in a variety of other compounds such as the cobalticinium ion and $(C_5H_5)Mn(CO)_3$.

Iron in the I oxidation state is rare but there is a well-developed chemistry of iron(0). Three carbonyl compounds are known, $Fe_2(CO)_9$ and $Fe_3(CO)_{12}$ as well as the penta-carbonyl, and a fair number of compounds exist where the carbonyls are replaced by other π-bonding ligands, as in $(Ph_3P)_2Fe(CO)_3$. Trifluorophosphine behaves in a very similar manner to carbonyl and analogues of most carbonyl compounds exist. Thus, $Fe(PF_3)_5$ and $Fe_2(PF_3)_9$ are found, as are all the mixed carbonyl-phosphine analogues of the pentacarbonyl such as $(PF_3)Fe(CO)_4$ and $(PF_3)_3Fe(CO)_2$.

13.7 Cobalt, $3d^7 4s^2$

Figure 13.22 shows the oxidation state energy diagram for cobalt. Only the II and III states have any stability in water

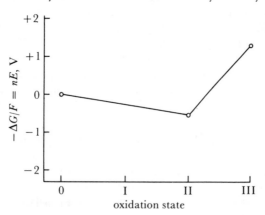

FIGURE 13.22 *Oxidation state free energy diagram for cobalt*
In aqueous acid conditions, cobalt(II) is much more stable than cobalt(III).

Cobalt metal is usually found in association with nickel in arsenical ores. It is relatively unreactive and dissolves only slowly in acids. It does not combine with hydrogen or nitrogen but it does react with carbon, oxygen, and steam at elevated temperatures, giving CoO in the latter cases.

The highest oxidation state shown by cobalt is IV. An ill-defined, hydrated CoO_2 is formed when cobalt(II) solutions are oxidized in alkaline media, and there is a report of Ba_2CoO_4. In addition, cobalt(IV) seems to be contained in a peroxy complex. When the binuclear cobalt(III) complex, $(H_3N)_5CoO_2Co(NH_3)_5^{4+}$, is oxidized, it appears that half the cobalt goes to cobalt(IV),

$$(H_3N)_5Co-O-O-Co(NH_3)_5^{4+} \longrightarrow$$
brown
$$(H_3N)_5Co-O-O-Co(NH_3)_5^{5+}$$
green

The magnetic moment corresponds to a $Co(III)-Co(IV)$ compound and its electron spin resonance spectrum indicates that the electron spends its time equally on both cobalts, passing readily across the peroxide bridge.

Cobalt(III) is strongly oxidizing in its simple compounds, and as the hydrated ion, but it forms a wide variety of stable octahedral complexes. It is the trivalent d^6 ion, therefore ΔE is large and spin-pairing is expected, to take advantage of the large CFSE of this configuration in the low-spin state, t_{2g}^6. Spin pairing appears to occur in most complexes, although CoF_6^{3-} is high-spin. The situation in the hydrate, $Co(H_2O)_6^{3+}$, is interesting. This is low-spin and diamagnetic but the CFSE gain appears to be only slightly greater than the loss of exchange energy, and the hydrated ion readily reduces to $Co(H_2O)_6^{2+}$, which is high-spin d^7. As the CFSE gain in d^6 is so high, it only takes a small increment in the ligand field to make the cobalt(III) state the more stable. Thus the hexammine, $Co(NH_3)_6^{3+}$, is prepared by air oxidation of cobalt(II) in aqueous ammonia. A still higher ligand field gives complexes in which the cobalt(II) state becomes strongly reducing. This is shown by the redox potentials for Co^{III}/Co^{II} complexes of the ligands shown below (all for octahedral complexes in both states):

ligand	H_2O	NH_3	CN^-
redox potential (V)	$+1\cdot84$	$+0\cdot1$	$-0\cdot8$

In the hydrates, cobalt(III) decomposes water by oxidizing it; in the ammines, cobalt(III) is stable to water; and in the cyanides, cobalt(II) decomposes water by reducing it. This range of activity resembles that already quoted for iron, but is more extreme because of the large ligand field effects.

Cobalt(III) complexes are extremely numerous and they undergo substitution only slowly so that a large variety of them are readily prepared and handled. It was the study of cobalt(III) and platinum(IV) complexes (both d^6), together with chromium(III) and square planar platinum(II) compounds which first led to the development of the ideas of complex chemistry at the start of this century by Werner and his school. By distinguishing different types of isomers and by proving the constancy through a series of chemical changes

and the II state is far more stable than Co^{III}, except in the presence of complexing ligands. Hydrated Co^{3+} decomposes water. In the solid state, the only trihalide is CoF_3, which is a strong oxidizing and fluorinating agent.

of certain groupings of atoms, Werner was able to formulate the idea of definite complex species and to determine the coordination numbers and shapes—all this before the development of any of the powerful modern techniques of structural determination. The variety and types of cobalt(III) complexes are best illustrated by some of these classical sequences of preparations and interactions in the field of cobalt-ammonia compounds, Figure 13.23. A number of other examples, including ethylenediamine complexes, are discussed in Chapter 12.

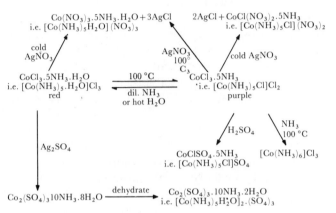

FIGURE 13.23 *Interconversions of cobalt(III) ammonia complexes*
This is an example of one of the series of interconversions which led Werner to the formulation of the concept of a complex.

An example of a natural cobalt(III) complex is provided by vitamin B_{12}. This has a cobalt(III) ion in a situation rather similar, though not identical, to that of iron in haemoglobin. The cobalt is in the middle of a porphyrin-type structure, and coordinated by the four ring nitrogens and by a fifth nitrogen from a side-chain group. The sixth site, completing the octahedron, is the active site and a number of derivatives are known with different groups occupying this site, including CN^-, hydrogen, and compounds containing a direct sigma cobalt−aliphatic carbon bond.

Although nitrogen is probably the commonest donor atom in cobalt complexes, a variety of oxygen complexes exist with ligands of the type of oxalate and acetylacetone, $Co(C_2O_4)_3^{3-}$ or $Co(acac)_3$. Cobaltinitrite, $Co(NO_2)_6^{3-}$, is coordinated through nitrogen of course, but there is some evidence for the existence of the isomeric nitrito-form, $Co-O-N-O$, in solution in equilibrium with the nitro-compound. Essentially all cobalt(III) compounds are octahedral.

Cobalt trifluoride is used as a fluorinating agent. CoF_2 reacts readily with F_2 to give CoF_3 and the latter is a strong fluorinating agent, though less reactive than fluorine. It provides a suitable way of moderating fluorination reactions. The compound to be fluorinated is streamed over CoF_3, giving the desired product and CoF_2. CoF_3 can then be regenerated by passing fluorine over the cobalt difluoride and the process may be continued in a cyclic manner.

One example of a simple cobalt(III) derivative is the anhydrous nitrate, $Co(NO_3)_3$. This is prepared in non-aqueous solvents and has bidentate nitrate groups giving octahedral CoO_6 coordination.

Although cobalt(III) exists largely in complexes and has unstable simple compounds, cobalt(II) is just the reverse. It is perfectly stable in simple compounds and salts and forms a number of complexes with ligands of relatively weak ligand field. However, with ligands further along the spectrochemical series than water, the CFSE gain on achieving the low-spin d^6 configuration is sufficiently great to make cobalt(III) the preferred state.

Cobalt(II) oxide, halides, and sulphide are well-known and may be made by normal methods. Red or pink hydrated cobalt salts of all the common anions are known. On addition of base to Co^{II} solutions, the pink (occasionally blue) hydroxide is precipitated and this dissolves in concentrated alkali to give the deep blue $Co(OH)_4^{2-}$ anion. The latter may be precipitated as the sodium or barium salt.

Cobalt(II) complexes with tetrahedral, octahedral, and square planar configurations are known. The hydrate is octahedral, as is $Co(NH_3)_6^{2+}$. This, and a number of related complexes, may be prepared as long as an inert atmosphere is maintained to stop oxidation. The deep blue $CoCl_4^{2-}$, formed by addition of excess HCl to the pink hydrated solutions, is tetrahedral. The other halides form similar, blue anions as does thiocyanate, $Co(CNS)_4^{2-}$. These tetrahedral ions generally have to be precipitated from solution as salts of large cations. A related compound is the mercury complex, $CoHg(CNS)_4$, which is used as a calibrant in magnetic measurements. This contains cobalt(II) tetrahedrally coordinated by the nitrogen atoms of the thiocyanate groups while the sulphur atoms tetrahedrally coordinate mercury(II) ions to give a polymeric solid. Square planar cobalt(II) complexes are found for some chelating ligands, such as dimethylglyoxime and salicylaldehyde-ethylenedi-imine, which form stable planar complexes in general (compare nickel dimethylglyoxime).

The tetrahedral complexes of cobalt(II) have three unpaired electrons and the square planar ones have only one, both as expected for d^7. The octahedral complexes include both high-spin and low-spin cases, the former with three and the latter with one unpaired electron. The change-over appears to come between ammonia and nitro anion, NO_2^-, in the spectrochemical series.

Cobalt occurs in a few compounds in the I, 0, and −I state, all with π-bonding ligands. The carbonyl, $Co_2(CO)_8$, gives the anion, $Co(CO)_4^-$ quite readily and also reacts with organic isonitriles to give the ion, $Co(CNR)_5^+$, which contains cobalt(I). This cation can also be prepared by reduction of the cobalt(II) compound, $Co(CNR)_4X_2$, by a metal or hydrazine. Cobalt(0) also occurs in $K_4Co(CN)_4$ which is the reported product of the reduction of the cobalt(III) hexacyanide with potassium in ammonia. The cobalt(I) isonitrile has been shown to be a trigonal bipyramid.

Another interesting cobalt(I) species is the compound $CoH(N_2)(PPh_3)_3$, hydridodinitrogentris(triphenylphosphine)cobalt(I). This is a representative of a growing class of compounds which have N_2 as a ligand (compare under

Ruthenium, section 14.6) and the cobalt species is one of the few whose structure is known. This is shown in Figure 13.24, with trigonal bipyramidal coordination of the cobalt. The nitrogen molecule is bonded end-on with the $N-N-Co$

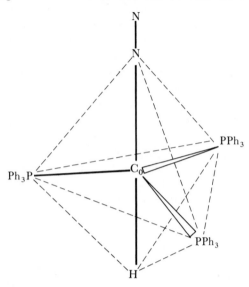

FIGURE 13.24 *The structure of the cobalt(I) nitrogen compound,* $HCo(N_2)(PPh_3)_3$.

The $Co-N_2$ unit is linear with bond length 180 ± 3 pm, the $Co-H$ distance is 160 pm and the equatorial P atoms are bent slightly towards the hydrogen position as is usual in hydrido-complexes) with $N-Co-P$ angles of $97 \pm 2°$.

angle linear or just slightly less. Thus the N_2 unit is bonded similarly to a terminal CO group, with which it is isoelectronic. The $Co-N$ bond length of 180 pm suggests multiple bonding, which would be similar to that indicated for carbonyl-metal in Figure 12.31, although the sigma donation from nitrogen is probably weaker than from carbon. $N-N = 111 \pm 2$ pm compared with 109.8 pm in free N_2.

Interesting cluster compounds are found among the more complex cobalt carbonyl anions. For example, in $Na_4Co_6(CO)_{14}$, the $Co_6CO_{14}^{4-}$ anion consists of an octahedron of Co atoms, with each of the eight triangular faces bridged by a CO. The remaining six CO units are normal terminal groups, one on each cobalt. In other words, the structure is like that of $Mo_6Cl_8^{4+}$ shown in Figure 14.17, with Co in place of Mo, carbonyl in place of the chlorines together with six terminal CO groups.

Cobalt carbonyl hydride, $HCo(CO)_4$, has been shown to be the active species in the OXO process for the conversion of alkenes to alcohols in presence of a cobalt catalyst

$$RCH=CH_2 + CO + 2H_2 = RCH_2CH_2CH_2OH$$

The reaction proceeds by initial insertion of $H-Co$ into the double bond (an example of a *hydrometallation* reaction) followed by CO insertion into the reactive $Co-C$ bond

$$RCH=CH_2 + HCo(CO)_4 =$$

$$RCH_2CH_2Co(CO)_4 \xrightarrow{+CO} RCH_2CH_2COCo(CO)_4$$

It is thought that the acyl-cobalt compound then loses CO

and the resulting tricarbonyl is cleaved by hydrogen and the product aldehyde is reduced to the alcohol.

$$H_2 + RCH_2CH_2COCo(CO)_n =$$
$$(n = 3?)$$
$$RCH_2CH_2CHO + HCo(CO)_n$$
$$\searrow^{CO}$$
$$HCo(CO)_4$$

13.8 Nickel, $3d^8 4s^2$

The chemistry of nickel is much simpler than that of the other first row elements. The only oxidation state of importance is nickel(II) and these compounds are stable. Nickel(II) is the d^8 ion and is able to form stable square planar complexes as well as octahedral ones. Ligands with large crystal field favour square planar coordination, because of the more favourable CFSE. The heavier elements in the nickel Group are exclusively square planar in the II state.

The element occurs largely in sulphide or arsenic ores and is extracted by roasting to NiO and then reducing with carbon. Pure nickel is made by the Mond process in which CO reacts with impure Ni at 50 °C to give $Ni(CO)_4$. This is decomposed to Ni and CO at 200 °C to yield metal of 99.99 per cent purity. The metal resists attack by water or air and is used as a protective coating. It has good electrical conductivity. The metal dissolves readily in dilute acids.

Oxidation states above II are represented by obscure oxides and some complex compounds. Oxidation of $Ni(OH)_2$, suspended in alkali, by a moderately strong oxidizing agent like Br_2 gives a black solid which can be dried to the composition, $Ni_2O_3.2H_2O$. Attempts at further dehydration lead to decomposition to NiO. The action of powerful oxidizing agents in an alkaline medium give impure, hydrated products of approximate composition NiO_2. This is a powerful oxidant, forming permanganate from Mn^{2+} in acid, and decomposing water. If nickel is heated in oxygen in fused caustic soda, sodium nickelate(III), $NaNiO_2$, results.

The high oxidation states may be stabilized in complexes, such as K_2NiF_6, and with oxidizing anions, as in the iodine(VII) purple compound, $KNiIO_6.nH_2O$. Nickel forms diarsine complexes, $Ni(diars)_2Cl_2^{2+}$, which were thought to contain nickel(IV) ($n = 2$) and nickel(III) ($n = 1$) but this formulation has recently been queried and one arsine may be positively charged.

Nickel(III) is found in the green $(NiCl_2en_2)^+Cl^-$ complex which has a magnetic moment corresponding to one unpaired electron. Nickel(III) is also found in the phosphine complex $Ni(PEt_3)_2Br_3$ which is probably a square pyramid with Br at the apex and *trans* bromine and phosphine groups in the base plane.

A wide variety of simple compounds of nickel exist, including all the halides and all the oxygen Group compounds. Ni^{2+} forms salts with even strongly oxidizing anions such as chlorite, and with relatively unstable ions like carbonate. The relative stabilities of the II and III states of iron, cobalt, and nickel, as shown by the compounds formed with the oxychlorine anions are discussed in section 7.6. As chlorine in alkali oxidises nickel(II) to nickel(IV), it is to be assumed that the extremely powerfully oxidizing ClO^- anion will not form even a nickel(II) salt. The complex ions with ligands such as water and ammonia are octahedral, the green $Ni(H_2O)_6^{2+}$ ion being responsible for the typical colour of hydrated nickel salts. The hexammines and related complexes such as $Ni(en)_3^{2+}$ are generally blue. All possible mixed forms occur such as $Ni(H_2O)_2(NH_3)_4^{2+}$.

With ligands of high field strength, square planar, diamagnetic complexes are formed, such as the cyanide, $Ni(CN)_4^{2-}$ and the dimethylglyoxime complex used in quantitative analysis for nickel (Figure 13.25). In the latter, there may be interaction between nickel atoms in the crystal, where the flat molecules are stacked vertically above each

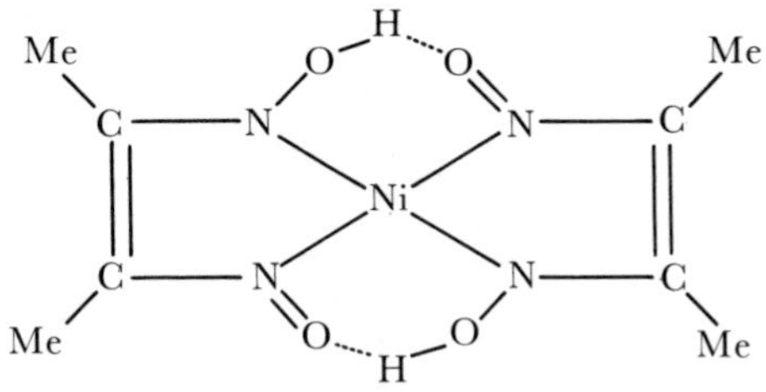

FIGURE 13.25 *Nickel dimethylglyoxime*
The dimethylglyoxime loses a proton and coordinates to Ni giving a five-membered ring. The complex is further stabilized by hydrogen bonding.

other, so that the compound could be considered as a very distorted octahedron.

There are several examples of tetrahedral nickel(II) complexes. These involve halogen compounds, either as anions with large cations, as in $(Ph_4As)_2^+(NiCl_4)^{2-}$, or in neutral complexes with phosphine or phosphine oxide and related ligands, as in $(PPh_3)_2NiI_2$ or $(Ph_3AsO)_2NiBr_2$. Such complexes are characteristically intensely blue due to a relatively intense absorption in the red end of the spectrum. These intense spectral lines distinguish tetrahedral nickel(II) from octahedral complexes, where the absorptions are relatively weak.

Since nickel(II) is found in octahedral, square planar and tetrahedral environments, and there is no large energy difference between these, there are many cases where more than one form of a complex occurs, differing in stereochemistry, or where an equilibrium exists between two forms in solution. The salicylaldimato complexes (compare Appendix B) of the type shown in Figure 13.26, provide examples of all three stereochemistries. When R = methyl or isobutyl,

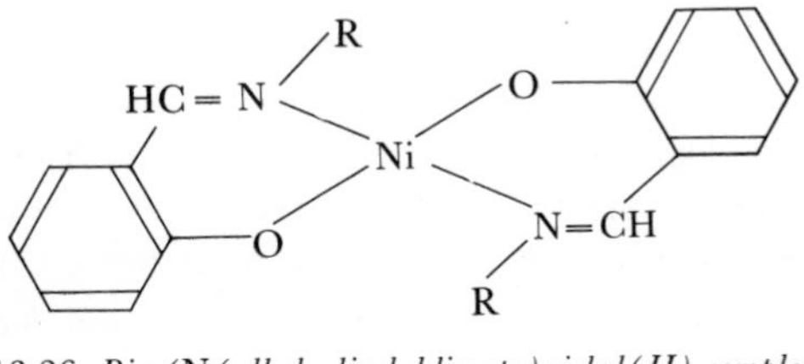

FIGURE 13.26 *Bis-(N(alkylsalicylaldimato)nickel(II) complexes*

the complex of Figure 13.26 is square planar, but when R = isopropyl, the coordination about the nickel becomes tetrahedral. Further, if the diamagnetic square planar complex (e.g. R = Me) is dissolved in pyridine, it becomes paramagnetic with the moment expected for an octahedral complex, and an octahedral dipyridine derivative may be isolated.

When a ligand atom is a moderately weak donor, it is possible to isolate complexes with the same formula but different coordinations. For example, NiL_2Cl_2 where

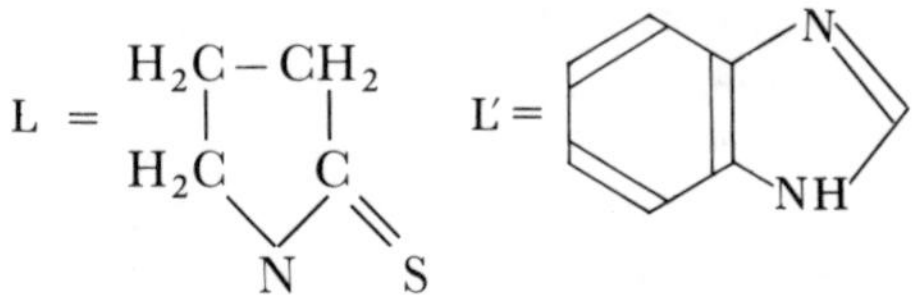

may be found in a blue tetrahedral form or in a yellow octahedral one. Similarly, NiL_4Br_2, where L = benzimidazole, is found in an orange square planar form and a yellow octahedral one.

In the above cases, solids may be isolated and characterized, but often the two forms coexist in solution, even though only one is isolable. This is true of a wide range of complexes NiL_2X_2 which show tetrahedral-octahedral equilibrium (usually as blue-yellow changes) in solution, with the tetrahedral form most favoured when X = I. Such equilibria may be studied spectroscopically, and when L = phosphine with bulky groups, the exchange is slow enough to be followed by nmr at low temperatures.

One particularly striking example of the ease of interconversion of these stereochemistries is provided by the compound $Ni(PRR_2')_2Br_2$ where the substituents on phosphorus in the phosphine are R = benzyl ($C_6H_5CH_2$) and R' = phenyl (C_6H_5). In the crystal of this compound, both square planar and tetrahedral nickel environments are found. Thus any small energy difference between the two coordinations is compensated for by the readier packing of the two different sorts of molecule.

Nickel(II) also forms a range of five-coordinate complexes (compare section 12.6), both in the square pyramidal configuration like NiX_2terpy (terpyridyl) and trigonal bipyramidal, as in many Schiff's base complexes (compare Appendix B). There is even the case, quoted in section 12.6 of the $Ni(CN)_5^{3-}$ ion adopting both coordinations in the same compound.

Low oxidation states are represented by the cyanide complexes, $K_4Ni_2^I(CN)_6$ and $K_4Ni^0(CN)_4$, which result from the reduction of the $Ni^{II}(CN)_4^{2-}$ species with potassium in liquid ammonia. The nickel(I) compound may also be made by reduction with hydrazine in an aqueous medium. It is relatively stable but the nickel(0) complex is extremely reactive and not well characterized. Both are oxidized in air or water. Nickel(I) is also represented by the phosphine complexes $NiX(PPh_3)_3$, while nickel(0) is found in the carbonyl, $Ni(CO)_4$, in $Ni(PF_3)_4$, and in all the mixed trifluorophosphine-carbonyls. The carbonyl anion, $Ni_2(CO)_6^{2-}$ contains nickel(−I). Although nickel carbonyl is one of the best-known carbonyls, the stability of such compounds is relatively low at this end of the d block and nickel has a much less rich carbonyl chemistry than the earlier elements.

13.9 Copper, $3d^{10}4s^1$

There are three possible oxidation states in the copper Group, the I state corresponding to d^{10}, the II state which is common to the whole transition series, and the III state corresponding to d^8 which would be stable in a square planar environment. The elements in this Group show these states, with II the stable state of copper, I the stable state of silver, and III the stable state of gold. Copper also exists in the I state, which is quite stable in solids, and one or two copper(III) compounds are reported. This wide variation in stabilities in this Group is probably a resultant of size, exchange energy, and ligand field effects, and is discussed a little later.

Copper is found in sulphide ores and as carbonate, arsenide, and chloride. Extraction involves roasting to the oxide, reduction, and purification by electrolysis. Pure copper has an electrical conductivity second only to that of silver and its major application is in the electrical industry. The element is inert to non-oxidizing acids but reacts with oxidizing agents. With oxygen, it combines on heating to give CuO at red heat, and Cu_2O at higher temperatures. It also reacts with halogens and dissolves in hot nitric acid or hot sulphuric acid.

Copper(II) in alkali is oxidized and the $KCuO_2$ compound of copper(III) may be isolated. Fluorination of mixed KCl and $CuCl_2$ gives the copper(III) complex, K_3CuF_6. A third representative of copper(III) is the periodate, $K_7Cu(IO_6)_2.7H_2O$. It will be noticed, if the higher oxidation states of iron, cobalt, nickel, and copper are compared, that the methods of preparing such compounds tend to be similar, for example, by oxidation in alkaline media and the use of oxidizing anions.

Copper(II) is the main state in aqueous solution and the compounds are paramagnetic with one unpaired electron. The hydrated ion may be written $Cu(H_2O)_6^{2+}$ although the structure is not regular. There are four near ligands in a square plane and the other two are further away as a result of the unequal occupation of the two e_g orbitals. This distorted shape is common for all copper(II) compounds as the examples of bond lengths below show:

Compound	Distances, pm	
	shorter	longer
CuF$_2$	4 of 193	2 of 227
CuCl$_2$	4 of 230	2 of 295
CsCuCl$_3$	4 of 230	2 of 265
CuCl$_2$.2H$_2$O	2 of 231 (Cl)	2 of 298
	2 of 201 (O)	
K$_2$CuF$_4$	2 of 195	4 of 208

(note inversion to two short and four long bonds)

Mixed complexes of water and ammonia are found up to $Cu(H_2O)_2(NH_3)_4^{2+}$, but replacement of the last two water molecules is impossible in aqueous solution and $Cu(NH_3)_6^{2+}$ can only be prepared in liquid ammonia. It is similarly

possible to form $Cu(H_2O)_4en^{2+}$ and $Cu(H_2O)_2en_2^{2+}$ but formation of $Cuen_3^{2+}$ is difficult. These, and similar amine complexes, are all a much deeper blue than the hydrated ion.

Halide complexes are also distorted but are tetrahedral with a flattened structure, for example, $CuCl_4^{2-}$, which can be precipitated from a chloride medium by large cations. Such tetrahedral complexes are generally green or brown.

Among the salts, mention must be made of the unusual structures of copper(II) acetate and of anhydrous copper nitrate. The acetate is dimeric and hydrated, $Cu_2(CH_3COO)_4.2H_2O$, and has the structure of Figure 13.27. The copper atoms are surrounded by a square plane of oxygen atoms and the acetate groups bridge the planes together. Similar structures are found for copper derivatives of other carboxylic acids.

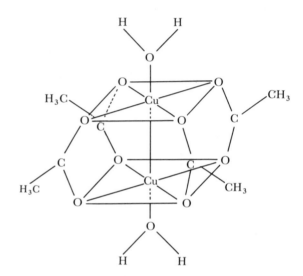

FIGURE 13.27 *Structure of hydrated copper(II) acetate*
The two square planar CuO_4 units are linked by the acetic acid residues and also by $Cu-Cu$ bonding.

Anhydrous copper nitrate cannot be made by dehydrating the hydrated salt as this decomposes to the oxide. However, copper metal dissolves in liquid N_2O_4/ethyl acetate mixture to give $Cu(NO_3)_2.N_2O_4$, and $Cu(NO_3)_2$ results when the solvating molecule is pumped off. The structure in the solid is shown in Figure 13.28. The structure consists of chains of copper atoms bridged by NO_3 groups and cross-linked by other nitrate groups. Anhydrous copper nitrate is slightly volatile as single molecules.

There are some apparently three-coordinated copper compounds but all are of more complex structure. For example, $KCuCl_3$ contains dimeric planar $Cu_2Cl_6^{2-}$ ions. In the lithium salt, these dimeric ions are joined into longer chains by long chloride bridges.

As with nickel, copper(II) appears in a variety of geometries and examples exist of the same compound occurring in two different structural modifications. One example is provided by complexes $CuCl_2L_2$, where L is an N-oxide such as pyridine-N-oxide, C_5H_5NO. These occur in a green form, which is thought to be *trans*-square planar and also in a yellow form in which the coordination is tetrahedral.

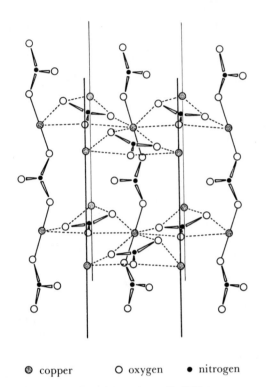

 copper   O oxygen   • nitrogen

FIGURE 13.28 *Solid anhydrous copper nitrate*, $Cu(NO_3)_2$
This structure consists of cross-linked chains of nitrate groups bonded to two copper atoms. A second crystal form of copper nitrate has been discovered recently.

Copper(II) is found in five-coordination which is usually square pyramidal although $CuCl_5^{3-}$ and $CuCl_2Br_3^{3-}$ are trigonal bipyramids (with axial Cl in the latter).

As many of these structures are distorted, and often show a variety of bond lengths to the same ligand atom, copper stereochemistry presents an extremely difficult field. Normal spectroscopic methods are inadequate to distinguish distorted shapes from those of lower coordination number, and even X-ray crystallography may give inconclusive results because of the problems of correlating several different interatomic distances with short, long, or non-existent bonds.

The I state of copper is represented largely by solid compounds which are insoluble in water. For example, if iodide is added to a copper(II) solution, the cupric iodide initially formed rapidly decomposes to give a precipitate of CuI. Similarly, one method of determining copper quantitatively is by precipitation of CuSCN. The stability of copper(I) in solution is low and the redox potentials

$$Cu^+ + e^- = Cu \quad 0.52\ V$$
$$Cu^{2+} + e^- = Cu^+ \quad 0.15\ V$$

show that Cu^+ is unstable to disproportionation:

$$2Cu^+ = Cu + Cu^{2+},\ E = 0.37\ V,\ K = \frac{[Cu^{2+}]}{[Cu^+]^2} = 10^6$$

The reason for the marked instability of Cu^+ in water is not completely clear, as the d^{10} configuration, with its very high exchange energy, would be expected to be reasonably favoured. One possible explanation lies in the low hydration energy which is likely for the Cu^+ hydrated ion, as compared

with that for the Cu^{2+} ion. The cuprous ion is larger and has only half the charge so that its charge density is markedly lower, and hence the energy of interaction with the water dipole is less. In addition, in known complexes of Cu^+, especially the ammine, Cu^+ is only two-coordinated as in $Cu(NH_3)_2^+$. If the hydrate is reasonably supposed to have the same formula, then there would be only two interactions instead of the four strong and two weaker interactions in the Cu^{2+} hydrate. Thus Cu^+ has about half the interaction energy with half the number of water molecules compared with Cu^{2+}. When this case is compared with that of Ag^+, where the behaviour is quite the reverse and Ag^{2+} is quite rare and unstable, it will be seen that the greater size of silver tends to decrease these differences between the two oxidation states and the exchange energy probably then becomes the dominant term. Passing to gold, this trend again alters with the size, and the large CFSE term for a trivalent species of the third transition series helps to make square planar Au(III), with a d^8 system, the preferred state, although Au(I) also occurs. It is probable that the marked variation in chemistry in this Group, where the three elements differ markedly more than in any other Group, is a function of the increase in size coupled with the existence of stable electronic configurations on either side of the divalent d^9 state.

Apart from the insoluble CuCl, CuBr, CuI, and the cyanide and thiocyanate, copper(I) gives soluble complexes with these groups as ligands. Two cyanide complexes exist, soluble $Cu(CN)_4^{3-}$, and the compound $KCu(CN)_2$ which has a chain structure containing three-coordinate Cu^I:

$$- - - - Cu - C - N - Cu(CN) - C - N - Cu - - - -$$

The red cuprous oxide, Cu_2O is well-known. This is the compound precipitated in Fehling's test for sugars; it is produced by the reduction of the blue cupric tartrate complex by glucose or related molecules.

It is still uncertain whether copper can exist in the 0 state as a number of reported compounds have been reformulated. One possible example is the diamagnetic species

$$Cu(NC-(CH_2)_4-CN)_2.$$

Copper is one of the more important trace elements in biological systems. It is particularly concerned in the uptake of inorganic sulphur into organic molecules and copper(I)/copper(II) changes are involved in redox transfer systems in cytochromes in a similar way to iron(II)/(III) ones. Intensely blue copper(II)-containing proteins are typical, and are involved in oxidation steps while copper(I) is found in haemocyanins.

13.10 The relative stabilities of the dihalides and trihalides of the elements of the first transition series

Although, at present, the factors affecting the stabilities of oxidation states are incompletely understood, a recent paper makes it clear that answers can be obtained in a rigorous manner from the thermodynamic parameters involved, providing that the problem is precisely formulated. The case discussed was the thermal stabilities of the trihalides (excluding fluorides) of the first row transition elements to decomposition according to the equation

$$MX_{3(solid)} \rightarrow MX_{2(solid)} + \tfrac{1}{2}X_2$$

at 25 °C. This problem is only a small part of the question of the relative stabilities of the II and III states of the first row elements but its solution points the way for further work. Other compounds, other conditions, and even other decomposition routes of the trihalides, would have to be considered to extend our understanding of the general problem.

The observed stabilities show a fairly regular trend from scandium to zinc, with the trihalide stable relative to the dihalide at the left hand end of the transition series, and unstable with respect to the dihalide for the elements to the right, especially nickel, copper and zinc. There is an anomalous order of stability in the middle of the series with Mn < Fe > Co in the stability of the trihalide relative to the dihalide. This order holds for X = Cl, Br, and I. In the detailed analysis, it turns out that the change of entropy in the decomposition reaction is small and nearly the same for all the elements. Thus the free energy change in the reaction depends on the changes in enthalpy. These are best analyzed by breaking the reaction down into a number of simpler steps in a Haber cycle:

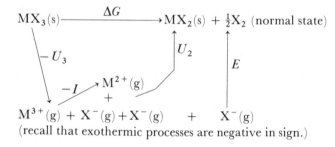

(recall that exothermic processes are negative in sign.)

Thus $\Delta G = -U_3 - I + U_2 + E - T\Delta S$, where ΔG is the free energy of the decomposition reaction, U_2 and U_3 are the lattice energies of the di- and tri-halide, $T\Delta S$ is the entropy energy, and E and I are the appropriate electron affinity and the third ionization potential of the metal. As the changes in the chlorides, bromides and iodides are parallel, attention may be focussed on the chlorides.

As we are comparing the chlorides of one element with the next, the value of E is a constant for the series, and as $-T\Delta S$ is found to vary only slightly, we can take the sum of E and $-T\Delta S$ as a constant along the series. Thus the *variation* in ΔG along the series is a consequence of the variations in the terms $-I$, $-U_3$ and U_2. The values of the ionization energies, I, are well established (Table 2.8) and the lattice energies, U, may be evaluated by Born-Haber cycles, as discussed in Chapter 4, in terms of atomic properties.

The variations across the series of $-I$, U_2, and $-U_3$ are plotted in Figure 13.29, together with the resultant value of ΔG which is the combination of the constant terms E and

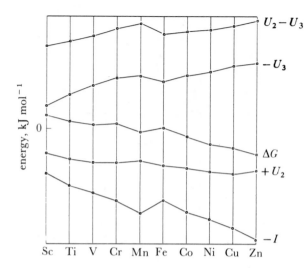

FIGURE 13.29 *The variation in the principle energy terms involved in the process* $MX_3 = MX_2 + \tfrac{1}{2}X_2$
The resultant ΔG follows the $-I$ curve, moderated by the lattice energy difference $U_2 - U_3$. For clarity, the zeros of the different curves have been shifted: the $-I$ curve ranges from -2430 to -3850 kJ mol^{-1}, the U_2 curve from -2340 to -2720 kJ mol^{-1}, the ΔG curve from $+375$ to -630 kJ mol^{-1}, and the $-U_3$ curve from $+5270$ to $+5860$ kJ mol^{-1}. (Recall that lattice energies are exothermic as defined so that $+U_2$ is exothermic and therefore has negative values while $-U_3$, the energy for the reverse process to forming the trihalide lattice, is endothermic and has positive values.)

$-T\Delta S$ with $(U_2 - U_3 - I)$. The lattice energy curves, U_2 and U_3, are similar to those shown in Figures 12.24 and 12.25 and clearly reflect ligand field stabilization energy changes. In particular, the d^5 species—$MnCl_2$ and $FeCl_3$—are less stable than their neighbours. (Note that U_2 and U_3 are plotted with opposite signs.) However, for the difference $(U_2 - U_3)$, these variations partly cancel (as shown in the top curve of Figure 13.29) as they occur one element later for the trihalides than for the dihalides. It is clear that the most important single factor is the third ionization potential of the metal, which is exothermic for the cycle step shown

$$M^{3+}(g) + e^- \rightarrow M^{2+}(g).$$

The free energy curve ΔG is clearly seen in the Figure to be dominated by the value of $-I$, moderated to some extent by the variation in $(U_2 - U_3)$. The change in free energy of the reaction

$$MCl_{3(solid)} \rightarrow MCl_{2(solid)} + \tfrac{1}{2}Cl_{2(gas)}$$

is positive for the elements at the left of the transition series and varies through near-zero values to negative at the right hand end. That is, the trichlorides (and similarly for the tribromides and triiodides) are stable with respect to decomposition at the left of the series while the decomposition to the dihalide is favoured towards the right. The variation of I, in turn, may be analyzed as a general increase across the transition series as the nuclear charge increases (that is, $-I$ becomes more negative) as expected from the incomplete shielding effect of the d electrons in the same shell. To this is added the major break between $Mn(d^5 \rightarrow d^4)$ and $Fe(d^6 \rightarrow d^5)$ which again reflects the exchange energy effect in the d^5 configuration.

Thus, although this decomposition reaction depends on the total effect of a large number of factors, it is seen that the pattern is set by exchange energy effects in the d^5 state as reflected in the value of I moderated by CFSE effects on the lattice energies of the halides.

14 The Elements of the Second and Third Transition Series

14.1 General properties

This Chapter deals with the remaining elements of the transition block, those in which the $4d$ or $5d$ level is filling. The effect of the lanthanide contraction is to make the chemistry of the heavier pair of elements in the same Group very similar. A summary of the oxidation states of these elements, in relation to those of the first series, is given in Table 12.2, while Tables 12.3, 4, and 5 give the known oxides and halides. Other properties already listed include electronic configurations, atomic weights and numbers (Table 2.5), and Ionization Potentials (Table 2.8).

In these elements, higher oxidation states are more stable, in general, than in the first series. This is shown both by the relatively non-oxidizing behaviour of states like Re(VII) and Mo(VI) or W(VI), and by the existence of high oxidation states to the right of the series, which are not found for the first row elements: examples include Os(VIII) and Ir(VI). Complementary to this increased stability of high oxidation states is a decrease in the stability of many lower oxidation states, as shown by such examples as the strong reducing properties of Zr(III) or Nb(IV) and the virtual non-existence of compounds of W(II). Some similarities do, of course, exist between the lighter and heavier elements, especially in the properties of the higher oxidation states to the left of the d block and in those of the lower oxidation states at the right hand end. The maximum range of oxidation states comes in the middle of the block, rising to the VIII state of osmium and ruthenium.

Some of these elements are rather rare or inaccessible, especially those in the Groups immediately following the lanthanides, which are difficult to separate. Both hafnium and niobium, which occur along with their more abundant congeners, zirconium and tantalum, are not well studied and there are gaps in their chemistries where they are presumed to be similar to their congeners but without full proof. Other rare elements are technetium, which has only unstable isotopes and has to be made artificially, and the platinum metals and gold which, though accessible, are expensive to work with. The elements become less reactive towards the right of the transition block, and the tendency to reduction

to the metal is an important characteristic especially of the platinum metals and gold, and also, to a lesser extent, of rhenium and silver. Few of these elements find large scale applications, though molybdenum and tungsten are components of highly resistant steels. The precious and semi-precious metals, apart from their uses in coinage and jewellery, find some application in precision instruments, electrical apparatus, and in surgery. Platinum and palladium are used as catalysts in a number of industrial processes. Zirconium and niobium, the former especially, find use as 'canning' materials for the fuel in nuclear reactors. For this purpose, they must be separated from their congeners, hafnium and tantalum, which 'poison' the reactors by capturing neutrons. Most elements are extracted by carbon or metal reduction of the oxides or chlorides.

Many of these elements form non-stoichiometric carbides, nitrides, and hydrides, similar to those already discussed for the first series elements.

As these elements are larger than the first row members, higher coordination numbers are found. Although six-coordination to singly-bonded ligands is still common, seven- and eight-coordination are found, as in ZrF_7^{3-} or $Mo(CN)_8^{2-}$. To pi-bonding ligands such as oxygen, six becomes a common coordination number as well as four; compare the polymeric oxyanions of Mo and W, which are based on MO_6 groups, with the vanadates and chromates, which contain MO_4 units.

The increase in ligand field splittings which is shown in these larger atoms means the CFSE values increase markedly in all configurations, and the normal electronic configurations are the low-spin ones. Evidence for high-spin complexes is very rare.

14.2 Zirconium, $4d^25s^2$, and hafnium, $5d^26s^2$

Only the IV oxidation state is stable for these elements and potential data are not available for the lower states. The M/M^{IV} values for the two elements are similar and differ from that of titanium by a factor of about two; the elements being better reducing agents than titanium. $Ti^{IV}/Ti =$

-0.89 V, $Zr^{IV}/Zr = -1.56$ V, and $Hf^{IV}/Hf = -1.70$ V. The IV state occurs in solution and in a variety of solid compounds, but the III and II states decompose water and are only found in solid products.

Zirconium ores occur, including the oxide and zircon, $ZrSiO_4$, but there are no discrete sources of hafnium which is always found as a minor component in zirconium minerals. Zirconium was known for nearly a hundred and fifty years before hafnium was discovered in its ores and compounds. The elements are extracted as tetrahalides and reduced with magnesium in a process similar to that for titanium. Here also, argon has to be used to provide an inert atmosphere, as the metals combine with nitrogen.

The covalent radii, 145 pm and 144 pm respectively for Zr and Hf, and also the radii of the hypothetical ions Zr^{4+} (74 pm) and Hf^{4+} (75 pm), are so close that the chemistry of this pair of elements is virtually identical. Separation is even more difficult than for the lanthanides but ion exchange and solvent extraction procedures are now available. One example of these is the separation of the tetrachlorides dissolved in methanol on a silica gel column. Elution is by an anhydrous HCl/methanol solution, with the zirconium coming off the column first.

In the IV state, these elements show a general resemblance to titanium(IV) but differ in the acidity of the dioxides and the solvolytic behaviour of the oxycation. The dioxides are soluble in acid solution and addition of base precipitates gelatinous hydrated oxide, $ZrO_2.nH_2O$. No true hydroxide exists. The hydrated oxide gives ZrO_2 (or HfO_2) on heating and these are hard, white, insoluble, unreactive materials with very high melting points (above 2500 °C). The more abundant zirconium dioxide is used for high temperature equipment, such as crucibles, because of these properties. The hydrated oxide is quite insoluble in alkali and the dioxides have therefore no acidic properties. In solution, zirconium(IV) and hafnium(IV) hydrolyse less than titanium(IV). The main species in solution is frequently written as the oxyion ZrO^{2+} (zirconyl) or HfO^{2+} (hafnyl) but it is doubtful whether these simple species exist. The species in solution are probably $M(OH)_n^{(4-n)-}$ and trimeric or tetrameric hydroxy species. Tracer experiments indicate the presence of Zr^{4+} ions in very dilute solution in perchloric acid. Compounds containing the MO group are common, for example $ZrO(NO_3)_2.2H_2O$ or $HfO(OOCCH_3)_2$, but their structures may not be simple. Thus, $ZrOCl_2.8H_2O$ contains the cation $[Zr_4(OH)_8(H_2O)_{16}]^{8+}$ which has dodecahedral coordination around the zirconium. There is also evidence for the, possibly polymeric, ion $Zr_2O_3^{2+}$ in solution. The complexity of these hydrolysis products is typical of the solution chemistry of this part of the Periodic Table, although the larger atoms mean that there is less extensive hydrolysis than for titanium(IV). However, the elements are not large enough to allow the formation of the M^{4+} ions. Simple zirconium or hafnium compounds, such as the tetrahalides or the tetraacetate, $Zr(OOCCH_3)_4$, are covalent compounds, not salts.

The halides may be made by standard methods and show the expected reactions:

$$MO_2 + C + Cl_2 \longrightarrow MCl_4 + CO_2$$

$$M + X_2 \longrightarrow MX_4$$

(also MF_7^{3-} and (stable)
$M_2'^{II}MF_8$ in the case X = F)

The tetrahalides, except the fluoride, are volatile solids. In the vapour phase, the structure is monomeric and tetrahedral, while in the solid the Zr or Hf atoms are in octahedral coordination. The structure of one form of $ZrCl_4$ is a chain of linked octahedra ($ZrCl_2Cl_{4/2}$) like the NbI_4 structure shown in Figure 14.7. The halides hydrolyse vigorously to the oxyhalide, which is stable to further hydrolysis. The most stable complex halides are the fluorides, and the hexa-, hepta-, and octa-fluorides are known. The octahedral ZrF_6^{2-} ion is found for example in Li_2ZrF_6. In the formally similar K_2ZrF_6, the structure involves bridging fluorines and ZrF_8 coordination. In the mixed potassium-cupric hexafluorozirconate, the structure is even more complex and is formulated as $K_2Cu(H_2O)_6[Zr_2F_{12}]$. The $(Zr_2F_{12})^{4-}$ anion has the unusual pentagonal bipyramidal coordination around each zirconium, and the two bipyramids share an edge—$F_5ZrF_{2/2}ZrF_5$. The pentagonal bipyramid is also found in the sodium salts Na_3ZrF_7 and Na_3HfF_7. These contain pentagonal bipyramid MF_7 groups, as in IF_7, but the ammonium salt of the heptafluorides has the same structure as the isoelectronic niobium and tantalum heptafluorides. This is shown in Figure 14.1 and may be described as a trigonal prism with the seventh fluoride added beyond the

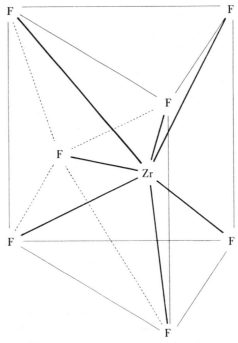

FIGURE 14.1   *The structure of the ZrF_7^{3-} ion in $(NH_4)_3ZrF_7$*
This species is basically a trigonal prism with an extra ligand attached to one face. The NbF_7^{2-} and TaF_7^{2-} ions adopt the same structure.

centre of one of the rectangular faces. In the MF_8 groups, the eight fluorines adopt the bisdisphenoid configuration shown in Figure 14.2, and described in section 12.6. If the cation is $Cu(H_2O)_6^{2+}$, in place of the alkali metals, the antiprismatic form is found for ZrF_8^{4-}, like the octafluorotantalate (Figure 14.4b).

The lower oxidation states of zirconium and hafnium are very strongly reducing. Mixtures of di- and tri-halides result from reduction by the element, or by reduction with H_2 at 400–500 °C. It may be noted that treatment of a mixture of $ZrCl_4$ and $HfCl_4$ with zirconium metal gives $ZrCl_3$, which is

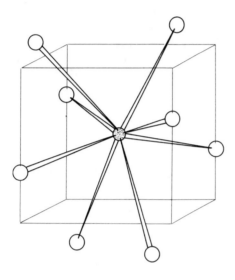

◎ Zr or Hf
◯ F

FIGURE 14.2 *The structure of* ZrF_8^{4-} *and* HfF_8^{4-}
This form of eight-coordination is similar to that found in the octacyanomolybdate ion (Figure 14.16). Contrast this with the structure of the octafluoride of tantalum (Figure 14.4b).

involatile, and leaves $HfCl_4$ unreacted. The latter may then be sublimed out of the reaction mixture, providing a method of separating the elements. The III and II states do not exist in solution. One striking example of the reducing power of the III state is provided by the production of the blue potassium solution in liquid ammonia, in the following reaction:

$$ZrCl_3 + 4KNH_2 \xrightarrow{NH_3} Zr(NH_2)_4 + K + 3KCl$$

$HfCl_4$ gives $HfCl_3$ when heated with hafnium metal, and the trichloride is stable to 350 °C in the presence of $HfCl_4$. Hafnium dichloride is said to disproportionate to $HfCl$ and $HfCl_4$ when heated to 627 °C. In contrast there is no lower hafnium iodide with iodine content less than $HfI_{3.2}$.

One compound of zero-valent zirconium is known. This is the violet dipyridyl complex, $Zr(dipy)_3$, formed, like the corresponding Ti compound, by reduction of $ZrCl_4$ with Li in ether in presence of dipyridyl.

The recently-announced element 104 is presumably the heaviest representative of this group. If its chemistry is ever explored, it is expected to be predominantly that of the IV state.

14.3 Niobium, $4d^45s^1$, and tantalum, $5d^36s^2$

These two elements resemble zirconium and hafnium in the very close similarity of their chemistries, although here it is the lighter niobium which is the rarer element. The different ground state electronic configurations appear to have no effect on the chemistry in the valency states. Rather more is known of the lower oxidation states of these elements than was the case with zirconium and hafnium, but the V state is by far the most stable and well-known. The element/M^{5+} potentials show the same trend as in the preceding Group:

$$V^V/V = -0.25 \text{ V}, \quad Nb^V/Nb = -0.65 \text{ V}, \quad Ta^V/Ta = -0.85 \text{ V}$$

A potential of about -1.1 V has been estimated for Nb^{III}/Nb and one of about -0.1 V for Nb^V/Nb^{III}, both in sulphuric acid solution where complex species are probably formed in both states.

Niobium and tantalum generally occur together and are separated by fractional crystallization of fluoro-complexes:

$$ore \rightarrow M_2O_5 + KHF_2/HF \rightarrow K_2TaF_7 \xrightarrow[\text{less soluble}]{\text{electrolysis}} Ta$$

$$\searrow$$

$$K_2NbOF_5 \xrightarrow{Al} Nb$$
$$\text{more soluble}$$

The metals are very resistant to acid attack but will react slowly with fused alkalis and with a variety of non-metals at high temperatures. They have very high melting points (Ta above 3000 °C) and find some use in high temperature chemistry. Tantalum is also used in surgery as it can be inserted in the body, as in fracture repair, without causing a 'foreign body' reaction.

In the V state, the oxides, Nb_2O_5 and Ta_2O_5, may be prepared by igniting the metals, their carbides, sulphides, or nitrides, or any compound with a decomposable anion. The oxides are inert substances which are generally brought into solution by alkali fusion, or treatment with concentrated HF. The oxides are therefore amphoteric but the acidity is very slight and the niobates are decomposed, even by as weak an acid as CO_2. The product of alkali fusion contains one metal atom and is written as NbO_4^{3-} (or TaO_4^{3-}) and termed orthoniobate (or orthotantalate). A monatomic metaniobate, $NaNbO_3$, is also known which has the perovskite structure. A number of more complex species are also known which contain 2, 5, or 6 metal atoms, for example, $M_4^INb_2O_7$ or $M_8^ITa_6O_{19}$. The latter $Ta_6O_{19}^{8-}$ ion has also been shown to be present in solution. The structure of none of these species is known, nor are the formulae unambiguous but may include water molecules and hydroxyl ions. It does seem clear, though, that niobium and tantalum share with vanadium the tendency to form polymeric oxyanions. The neutralization of these niobate or tantalate solutions with acid leads to the precipitation of white gelatinous precipitates of the hydrated pentoxides. These dissolve in hydrofluoric acid, probably as fluoro-complexes, but there is no evidence of cationic forms

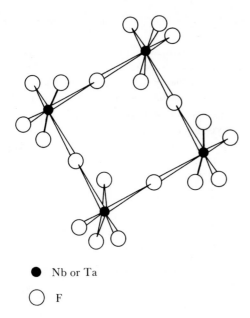

● Nb or Ta

○ F

FIGURE 14.3 *The structure of* NbF$_5$ *or* TaF$_5$
Most other pentafluorides of the transition metal adopt similar, tetrameric, structures.

of niobium or tantalum in solution analogous to the vanadium oxycations.

The pentafluorides may be prepared by the reaction of fluorine on the metal, pentoxide, or pentachloride, or by HF on the pentachloride. Both are volatile white solids melting below 100 °C and boiling near 230 °C. The structures are tetrameric in the solid with MF$_6$ units and one bridging fluoride between each pair of metal atoms (Figure 14.3), isostructural with MoF$_5$. In the liquid phase, there is evidence for similar *cis*-bridging, but forming polymers rather than tetramers.

The pentoxide dissolves in HF with formation of fluoro-complexes and these are also formed by the pentafluorides and F$^-$. Crystallization from solutions of moderate F$^-$ concentration gives the salts M^{I}Nb(or Ta)F$_6$ containing octahedral MF$_6^-$ ions. In the presence of excess F$^-$, TaF$_7^{2-}$ and NbOF$_5^{2-}$ are formed. A larger excess of F$^-$ leads to the formation of NbF$_7^{2-}$ and, at very high F$^-$ concentrations, NbOF$_6^{3-}$ and TaF$_8^{3-}$ are formed. The octafluoroniobate has not been reported. The structures of the MF$_7^{2-}$ ions are those based on the trigonal prism shown in Figure 14.1. NbOF$_5^{2-}$ has an octahedral structure, while NbOF$_6^{3-}$ illustrates another form of seven-coordination. This, Figure 14.4a, is based on the octahedron with the seventh ligand placed at the centre of one triangular face. The TaF$_8^{3-}$ ion is the square antiprism (Figure 14.4b) which reduces interactions between ligands to a minimum in eight-coordination. This corner of the *d* block gives a variety of seven- and eight-coordinated structures which is not found elsewhere in the Periodic Table.

The other pentahalides all exist and can be prepared by standard methods. All six compounds are volatile covalent solids with boiling points below 300 °C. The vapours are all monomeric and electron diffraction results indicate the

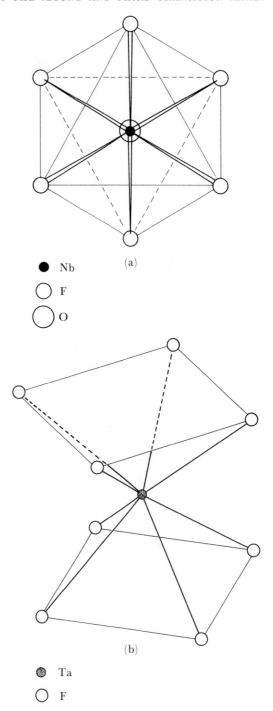

● Nb

○ F

◯ O

(a)

(b)

◍ Ta

○ F

FIGURE 14.4 *Higher coordination numbers* (a) NbOF$_6^{3-}$ *and* (b) TaF$_8^{3-}$

structures are probably the expected trigonal bipyramids. In the solid, an X-ray study shows that NbCl$_5$ is a dimer,

$$\text{Cl}_4\text{Nb}\underset{\text{Cl}}{\overset{\text{Cl}}{\diamondsuit}}\text{NbCl}_4,$$ with two bridging chlorides giving an

octahedral configuration around each niobium. In solution in non-donor solvents, such as CCl$_4$, the dimeric form is retained. Tantalum pentachloride and both bromides have the same solid structure but the pentiodides are of a different, and unknown, structure. Much less is known about com-

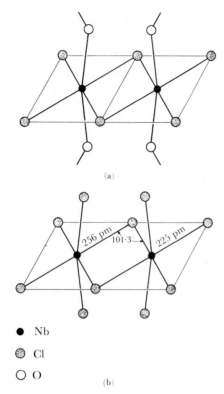

FIGURE 14.5 *The structures of* (a) NbOCl₃ *and* (b) NbCl₅

• Nb
◉ Cl
○ O

into long chains by oxygen bridges between the niobium atoms.

Apart from the oxides and halides, there are few important compounds of these elements in the V state. One interesting one is the cyclopentadienyl hydride, $(\pi\text{-}C_5H_5)_2TaH_3$, which provides another example of the stabilization of transition metal-hydrogen bonds by the presence of a pi-bonding ligand. The structure proposed is a distorted trigonal bipyramid shown in Figure 14.6. The C_5H_5 rings

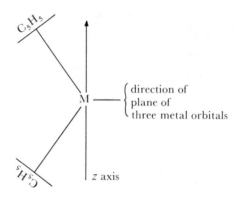

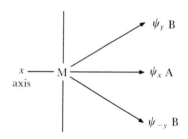

FIGURE 14.6 *Bonding in dicyclopentadienyltantalumtrihydride,* $\pi\text{-}(C_5H_5)_2TaH_3$
This is a diagrammatic representation of the three orbitals which become available on the central atom when the $C_5H_5-Ta-C_5H_5$ angle is reduced from the linear arrangement in ferrocene. The two H atoms bonded to the B orbitals are identical and differ from the central A hydrogen.

plexes of these heavier halides but there is good evidence for the existence of MCl_6^- species.

The pseudohalogen analogues, $M(NCS)_5$ and $M(NCS)_6^-$ are also known both for M = Tb and M = Ta, the penta-thiocyanates are dimers like the pentachlorides.

The halides all hydrolyse readily to the hydrated pent-oxides and some oxyhalides occur as intermediate hydrolysis products. These are $NbOCl_3$, $TaOCl_3$, and the correspond-ing oxybromides. A better preparation is by reaction be-tween the pentahalide and oxygen. The oxyhalides are covalent but less volatile than the pentahalides. The vapours appear to be monomeric and tetrahedral but the solids are polymeric. The structure of $NbOCl_3$ is shown in Figure 14.5, with the pentahalide for comparison. The oxyhalide contains

planar, binuclear $Cl_2Nb \begin{array}{c} Cl \\ \diagup\diagdown \\ \diagdown\diagup \\ Cl \end{array} NbCl_2$ groups which are linked

are on the vertical axis but bent towards each other so that the ring$-$Ta$-$ring angle is less than 180°. The three hydro-gens are bunched together with H$-$Ta$-$H angles of less than 120° on the opposite side from the rings. Nmr studies

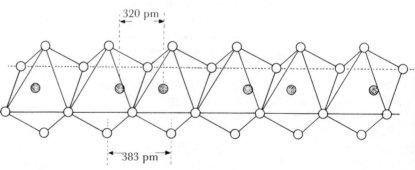

FIGURE 14.7 *The structure of niobium or tantalum tetraiodide*

◉ Nb or Ta
○ I

show two different types of H atom in the ratio 2:1 and the coupling to the ring protons fits with the angles given.

The IV state is represented by NbO_2 and the tetrahalides, MX_4, but TaO_2 probably does exist. The state is unstable, reducing, and shows some tendency to disproportionate. The dioxides are prepared by high temperature reduction of the pentoxides. They dissolve only in hot alkali, with reduction of the solvent.

All the tetrahalides exist except the tetrafluorides: the iodides are best known. NbI_4 results from prolonged heating of NbI_5 at 270 °C, when iodine sublimes off leaving the tetraiodide. TaI_4 is most easily made by heating TaI_5 with the metal. These iodides are diamagnetic solids, volatile at 300 °C. The diamagnetism arises because the metal atoms are linked by long bonds in dimers, as shown in Figure 14.7, thus pairing the single electron on each. The structure consists of chains of MI_6 octahedra, each joined by the edges to its neighbours, and with the metal atoms placed unsymmetrically in the octahedra and linked in pairs. The structures of the other tetrahalides are unknown. The tetrachlorides are made by reduction of the pentachlorides and both $NbCl_4$ and $TaCl_4$ disproportionate,

e.g. $TaCl_4 \xrightarrow{400\ °C} TaCl_3 + TaCl_5\uparrow$

Similarly, on hydrolysis, $TaCl_4$, gives a precipitate of tantalum(V) oxide and a green solution of the trichloride, which is fairly stable unless heated. The IV state is also reported from the electrolytic reduction of niobium(V) in 13M hydrochloric acid. An orange Nb^{IV} solution results which probably contains the oxyhalide ion, $NbOCl_4^{2-}$. This solution disproportionates to niobium(III)(+V).

In the III state, all the halides are known, including the trifluorides, but with the exception of TaI_3. Niobium tri-iodide is formed by heating the pentiodide at 430 °C, while the other niobium trihalides are formed by reducing the pentahalide with hydrogen at about 500 °C. Tantalum trichloride is obtained from the tetrachloride as above, while the tribromide is made by hydrogen reduction. Most of the trihalides are brown or black and strongly reducing, although the reactivity depends greatly on the thermal history of the sample. In strong hydrochloric acid, electrolytic reduction of niobium(V) to niobium(III) is reported. The III state solutions are yellow or blue, depending on the conditions used.

$TaCl_2$ is the only known tantalum dihalide. It is a non-stoichiometric, green-black solid resulting from the disproportionation of $TaCl_3$ at 600 °C. It is much more reactive than the trichloride, and attacks water under all conditions. Small quantities of $NbBr_2$ have been produced from the reaction of the pentabromide with hydrogen in an electric discharge. Nothing is known of its properties. Electrolytic reduction of niobium in 10 M HCl gives a violet solution colour attributed to niobium(II) but this state is not otherwise reported in solution.

Apart from the tetraiodides, nothing is known of the structures or degrees of polymerization of these other halides. They are all relatively involatile and the volatility drops

from the tetrahalide to the trihalide to the dihalide; the structures are thus probably polymeric, although the ability of $TaCl_3$ to dissolve in water must be recalled.

In addition to these halides of simple stoichiometry, niobium and tantalum form compounds of formula M_6X_{14} for X = Cl, Br, and I. Related compounds Nb_6F_{15}, Ta_6Cl_{15}, Ta_6Br_{15} and Ta_6Br_{17} are also found. These result from sodium amalgam reduction of the pentahalides (a better preparation is by reduction with cadmium metal at red heat followed by precipitation of CdS). The compounds are soluble in water and alcohol and the action of Ag^+ on the compounds M_6X_{14} precipitates only one-seventh of the halide as AgX. The compounds are therefore salts of the complex cation $(M_6X_{12})^{2+}$. The structure of this ion has been determined by X-rays. The metal atoms (Figure 14.8)

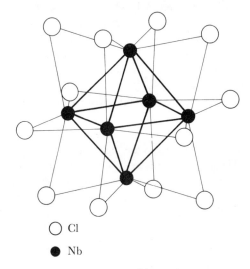

○ Cl

● Nb

FIGURE 14.8 *The structure of* $Nb_6Cl_{12}^{2+}$
The octahedron of niobium atoms is bonded, not only by the bridging chlorines, but by metal-metal bonding within the cluster.

form an octahedron and are bridged in pairs along the octahedron edges by halogen atoms. (Compare with the structure, of the dihalides of molybdenum and tungsten.) The octahedron of metal atoms is an example of the *metal cluster* compounds which are attracting current attention. It is held together by poly-centred metal−metal bonding as well as by the bridging halogen atoms.

A variety of related clusters are reported which are related to the $M_6X_{12}^{2+}$ structure, either by showing charges of +3 or +4, by containing terminal M−X groups in addition as in $M_6X_{18}^{4-}$, or in consisting of such clusters linked together by halogen bridges.

The lowest oxidation states are limited to a few carbonyl and organometallic species. For example, the −I oxidation state is found in the carbonyl anion, $M(CO)_6^-$, which is formed by both elements (and also by vanadium). These elements do not, however, form a simple carbonyl as does vanadium.

It is thus seen that this Group is quite similar to the titanium one. The heavier elements are broadly similar to the lighter ones in the Group oxidation states of IV and V respectively, but the oxides are less hydrolysed in solution and less amphoteric. The heavier elements form complexes with fluorine of high coordination numbers. The vanadium Group has a wider range of oxidation states and the lower ones are not quite so unstable for niobium and tantalum as they are for zirconium and hafnium, but they are much less stable than the Group state and are strongly reducing.

14.4 Molybdenum, $4d^5 5s^1$, and tungsten, $5d^4 6s^2$

These two elements, although there is still a close resemblance, show much more distinct differences in their chemistries than the two earlier pairs in these series. This is illustrated by the ready separation of the two elements in qualitative analysis, where tungsten appears in Group I of the conventional scheme and molybdenum in Group II. This is because tungsten(VI) which is soluble in neutral and alkaline solution precipitates an insoluble hydrated oxide, $WO_3.nH_2O$, in acid solution. Molybdenum(VI) oxide, on the other hand, dissolves again due to formation of chloro-complexes and molybdenum is first reduced by H_2S and then precipitates as MoS_2.

Examples of all oxidation states from VI to −II are found for these elements. The Group state of VI is the most stable, but the V and IV states are well-represented and occur in aqueous solution. Strong reducing properties are not shown until the III and II states are met. This behaviour is illustrated by the occurrence of the various halides and oxides (Tables 12.3 to 5) and by the free energy changes in solution

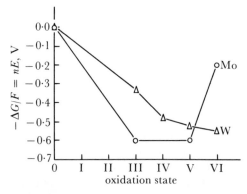

FIGURE 14.9 *Oxidation state free energy diagrams for molybdenum and tungsten*
This illustrates the greater relative stability of the lower oxidation states of molybdenum compared with those of tungsten.

shown in Figure 14.9. This diagram shows the differences in behaviour of the two elements, in particular the greater relative stability of Mo^V and the similar effect at W^{IV}, both of which lie at shallow minima. These two elements show little resemblance to chromium: the stability of the VI state contrasts with the strong oxidizing nature of chromium(VI) while the III state, which is so stable for chromium, is very much less stable for molybdenum and tungsten.

The elements occur as oxyanions and molybdenum is also found as the sulphide, MoS_2, which has found recent popularity as a solid lubricant, as it has a layer structure which allows the planes of atoms to slide over each other easily. The

ores are converted to the trioxides *via* the oxyanions and these are reduced to the metals by reduction with hydrogen (carbon cannot be used as the very stable carbides would result). Both metals are relatively inert, with very high melting points and with fairly high electrical conductivity. The metals are most readily attacked by alkaline peroxide or other fused, oxidizing alkaline medium. Attack by aqueous alkali and by acids is only slight.

In the VI state, the oxides, the hexafluorides, and the hexachlorides all occur and tungsten also forms WBr_6. Uranium is the only other element to form a hexachloride, and hexabromides are unknown apart from the tungsten compound. There is also an extensive aqueous chemistry of Mo^{IV} and W^{VI} and a vast collection of polymeric oxyanions.

The oxides, MoO_3 and WO_3, are the final products of igniting molybdenum or tungsten compounds. Both oxides are insoluble in water but dissolve in alkali to give oxyanions. MoO_3 also dissolves in acids to give oxy-cations or related hydrolysed species but WO_3 is insoluble in acid. The simplest product of the solution in alkali is the MoO_4^{2-} ion (molybdate or tungstate) which is tetrahedral. The molybdates and tungstates of most metals except the alkalis, NH_4^+, Mg^{2+}, and Tl^+, are insoluble. The neutralization of the alkaline solutions leads to the precipitation of the hydrated oxides, $MoO_3.2H_2O$ which is yellow, and $WO_3.2H_2O$ which is nearly white. These are definitely hydrates, as written above, not the hydrated acids, $H_2MoO_4.H_2O$. As with the anhydrous oxides, which are derived from the hydrates by ignition, the hydrated oxides differ in their behaviour to acid, the molybdenum compound dissolving and the tungsten compound being relatively insoluble.

These simple, mononuclear oxyanions are found only in strong alkali. At lower pH values these species show a strong tendency to condense to poly-oxyanions, either alone in the *isopolyacids* and their salts, or together with other oxyanions such as silicate or phosphate in the *heteropolyacids* and their salts. This gives a rich and varied field of condensed structures, probably only exceeded in complexity by the silicates. It is not possible to deal with this group of compounds in anything like full detail, and many structures are not fully determined, but Table 14.1 lists some typical formulae. The polyacids are built up from MoO_6 octahedra, Figure 14.10, and most of the structures so far determined are rings, double-rings, or clusters. On the basis of older experimental results, it was thought that Mo and W behaved similarly and passed, on increasing condensation, through stages which were termed ortho-, para-, and meta-acids (ortho being the normal MoO_4^{2-}). It is now known that some of these stages correspond to a number of different compounds which are not represented by the same formulae for Mo as for W. The ions or other species are most conveniently named on the basis of the number of Mo or W atoms. The structures of a number of ions are known, and ions with the same number of metal atoms appear to exist in solution, but there is no final proof that the solution equilibria do correspond to the solids found, and there is certainly evidence for a number of intermediate species in solution. Problems

TABLE 14.1 Iso- and hetero-poly molybdates and tungstates

Iso-compounds

(a) Molybdenum in solution.

$$MoO_4^{2-} \xrightarrow{pH = 6} Mo_7O_{24}^{6-} \xrightarrow{\text{lower pH}} Mo_8O_{26}^{6-} \xrightarrow{\text{strong acid}} Mo_n \text{ species?}$$

(7– and 8– ions may be protonated and hydrated.)

(b) Solid molybdenum compounds (all except the 1– and 2– species are heavily hydrated).

MoO_4^{2-}	Tetrahedron.
$(Mo_2O_7^{2-})_n$	Chain of MoO_6 octahedra sharing opposite corners and linked in pairs through adjacent corners by MoO_4 tetrahedra (Figure 14.11).
$Mo_7O_{24}^{6-}$	Linked octahedra (Figure 14.12).
$Mo_8O_{26}^{6-}$	Linked octahedra.

(c) Tungsten in solution.

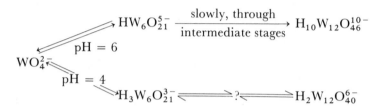

$$WO_4^{2-} \underset{pH = 4}{\overset{pH = 6}{\rightleftharpoons}}$$

$$HW_6O_{21}^{5-} \xrightarrow[\text{intermediate stages}]{\text{slowly, through}} H_{10}W_{12}O_{46}^{10-}$$

$$H_3W_6O_{21}^{3-} \rightleftharpoons ? \rightleftharpoons H_2W_{12}O_{40}^{6-}$$

(d) Solid tungsten compounds (all are heavily hydrated as salts except the 1– and 2– species).

WO_4^{2-}	Tetrahedron.
$(W_2O_7^{2-})_n$	Linked tetrahedra and octahedra: see Mo compound.
W_6 species	Structures not known.
$W_{12}O_{46}^{20-}$	See Figure 14.13, linked octahedra.
$W_{12}O_{40}^{8-}$	Linked octahedra; isomorphous with the 12-hetero-acids (q.v.).

Hetero-compounds

(a) MO_6 (M = Mo, W) octahedra surrounding a central (hetero) $M'O_6$ octahedron (all compounds are heavily hydrated).

$M'M_6O_{24}^{(12-x)-}$ (x = formal positive charge on hetero-atom M')	$M' = I^{7+}$, Te^{6+}, or a number of trivalent ions such as Co^{3+}, Al^{3+} or Rh^{3+}. Structure is a ring of six linked MO_6 octahedra with a central octahedral site for M' (Figure 14.14).
$M'M_9O_{32}^{(10-x)-}$	$M' = Mn^{4+}$, Ni^{4+}. Structure consists of three sets of clusters of three MO_6 groups giving a central octahedral hole for M'.
$M'M_{12}O_{40}^{(8-x)-}$	$M' = Ce^{4+}$, Th^{4+}, Sn^{4+}? Structure unknown.

(b) MO_6 octahedra surrounding a central (hetero) $M'O_4$ tetrahedron.

$M'M_{12}O_{40}^{(8-x)-}$	$M' = P^{5+}$, As^{5+}, Si^{4+}, Ge^{4+}, Ti^{4+}, Zr^{4+} (and perhaps the Sn^{4+} given in (a) above). Structure contains four sets of three MO_6 octahedra joined by edges and defining a central tetrahedral site. The structure of $W_{12}O_{40}^{8-}$ is the same as these with the central site empty.

Other heteroacids with tetrahedral $M'O_4$ groups occur with M'/M ratios of 1/11, 1/10, 2/18 and 2/17 and containing mainly P^{5+} or As^{5+}. The structures are unknown except that of compounds related to:

$P_2Mo_{18}O_{62}^{6-}$	This is the 12-acid structure as above with the three MO_6 octahedra at the base removed to give the 'half-unit' PM_9O_{34} and two of these are linked, sharing six oxygens, to give the P_2M_{18} unit.

(c) More complex hetero-compounds.

Much greater complication is possible as illustrated by the recent studies on phosphorus-molybdenum compounds, two of which are included:

$(MoP_2O_{11})_n$	Chains of MoO_6 octahedra formed by linking corners. The chains are cross-connected by P_2O_7 groups into a three-dimensional structure.
$(MoP_2O_8)_n$	Layers formed by MoO_6 octahedra sharing oxygens and these layers linked up by $(PO_3)_n$ chains which run perpendicular to the planes of the layers.

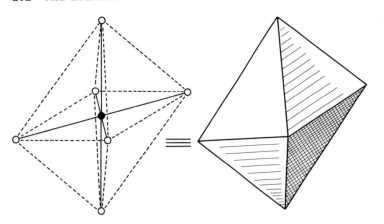

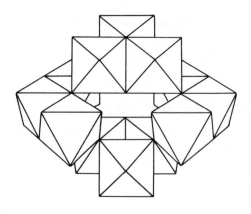

FIGURE 14.13   *The* $W_{12}O_{46}^{20-}$ *structure*

● molybdenum

○ oxygen

FIGURE 14.10   *The* MO_6 *octahedron*
The unit from which the polymolybdates and tungstates are built up.

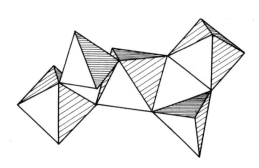

FIGURE 14.11   *The* $Mo_2O_7^{2-}$ *structure*

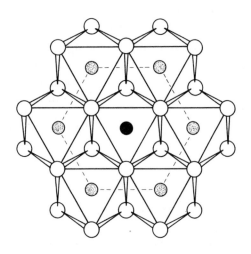

○ O

● M′ e.g., Te, I, Rh

▨ Mo or W

FIGURE 14.14   *The* $M'M_6O_{24}^{(12-x)-}$ *structure*

This consists of a ring of six linked MO_6 octahedra forming a central octahedral site for the M′ metal ion.

arise as it is difficult to obtain unambiguous analyses for alkali metals and Mo or W in the heavier species, and because mixed crystals are readily formed.

The polyacids, and especially the heteropolyacids, form a variety of unusual environments and are at present attracting attention as a means of studying unusual coordinations or unusual oxidation states such as Ni^{IV}.

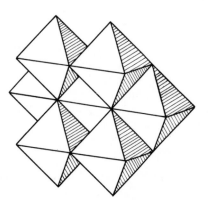

FIGURE 14.12   *The* $Mo_7O_{24}^{6-}$ *structure*
This structure is more compact than that found in the similar heteropolyion (Figure 14.14).

A recently-reported structure, of $CeMo_{12}O_{42}^{8-}$, contains twelve-coordinated cerium(IV) and the novel structure of MoO_6 octahedra linked in pairs by sharing a face, that is, with $Mo(O)_3Mo$ bridges.

Apart from these oxygen compounds, the VI state is found in halides, oxyhalides, and complexes related to these. Both molybdenum and tungsten form MF_6, MOF_4, and MO_2F_2 compounds but molybdenum has a much lower affinity for the heavier halogens than has tungsten, as Table 14.2 shows.

TABLE 14.2 Halides and oxyhalides of molybdenum(VI) and tungsten(VI)

MX_6		MOX_4		MO_2X_2	
MoF_6	WF_6	$MoOF_4$	WOF_4	MoO_2F_2	$WO_2F_2(?)$
$MoCl_6$	WCl_6	$MoOCl_4$	$WOCl_4$	MoO_2Cl_2	WO_2Cl_2
	WBr_6		$WOBr_4$	MoO_2Br_2	WO_2Br_2
					WO_2I_2

The hexahalides are made by direct reaction. They are volatile and unstable to oxygen and moisture, with which they react readily to give oxyhalides. Tungsten hexabromide is thermally unstable and decomposes on gentle warming. The oxyhalides are also volatile, covalent compounds which hydrolyse to the trioxides, the MOX_4 type more rapidly than the MO_2X_2 compounds. WO_2Cl_2, which is yellow, disproportionates above 200 °C to WO_3 and red $WOCl_4$. Molybdenum also forms a compound of formula $MoO_2Cl_2.H_2O$, which may well be the hydroxy compound, $MoO(OH)_2Cl_2$.

The hexahalides are octahedral, and WOF_4 has the tetrameric structure of NbF_5 (Figure 14.3). In contrast, $MoOF_4$ is formed into chains of octahedral units, linked by shared fluorines in both structures.

Molybdenum and tungsten form a variety of complex halides in the VI state, of which the fluorine compounds are the most widely represented. Examples include $M_2^IWF_8$ and M^IWF_7, and oxyhalides of both elements of the types MOF_5^-, $MO_2F_4^{2-}$, and $MO_3F_3^{3-}$. The oxychlorides $MO_2Cl_4^{2-}$ are also found. Other oxy-species are known, including molybdenyl sulphate, MoO_2SO_4, formed from molybdenum trioxide and sulphuric acid. Such compounds are probably molecular, rather than salts of the oxycation.

Molybdenum and tungsten are sufficiently stable in the VI state to form sulphides, MS_3. These are precipitated as hydrated compounds when H_2S is passed through slightly acid M^{VI} solutions. In stronger acid, the H_2S reduces the VI state to the IV state, giving the disulphides MS_2.

The VI state is also sufficiently stable to allow the preparation of a polyhydrido-complex, $WH_6(PR_3)_3$. This shows nine-coordination and multiple substitution by hydrogen as in the similar rhenium hydrides discussed on page 219.

Two interesting types of compound are found which fall between the VI and V states. Mild reduction of the trioxides gives intensely blue oxides whose composition is intermediate between M_2O_5 and MO_3. These blue oxides appear to contain both M^V and M^{VI} in an oxide lattice, and the intense colour arises from the existence of two oxidation states in the

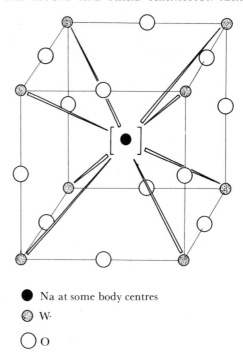

● Na at some body centres

◎ W·

○ O

FIGURE 14.15 *Tungsten bronze: the relation between the structures of rhenium trioxide, tungsten bronze, and perovskite*
The figure shows the ReO_3 structure, with the black circle indicating the site occupied, at random, by sodium ions in the bronze. If all the black circles are occupied, the structure is that of perovskite.

same compound. Similarly intense colours are observed in other cases like this, for example in magnetite, Fe_3O_4, and in Prussian blue. The second compound is the product of reduction of sodium tungstate, of formula Na_nWO_3 (n lying between 0 and 1), called tungsten bronze. The colour varies from yellow to blue-violet as n varies from about 0·9 to 0·3. The structure of these bronzes is based on the ReO_3 structure shown in Figure 14.15. Here, the metal atoms lie at the corners of a cube and the oxygens are at the mid-points of the edges. If a M^I ion lies at the centre of each cube, the structure is that of perovskite, $M'MO_3$. The sodium bronzes are compounds where sodium ions appear at random in the cube centres. The metallic appearance and conductivity of the bronzes arises as the sodium valency electron is delocalized over the structure. It may be noted that the structure of WO_3 itself is a distorted form of the ReO_3 structure.

The V state

The V state is prepared by mild reduction of the VI compounds and is coloured, usually green or red. The oxides are not acidic. Although W^V does exist, there are fewer tungsten than molybdenum compounds in this state.

Mo_2O_5 may be made by reacting MoO_3 with Mo at 750 °C. It is violet in colour and insoluble in water and dilute acids. It now seems likely that W_2O_5 does not exist and reports of its preparation apply to oxygen-deficient WO_3. Mo_2S_5 is also known from the reaction of H_2S on Mo^V solutions.

Among the halides, MF_5, MCl_5 (M = Mo, W), and WBr_5 are found. The existence of WBr_5 parallels the behaviour in the VI state. WF_5 is unstable and disproportionates to $WF_6 + WF_4$ above 50 °C. Molybdenum pentafluoride is prepared by reduction with the carbonyl:

$$5MoF_6 + Mo(CO)_6 = 6MoF_5 + 6CO$$

The pentachloride is the product of direct reaction between Mo and Cl. The tungsten pentahalides are formed by mild reduction of the hexahalides. All the pentahalides are covalent, relatively volatile solids. Solid MoF_5 has a tetrameric structure like NbF_5 (Figure 14.3) while $MoCl_5$ and WCl_5, in the solid, are dimeric like $NbCl_5$. Among the oxyhalides of the V state are $MOCl_3$ (M = Mo, W), $WOBr_3$ and WO_2I.

Most complexes of the V state are halogen or oxy-halogen compounds or contain pseudohalogen groups like CN or CNS. Reaction of $M(CO)_6$ in liquid IF_5 in the presence of alkali halides gives the two formula types $M^IMo(W)F_6$ and $M^I_3Mo(W)F_8$. The former are precipitated as salts of large cations and contain MF_6^- units. It is not known if the latter contain MF_8 units. Oxyhalide complexes result from the reduction of solutions of M^{VI}. The commonest types are MOX_5^{2-} and MOX_4^-, where X includes Cl, Br, CNS, and the Mo compounds are found in greater variety than the W ones.

The M^V octacyanides, $Mo(CN)_8^{3-}$ and $W(CN)_8^{3-}$ are obtained from strong oxidation of the M^{IV} octacyanides and have similar structures. These are discussed below.

The IV state

The two elements are quite similar in the IV state. This is usually formed by stronger reduction from the VI state than is used to reach the V state. The two dioxides, the disulphides and all the tetrahalides are known. Complexes are not very numerous and are similar in type to the complexes of the V state.

The dioxides, MO_2, are prepared by reduction of the trioxides by hydrogen or by careful oxidation of the metals. They are readily oxidized by halogens or oxygen and are reduced to the metal in hydrogen at temperatures above 500 °C. The dioxides are insoluble in nonoxidizing acids, but dissolve in nitric acid with oxidation to M^{VI}.

Both elements form the disulphide, MS_2. Molybdenum disulphide is an important naturally-occurring form of molybdenum. It has a layer lattice in which Mo atoms are surrounded by six S atoms at the corners of a trigonal prism. The outer S planes of neighbouring layers are only weakly cross-linked and the material is a solid lubricant.

All the tetrahalides are known, although the bromide and iodide are poorly characterised. The IV halides tend to disproportionate, thus:

$$MoO_2 + Cl_2 \rightarrow MoCl_4 \rightarrow MoCl_3 + MoCl_5$$

$$WCl_6 + H_2 \rightarrow WCl_4 \xrightarrow{\text{heat}} W_6Cl_{12} + WCl_5$$

The WX_4 compounds are slightly more stable than the MoX_4 ones. All are coloured, involatile solids which are readily oxidized. The tetrachlorides have polymeric struc-

tures in which the metal coordination is octahedral and where the M—M distances indicate some metal-metal bonding. $MoCl_4$ exists in a second form where there are four shared chlorines giving octahedra linked into chains by edge sharing.

Tungsten(IV) can be produced in solution by reduction of tungstate by tin and HCl. From the dark green solution the salt $K_2[W(OH)Cl_5]$ can be crystallized. This was formulated more recently as a dimer with bridging oxygen: $K_4[Cl_5W(O)WCl_5]$ and presumably hydrated. Some octahedral MX_6^{2-} complexes are also found, including the fluorides, chlorides, and thiocyanates. Also in the IV state is the hydride-cyclopentadienyl class of compounds, $(C_5H_5)_2Mo(W)H_2$ and $(C_5H_5)_2WH_3^+$. The latter is isoelectronic with the tantalum trihydride and appears to have the same kind of distorted trigonal bipyramidal structure. The dihydrides are probably distorted tetrahedra.

The best-known complex ion of the IV state is the octacyanomolybdate or tungstate, $M(CN)_8^{4-}$. This is an extremely stable grouping and is attacked only by permanganate or ceric which oxidize it only as far as the corresponding V octacyanide ion. The structure in the solid is dodecahedral (compare section 12.6) and is shown in Figure 14.16. A similar arrangement is shown in Figure 14.2. This arrangement of ligands stabilizes the d_{xy} orbital relative to all the rest, as the energy level diagram (14.16b) shows. The configuration is thus suitable for d^1 and d^2 arrangements with large ligand fields and the CFSE must be a major factor in the stability of these octacyanides.

The $M(CN)_8^{3-}$ ion, which results from oxidation of $M(CN)_8^{4-}$, is also found in the dodecahedral configuration. However, as we have seen in other cases of high coordination, the energy difference between alternative structures is small, and the octacyanides are also found in square antiprismatic coordination (as in Figure 14.4b) when the cation is large. Examples include $Mo(CN)_8^{4-}$ in the $Cd(N_2H_4)_2^{2+}$ salt, and both $M(CN)_8^{3-}$ ions when the cation is $Co(NH_3)_6^{3+}$.

The lower oxidation states

The lower oxidation states are unstable and strongly reducing. There is no evidence for simple cations in the III or II state. In the III state, tungsten is quite unstable and only a few compounds are known. These are WCl_3 and WBr_3, both recently reported, and polynuclear complexes such as $K_3W_2Cl_9$. The structure of WCl_3 consists of a cluster $W_6Cl_{12}^{6+}$ which is isostructural with $Nb_6Cl_{12}^{2+}$ (Figure 14.8). In contrast, WBr_3 contains the $W_6Br_8^{6+}$ cluster which is related to the $Mo_6Cl_8^{4+}$ structure (Figure 14.17) which is found in the dihalides of these elements. By contrast, Mo^{III} is relatively stable and well-represented. The oxide Mo_2O_3 is not known, but the sulphide exists. All the trihalides are reported, being derived from the higher oxidation states by strong reduction, or disproportionation. $MoCl_3$ is fairly stable, being only slowly oxidized in air and slowly hydrolysed by water. In solution, in presence of excess chloride, the complex anion $MoCl_6^{3-}$ may be prepared. A considerable number of other representatives of the type MoX_6^{3-} are

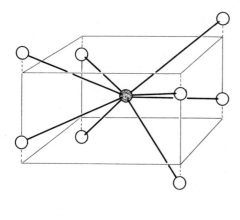

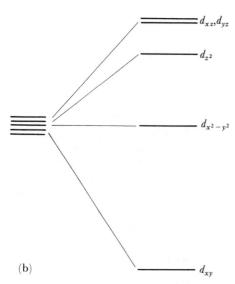

○ CN

◉ Mo   (a)

d_{xz}, d_{yz}

d_{z^2}

$d_{x^2-y^2}$

(b)

d_{xy}

FIGURE 14.16 (a), the structure of the octacyanomolybdate(IV) ion, $Mo(CN)_8^{4-}$, (b) the energy level diagram for this structure
(b) shows how this structure stabilizes one d orbital relative to the rest.

known including the fluoride and thiocyanate. These are all octahedral complexes with three unpaired electrons, as expected. Some complex cations of Mo^{III} are known, including $Mo(dipy)_3^{3+}$ and $Mo(phen)_3^{3+}$, and neutral compounds like $Mo(acac)_3$ also occur. All these compounds are probably octahedral.

The II state is represented by the complex halides $M_6X_{12}(M = Mo, W: X = Cl, Br, I)$ and by a few complexes with π-bonding ligands. The dihalides are relatively inert and insoluble materials. It was soon found that only a third of the halogen could be replaced, for example with $AgNO_3$ or by OH^-, and structural evidence has led to the formulation of the dihalides as hexamers:

$$\text{`}MX_2\text{'} = M_6X_{12} = (M_6X_8)^{4+}X_4^- \xrightarrow{Y^-} M_6X_8Y_4$$

The structure of the $Mo_6Cl_8^{4+}$ group is shown in Figure 14.17; the other compounds are isomorphous. This structure

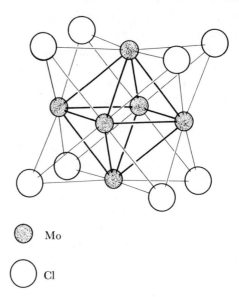

◉ Mo

○ Cl

FIGURE 14.17 The structure of the ion, $Mo_6Cl_8^{4+}$
This structure also contains metal-metal bonding in the Mo_6 cluster.

resembles that of the $Ta_6Cl_{12}^{2+}$ ion already discussed. The six metal atoms are again in an octahedron and are bridged by the chlorines. In the case of the molybdenum compound the 8Cl atoms lie approximately at the corners of the cube circumscribing the metal atom octahedron and each Cl bridges three Mo. In the tantalum compound, the Cl atoms bridge two Ta atoms along the octahedron edge. In both these types of compound, there is polycentred metal-metal bonding as well as the bonding involving the bridging chlorines.

Most other compounds of the divalent elements are carbonyl complexes such as $(C_5H_5W CO)_3Cl$ and $Mo(diars)_2(CO)_2X^+$. There is also an involatile acetate $Mo(CH_3COO)_2$, formed from $Mo(CO)_6$ and glacial acetic acid. A recent structural study shows that this compound is a dimer like the copper (Figure 13.27) and chromium (Figure 13.16) analogues. In the molybdenum compound, the metal-metal distance is short indicating that metal-metal bonding occurs in this case, as in the copper and chromium analogues. However, the molybdenum-molybdenum distance is so much shortened (211 pm compared with a 'single bond' distance of 290 pm) that an unusually high bond order is indicated which may include δ bonding. This suggestion is reinforced by the recent characterization of the red crystalline Mo(II) chloride complex $K_4Mo_2Cl_8.2H_2O$. This contains the $Mo_2Cl_8^{4-}$ ion which is isostructural with the rhenium analogue, discovered a few years earlier, Figure 14.20b. These species are formulated with a quadruple M−M bond containing one sigma, two pi, and one delta components. The Mo−Mo distance is 213·8 pm which is very similar to that in the acetate above.

The I state occurs in π-bonded compounds such as $Mo(C_6H_6)_2^+$ (compare the analogous Cr compound) or $[C_5H_5Mo(CO)_3]_2$.

The 0 state is represented by the octahedral and rather stable carbonyls, $Mo(CO)_6$ and $W(CO)_6$, which are white solids. Related compounds are also quite common, for example, $Mo(CO)_5I^-$, $W(PF_3)_6$, or $py_3Mo(CO)_3$. The $-II$ state is shown by the carbonyl anion $M(CO)_5^{2-}$.

14.5 Technetium, $4d^65s^1$, and rhenium, $5d^56s^2$

The comparative chemistry of these two elements is somewhat obscured by the inaccessibility of technetium. This occurs only as radioactive isotopes and had not been thoroughly studied until quite recently. These elements, in marked contrast to manganese, are particularly stable in the VII state, with little oxidizing properties. The lower states are also quite stable, especially Tc(IV) and Re(III) and (IV). The V and VI states are relatively unknown and tend to disproportionate. Low oxidation states are strongly reducing and not well-known, especially the II state which was so stable for Mn. These characteristics are illustrated by the oxides and halides in Tables 12.3–12.5 and by the oxidation state free energy diagrams in Figure 14.18.

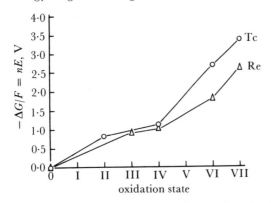

FIGURE 14.18 *The oxidation state free energy diagram for technetium and rhenium*

The longest-lived isotopes of technetium, ^{97}Tc and ^{98}Tc, have half-lives of the order of two million years. These are prepared by neutron bombardment of molybdenum isotopes. However, the most accessible source is nowadays the isotope ^{99}Tc which is one of the fission products of uranium and thus occurs to a considerable extent in spent fuel elements. This has a half-life of 212 000 years and is a weak β-emitter. It is therefore only mildly radioactive and relatively easy to handle.

Rhenium is a very rare element in the earth's crust and does not occur in quantity in any ore. However, it is found in molybdenum ores and can be quite readily recovered from these so that it is not too inaccessible and its chemistry is quite well known. It is left in oxidized solution as the perrhenate ion, ReO_4^-, whence it may be precipitated as the insoluble potassium salt. The metal is obtained by hydrogen reduction.

Technetium is separated from uranium and fission products by oxidation and distilled out as the volatile heptoxide, Tc_2O_7. It may be separated from rhenium, whose heptoxide is also volatile, by fractional precipitation of the sulphides, M_2S_7. At an acid concentration above 8 M HCl, Re_2S_7 is precipitated but Tc_2S_7 is not.

The Group oxidation state is represented by the hept-

oxides, the acid and salts of the MO_4^- ion, by the heptasulphides, by oxyhalides and by the MH_9^{2-} complexes discussed below. Rhenium alone forms the heptafluoride, ReF_7.

The heptoxides result from heating the metals in oxygen or air. Tc_2O_7 is a yellow solid melting at 120 °C and boiling at 310 °C. It is stable up to the boiling point. Re_2O_7 is also yellow and the solid sublimes; the calculated boiling point is 360 °C. In the vapour, the structure is $O_3M-O-MO_3$, with tetrahedral coordination. The crystal structures of Tc_2O_7 and Re_2O_7 differ, with the technetium oxide having the same structure as the vapour, with a linear $Tc-O-Tc$ bridge. The Re_2O_7 solid contains regular ReO_4 tetrahedra and ReO_6 octahedra with three short and three long bonds; the overall arrangement indicates that $ReO_3^+ReO_4^-$ (compare N_2O_5 in the solid) may be a reasonable formulation. Technetium heptoxide is somewhat more oxidizing than the rhenium compound and these two heptoxides differ strikingly in stability from Mn_2O_7 which is also volatile but which rapidly decomposes at room temperature.

The heptoxides dissolve in water to give colourless solutions of the acids. Pertechnic acid, $HTcO_4$, is produced as dark red crystals on evaporation, but perrhenic acid, $HReO_4$, cannot be isolated although the colour of the solution changes to yellow-green on concentration and lines due to the acid appear in the Raman spectrum of the concentrated solution. These colours are due to the lowering of the symmetry on passing from the tetrahedral anion ReO_4^- to the acid $(HO)ReO_3$ on concentration. Perrhenic acid resembles periodic acid in having a second form H_3ReO_5 (compare HIO_4 and H_5IO_6), and salts derived from this are readily prepared. The normal perrhenates are formed in dilute solution while salts of the tribasic form of the acid are formed in media of higher basicity:

$$ReO_4^- + K^+ = KReO_4 \text{ (yellow)}$$
$$ReO_4^- + K^+ + OH^- = K_3ReO_5 \text{ (red)}$$
or $$Ba(ReO_4)_2 \underset{H_2O}{\overset{Ba(OH)_2}{\rightleftharpoons}} Ba_3(ReO_5)_2$$

The peracids are strong acids, with perrhenic acid lying between perchloric acid and periodic acid in strength. Permanganic acid is stronger than perrhenic acid, so it is likely that pertechnic acid is also stronger. A third anion is found in $Ba_5(ReO_6)_2$ which probably contains isolated ReO_6^{5-} ions. The technetium analogue of the tribasic anion has not been isolated but TcO_5^{3-} may exist in fused sodium hydroxide. Some reactions of perrhenic acid are shown in Figure 14.19. Reactions of $HTcO_4$ are similar, as far as is known, apart from the lack of H_3TcO_5 and some differences in solubilities. Thus $KTcO_4$ is twice as soluble as $KReO_4$ and does not precipitate so readily. Both these potassium salts are very stable and can be distilled at temperatures of over 1000 °C without decomposition. $KMnO_4$, on the other hand, loses oxygen above 200 °C. The perrhenates, even of organic bases such as strychnine, may be isolated while permanganate readily oxidizes such compounds.

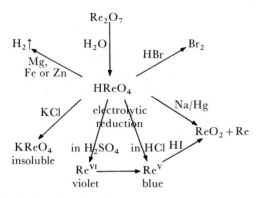

FIGURE 14.19 *Reactions of perrhenic acid*

The VII state is found in halides and oxyhalides as shown in Table 14.3.

TABLE 14.3 Halides and oxyhalides of the VII state

Re	Tc	Mn
ReF_7		
$ReOF_5$		
ReO_2F_3		
ReO_3F	TcO_3F	MnO_3F
ReO_3Cl	TcO_3Cl	MnO_3Cl ?

Technetium heptafluoride is not formed by direct combination at 400 °C, the reaction which gives ReF_7, but TcF_6 is found instead. The halides and oxyhalides are all colourless or pale yellow compounds which are either liquids or low-melting solids. The oxyhalides result from halogenation of the oxide or oxyion, or from the action of oxygen or water on the

fluoride in the case of the rhenium oxyfluorides.

Few complexes of the VII state are known, but oxidation of the octacyanide of Re(VI) gives salts of what appears to be the $Re(CN)_8(OH)_2^{3-}$ ion, which contains Re(VII) in ten-coordination. Unfortunately, its structure is not yet known.

The VI state is represented by a number of compounds but shows a marked tendency to disproportionate, especially in aqueous solution. In contrast, in concentrated sulphuric acid, Re(VII) may be reduced to pink Re(VI) and then to blue Re(V) in distinguishable steps. ReO_3 is well-known and there is one report of an oxide of composition $TcO_{3\cdot05}$. The rhenium trioxide is made by reaction of rhenium on the heptoxide and is red. In vacuum at 300 °C it disproportionates to ReO_2 plus Re_2O_7. It is inert to acids and bases in a non-oxidizing medium, and rhenates, ReO_4^{2-}, have to be made by fusing mixtures of perrhenate and rhenium dioxide. Technates exist in alkaline solution and pink $BaTcO_4$ may be precipitated, indicating less tendency to disproportionate than for rhenium(VI).

The halides, oxyhalides, and related complexes, of the VI state are given in Table 14.4. The VI state of Re is well-represented but only the fluoride, oxyfluoride and chloride are found for Tc. The halides are prepared by direct combination: ReF_6 may be purified from ReF_7 by heating with Re metal. $ReOF_4$ results from the reaction of ReF_6 with rhenium carbonyl. $TcOF_4$ exists in two forms: the blue variety contains infinite chains of octahedra while the green form contains trimers of octahedra. In both cases the octahedra are formed by sharing fluorines, $(TcOF_3F_{2/2})_n$. The other oxyhalides result from the reaction of the halides with air.

TABLE 14.4 Halogen compounds of the lower oxidation states of technetium and rhenium

VI		V		IV		III	
ReF_6	TcF_6	ReF_5	TcF_5	ReF_4		$[ReF_3]$	
$ReCl_6$	$TcCl_6$	$ReCl_5$		$ReCl_4$	$TcCl_4$	Re_3Cl_9	
		$ReBr_5$		$ReBr_4$		Re_3Br_9	
				ReI_4		Re_3I_9	
Also ReF_2, ReX_2 in complexes, and a reported compound of formula ReI							
$ReOF_4$	$TcOF_4$	$ReOF_3$					
$ReOCl_4$	$(TcOCl_4)$		$TcOCl_3$				
$(ReOBr_4)$			$TcOBr_3$				
ReO_2Br_2 ?							
ReF_8^{2-}	TcF_8^{2-}	ReF_6^-		ReF_6^{2-}	TcF_6^{2-}	$ReCl_6^{3-}$	
	TcF_7^-			$ReCl_6^{2-}$	$TcCl_6^{2-}$	$Re_3X_{12}^{3-}$	
				$ReBr_6^{2-}$	$TcBr_6^{2-}$	$Re_3X_{11}^{2-}$	
				ReI_6^{2-}	TcI_6^{2-}	$Re_3X_{10}^-$	
				$Re_2X_9^-$		$Re_2X_8^{2-}$	$Tc_2Cl_8^{2-}$
				(X = Cl, Br)		(X = Cl, Br)	
$ReO_2F_4^{2-}$?							
$ReOCl_6^{2-}$		$ReOCl_5^{2-}$	$TcOCl_4^-$?	$Re(OH)Cl_5^{2-}$	$Tc(OH)Cl_5^{2-}$		
$ReOCl_5^-$		$ReOBr_5^{2-}$	$TcOCl_5^{2-}$	$Re_2OCl_{10}^{4-}$			
		$ReOX_4^-$					
		(X = Cl, Br, I)					

The V state is also relatively unstable and disproportionates:

$$3M^V = 2M^{IV} + M^{VII}$$

For example, $ReCl_5$ reacts with HCl to give ReO_2, perrhenic acid, and $ReCl_6^{2-}$. $ReCl_5$ exists as a dimer with octahedral rhenium, like other pentahalides such as $NbCl_5$ (Figure 14.5).

The complex of the V state which is of most interest is the cyanide, $Re(CN)_8^{3-}$. This is the d^2 octacyanide, isoelectronic with the Mo^{IV} compound, and it appears to have the same dodecahedral structure. On oxidation it gives the d^1 compound of Re^{VI}, $Re(CN)_8^{2-}$. Related compounds are produced from the dioxides in alkaline cyanide solutions. The complex ions $Tc(OH)_3(CN)_4^{3-}$ and $Re(OH)_4(CN)_4^{3-}$ are reported to be formed under similar conditions. It is interesting that Re is oxidized in this system while Tc remains in the IV state.

The diarsine ligand already encountered (page 194) appears to stabilize the relatively unstable II, III, and V states of Tc and Re. Both metals form the $M(diars)_2Cl_4^+$ ion in which the element is in the V oxidation state and eight-coordinate. The $M^{II}(diars)_2Cl_2$ and $M^{III}(diars)_2Cl_2^+$ compounds are also known.

The IV state is the second most stable oxidation state for both elements. The dioxides, MO_2, disulphides, MS_2, and most of the halides are known. A number of complexes exist including the very important class of hexahalides, MX_6^{2-}. These are formed by dissolving the dioxides in the hydrohalic acid and are a most useful starting point for preparations.

The dioxides are formed by reducing the heptoxides, or from the metal by controlled oxidation. Apart from dissolving in the hydrogen halides, ReO_2 dissolves in fused alkali to give the oxyanion, rhenite, ReO_3^{2-}. This precipitates the dioxide when treated with water. It does not appear that the technetium dioxide is soluble in alkali.

A further example of the smaller tendency of Tc to enter the V state is provided by the reaction of the MI_6^{2-} complexes with KCN. TcI_6^{2-} reacts in methanol with KCN to give the Tc^{IV} cyanide complex, $Tc(CN)_6^{2-}$, but ReI_6^{2-} undergoes oxidation in the course of the reaction to the Re^V octacyanide $Re(CN)_8^{3-}$. It is easy to see that, if the octacyanide is to be formed, the M^V (d^2) compound will be more stable than the M^{IV} (d^3) compound, as the third electron is in an unstable orbital in the dodecahedral field (Figure 14.16b). It is therefore likely that this difference in behaviour is to be ascribed to a steric effect favouring eight-coordination for Re rather more than for Tc.

In the III state, technetium is quite unstable but rhenium forms the oxide, $Re_2O_3.nH_2O$, and the heavier halides. The oxide is formed by hydrolysis of the rhenium(III) chloride.

Rhenium trichloride and tribromide, formed by heating the pentahalides in an inert gas, are dark red solids which have been shown to be trimers, Re_3X_9. The structure is similar to that of the $Re_3Cl_{12}^{3-}$ ion shown in Figure 14.20a. In these species, a triangle of Re atoms is bridged along each

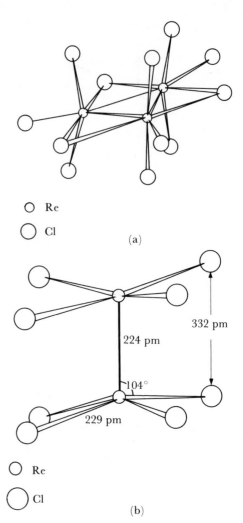

O Re

O Cl

(a)

332 pm

224 pm

104°

229 pm

O Re

O Cl

(b)

FIGURE 14.20 (a) *The structure of the ion,* $Re_3Cl_{12}^{3-}$, (b) *structure of the* $Re_2Cl_8^{2-}$ *ion*

edge by a halogen atom, and there are two more halogen atoms on each rhenium situated above and below the Re_3 plane giving a basic Re_3Cl_9 unit. In the ion $Re_3Cl_{12}^{3-}$, the structure is completed by adding a terminal halogen, in the plane of the triangle, to each rhenium. The related ions (see Table 14.4) $Re_3Cl_{11}^{2-}$ and $Re_3Cl_{10}^-$ lack one and two, respectively, of these in-plane terminal chlorines. The corresponding bromides have similar structures. In the parent trihalide, the basic Re_3Cl_9 units are joined together by forming $Re(Cl)_2Re$ bridges, using the in-plane positions and one of the Cl atoms which are perpendicular to the plane from each Re. A range of related complexes are also found, of type $Re_3X_9L_3$ where neutral ligands occupy the in-plane terminal positions. Earlier reports of dimeric forms of rhenium trihalides have been shown to be unfounded. ReI_3 is a black solid, which results from heating the tetraiodide, and further heating leads to another black compound of formula ReI. The structure is also trimeric and similar to that of the trichloride. The overall linking of Re_3 units gives long chains in the iodide, while the chloride and bromide have layer structures.

A further type of complex halide ion of the III state of rhenium is formed by reduction of perrhenate in acid solution and contains dimeric units, $(Re_2Cl_8)^{2-}$. The corresponding bromide is also known. The structure of these units has been determined and is shown in Figure 14.20b. This consists of two square-planar $ReCl_4$ units linked face-to-face by a $Re-Re$ bond so that the chlorines lie approximately at the corners of a cube. This eclipsed configuration, together with the very short Re-Re distance of 224 pm, is taken to indicate unusual bonding. The structure has been explained in terms of a quadruple $Re-Re$ bond made up of a σ, two π, and a δ interaction. If the $Re-Re$ axis is the z axis, and the square $ReCl_4$ part is bonded by $s+p_x+p_y+d_{x^2-y^2}$ components, the quadruple bond may be described in simple terms as follows. The sigma bond will have contributions from the metal d_{z^2} and p_z orbitals which lie along the $Re-Re$ axis. Then the metal d_{xz} and d_{yz} orbitals are aligned correctly to form two π bonds, leaving the d_{xy} orbital (in the $ReCl_4$ plane and lying between the $Re-Cl$ directions) on each Re to combine to form a delta bond (compare Figure 3.15). These four orbitals accommodate the eight electrons arising from the d^4 configuration of rhenium(III). Since, there is no other obvious reason for the eclipsed configuration, this is perhaps the best evidence to date for the existence of a δ bond.

Similar structures are found for $Mo_2Cl_8^{4-}$, $Re_2Br_8^{2-}$ and for a number of derivatives such as $Re_2Cl_6[P(Et_3)]_2$. The delta-bonded form has also been proposed for molybdenum diacetate (section 14.4) to account for the very short bond although the presence of bridging groups means that the eclipsed form of the MoO_4 units is now to be expected.

Other complexes of the trichlorides with donor ligands are found, for example, $(Ph_3PO)_2ReCl_3$. Rhenium(III) also gives hydrides of formula $(C_5H_5)_2ReH$ and $(C_5H_5)_2ReH_2^+$.

Representatives of lower oxidation states include complexes of MX_2 with donor ligands such as py_2ReI_2 and $(diars)_2TcCl_2$.

Both elements form a cyanide in the I state corresponding to the manganese compound, $K_5M(CN)_6$. The I state is also found in carbonyl halides and similar species, for example, in $Re(PF_3)_4Cl$.

The 0 state appears in the carbonyls, $Tc_2(CO)_{10}$ and $Re_2(CO)_{10}$, which have the same structure as the manganese carbonyl. The anion, $Re(CO)_5^-$, is formed with the alkali metals and contains $Re(-I)$.

Finally, mention must be made of the hydrogen complexes, K_2ReH_9 and K_2TcH_9. It has long been known that treatment of perrhenate solutions with potassium/ethylenediamine/water gave a water-soluble reactive rhenium species. This was originally identified as a 'rhenide' anion, analogous to the halide ions in the corresponding Main Group. Recently, evidence from infra-red and nmr spectra showed the presence of $Re-H$ bonds in the compound and, after various other formulations had been suggested, it has been shown to contain Re(VII) coordinated to nine hydrogen atoms. The compound is colourless and diamagnetic, which accords with Re(VII). It was soon afterwards shown that a similar Tc compound existed.

The structure of the ion is shown in Figure 14.21. The

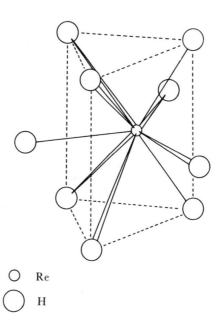

◯ Re
◯ H

FIGURE 14.21   *The structure of the enneahydridorhenium (VII) ion,* ReH_9^{2-}

positions of the metal atoms were determined by X-rays and then the hydrogen positions were found from neutron diffraction. The $Re-Re$ distances rule out the possibility of any metal-metal bonding. The Re atoms are at the centre of trigonal prisms of six H atoms with the other three H atoms beyond the rectangular faces. This lies inside a similar prism of K atoms. The Tc compound is isostructural. The closely related hydride, $ReH_7(PR_3)_2$ is reported as is the rhenium(V) species, $ReH_5(PR_3)_3$. The existence of such compounds, like the $WH_6(PR_3)_3$ species indicated in the last section, highlights the lack of oxidizing tendency in the high oxidation states of these heavy elements.

14.6 Ruthenium, $4d^75s^1$, and osmium, $5d^66s^2$

These are the first two of the six platinum metals, i.e. the six heavier members of the iron, cobalt, and nickel Groups. These elements, together with rhenium and gold, are broadly similar in that the element is fairly unreactive—'noble'—and decomposition of compounds to the element is fairly ready. The platinum metals, gold, and silver are commonly found together and a number of schemes are in current use for their separation. One method involves extracting the mixed metals with aqua regia and then treating the soluble and insoluble portions as in Figure 14.22. Osmium may occur in either fraction and is removed as the volatile tetroxide, while ruthenium ends up in the VI state in fused alkali.

These two elements share with xenon the highest observed oxidation state of VIII, and their oxidation states range downwards to $-$II. The VI and IV states are stable while the VII and V states are poorly represented and tend to disproportionate. Osmium is most stable in the IV state and ruthenium in the III state.

Some inter-relations among the oxides and oxyions of the different oxidation states are shown in the reaction diagram of Table 14.5.

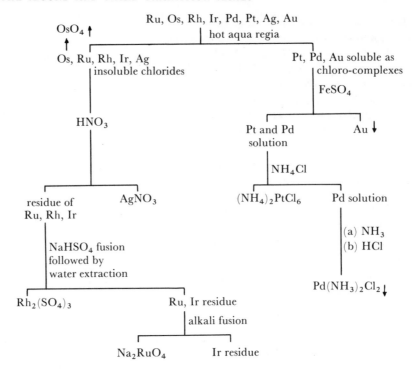

FIGURE 14.22   *A reaction scheme for the separation of the platinum metals, gold, and silver*

Note that osmium is more stable in the VIII state than ruthenium, the metal being oxidized directly to the tetroxide. One or two osmium(VIII) complexes occur, including the oxyfluoride $OsO_4F_2^{2-}$ and the corresponding hydroxide. Only mild reduction of osmium tetroxide is required to give the dioxide, and ruthenium burns in air directly to the dioxide. The stability of the IV state is shown by the existence of the disulphide and the compounds of the other oxygen Group elements. The dioxides are stable compounds with the rutile structure, while the tetroxides are volatile (b.p. about 100 °C), tetrahedral, covalent molecules which are strongly oxidizing.

The inter-relations among the halides, and halogen complexes, are similar, except that no compounds of the VIII or VII state exist (older reports of the octafluorides have been disproved). The VIII state is also found in the oxyfluoride, OsO_3F_2 which gives the Os(VII) and (VI) species on reaction with fluorine

$$OsO_3F_2 + F_2 \rightarrow OsOF_5 + OsF_6 + OF_2$$

That is, osmium(VIII) is reduced by the action of fluorine gas, although the reaction also involves the oxidation of $O(-II)$ to $O(II)$.

The pentafluorides are the only examples of the unstable V state. RuF_5 is a tetramer with non-linear $Ru-F-Ru$ bridges similar to the niobium compound. This tetrameric form, with bridging fluorides (Figure 14.3) is adopted, with minor variations, by most of the pentafluorides of the heavier transition metals. Molybdenum, osmium and platinum pentafluorides are all tetramers. The stable states of osmium (IV) and ruthenium(III) again show clearly. The II state is represented only by the osmium compounds which are polymeric reactive solids.

Osmium(IV) is also found in the polyhydride, $OsH_4(PR_3)_3$, which continues the sequence of the W(VI) and Re(V) polyhydrides. An example of Ru(IV) is the oxychloride, $Ru_2OCl_{10}^{4-}$, which has the oxygen-bridged structure $Cl_5Ru-O-RuCl_5^{4-}$.

Apart from those given above, these elements form a variety of complexes in the IV, III, and II states. In the IV state, osmium complexes are more extensive and stable than the ruthenium ones. The commonest are the halide, hydroxy-halide, amine, and diarsine complexes. In the III state, the situation is reversed and there are more ruthenium species, mainly octahedral. Both elements give hexammines, $M(NH_3)_6^{3+}$, and ruthenium gives the whole range of mixed halogen-ammonia complexes down to $Ru(NH_3)_3X_3$. A variety of other ruthenium complexes occurs, including those with substituted amines, and complex chlorides with 4, 5, 6, and 7 chlorine atoms in the anion.

Although there are few simple compounds of the II state, there are a variety of complexes of both ruthenium and osmium. All are formed by reduction of metal solutions, in the IV or III states, in presence of the ligands. Examples include $M(dipy)_3^{2+}$, the very stable $M(CN)_6^{4-}$, and a variety of ammine and arsine complexes. As the II state is the d^6 configuration, there will be a large CFSE contribution to octahedral complexes of these heavy elements. An important group of compounds are those containing NO bonded to ruthenium(II), such as $Ru(NO)X_5^{2-}$, where X = halogen, OH, CN, and many more.

In lower oxidation states, there are a number of compounds including $Os(NH_3)_6Br$ and $Os(NH_3)_6$—Os(I) and Os(0) respectively—formed by reduction with potassium in liquid ammonia. Other representatives of the 0 state include the

TABLE 14.5

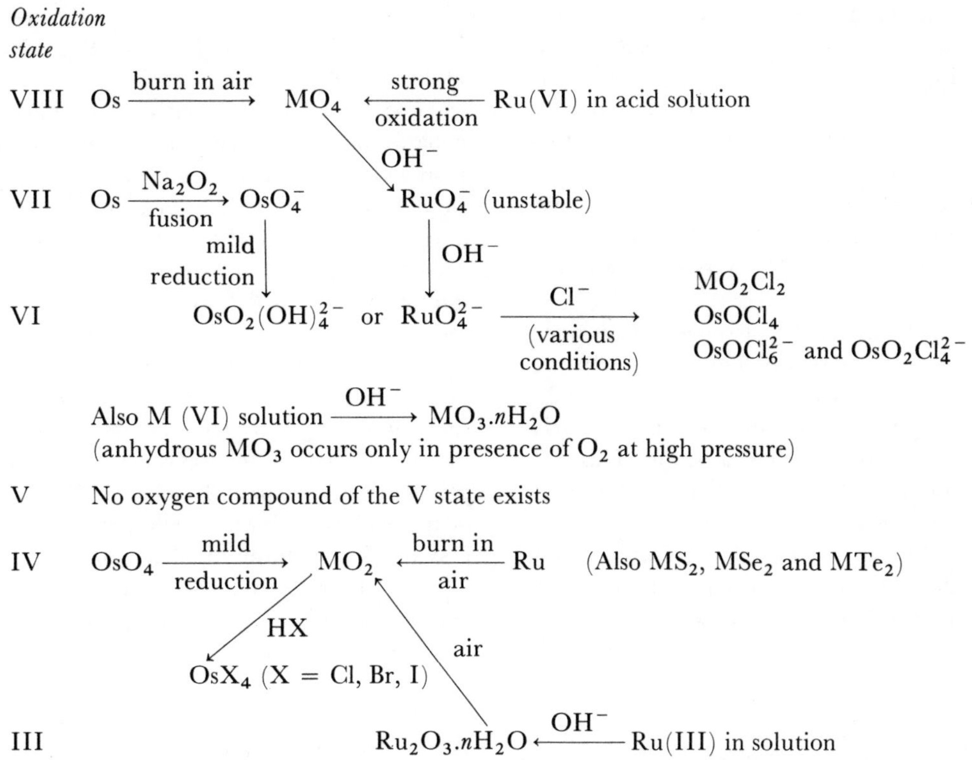

*Oxidation
state*

VIII $Os \xrightarrow{\text{burn in air}} MO_4 \xleftarrow[\text{oxidation}]{\text{strong}} Ru(VI)$ in acid solution

VII $Os \xrightarrow[\text{fusion}]{Na_2O_2} OsO_4^- \qquad RuO_4^-$ (unstable)

$\quad\quad\quad \downarrow \begin{array}{c}\text{mild}\\\text{reduction}\end{array} \qquad\qquad \downarrow OH^-$

VI $\quad\quad\quad OsO_2(OH)_4^{2-}$ or $RuO_4^{2-} \xrightarrow[\substack{\text{(various}\\\text{conditions)}}]{Cl^-} \begin{array}{l} MO_2Cl_2 \\ OsOCl_4 \\ OsOCl_6^{2-} \text{ and } OsO_2Cl_4^{2-} \end{array}$

Also M (VI) solution $\xrightarrow{OH^-} MO_3.nH_2O$
(anhydrous MO_3 occurs only in presence of O_2 at high pressure)

V No oxygen compound of the V state exists

IV $OsO_4 \xrightarrow[\text{reduction}]{\text{mild}} MO_2 \xleftarrow[\text{air}]{\text{burn in}} Ru \quad$ (Also MS_2, MSe_2 and MTe_2)

$\quad\quad\quad \swarrow HX \qquad\qquad \nwarrow \text{air}$

$\quad\quad OsX_4$ (X = Cl, Br, I)

III $\quad\quad\quad\quad\quad\quad\quad Ru_2O_3.nH_2O \xleftarrow{OH^-} Ru(III)$ in solution

---

TABLE 14.6

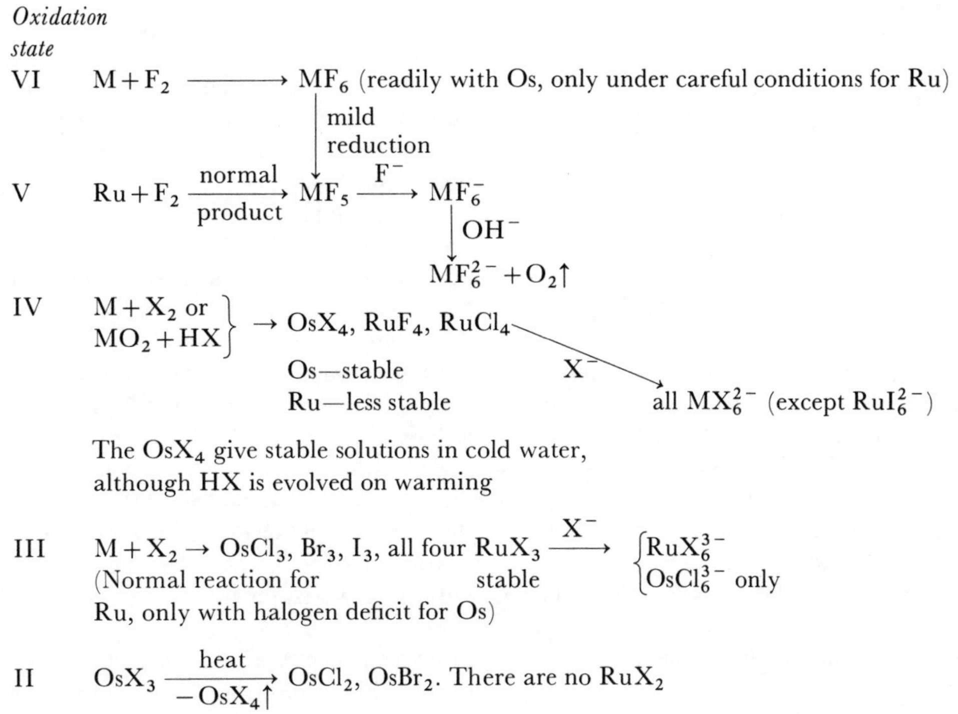

*Oxidation
state*

VI $M + F_2 \xrightarrow{\quad\quad} MF_6$ (readily with Os, only under careful conditions for Ru)

$\quad\quad\quad\quad \downarrow \begin{array}{c}\text{mild}\\\text{reduction}\end{array}$

V $Ru + F_2 \xrightarrow[\text{product}]{\text{normal}} MF_5 \xrightarrow{F^-} MF_6^-$

$\quad\quad\quad\quad\quad\quad\quad\quad \downarrow OH^-$

$\quad\quad\quad\quad\quad\quad\quad\quad MF_6^{2-} + O_2\uparrow$

IV $\left.\begin{array}{l} M + X_2 \text{ or} \\ MO_2 + HX \end{array}\right\} \rightarrow OsX_4, RuF_4, RuCl_4$

$\quad\quad\quad\quad Os\text{—stable} \qquad\qquad\qquad \searrow X^-$

$\quad\quad\quad\quad Ru\text{—less stable} \qquad\qquad\qquad$ all MX_6^{2-} (except RuI_6^{2-})

The OsX_4 give stable solutions in cold water,
although HX is evolved on warming

III $M + X_2 \rightarrow OsCl_3, Br_3, I_3,$ all four $RuX_3 \xrightarrow{X^-} \begin{cases} RuX_6^{3-} \\ OsCl_6^{3-} \text{ only} \end{cases}$
(Normal reaction for $\quad\quad\quad\quad\quad$ stable
Ru, only with halogen deficit for Os)

II $OsX_3 \xrightarrow[-OsX_4\uparrow]{\text{heat}} OsCl_2, OsBr_2.$ There are no RuX_2

carbonyls, $M(CO)_5$ and $M_3(CO)_{12}$. Among the more complex carbonyls, there is the very interesting derivative $Ru_6C(CO)_{17}$. This contains an isolated carbon atom which is situated at the centre of a distorted octahedron of Ru atoms giving a CRu_6 environment which is also found in most interstitial metallic carbides (section 15.5). Four of the Ru atoms carry three terminal CO groups and the other two have two terminal carbonyls and are bridged by the seventeenth one. The Ru_6C cluster involves delocalised polycentred bonding. There is a reported anion, $Ru(CO)_4^{2-}$, which would contain ruthenium($-II$) and be isoelectronic with nickel carbonyl.

Nitrogen complexes. A little after the first reports of nitrogen fixation in the systems metal halide/organometallic reducing agent which were discussed under *Titanium* (section 13.2), there was reported the first compound in which it was clearly shown that nitrogen was coordinated as a ligand. This was the ruthenium(II) complex $[Ru(NH_3)_5N_2]^{2+}$. This compound was first made by the action of hydrazine (which was the source of the coordinated N_2) on the trichloride, but a number of other routes have since been reported including the reaction of a ruthenium(II) complex with gaseous nitrogen. Some of these preparations and some reactions are summarized in Figure 14.23.

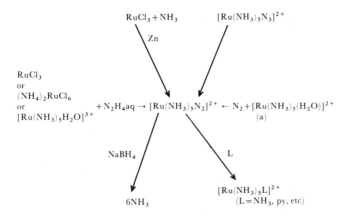

(a) The ruthenium(II) complex is made *in situ* by zinc amalgam reduction of ruthenium(III) pentammine chloro complex.

FIGURE 14.23 *Preparations and some reactions of* $[Ru(NH_3)_5N_2]^{2+}$

This series of reactions shows that the ligand nitrogen may be derived from a number of sources, including the direct fixation of atmospheric nitrogen. The coordinated nitrogen is replaceable by a range of ligands, presenting a useful route to otherwise unobtainable ruthenium(II) complexes. In these cases, the nitrogen comes off as N_2 gas, so that the cycle of reactions does not lead to fixation of N_2 as a compound. The only indication of the latter process is in the reaction with $NaBH_4$ which yields six volumes of ammonia, that is, one N atom out of the two in the coordinated group is converted to NH_3.

A number of other metals in this part of the periodic table have been shown to form nitrogen complexes. Most of these are of low oxidation state phosphine complexes, often containing hydride, like the cobalt complex of Figure 13.24, $ClIr(N_2)(PPh_3)_2$, $(R_3P)_2Ni(N_2)Ni(PR_3)_2$ and $H_2Fe(N_2)(PR_3)_3$. So far, nitrogen complexes have been isolated for all the elements of the manganese, iron, and cobalt groups, together with Mo and Ni. Osmium forms the analogue of the ruthenium complex, $Os(NH_3)_5N_2^{2+}$, by similar routes. The osmium complex is much more stable and resists replacement of the nitrogen.

The coordinated nitrogen is characterized by a stretching frequency around 2100 cm^{-1}, compared with a value of about 2300 cm^{-1} for free nitrogen. In those species whose structure has been determined, the nitrogen is coordinated linearly with the $M-N-N$ angle equal to 180°, or nearly so. Thus the N_2 molecule is bonded to the metal in a similar way to CO, with which it is isoelectronic, and isolatable complexes result most readily for transition metals to the right of the *d* block in low oxidation states. As the N_2 molecule is symmetric, with no dipole, sigma donor bonding from N to the metal is likely to be less important than is the sigma component of $M-CO$ bonding. Thus the main interaction in the $M-NN$ bond is probably the back donation from the metal into π orbitals on nitrogen.

A limited number of compounds are known, like $(NH_3)_5Ru(N_2)Ru(NH_3)_5^{4+}$ and the nickel compound above, where the NN unit bridges two metal atoms. Structural studies on these compounds indicate a linear $M-N-N-M$ bridge, in contrast to bridging carbonyl which bonds to both metals through carbon in a non-linear unit.

It should be noted that nitrogen fixation reactions, that is, activation of coordinated N_2 so that products such as NH_3 may be formed from it, are rare and inefficient for the compounds of those elements to the right of the periodic table which are stable enough to be isolated. In contrast, those species formed by reaction of metal halide plus reducing agent plus N_2, which do yield ammonia (usually on hydrolysis) do so via intermediates which cannot be isolated and which are commonly formed by elements of the titanium, and chromium groups. Thus, it appears at present that where the metal-nitrogen interaction is strong enough to allow the isolation of compounds, the nitrogen is usually displaced as N_2 gas, whereas those systems which allow reaction of the nitrogen to form ammonia or other compounds, do not bind the nitrogen sufficiently firmly to allow derivatives to be characterized. It may be that these active intermediates no longer contain $N-N$ units.

14.7 Rhodium, $4d^85s^1$, and iridium $5d^96s^0$

In this Group, the Group oxidation state of IX is not shown and a strong tendency for lower oxidation states to become stable is seen, as compared with the previous Groups of heavy transition elements. The highest state found is VI, while the most stable states are iridium(IV) and (III) and rhodium(III). In the separation from the rest of the platinum metals, Figure 14.22, both elements are found in the fraction insoluble in aqua regia. Iridium remains as the insoluble residue, after alkaline fusion, while rhodium is extracted in aqueous solution as sulphate after fusion with sodium bisulphate.

The highest oxidation state is represented by the hexafluorides, MF_6, formed by direct reaction. As in the previous Group, the heavier element is more stable in the high oxidation states and IrF_6 is much more stable than RhF_6. No oxygen species exist in the VI state: the reported IrO_3 is possibly a peroxide rather than an iridium(VI) compound.

The V state is also unstable and is found only in the IrF_6^- ion, whose salts are prepared by fluorinating mixtures of iridium trihalides and alkali halides in bromine trifluoride solution.

The IV state is fairly stable for iridium but represented by only a few rhodium compounds. The products formed on igniting the metals in air are IrO_2 and Rh_2O_3, respectively, and RhO_2 can be made only by strong oxidation of rhodium(III) solutions. It can be dehydrated without reverting to rhodium(III) only by heating under a high pressure of oxygen. Both elements form the tetrafluoride, MF_4, and $IrCl_4$ is also reported. Rhodium gives only the halogen complexes, RhF_6^{2-} and $RhCl_6^{2-}$, but a wider variety of iridium(IV) complexes occur. Fairly stable hexahalides, IrX_6^{2-}, are formed for X = F, Cl, and Br, and other complexes include the oxalate, $Ir(C_2O_4)_3^{2-}$, which can be resolved into optical isomers.

In the III state both elements form all four trihalides and the oxides, M_2O_3. Ir_2O_3 can only be formed, by addition of alkali to iridium(III) solutions, as a hydrated precipitate in an inert atmosphere. Attempts to dehydrate it lead to decomposition, and it oxidizes in air to IrO_2. Rhodium trifluoride results, along with some tetrafluoride, from fluorination of $RhCl_3$, but IrF_3 can only be prepared by reduction of IrF_6 with Ir, as fluorination of iridium(III) compounds gives the tetra- or hexa-fluoride. The other trihalides are made by direct reaction with halogen. They are insoluble in water and probably polymeric.

A considerable number of rhodium(III) complexes are known, and these are generally octahedral. The hydrate, $Rh(H_2O)_6^{3+}$, the ammine, $Rh(NH_3)_6^{3+}$, and a large variety of partially substituted hydrates and ammines are found. For example, hydrates containing 1, 2, 3, 4, and 5 chloride ions in place of water molecules are formed, as well as $RhCl_6^{3-}$ and other hexahalides. Other halide complexes include RhX_5^{2-}, and RhX_7^{4-}, especially for the bromides. An interesting example of a new type of metal-metal bond is provided by the recently discovered compound, $(R_3As)_3Rh^{III}(HgCl)^+Cl^-$, which contains the Rh—Hg bond. A similar compound is the dimeric iridium(III) species $[Ir_2Cl_6(SnCl_3)_4]^{4-}$ which contains Ir—Sn bonds. Rhodium also forms a compound with $SnCl_3^-$ as a ligand, this time in

the I state but again a dimer, $[Rh_2Cl_2(SnCl_3)_4]^{4-}$. In such compounds, the $SnCl_3^-$ species behaves rather like a halide ion. Such compounds with bonds from transition to non-transition metals serve to link the long-established Main Group metal-metal bonded compounds with the newer field of interest in bonds between transition metal atoms.

The complexes of rhodium(III) are relatively inert to substitution so that isomers may be isolated, as in the resolution into optical isomers of the cis-$Rh(en)_2Cl_2^+$ ion. There is also an extensive series of iridium complexes which are generally similar to the rhodium compounds. Iridium also gives the hydrides, $(R_3P)_3Ir(H)_n(Cl)_{3-n}$, for $n = 1$, 2, and 3. Rhodium forms only the analogue with $n = 1$.

These elements form only a few compounds in the II state. Simple compounds are restricted to polymeric $IrCl_2$ and the reported oxide, RhO, which is not well-established. There are also a few complexes such as $Ir(CN)_6^{4-}$ and $Ir(NH_3)_4Cl_2$.

Low oxidation state compounds include the carbonyl monohalides of the I state, the carbonyl anion, $Rh(CO)_4^-$, in the $-I$ state. These are paralleled by trifluorophosphine analogues like $KIr(PF_3)_4$ which is oxidized from the $-I$ state to the $+I$ state by iodine to yield $Ir(PF_3)_4I$. Compounds in the 0 state include $Ir(NH_3)_5$ and $Ir(en)_3$ which are formed by reduction in liquid ammonia, like the similar osmium compounds. The 0 state is also represented in the carbonyls which include $M_2(CO)_8$ and the interesting polymeric molecule $Rh_6(CO)_{16}$. This has a structure with an octahedron of rhodium atoms, each with two terminal carbonyl groups, and the other four carbonyl groups are found in the middle of opposite faces. The structure is held together by delocalized metal-metal bonding in the Rh_6 cluster.

Among the most fully explored reactions are those of the square planar compounds of the I state, which add neutral molecules to give octahedral compounds of the III state in the process which has been called *oxidative addition*. For example,

$$IrCl(CO)(PPh_3)_2 + XY = XIrY(Cl)(CO)(PPh_3)_2$$

where XY = HCl, CH_3I, Cl—HgCl, etc. Molecules such as H_2, O_2 or SO_2 are also taken up and may be lost again in a reversible reaction. While H_2 gives two M—H bonds, the O_2 molecule is not split and the two O atoms are cis to each other at a distance O—O = 130 pm, suggesting coordination as superoxide, O_2^-.

The I state is also found in five-coordinate compounds, usually involving phosphine ligands. The structure of $HRhCO(PPh_3)_3$ is trigonal bipyramidal, with the P atoms in the equatorial plane.

14.8 Palladium, $4d^{10}5s^0$, and platinum, $5d^96s^1$

This pair of elements continues the trends already observed. The higher oxidation states are unstable and the VI and V states are represented only by a few platinum compounds. The IV state is stable for platinum and well-represented for palladium. This corresponds to d^6 and many octahedral Pt^{IV} complexes occur with high CFSE. The II state is the other common oxidation state. In this palladium and platinum are almost invariably four-coordinated and square planar. Palladium continues the trend, which probably starts at technetium or molybdenum and is clearly seen for ruthenium and rhodium, in being less stable than the heaviest element of the Group in the higher oxidation states and in having a well-developed lower oxidation state.

Only platinum forms a hexafluoride, PtF_6, and attempts to isolate the palladium compound have failed. The VI state of platinum may also occur in the reported oxide, PtO_3. There is also a compound of unknown structure which may contain $Pt(VIII)$. This is $PtF_8(CO)_2$ which is reported to be formed by the reaction of CO under pressure on PtF_4. It is difficult to see why such a system should be oxidizing, but spectroscopic evidence shows no bridging carbonyl groups and no adduct molecules such as F_2CO.

The V state is found in PtF_5 and in the PtF_6^- ion. The latter was first found as a product of the reaction of O_2 and PtF_6 which yielded the unexpected oxygen cation in $O_2^+PtF_6^-$. The now famous first report of a rare gas compound was of the $Xe^+PtF_6^-$ complex and a number of other compounds of the anion have since been made.

The highest state for palladium, and the first stable state for platinum, is the IV state. This is represented by all four PtX_4 halides and by PtO_2. Palladium forms PdF_4 and PdO_2. The latter is found as the poorly characterized hydrated oxide but PtO_2 is the most stable oxide of platinum. It is obtained as a hydrated precipitate from the action of carbonate on Pt^{IV} solution and it is soluble in acid and alkali in this condition. It can be dehydrated by careful heating when it becomes insoluble. On heating to 200 °C, it decomposes giving platinum metal and O_2.

PtF_4 is the major product of fluorination of platinum, although PtF_5 and PtF_6 also result from the direct reaction. PdF_4 is also formed by direct fluorination though here the main product is PdF_3. The other platinum tetrahalides are formed by direct halogenation and $PtCl_4$ also results when H_2PtCl_6, the product from the aqua regia solution of the metal, is heated. These heavier halides of platinum are quite stable, even PtI_4 does not decompose until about 180 °C, when it goes to PtI_2 and iodine.

Palladium(IV) complexes are only a little more stable than the simple Pd^{IV} compounds. The common examples are the halides, PdX_6^{2-}, all of which are known except the iodide, and the tetrahalide amine, $Pd(amm)_2X_4$, where amm = ammonia, pyridine or related ligands.

By contrast, platinum(IV) forms a large number of complexes which are always octahedral, and are stable and inert in substitution reactions. As platinum(II) also gives a wide range of complexes, platinum is probably the most prolific complex-forming element of all. All the common types of complex are found, for example all members of the set between $Pt(NH_3)_6^{4+}$ and PtX_6^{2-} are known for a variety of amines as well as ammonia, and for X = halogen, OH, CNS, NO_2, etc. However, fluoride is found only in PtF_6^{2-}.

The Pt^{IV} octahedral complexes may readily be obtained from the Pt^{II} square planar complexes if the attacking ligand

is also an oxidizing agent. Thus $Pt^{II}L_4 + Br_2 \rightarrow$ *trans*-$Br_2Pt^{IV}L_4$ for a variety of ligands (L). The halogen atoms simply add on opposite sides of the square plane.

An interesting class of complexes is the deeply coloured, usually green, type of compound which apparently contains trivalent platinum, such as $Pt(en)Br_3$ or $Pt(NH_3)_2Br_3$. These are not Pt^{III} compounds at all but chains made up of alternate $Pt^{II}(NH_3)_2Br_2$ and $Pt^{IV}(NH_3)_2Br_4$ units as shown in Figure 14.24. The Pt—Br distances on the vertical axis are 250 pm for Pt^{IV}—Br and 310 pm for the weak Pt^{II}—Br interaction.

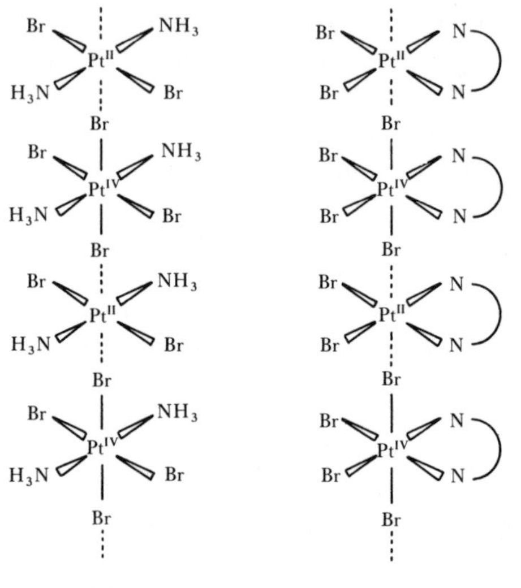

FIGURE 14.24 *Structures of* (a) $Pt(NH_3)_2Br_3$, (b) $Pt(en)Br_3$ These structures consist of chains of alternate square planar platinum(II) units, PtN_2Br_2, and octahedral platinum(IV) units, PtN_2Br_4. (N = NH_3 or half an **ethylenediamine molecule**).

The existence of compounds of these two elements in the III state is dubious. The well-known PdF_3 has recently been shown to be $Pd^{2+}PdF_6^{2-}$, a mixed-valency compound of palladium(II) and palladium(IV). It is likely that the reported platinum trihalides are similar. Hydrated oxides, M_2O_3, are reported but their identity is not proven.

Like nickel, these elements have a very stable II state, but occur only in the square-planar configuration in this state. All the dihalides except PtF_2 are found. Palladium forms a stable PdO while platinum gives a hydrated PtO which readily oxidizes to PtO_2. Palladium dichloride has a chain structure:

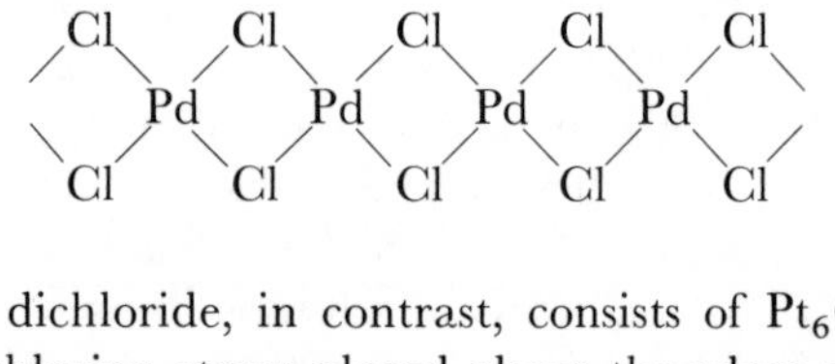

Platinum dichloride, in contrast, consists of Pt_6Cl_{12} units with the chlorine atoms placed above the edges of an octahedron of platinum atoms. This structure is reminiscent of the tantalum(II) and molybdenum(II) halide complex ions and is a further example of a metal cluster compound.

The complexes of palladium(II) and platinum(II) are abundant and include all common ligands. Some examples have already been given in Chapter 12 (see Figure 12.21). Palladium(II) complexes are a little weaker in bonding and react rather more rapidly than the platinum(II) complexes, but are otherwise very similar. The commonest donor atoms are nitrogen (in amines, NO_2), cyanide, the heavier halogens and phosphorus, arsenic, and sulphur. The affinity for F and O donors is much lower.

One or two examples do exist of coordination numbers other than four in the II complexes: for example, $Pd(diars)_2 Cl^+$ is a trigonal pyramid and $Pt(NO)Cl_5^{2-}$ is octahedral (though the latter might be formulated as an NO^- compound of Pt^{IV} rather than as the NO^+ compound of Pt^{II}). The $SnCl_3^-$ complexes also provide examples. Platinum(II) forms a square-planar ion $PtCl_2(SnCl_3)_2^{2-}$ (as does ruthenium) but it also gives the ion $Pt(SnCl_3)_5^{3-}$ which has a trigonal bipyramidal $PtSn_5$ skeleton. This species reacts with hydrogen under pressure to form the hydride ion $HPt(SnCl_3)_4^{3-}$. Palladium forms the ion $PdCl(SnCl_3)_4^{2-}$, but it is not yet known if this is a monomer or forms a chloride-bridged dimer which would be square-planar around the palladium. Such compounds are commonly precipitated as salts of very large cations (like Ph_4As^+) from acid chloride solutions containing palladium, or platinum, and tin. A number of compounds containing the analogous $M-Ge$ bond have also been reported such as $(R_3P)_2M(GePh_3)_3$ for $M = Pd$ or Pt.

The I state is represented by the reported $Pd_2(CN)_6^{4-}$ analogous to the nickel complex. There are no carbonyls in the 0 state analogous to $Ni(CO)_4$, although the isoelectronic $M(CN)_4^{4-}$ ions are reported. There are also reports of $Pt(NH_3)_5$ and $Pt(en)_2$ analogous to the iridium compounds. It has also been shown that the phosphine complexes, $(Ph_3P)_4Pt$ and $(Ph_3P)_3Pt$, are probably derivatives of Pt^0 rather than hydrides of higher states as originally reported. The structure of $Pt(PPh_3)_3$ has been shown to contain the planar PtP_3 group.

14.9 Silver $4d^{10}5s^1$, and gold, $5d^{10}6s^1$

The major part of silver chemistry is that of the d^{10} state, Ag(I). Gold is most often found in Au(III) square planar complexes, but there are also a variety of Au(I) compounds, mostly complexes. Simple compounds of gold readily give the element, and silver is also reduced to the element fairly readily. These elements differ quite widely in their chemistries, and also differ from copper which occurs in the II state. Reasons for this were discussed in the copper section.

The elements are found uncombined and in sulphide and arsenide ores. They may be recovered as cyanide complexes which are reduced to the metal, in aqueous solution, by the use of zinc. Gold is inert to oxygen and most reagents but dissolves in HCl/HNO_3 mixture (aqua regia) and reacts with halogens. Silver is less reactive than copper and is similar to gold, except that it is also attacked by sulphur and hydrogen sulphide.

It is easiest to treat the two elements separately as there is little resemblance in their chemistries.

There are one or two examples of Ag^{III} compounds. Fluorination of a mixture of alkali and silver halides gives M^IAgF_4 and oxidation in basic solutions gives $Ag(IO_6)_2^{7-}$ or $Ag(TeO_6)_2^{9-}$: copper(III) compounds are obtained similarly. Electrolytic oxidation appears to give an impure Ag_2O_3. All these compounds are unstable and strongly oxidizing.

When Ag_2O is oxidized by persulphate, black AgO is obtained. This is a well-defined compound which is strongly oxidizing. It is diamagnetic, which excludes its being an Ag^{II} compound which would contain the paramagnetic d^9 configuration. A neutron diffraction study has shown the existence of two units in the lattice, linear $O-Ag-O$ and square planar AgO_4 groups. This strongly suggests a formulation as $Ag^IAg^{III}O_2$. The square planar configuration is expected for the d^8 Ag^{III} while the linear one is common for d^{10} as in Ag^I (compare Ag^I with Cu^I or Au^I which both show linear configurations and compare the III state with Au^{III}). AgO is stable to heat up to 100 °C and it dissolves in acids giving a mixture of Ag^+ and Ag^{2+} in solution and evolving oxygen.

The only simple compound of silver(II) is the fluoride, AgF_2. This is formed from the action of fluorine on silver or AgF. It is a strong oxidizing or fluorinating agent and can be used in reversible fluorinating systems in a similar way to CoF_3.

Silver ions of the II state can exist in solution but only transiently. They are produced by ozone on Ag^+ in perchloric acid. The potential for $Ag^{2+} + e^- = Ag^+$ has been measured as 2·00 V in 4M perchloric acid. This makes Ag^{2+} a much more powerful oxidizing agent than permanganate or ceric. Silver also gives a number of complexes in the II state. These are usually square planar and paramagnetic corresponding to the d^9 state, and known structures, for example, of $Agpy_4^{2+}$, are isomorphous with the Cu^{II} analogues. Other examples include the cations $Ag(dipy)_2^{2+}$ and $Ag(phen)_2^{2+}$. These cations are stable in the presence of non-reducing anions such as nitrate, perchlorate or persulphate.

The normal oxidation state of silver is Ag^I and the chemistry of this state is already familiar, for example, from the precipitation reactions of Ag^+ in qualitative analysis. Salts are colourless (except for anion effects) and generally insoluble apart from the nitrate, perchlorate, and fluoride.

Silver (I) forms a wide variety of complexes. With ligands which do not π-bond, the most common coordination is linear, two-coordination as in $Ag(NH_3)_2^+$, while π-bonding ligands give both 2- and 4-coordinate complexes and 3-coordination is found for some strongly π-bonding ligands such as phosphines.

Gold(III) is the most common oxidation state of gold in compounds. The interrelations are illustrated in Figure 14.25, starting off from the solution in aqua regia. The simple compounds readily revert to the element but the

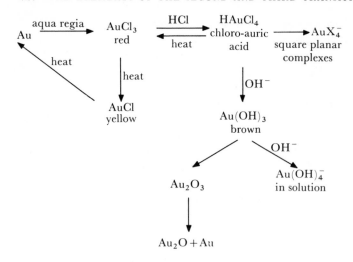

FIGURE 14.25 *Reactions of gold(III)*

complexes are more stable. Square planar four-coordination is the most common shape, but there is evidence for a few six-coordinate complexes such as $AuBr_6^{3-}$. Examples of the tendency to four-coordination are provided by the alkyls such as R_2AuX. These contain sigma $Au-C$ bonds which are among the most stable metal-carbon bonds in the transition block. The halides have dimeric structures:

$$\begin{array}{ccccc} R & & X & & R \\ & \diagdown & & \diagup & \\ & Au & & Au & \\ & \diagup & & \diagdown & \\ R & & X & & R \end{array}$$

while, when $X = CN$, tetrameric structures are found:

$$\begin{array}{ccc} R_2Au - CN - AuR_2 \\ | & & | \\ \begin{pmatrix} C \\ N \end{pmatrix} & & \begin{pmatrix} C \\ N \end{pmatrix} \\ | & & | \\ R_2Au - CN - AuR_2 \end{array}$$

Gold(I) is less stable but is represented by the halides and Au_2O and also by complexes which are usually linear. Examples include $Au(CN)_2^-$, and similar halides, as well as complexes like R_3PAuX. Among the latter are alkyls of Au^I, such as R_3PAuCH_3.

Some apparent gold(II) compounds exist, such as $CsAuCl_3$. These are all mixed compounds of Au^I and Au^{III}; e.g. $[(Au^ICl_2)(Au^{III}Cl_4)]^{2-}$. The recently isolated complexes $Au_5(PPh_3)_4Cl$ are thought to contain four Au^0 atoms and one Au^I atom. The structure is not established.

14.10 The zinc group

The Zinc Group does not fit the general picture of the transition Group, as developed in the last two chapters. It shares with beryllium the property of belonging to one block of the Periodic Table and having many of the properties characteristic of another. In this case, the three elements of

this Group resemble the three heavy elements of the Boron Group.

The elements, zinc, cadmium, and mercury, in this Group have the outer electronic configuration $d^{10}s^2$ and have the common oxidation state of II, corresponding to the loss of the two s electrons. In addition, mercury shows a well-established I state and cadmium and zinc form analogous but very unstable I compounds. Thus the heaviest element is more stable than the lighter ones in a low oxidation state, a characteristic of the Main Groups.

The persistence of the lanthanide contraction into this Group is in some doubt. The radii of the covalent species, and the set of ionic radii given by Goldschmidt, indicate that the lanthanide contraction persists, while the ionic radii given by Pauling show a similar increase in size between Cd^2 and Hg^{2+} as between Y^{3+} and La^{3+}:

	Zn	Cd	Hg	
covalent radii (pm)	125	141	144	
ionic radii (pm) Pauling	74	97	110	for M^{2+}
Goldschmidt	69	92	93	

The chemical behaviour does not support the close similarity in the values given for Cd^{2+} and Hg^{2+} by Goldschmidt, and the M^{II}/M redox potentials show a large difference between Cd and Hg which can be ascribed in part to the higher solvation energy of Cd^{2+}, which is an effect of smaller size. The potential values for $M^{2+} + 2e^- = M$ are:

$$Zn = -0.762 \text{ V}, \quad Cd = -0.402 \text{ V}, \quad Hg = 0.854 \text{ V}$$

These values show the relatively high electronegativity of zinc and cadmium and reflect the reducing power of these elements. By contrast, mercury is unreactive and 'noble'.

The elements are all readily accessible as they occur in concentrated ores and are easily extracted. Zinc and cadmium are formed by heating the oxides with carbon and distilling out the metal (boiling points are 907 °C for zinc and 767 °C for cadmium). Mercury(II) oxide is decomposed by heating alone, without any reducing agent, at about 500 °C, and the mercury distills out (b.p. = 357 °C). Mercury is the lowest melting metal with a melting point of −39 °C, while zinc and cadmium melt at about 420 °C and 320 °C respectively. Mercury is monatomic in the vapour, like the rare gases. The element, and many of its compounds, are very poisonous and, as mercury has a relatively high vapour pressure at room temperature, mercury surfaces should always be kept covered to avoid vaporization.

In the I state, mercury exists as the dimeric ion, Hg_2^{2+}. This has been demonstrated by a number of independent lines of evidence:

(i) The Raman spectrum of aqueous mercurous nitrate shows an absorption attributed to the $Hg-Hg$ stretching vibration.
(ii) The crystal structures of mercury(I) salts show the existence of discrete Hg_2 units: see Figure 14.26 for the case of Hg_2Cl_2.
(iii) All mercury(I) compounds are diamagnetic, whereas an Hg^+ ion would have one unpaired electron.
(iv) E.m.f. measurements on concentration cells of mercurous

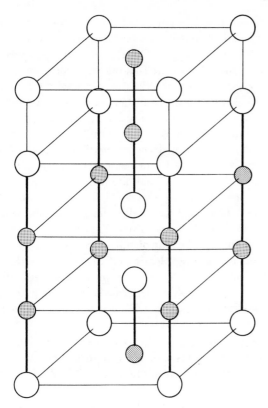

tetragonal unit cell of Hg_2Cl_2

○ Cl   ● Hg

FIGURE 14.26 *The crystal structure of mercurous chloride*, Hg_2Cl_2

salts show that two electrons are associated with the mercurous ion. For example, the e.m.f. of the cell:

| Hg | mercury(I) nitrate in 0·1M HNO₃ | (0·05M) | mercury(I) nitrate in 0·1M HNO₃ | (0·005M) | Hg |

was found to be 0·029 V at 25 °C. Now $E = \dfrac{RT}{nF}\ln(c_1)/(c_2)$,

where the concentrations are c_1 and c_2, n is the number of electrons, and the other constants have their usual significance. If values are put in for the constants and conversion to logarithms to the base ten is carried out, the equation becomes $E = (0·059/n) \log (0·05/0·005)$: hence $n = 2$.

(v) Conductivities and equilibrium constants also fit for a dimeric ion of double charge and not for Hg^+.

The redox potentials involving mercury(I) are

$$Hg_2^{2+} + 2e^- = 2Hg_{(liquid)} \quad E = 0·789 \text{ V}$$
$$2Hg^{2+} + 2e^- = Hg_2^{2+} \quad E = 0·920 \text{ V}$$

It follows that, for the disproportionation reaction:

$$Hg + Hg^{2+} = Hg_2^{2+},$$

the potential is $E = 0·131$ V. Then, as $E = (RT/nF) \ln K$.

where K is the equilibrium constant for the reaction:

$$K = [(Hg_2^{2+})]/[(Hg^{2+})] = \text{about } 170$$

In other words, in a solution of a mercury(I) compound there is rather more than $\frac{1}{2}$ per cent mercury(II) in equilibrium. Thus, if another reactant is present which either forms an insoluble mercury(II) salt or a stable complex, the mercury(II) ions in equilibrium are removed and the disproportionation goes to completion. If OH^- is added to a solution of Hg_2^{2+}, a grey precipitate of HgO mixed with Hg is formed. Similarly, sulphide precipitates HgS and CN^- ions give a precipitate of mercury and the undissociated cyanide of mercury(II), $Hg(CN)_2$.

The existence of this disproportionation reaction means that a number of mercury(I) compounds, such as the sulphide, cyanide or oxide, do not exist. The main compounds of mercury(I) are the halides and a number of salts. Hg_2F_2 is unstable to water giving HF and HgO plus Hg. The other halides are very insoluble. Mercurous nitrate and perchlorate are soluble and give the insoluble halides, sulphate, acetate and other salts by double decomposition reactions. The nitrate and perchlorate are isolated as hydrates which contain the hydrated ion $[H_2O-Hg-Hg-H_2O]^{2+}$.

The existence of a cadmium(I) ion, Cd_2^{2+}, has been conclusively proved in a molten halide system. $CdAlCl_4$ is obtained as a yellow solid and shown to be $Cd_2^{2+}(AlCl_4)_2^-$. This reacts violently with water to give cadmium metal and Cd^{2+} in solution. Cd^I may also exist in the deep red melts of Cd in cadmium(II) halides. The Cd_2^{2+} ion is quite unstable to water. A recent report indicates that zinc(I), Zn_2^{2+}, exists under similar conditions.

The most stable state for all three elements is the II state. In this, zinc and cadmium resemble magnesium and many of the compounds are isomorphous. Mercury(II) compounds are less ionic and its complexes are markedly more stable than those of zinc and cadmium. All three elements resemble the transition elements more than the Main Group elements in forming a large variety of complexes.

The halides of all three elements are known. All the fluorides, MF_2, are ionic with melting points above 640 °C. HgF_2 crystallizes in the fluorite lattice and is decomposed in contact with water. The structures of ZnF_2 and CdF_2 are unknown: these compounds are stable to water and are poorly soluble, due both to the high lattice energy and to the small tendency of the fluorides to form complex ions in solution. In this the zinc and cadmium fluorides resemble the alkaline earth fluorides. The chlorides, bromides, and iodides of zinc and cadmium are also ionic, although polarization effects are apparent and they crystallize in layer lattices. The structures are approximately close-packed arrays of the anions with Zn^{2+} in tetrahedral sites in the zinc halides while, in the cadmium halides, Cd^{2+} ions occupy octahedral sites. The zinc and cadmium halides have lower melting points than the fluorides and are ten to thirty times more soluble in water. This is due, not only to lower lattice energies, but also to the ready formation of complex ions in solution. A variety of species result, especially in the case of cadmium halides. Thus a 0·5M solution of $CdBr_2$ contains

Cd^{2+} and Br^- ions and, in addition, $CdBr^+$, $CdBr_2$, $CdBr_3^-$ and $CdBr_4^{2-}$ species, the most abundant being $CdBr^+$, $CdBr_2$ and Br^- (these species are probably hydrated). Hydrolysis also occurs and species such as $Cd(OH)X$ are observed. The tetrahalides, ZnX_4^{2-} and CdX_4^{2-} may be precipitated from solutions of the halides in excess halide by large cations. These are tetrahedral ions, as are all four-coordinated species in this group.

By contrast, $HgCl_2$, $HgBr_2$, and HgI_2 are covalent solids melting and boiling in the range 250 °C–350 °C. $HgCl_2$ is a molecular solid with two $Hg—Cl$ bonds of 225 pm and the next shortest $Hg—Cl$ distance equal to 334 pm, so that there is little interaction between the mercury atom and these external chlorines. In the bromide and iodide, layer lattices are formed, but in the bromide, two $Hg—Br$ distances are much shorter (248 pm) than the rest (323 pm) so that this is a distorted molecular lattice. HgI_2 forms a layer lattice in which there are HgI_4 tetrahedra with the $Hg—I$ distance equal to 278 pm. In the gas phase, the mercury halogen distances in the isolated HgX_2 molecules are, for $X = Cl$, 228 pm; $X = Br$, 240 pm; $X = I$, 275 pm. Thus the $Hg—Cl$ distances are the same in the solid and gas, underlining the molecular form of the solid. The $Hg—Br$ distance is a little longer in the solid while the $Hg—I$ distance is markedly longer in the solid, showing the increasing departure from a purely molecular solid on passing from the chloride to the iodide. Mercury also forms halogen complexes, and the same species are found for mercury as for cadmium. The stability constants for the mercury complexes are much higher than those for the zinc and cadmium species.

The oxides are formed by direct combination. ZnO is white and turns yellow on heating. CdO is variable in colour from yellow to black. The colours in both cases are due to the formation of defect lattices, where ions are displaced from their equilibrium positions in the crystal lattices to leave vacancies. These may trap electrons whose transitions give rise to colours in the visible region. HgO is red or yellow, depending on the particle size. Zinc oxide and the hydroxide are amphoteric. $Zn(OH)_2$ is precipitated by the addition of OH^- to zinc solutions and dissolves in excess alkali. $Cd(OH)_2$ is precipitated similarly but is not amphoteric and remains insoluble in alkali. Mercury(II) hydroxide does not exist; the addition of alkali to mercuric solutions gives a precipitate of the yellow form of HgO. The elements all form insoluble sulphides and these are well-known in qualitative analysis. ZnS is somewhat more soluble than CdS and HgS, and has to be precipitated in alkali rather than the acid conditions under which yellow CdS and black HgS precipitate.

Most of the oxygen Group compounds of Zn, Cd, and Hg have the metal in tetrahedral coordination in the zinc blende structure (see Figure 4.1c) or in the related wurtzite structure (Figure 4.4a). Both these are found for ZnS, with the wurtzite form the stable one at high temperatures. ZnO, ZnS, $ZnTe$, CdS, and $CdSe$ all occur in both the wurtzite and the zinc blende forms. $ZnSe$, $CdTe$, HgO, $HgSe$ and $HgTe$ are found in the zinc blende form only. HgS occurs in two forms, one is zinc blende and the other is a distorted $NaCl$ lattice. The only other example of six-coordination is CdO which forms a sodium chloride lattice. These structures again illustrate the strong tendency for these elements to form tetrahedral coordination.

This formation of four and six-coordinated species is also found in the complexes of these elements. Zinc occurs largely in four-coordination in complexes like $Zn(CN)_4^{2-}$ or $Zn(NH_3)_2Cl_2$ and is also found in rather unstable six-coordination, as in the hexahydrate and the hexammine, $Zn(NH_3)_6^{2+}$. Cadmium forms similar four-coordinated complexes but is rather more stable than zinc in six-coordination, due to the larger size. Mercury is commonly found in four-coordination, though a few octahedral complexes such as $Hg(en)_3^{2+}$ are also found.

All three elements are also found in linear two-coordination especially in their organo-metallic compounds, and in the halides and similar compounds. The organic compounds R_2M (M = Zn, Cd or Hg) are well-known and mercury also forms $RHgX$ compounds with halides. The so-called $RZnX$ and $RCdX$ species are, like the Grignard reagents, $RMgX$, more complex and their structures are not fully understood. They are polymeric with some evidence for MX_2 and MR_2 groups and are usually coordinated by the ether used in their preparation.

15 The Elements of the 'p' Block

15.1 Introduction and general properties

Those elements which have their most energetic electron in a p orbital lie in the Main Groups of the Periodic Table headed by B, C, N, O, F, and He. Some aspects of the chemistry of these elements have been discussed in the earlier chapters. The hydrides of the p elements are included in Chapter 8, the structures are discussed in Chapter 3, solid state structures, especially of the silicates, are treated in Chapter 4, while aspects of aqueous and non-aqueous chemistry, including properties of oxyacids, are covered in Chapter 5. There is now so much information available that it is impossible to give anything like a full treatment of these elements and the aim in this Chapter is limited to giving a valid general picture of the chemistry, illustrated by the properties of the more common compounds, and to discuss some of the more striking recent advances in the field. The choice of the latter topics is necessarily somewhat arbitrary and further information will be found in any of the standard textbooks quoted in the references.

The maximum oxidation state shown by a p element (the 'Group oxidation state'), is equal to the total of the valency electrons, i.e. to the sum of the s and p electrons, and is the same as the Group Number in the Periodic Table. In addition to this oxidation state, p elements may show other oxidation states which differ from the Group state by steps of two. Clearly, the number of possible oxidation states increases towards the right of the Periodic Table. The most important oxidation states in the various Groups of p elements are shown in Table 15.1. Where oxidation states other than these occur, they usually arise either from multiple bonding, as in the nitrogen oxides, or from the existence of element-element bonds as in hydrazine, H_2N-NH_2, ($N = -II$) or disilane, $H_3Si-SiH_3$, ($Si = III$).

As fluorine is the most electronegative element, it can show only negative oxidation states and always exists in the $-I$ state. Oxygen is also always negative, except in its fluorides, and is found in the $-I$ state in peroxides (due to the $O-O$ link) and in the $-II$ state in its general chemistry. The other

TABLE 15.1 Oxidation states among the p elements

Group headed by	B	C	N	O	F	He
Group oxidation state	III	IV	V	VI	VII	VIII
Other states	I	II	III	IV	V	VI
		$(-IV)$*	I	II	III	IV
			$-III$	$-II$	I	II
					$-I$	(I)†

*As the electronegativity of C lies between that of H and those of O, N, or halogen, carbon in CH_4 is $-IV$ while carbon in CF_4 is IV, and all intermediate cases occur. The cases of Si, Sn, and Pb are simple, as these elements are in the IV state in their hydrides, but the case of Ge is uncertain, as its electronegativity is the subject of dispute and is approximately that of hydrogen, so Ge in GeH_4 may be in the $-IV$ state. In general chemistry, apart from germanium hydrides and carbon compounds, the carbon Group elements show the II, and IV states only.
†In $XePt^VF_6$ and similar compounds.

highly electronegative elements, nitrogen, sulphur, and the halogens also show stable negative oxidation states in which they form anions, hydrides, and organic derivatives.

In the boron, carbon, and nitrogen Groups, the Group oxidation state is the most stable state for the lighter elements in the Group while the state two less than the Group oxidation state is the most stable one for the heaviest element in each Group. The relative stabilities of these two states varies down the Group; thus, in the carbon Group, lead(II) is stable and lead(IV) is strongly oxidizing, tin(II) and (IV) are about equal in stability, germanium(II) is represented by a handful of compounds only, and germanium(IV) is the stable state, while silicon shows only the IV state.

In the oxygen and fluorine Groups the position is more complicated because of the wider ranges of oxidation states. Oxygen and the halogens are most stable in the negative states. The VI state is stable for sulphur and falls in stability for the heavier elements of the oxygen Group, but both the IV and the II states are relatively stable for the heavy elements. Among the halogens, chlorine and iodine both show the Group state of VII in oxyions, and bromine(VII) was recently found. Iodine is also fairly stable in the V and III states. Thus, the general trend for the lower oxidation states to be more stable for the heavier elements is shown by elements such as iodine or tellurium, but more than one state is involved and no simple rule may be given. The most stable rare gas compounds are the tetrafluorides, which probably exist for radon and krypton as well as for xenon. The other oxidation states in Table 15.1 are shown by xenon.

The general pattern of oxidation state stabilities in the Main Group elements may be summarized: the common oxidation states vary in steps of two, with states lower than the Group state being the most stable for the heavier elements. This trend is opposite to that found among the transition elements.

This pattern of stabilities is reflected in the oxides, sulphides, and halides formed by the Main Group elements, as given in Tables 15.2a and b. A stable oxidation state will form the oxide, sulphide, and all four halides, while a state which is unstable and oxidizing, such as the Group states of the heavy elements, thallium, lead, and bismuth, will not form compounds of the readily oxidizable sulphide or heavy halide ions. Similarly, an oxidation state which is unstable and reducing, such as gallium(I), will form the sulphide and heavier halides, but not the oxide or fluoride.

A similar picture of the stabilities of the oxidation states, but this time in solution, is given by the oxidation state free energy diagrams of Figures 15.5, 15.18, 15.37 and 15.47. These diagrams show how the Group oxidation states become unstable with respect to lower oxidation states for the heavier elements in each Group, and they also indicate the instability, with respect to disproportionation, of the intermediate oxidation states in Groups, such as the halogens, which show a number of states.

The effects shown in Tables 15.2 and 15.3 and in the oxidation state free energies are also reflected in the ionization potentials (Table 2.8), the electron affinities (Table 2.9), and in the electronegativities (Table 2.13), of the p elements. It

TABLE 15.2a Oxides and sulphides of p elements

Elements in oxidation state = number of valency electrons

B_2O_3	B_2S_3	CO_2	CS_2	N_2O_5		
Al_2O_3	Al_2S_3	SiO_2	SiS_2	P_4O_{10}	P_4S_{10}	SO_3
Ga_2O_3	Ga_2S_3	GeO_2	GeS_2	As_2O_5	As_2S_5	SeO_3
In_2O_3	In_2S_3	SnO_2	SnS_2	Sb_2O_5	Sb_2S_5	TeO_3
Tl_2O_3		PbO_2		$(Bi_2O_5?)$		$(PoO_3?)$

Elements in an oxidation state lower by two

		CO		N_2O_3		
				P_4O_6		SO_2
	Ga_2S	GeO	GeS	As_4O_6	As_4S_6	SeO_2
$In_2O?$	In_2S	SnO	SnS	Sb_4O_6*	Sb_2S_3	TeO_2
Tl_2O	Tl_2S	PbO	PbS	Bi_2O_3	Bi_2S_3	PoO_2

Elements in an oxidation state lower by four

		N_2O		
				TeO
				PoO PoS

Other compounds

$(BO)_x$	C_3O_2	$NO, NO_2, N_2O_4,$	
		$(NO_3$ or $N_2O_6)$	
		N_4S_4	
		(PO_2)	S_2O
		P_4S_3, P_4S_5, P_4S_7	
		As_4S_3, As_4S_4	
		(SbO_2)	
	Pb_3O_4		

*Antimony trioxide also occurs as $(Sb_2O_3)_n$ chains.

TABLE 15.2b Halides of the elements

Fluorides of the Groups headed by				Chlorides of the Groups headed by				Bromides of the Groups headed by				Iodides of the Groups headed by			
B	C	N	O	B	C	N	O	B	C	N	O	B	C	N	O

elements in oxidation state = Number of valency electrons

MF_3	MF_4	MF_5	MF_6	MCl_3	MCl_4	MCl_5	MCl_6	MBr_3	MBr_4	MBr_5	MBr_6	MI_3	MI_4	MI_5	MI_6
B	C			B	C			B	C			B	C		
Al	Si	P	S	Al	Si	P		Al	Si	P		Al	Si		
Ga	Ge	As	Se	Ga	Ge			Ga	Ge			Ga	Ge		
In	Sn	Sb	Te	In	Sn	Sb		In	Sn			In	Sn		
Tl	Pb	Bi	Po?	(Tl)	(Pb)			(Tl)							

elements in oxidation state = Two less than the number of valency electrons

MF	MF_2	MF_3	MF_4	MCl	MCl_2	MCl_3	MCl_4	MBr	MBr_2	MBr_3	MBr_4	MI	MI_2	MI_3	MI_4
		N				(N)				(N)				(N)	
		P	S			P	(S)			P				P	
	Ge	As	Se	Ga?	Ge	As	Se	Ga?	Ge	As	Se	Ga?	Ge	As	
	Sn	Sb	Te	In	Sn	Sb	Te	In	Sn	Sb	Te	In	Sn	Sb	Te
Tl	Pb	Bi	Po	Tl	Pb	Bi	Po	Tl	Pb	Bi	Po	Tl	Pb	Bi	Po

Other compounds

MF	MF_2	MF_3	MF_4	MCl	MCl_2	MCl_3	MCl_4	MBr	MBr_2	MBr_3	MBr_4	MI	MI_2	MI_3	MI_4
B_2F_4		N_2F_2	OF_2	B_2Cl_4											
		N_2F_4	O_2F_2	B_4Cl_4											
		NF_2	O_3F_2	B_8Cl_8											
			O_4F_2												
	Si_2F_6	S_2F_2 S_2F_{10}		Si_nCl_{2n+2} (up to $n=10$)		P_2Cl_4	S_nCl_2 (up to n ca.100)		Si_2Br_6	S_2Br_2			Si_2I_6	P_2I_4	
Ga_2F_4		Se_2F_{10}		Ga_2Cl_4	Ge_2Cl_6		$SeCl_2$ Se_2Cl_2	Ga_2Br_4			$SeBr_2$ Se_2Br_2	Ga_2I_4			
InF_2?		Te_2F_{10}		In_2Cl_3			$TeCl_2$				$TeBr_2$				Sb_2I_4?
				BiCl*			$PoCl_2$				$PoBr_2$	$Tl^I(I_3)$			

*See section 15.6 for a discussion of this compound.

will be seen that the five elements in a p Group, although they have many properties in common, split into three sets when their detailed chemistry is examined. This division is:

(i) the lightest element
(ii) the three middle elements
(iii) the heaviest element

The lightest element shows the most marked differences from the rest of the Group, in properties which are discussed in detail in the next section. The heavier elements are discussed in the following section; the division between the heaviest element and the rest is not so marked as that between the first element and the Group, but it is quite distinctive and has been noted above in the stability of the lower oxidation states. It might also be noted that the middle element in the Group does not always fit between the second and fourth elements (compare electronegativities and chemical evidence such as the non-existence of $AsCl_5$), but these deviations are minor and there is a fairly regular trend of properties among the four heavy elements in each Group.

It is useful here to summarize some bond energies of the

p elements and hydrogen. Two different types of bond energy are encountered. First, if the bond between each pair of atoms in most molecules is independent of the rest of the molecule, we might hope to compile a list of the energies of individual bonds which would allow us to predict the energy of formation of the molecule. For example, the energy of the process of forming SiH_3F from its atoms

$$Si + F + 3H = SiH_3F$$

should be given by three times the $Si-H$ energy plus the $Si-F$ energy. Bond energies determined in this way are given the symbol E and are written $E(Si-H)$, etc. Experimentally, such energies are obtained from heats of formation and heats of atomization.

It is also possible to determine the energy required to break a particular bond, leaving two fragments in their ground state and this is called the bond dissociation energy and given the symbol D. Thus $D(HO-OH)$ is the energy required to break the $O-O$ bond in hydrogen peroxide to give two OH radicals. In general, D and E values differ, and D values for successive steps also differ: thus $D(SiH_2F-H)$ is not the same as $E(Si-H)$, or $D(SiHF-H)$ or $D(SiF-H)$. We can also define the average value of the $Si-H$ dissociation energy in SiH_3F, written $\bar{D}$, and this is usually fairly close in value to the corresponding E value. Clearly, for a diatomic molecule, D and E values are equal.

In Table 15.3 are listed bond energies for some diatomic molecules, which may also be used as E values for these bonds, together with a set of other E values chosen to reasonably reproduce known heats of formation. Also included are a few D values to emphasize the difference between the two. For prediction of heats of formation, the E values should be used, but the D values give a better indication of relative

TABLE 15.3 Some bond energies kJ mol^{-1}

(a) From diatomic molecules

H—H	436·0	H—F	640	F—F	157·8	F—Cl	257
O=O	497·3	H—Cl	431·4	Cl—Cl	242·1	F—Br	238
N≡N	948·9	H—Br	365·7	Br—Br	192·8	Cl—Br	222
C≡O	1074	H—I	298·7	I—I	150·9	Cl—I	211
NO	626·3					Br—I	199

(b) *E* values from polyatomic molecules

C—C	348	N—N	161	O—O	139	C—H	413	C—F	439
Si—Si	297*	P—P	215	S—S	213	Si—H	318	Si—F	565
Ge—Ge	260*	As—As	134	Se—Se	184	Ge—H	289	C—Cl	328
Sn—Sn	240*	Sb—Sb	126	Te—Te	138	Sn—H	251	Si—Cl	381
Si—C	301	Bi—Bi	105	O—F	185	N—H	391	Ge—Cl	339
Ge—C	270*	N—F	270	O—Cl	203	P—H	320	Sn—Cl	318
Sn—C	226	N—Cl	200	S—Cl	250	As—H	245	C—Br	272
C—N	292	P—F	490	S—Br	212	O—H	463	Si—Br	310
C—P	264	P—Cl	331			S—H	339	Ge—Br	276
C—O	343	P—Br	274	O—Si	369	Se—H	277	Sn—Br	272
C—S	259	P—I	215			Te—H	241	C—I	213
		As—F	466	S—Si	227			Si—I	234
		As—Cl	288					Ge—I	213
		As—Br	236					Sn—I	272
		As—I	174						

multiple bonds									
		C=C	619	C=N	615	C≡C	812	C≡O	1071
		N=N	418	C=O	707	N≡N	946	C≡N	879
		O=O	402	C=S	477				

(c) Some bond dissociation energies, *D*

O—ClO	243	O=CO	536	$F_3P=O$	544	H—CH$_3$	435
HO—Br	239	$H_2C=CH_2$	699	$Cl_3P=O$	511	H—NH$_2$	431
O—NO	306	HC≡CH	963	$Br_3P=O$	498	H—OH	498
O—NN	167	H—CN	540			H—CF$_3$	444
HO—OH	213	NC—CN	607			CH$_3$—F	452
$F_2N—NF_2$	83						

*These values are still being reassessed. The moderately high values for Group IV elements given here are superseding the older, much lower, values.

bond strengths. Note that values for first row elements forming single bonds, for example the X_2 energies for the halogens, are often unusually small compared with the heavier members. This is another effect distinguishing the lightest elements.

15.2 The first element in a *p* Group

In a *p* Group, the first element differs from the remaining members in two major respects. First, the size—and all properties which depend on size—changes most sharply between the first and second elements of a Group. Thus the lightest *p* elements show the same kind of differences as the lightest *s* elements, lithium and beryllium. The second important difference, which applies only to the *p* elements, arises from the effects of *d* orbitals in the valency shells of the heavier elements, and their lack in the first row elements.

The valency shell configuration of the first row elements of the *p* Groups is $2s^2 2p^n$, and the orbital next in energy is the $3s$ level. This is separated from the $2p$ level by a considerable energy gap and is not used in bonding by these elements. The first row elements are accordingly restricted to a maximum coordination number of four (using the $2s$ and the three $2p$ orbitals). In contrast, the second element of a *p* Group, with the configuration $3s^2 3p^n$, has the $3d$ orbitals lying between the $3p$ level and the $4s$ level in energy (compare Figure 7.2). The second row elements make use of these *d* orbitals to expand their coordination numbers above four. Compare, for example, the fluorine complexes of boron and aluminium where boron can form only BF_4^- while aluminium gives the AlF_6^{3-} ion. Similar behaviour is shown by the third and subsequent members of each *p* Group, which also have *d* orbitals available in the valence shell.

The presence of these *d* orbitals affects the chemistry of the heavier elements in a number of ways. The effect on co-ordination number is one of these, the others include the marked differences in stability between carbon and silicon compounds, and lesser but similar effects in the other Groups, and also effects on π bonding. These effects are discussed below. Although the *d* orbitals are available to the higher elements in the *p* Groups, it will be recalled from Figure 7.4 that there is a significant energy gap between the np and nd orbitals at the atomic number values corresponding to *p* elements. This means that strong bonds have to be formed before it is energetically favourable to use the *d* orbitals to attain high coordination numbers. Thus fluorine is more likely to form molecules or complexes with high coordination numbers than are the heavier halogens. Table 15.2b presents a number of instances, for example in the formation of the pentafluoride by all the heavier elements of the nitrogen Group, while only a limited number of pentachlorides exist and only one pentabromide. Nitrogen itself is prevented from forming any pentahalide by its lack of *d* orbitals, even though the V state is perfectly stable for nitrogen in species such as the nitrates.

A very striking example of differences in behaviour between first and second elements is provided by the behaviour of the hydrides and halides of the carbon Group. Carbon

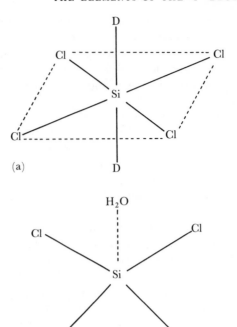

FIGURE 15.1   (a) *Structure of* $SiCl_4 2D$, (b) *possible structure for intermediate in hydrolysis of* $SiCl_4$

compounds, such as CH_4 and CCl_4, are stable in air and react only under vigorous conditions. In contrast, silicon hydrides inflame or explode in air and the silicon tetrahalides are hydrolysed violently on contact with water. Some part of this difference, particularly in the case of the hydrides, may be ascribed to the high $Si-O$ bond energy, but this cannot be the sole explanation. For example, the heats of the reaction:

$$MCl_4 + 2H_2O = MO_{2(aq)} + 4HCl$$

are exothermic and very similar in the two cases, $M = C$ and $M = Si$. A simple explanation of the difference arises if the mechanism of the reaction is examined. Carbon tetrachloride can react with water only if the strong $C-Cl$ bond is first broken, or considerably weakened, as there is no other way in which the water molecule can coordinate to the carbon. In contrast, silicon tetrachloride may be readily attacked by water if the $3d$ orbitals are used to expand the silicon co-ordination number above four, to give a reaction intermediate such as $Si(OH_2)Cl_4$ which then loses HCl. In support of this theory, silicon tetrachloride is known to form complexes, $SiCl_4.D$ or $SiCl_4.D_2$, with donor ligands, such as amines or pyridine (see Figure 15.1). Thus, the great difference in reactivity between carbon and silicon compounds may be ascribed, in considerable part, to the availability of a low-energy reaction path which makes use of the silicon *d* orbitals. Similar effects are probably present in the chemistry of the other *p* Groups, but, as boron has an empty *p* level, and nitrogen, oxygen, and fluorine have lone pairs, in their compounds other means for providing low-energy reaction intermediates are present, so differences are less marked in these Groups.

Both effects, of size and of availability of *d* orbitals, combine to effect considerable differences between first row elements and their congeners when π bonds are formed. The first row elements differ from the heavier *p* elements in their ability to form double bonds using only *p* orbitals (named $p_\pi - p_\pi$ bonds). One striking example is provided by carbon and silicon dioxides. Carbon dioxide is a volatile, monomeric molecule in which all the valency electrons are used in forming σ and π bonds between carbon and oxygen (section 3.9). In silicon dioxide, π-bonding between the silicon and oxygen does not occur. Instead, each silicon forms a single bond to four oxygen atoms, and each oxygen links two silicon atoms, to form a giant molecule with a three-dimensional structure of single $Si - O$ bonds.

Other examples are provided by the oxides of other first row elements, given in Table 15.2a. In each case (compare, for example, the electronic structure of NO in Figure 3.19), the oxide of the first row element, is a small molecule with $p_\pi - p_\pi$ bonds—CO, CO_2, NO, NO_2, N_2O_5, etc. The elements of the second, and subsequent, rows in the Main Groups rarely form $p_\pi - p_\pi$ bonds, probably because their greater size leads to a much weaker π overlap between *p* orbitals. The situation is illustrated diagrammatically in Figure 15.2 which shows that the sideways extension of the 3*p* orbital is insufficient to give good π overlap in association with the longer σ bond formed by the larger second row atom.

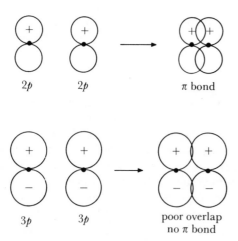

$$2p \qquad 2p \qquad \pi \text{ bond}$$

$$3p \qquad 3p \qquad \text{poor overlap}$$
$$\text{no } \pi \text{ bond}$$

FIGURE 15.2 *The poor overlap of 3p and higher p orbitals in π-bonding*
As the second row elements are larger than those of the first row, the sigma bond formed between them is longer. Thus the sideways extension of the 3*p* orbital is not sufficiently large to give good π overlap.

A second factor in this contrast between the tendency of first and second row elements to form $p_\pi - p_\pi$ bonds is the strength of the homonuclear *single* bonds of many of the first row elements. For example, in Table 15.3 it may be seen that the bond energies of $F - F$, $O - O$ or $N - N$ are distinctly smaller than, respectively, $Cl - Cl$, $S - S$ or $P - P$. It has been

suggested that this reflects repulsions between the nonbonding electrons on these atoms when they are brought close enough together to form the single bond. (Although the doubly-bonded atoms are even closer, some of the electrons are now used in the π bond.) Thus, it seems that there are two factors to be balanced, first that the $p_\pi - p_\pi$ bond is stronger for the first row elements, and second that the sigma bond may be weaker. Thus these elements are found to form a sigma and pi bond between them, rather than using the second orbital and electron to form a second sigma bond to a third atom. For the heavier element the converse case holds, and it is more stable forming two different sigma bonds rather than one sigma and one pi bond between the same pair of atoms.

The heavier elements do form π bonds, as in the oxyions, but this involves *d* orbitals (in $d_\pi - d_\pi$ or $d_\pi - p_\pi$ bonds). As the *d* orbitals have a considerable sideways extension, they can overlap with π orbitals on a second atom to form π bonds. However, as the *d* orbitals are of higher energy than the *p* orbitals, *d* orbitals are used in π bonding mainly to active ligands like oxygen, and π bonds involving *d* orbitals probably add less on average to the overall stability of molecules than does the $p_\pi - p_\pi$ bonding of the first row elements. As the second, and heavier, elements of a Group may use the *d* orbitals to contribute to π bonding, coordination numbers in oxyions are usually higher for these elements than for the oxyion of the first element in the same oxidation state. Thus, in the V state in the nitrogen Group, the ions are NO_3^- (with π bonding involving one nitrogen *p* orbital), PO_4^{3-}, and AsO_4^{3-} (four-coordination involving the *s* and three *p* orbitals and the *d* orbitals contributing to the π bonding), and, finally, $Sb(OH)_6^-$ (where the *d* orbital contributes to σ bonding to give six-coordination for the larger atom).

This discussion of π bonding may be summarized: strong π bonding using only *p* orbitals is confined to the first element of a *p* Group and often leads to small, volatile molecules, as with the carbon and nitrogen oxides. π bonding by the heavier elements involves their *d* orbitals, or possibly hybrid orbitals formed by *p* and *d* combinations.

The ability to form $p_\pi - p_\pi$ bonds, and the weakness of simple sigma bonds, are effects of the small size of the first row elements. Other properties, such as ionization potential and electron affinity, which depend in part on size, also change sharply between the first and second elements of a *p* Group. This shows up very clearly in the electronegativity values in Table 2.13, where there is a major drop in value between the first and second elements in each *p* Group, and then a much slower fall in values down the rest of the Group.

The size effect also shows up in the bond strengths, as long as π bonding plays no part. The larger atoms usually form weaker sigma bonds, so that bond strengths fall on passing down a *p* Group, and the largest relative changes come between the first two elements of the Group. This point is illustrated by the stabilities of the hydrides, discussed in Chapter 8, and by a similar fall in the stability of element-carbon bonds in the organic analogues of the hydrides. These changes are illustrated by the values of the element-hydrogen

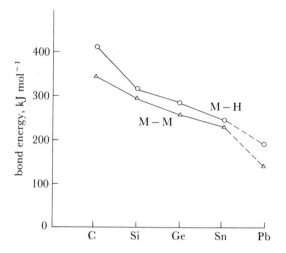

FIGURE 15.3 *The element—hydrogen and element—element bond energies in the carbon groups. Similar trends are found for the other Main Groups.*

and element-element bond energies shown in Figure 15.3.

The characteristic properties of the first element in each *p* Group may be summarized as:

(i) small size, with the greatest relative size change occurring between the first and second members of a Group

(ii) other properties which result from the size effect change in a similar way: for example, electronegativity and the tendency to form anions and negative oxidation states

(iii) the ability to form π bonds using the *p* orbitals is restricted to the first row elements, with the result that many simple compounds such as the oxides are small molecules instead of polymers

(iv) the first element has no low-lying *d* orbital and is limited to a maximum coordination number of four.

15.3 The remaining elements of the *p* Group

The last member of each *p* Group shows distinct differences from the other elements in the Group, though these differences are less pronounced than those which mark off the first member. The most obvious property of the last row of *p* elements is the stability of an oxidation state two less than the Group state in thallium, lead, and bismuth. The other two heavy elements, polonium and astatine, occur only as radioactive isotopes and their chemistry is less well known but, in polonium at least, the trend appears to continue and the IV, and also the II, states of polonium are more stable than the Group state of VI. This effect has been termed the inert pair effect as it corresponds to the valency which arises if two of the valency electrons are inactive. The stability of the lower state, and the oxidizing properties of the Group state, in these heavy elements is clearly shown in Tables 15.2 and 15.3 and in the oxidation state free energy diagrams. The heavy elements show the Group oxidation state only in their oxides and fluorides, while sulphides and other halides of the elements exist only for the lower oxidation states.

The heaviest element in each *p* Group is also the most metallic and least electronegative member of its Group (as the larger size means that the outer electrons are less tightly

held). Most of the heavy elements are metallic except polonium, which is metalloid, and astatine whose character is not clear. They all form basic oxides and appear in solution in cationic forms. The single bond strengths of these elements with hydrogen, organic groups, and halogens are lower than for the lighter elements in the *p* block and the very existence of some of the hydrides is dubious.

The chemistry of the middle three elements in a *p* Group follows a graded transition in properties from the small electronegative element at the head of the Group to the large, much more basic element at the foot. This is shown by the increasing stability of the oxidation state two less than the Group state, the increasing basicity of the oxides, and by the lower stability of bonds with hydrogen and organic groups as the Group is descended. The transition in properties between the middle three elements and the heaviest element is smooth and continuous, but there is a marked discontinuity between the properties of the first element and the middle elements in the Group for reasons discussed in the last section. This discontinuity between first and second element is most marked in the middle Groups of the *p* block. Boron and aluminium have each an empty *p* orbital in the valence shell in their trivalent compounds and thus have acceptor properties. The fact that aluminium also acts as an acceptor by using its 3*d* orbitals makes a difference in degree, but not in kind of behaviour so that boron and aluminium resemble each other reasonably closely. In sharp contrast, the ability of silicon, phosphorus, and sulphur to use the 3*d* orbitals, and the ability of carbon, nitrogen and oxygen to form *p*ₚ bonds combine to create marked differences in properties between the first and second members of these Groups. At oxygen, and still more at fluorine, the tendency to enter the negative oxidation states as ions or covalent molecules becomes more important and re-introduces some resemblance in the general chemistry, especially between fluorine and chlorine. These trends may be linked with those in the *s* block elements so that, in the Main Groups as a whole, the first row element differs from the rest of its Group. This difference is least at the ends of the Periods in the lithium and fluorine Groups, and is most marked in the centre, in the carbon and nitrogen Groups.

Size effects in the chemistry of the heavier elements are much less distinctive than the changes observed at the head of the Group. There is a general tendency for the heavier elements to show higher coordination numbers but this is offset to some extent by the weakening of element-ligand single bonds already noticed. In oxyacids, antimony, tellurium, and iodine give compounds in the Group oxidation state where the coordination to oxygen is six-fold in place of the four-coordinated oxyions of the lighter elements. In halogen compounds there is evidence for TeF_7^- or TeF_8^{2-} species and IF_7 exists as may XeF_8. It is not known whether this trend continues for polonium and astatine.

One further feature of the chemistry of the *p* Groups is the slight discontinuity in properties observed at the middle element in each Group (i.e. in the Period from gallium to bromine) which was mentioned above. These middle elements do not always have properties which interpolate between those of the second and fourth element. These effects are ascribed to the insertion of the first d shell immediately preceding this row of Main Group elements—the effect is reminiscent of the lanthanide contraction following the appearance of the first f shell and has even been termed the 'scandide contraction'. The effects are much less striking however than in the case of the lanthanide contraction. The general effect is most clearly seen if contours of equal electronegativity are drawn across the Periodic Table. In the chemistry of the elements, the effect shows up in minor anomalies such as the non-existence of $AsCl_5$ while phosphorus and antimony do form the pentachloride. Similarly, GeH_4 is found to be stable to dilute alkali while SiH_4 and SnH_4 are rapidly attacked. This is not a major effect, and most of the chemistry of the second, third, and fourth elements in any p Group follows a smooth trend, but the second order anomaly clearly exists.

These points may be summarized:

(i) The heaviest element in a p Group differs from the rest in the stability of oxidation states lower than the Group state, in its more metallic character with more basic oxides, and in its weak bonding to hydrogen and related ligands.

(ii) The properties of the middle three elements in a p Group form a smooth transition to those of the heaviest element, with the lower oxidation states becoming more stable and the oxides more basic, etc.

(iii) There is a sharp discontinuity in properties with the first element, especially in the carbon, nitrogen, and oxygen Groups.

(iv) The heavier elements tend to show higher coordination numbers.

(v) Minor anomalies are observed in the chemistry of the elements of the middle row of the p block as compared with the properties of the rows above and below.

The p Groups will now be discussed in turn.

15.4 The boron group, ns^2np^1

References to the properties of the boron Group elements given in earlier chapters include:

Ionization potentials	Table 2.8 and Figure 7.7
Electronegativities	Table 2.13
Hydrides	Chapter 8, especially electron-deficient hydrides, section 8.6

Table 15.4 lists some properties of the elements and Figures 15.4 and 15.5 give the variations of certain properties with Group position. It will be seen from the free energy diagram, Figure 15.5, that the III state is the most stable one, except for thallium.

Aluminium, the most common metallic element in the earth's crust, is extracted from the hydrated oxide, bauxite, by electrolysis of the oxide (after purification by alkaline treatment) dissolved in molten cryolite, sodium hexafluoro-aluminate. Boron is found in concentrated deposits as borax, and the element is formed by magnesium reduction of the

TABLE 15.4 Properties of the elements of the boron group

Element	Symbol	Oxidation states	Common coordination numbers	Availability
Boron	B	III	3, 4	common
Aluminium	Al	III	3, 4, 6	common
Gallium	Ga	(I), III	3, 6	rare
Indium	In	I, III	3, 6	very rare
Thallium	Tl	I, (III)	3, 6	rare

All the elements have high boiling points (above 2000 °C), but gallium has an unusually low melting point at 29·8 °C which gives it the longest liquid range of any element.

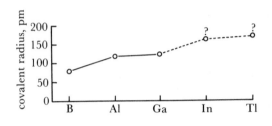

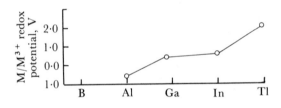

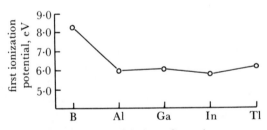

FIGURE 15.4 *Some properties of the boron Group elements*
The figure shows the covalent radii, the first ionization potentials, and the oxidation potentials as functions of Group position.

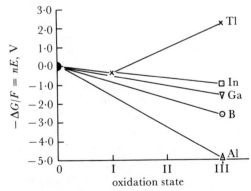

FIGURE 15.5 *The oxidation state free energy diagram for the boron Group elements*

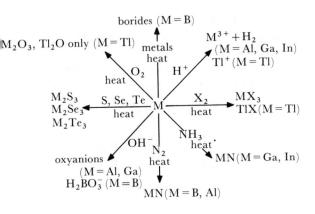

FIGURE 15.6 *Reactions of the elements of the boron Group*

oxide. The other three elements are found only in the form of minor components of various minerals, and the elements are produced by electrolytic reduction in aqueous solution. Gallium, indium, and thallium are relatively soft and reactive metals which readily dissolve in acids. Aluminium is also a reactive metal but is usually found with a protective, coherent oxide layer which renders it inert to acids, although it is attacked by alkalies. Boron is non-metallic and the crystalline form is very hard, inert, and non-conducting. The amorphous form of boron, which is more common than the crystalline variety, is much more reactive. Figure 15.6 shows some typical reactions of the boron Group elements.

Boron reacts directly with most metals to give hard, inert, binary compounds of various formulae. These borides some-

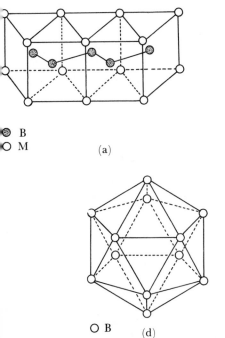

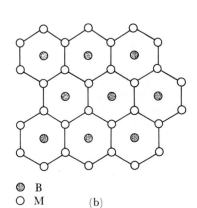

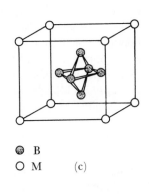

FIGURE 15.7 *Examples of the structures of boron units in borides*
(a) shows the configuration of the boron chain in MB compounds, (b) the sheet structure found in MB_2 and (c) the B_6 cluster in MB_6 compounds. The B_{12} icosahedron shown in (d) is found in boron itself, in some of the boron hydrides, and in BeB_{12} and AlB_{12}. The other MB_{12} species have a related structure. In the tetraborides, hexaborides, and **dodecaborides**, the boron clusters (such as the B_6 unit in Figure (c)) are themselves linked so that a bonded boron framework extends completely through the compound.

TABLE 15.5 Some typical borides

Formula	Boron atom structure	Boride examples
M_2B	single atoms	Be, Cr, Mn, Fe, Co, Ni, Mo, Ta, W
M_3B_2	pairs	V, Nb, Ta
MB	single chain (Figure 15.7a)	V, Cr, Mn, Fe, Co, Ni, Nb, Ta, Mo, W
M_3B_4	double chains	V, Cr, Mn, Ni, Nb, Ta
MB_2	sheets (Figure 15.7b)	Be, Mg, Al, Sc, Ti, V, Cr, Mn, Y, Lu, U, Pu, Zr, Hf, Nb, Ta, Tc, Re, Ru, Os, Ag, Au
MB_4	sheets linking B_6 octahedra	Mg, Ca, Mn, Y, Ln, Th, U, Pu, Mo, W
MB_6	B_6 octahedra (which occupy Cl^- positions in a CsCl structure with the metals (Figure 15.7c))	Na, K?, Be, Mg, Ca, Sr, Ba, Sc, Y, La, Th, Pu
MB_{12}	three-dimensional lattice consisting of linked B_{12} clusters and with metal atoms in the middle of each cluster	Be, Mg, Al, Sc, Y, Ln from Tb to Lu, U, Zr

Other more unusual species include CuB_{22}, $B_{13}M_2$ (M = P, As) and $Ru_{11}B_8$.

what resemble the interstitial carbides and nitrides. Table 15.5 lists some typical formulae, and some structures, which all involve chains, sheets, or clusters of boron atoms, are shown in Figure 15.7. The binary compound, boron nitride, BN, is interesting as it is isoelectronic with carbon and occurs in two structural modifications. One has a layer structure, like graphite (but is light in colour) and is soft and lubricating, while the other, formed under high pressure, has a very hard, stable, tetrahedral structure as in diamond.

The III state

The Group state of III is shown by all the elements, and is the most stable state for all except thallium. It is represented by a wide variety of compounds, of which the oxygen and halogen compounds are typical. There is no evidence for a free M^{3+} ion, either in the solid or in solution. A number of solids, especially fluorides and oxides, are high-melting and strongly bonded, but the bonds are intermediate between ionic and covalent and the stabilities of the solids are due to the formation of giant molecules with uniform bonding. For example, aluminium chloride, bromide, and iodide are volatile, covalent solids, while aluminium trifluoride is high-melting and a giant molecule. Similar effects are seen for the other trifluorides, except for BF_3, and for the oxides. In solution, extensive hydration and hydrolysis occur and ionic species (though often written as cations for convenience) are actually much more complex, e.g. $Al(OH)(H_2O)_5^{2+}$ has been shown to occur in 'Al^{3+}' solutions.

All the elements form the trioxides, usually as hydrated species by precipitation from solution or by hydrolysis of the trihalides. Chemical and structural properties are given in Table 15.6.

Oxides of the I state are treated later.

Hydration of the oxides gives a variety of hydrates and hydroxy-species. Boric oxide gives boric acid, $B(OH)_3$, on hydration which forms crystals in which the $B(OH)_3$ units are linked together by hydrogen bonding (Figure 8.18). When boric acid is heated, it dehydrates first to metaboric acid, HBO_2, and ultimately to boric oxide:

$$B(OH)_3 \xrightleftharpoons[+H_2O]{-H_2O} HBO_2 \xrightleftharpoons[+H_2O]{-H_2O} B_2O_3$$

Metaboric acid exists in three crystalline forms, one of which contains the cyclic unit shown in Figure 15.8. The structures of the other two are not known with certainty but appear to contain chains of BO_3 and BO_4 units. The cyclic anion is also found in sodium and potassium metaborates. A wide

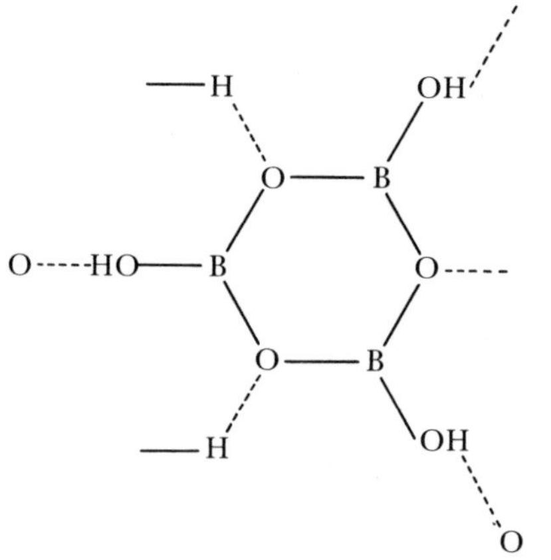

FIGURE 15.8 *The cyclic form of metaboric acid*

TABLE 15.6 Oxides of the III state of the boron Group elements

Oxide	Properties	Structure
B_2O_3	Weakly acidic Many metal oxides give glasses with B_2O_3 as in the 'borax bead' test.	Glassy form—random array of planar BO_3 units with each O linking two B atoms. Crystalline form—BO_4 tetrahedra linked in chains.
Al_2O_3 and Ga_2O_3	Amphoteric	α-form—inactive, high-temperature form. Oxide ions ccp with metal ions distributed regularly in octahedral sites. γ-form — low-temperature form, more reactive. Metal ions arranged randomly over the octahedral and tetrahedral sites of a spinel structure.
In_2O_3 and Tl_2O_3	Weakly basic Tl_2O_3 gives O_2 and Tl_2O on heating to 100 °C	The structure has the metal ions in irregular six-coordination, and four-coordinated oxygens. The same structure is adopted by most oxides of the lanthanide elements, Ln_2O_3.

Other Oxides: $(BO)_x$ formed by heating $B + B_2O_3$ at 1050 °C. This probably contains both $B-O-B$ and $B-B$ links as it reacts with BCl_3 to give B_2Cl_4.

gem forms of alumina: ruby—Al_2O_3 + traces of Cr^{3+}

blue sapphire—Al_2O_3 + traces of Fe^{2+}, Fe^{3+} or Ti^{4+}

white sapphire—this is the gem form of alumina itself

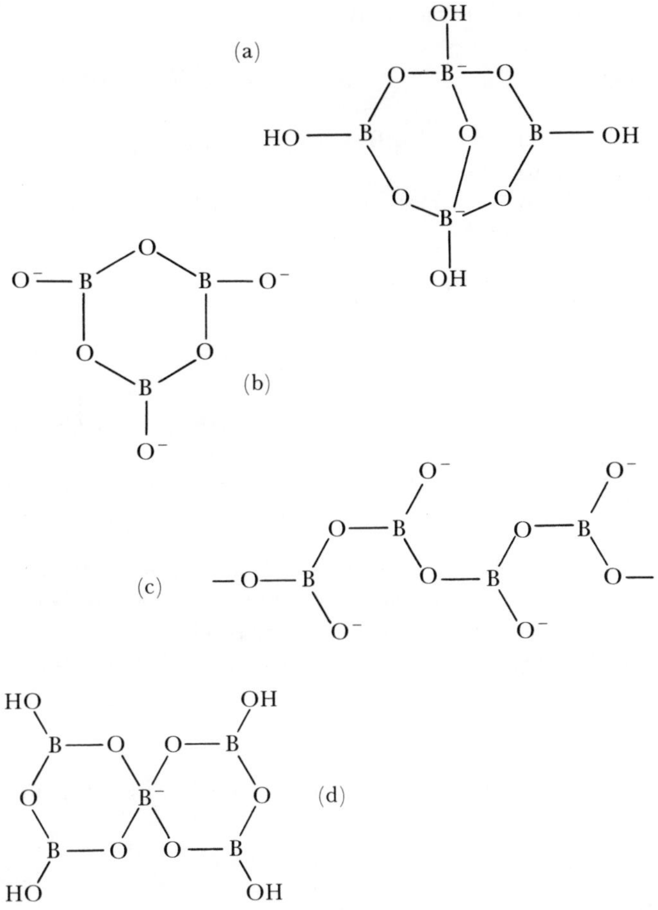

OH

FIGURE 15.9 *Examples of borate structures:* (a) *in borax,*
Na$_2$B$_4$O$_7$.10H$_2$O, (b) *in metaborates,* M$_3$B$_3$O$_6$ *(cyclic anion),*
(c) *in linear metaborates,* CaB$_2$O$_4$, (d) B$_5$O$_{10}$H$_4^-$

Borax, Na$_2$B$_4$O$_7$.xH$_2$O ($x = 5$ or 10), and other borates find uses in water treatment, and in preserving timber from insect attack. Large amounts of sodium or calcium borates, and of boric acid or oxide, are used in glass manufacture. Sodium perborate, of approximate overall composition NaBO$_3$.4H$_2$O, is widely used as a bleaching agent in detergents. This material was once formulated as borate with hydrogen peroxide of crystallization, but the material is now accepted as a true peroxyborate with the B—O—O linkage well-established.

All the trihalides of all the boron Group elements exist and all correspond to the III state, except TlI$_3$ which is the tri-iodide, I$_3^-$, of Tl$^+$. The normal trihalides are planar molecules which have an empty p orbital in the valence shell. Most of the trihalides make use of this empty orbital, both in the structure of the trihalide, and in the formation of complexes of the form MX$_3$.D, where D is a lone pair donor. Table 15.7 lists these applications for the halides and for the halide complexes. Aluminium, and the heavier elements, also use their d orbitals to become six-coordinate. It will be seen that the formation of a $p_\pi - p_\pi$ bond in BF$_3$, and the use of d orbitals, especially in the fluorides, mirrors the discussion of these effects in section 15.2.

Most of the trihalides react with water to give the hydrated oxides, but boron trifluoride gives 1:1 and 1:2 adducts, BF$_3$.H$_2$O and BF$_3$.2H$_2$O, which are not ionized in the solid state. The 1:1 adduct has the expected donor structure, F$_3$B.OH$_2$, but the structure of the second is unknown. When these adducts are melted, they each ionize:

$$2BF_3.H_2O = (H_3O.BF_3)^+ + (HO.BF_3)^-$$
and $$BF_3.2H_2O = H_3O^+ + (HOBF_3)^-$$

Among the complex halides, all the MX$_4^-$ species are tetrahedral while the MX$_6^{3-}$ ones are octahedral. All the boron Group trihalides act as catalysts in the Friedel-Crafts reaction, where their function is to abstract a halide

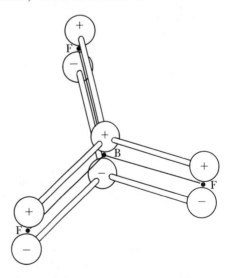

FIGURE 15.10 *Internal p_π-p_π bonding in boron trifluoride*
The p orbital on the boron which is not used in the sigma bonds, accepts electrons from the three corresponding fluorine orbitals to give internal pi-bonding in BF$_3$.

variety of other oxyanions of boron exists with very varied structural types. Not only are discrete ions, rings, chains, sheets, and three-dimensional structures found—as with the silicates—but boron occurs both in planar BO$_3$ units and in tetrahedral BO$_4$ units, and many borates contain OH groups. It is impossible to discuss all the borates, but Figure 15.9 gives a few representative borate structures.

Aluminium and gallium form hydrated oxides of two types —MO(OH) and M(OH)$_3$. These are precipitated from solution by, respectively, ammonia and carbon dioxide. Indium gives a hydrated oxide, In(OH)$_3$. In these compounds the metal is six-coordinated to oxygen.

Hydroxy species also occur in solution. Thus boric acid accepts an OH$^-$ group in dilute solution and polymerizes in more concentrated solutions:

$$B(OH)_3 + 2H_2O = B(OH)_4^- + H_3O^+$$
$$3B(OH)_3 = B_3O_3(OH)_4^- + H_3O^+ + H_2O$$

The hydrates of the other elements of the Group behave similarly. For example:

$$M(H_2O)_6^{3+} \rightleftharpoons M(H_2O)_5(OH)^{2+} \rightleftharpoons \text{intermediate stages} \rightleftharpoons M(OH)_6^{3-}$$

and a compound, Ca$_3$[Al(OH)$_6$]$_2$ has been isolated.

TABLE 15.7 *Acceptor and structural properties of the trihalides of the boron Group elements*

Halide	Structural use of empty p orbital	Halide complex
BF_3	Internal $p_\pi - p_\pi$ bonding: see note (1) and Figure 15.10.	BF_4^-
BX_3	Possibly slight π bonding in BCl_3, otherwise none.	BX_4^-
AlF_3	Accepts lone pair from fluorine (as do two d orbitals) to give AlF_6 units in a highly polymerized solid. M.p. above 1000 °C.	AlF_6^{3-}
AlX_3 (Note 2)	Accept one lone pair from a halide to give Al_2X_6 dimer (Figure 15.11). m.p. 100–200 °C.	AlX_4^-
MF_3 (M = Ga, In, Tl)	As AlF_3 (m.p. about 1000 °C).	MCl_6^{3-}, MBr_6^{3-},
GaX_3, InX_3	As AlX_3 (m.p. 100–600 °C).	MCl_5^{2-}, MCl_4^-
		MBr_4^- (M = Ga, In)
		$TlCl_5^{2-}$
TlX_3	As AlX_3 (see note 2)	

In all cases above, X = Cl, Br and I.

Note (1) The evidence for internal π bonding in BF_3 derives from two sources. First, the $B-F$ bond length is shortened compared with that in the BF_4^- ion, 130 pm compared with 142 pm. Second, the order of acceptor strengths for the boron trihalides (forming $BX_3.D$) is $BBr_3 > BCl_3 > BF_3$. As ability to accept an electron-pair depends on the electron density at the boron, the strongly electronegative fluoride would be the strongest acceptor unless other effects intervene.

Note (2) For X = Cl, the dimer is found in the vapour but solid $AlCl_3$ exists as a slightly deformed $CrCl_3$ layer lattice structure (compare Table 4.3) with 6-coordination of aluminium.

Note (3) $TlBr_3$ decomposes to $TlBr$ and Br_2 at room temperature, and $TlCl_3$ loses chlorine similarly at 40 °C.

Note (4) A more complex chloro-gallate, $Ga_3Cl_{10}^-$, has been reported recently. This may have an extended bridge structure with both 4- and 6-coordinate gallium:

$$\left[\begin{array}{ccccc} Cl & & Cl & Cl & Cl & & Cl \\ & Ga & & Ga & & Ga & \\ Cl & & Cl & Cl & Cl & & Cl \end{array} \right]^-$$

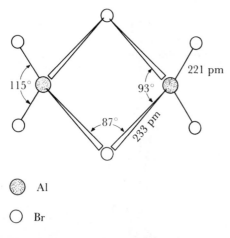

Al

Br

FIGURE 15.11 *The structure of the aluminium tribromide dimer*
The aluminium makes use of its empty p orbital to accept a lone pair from a bromine atom in a second $AlBr_3$ molecule, giving Al_2Br_6.

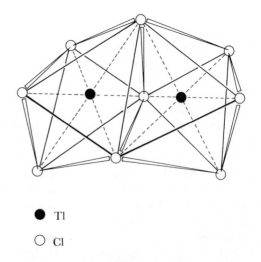

Tl

Cl

FIGURE 15.12 *The structure of the ion,* $Tl_2Cl_9^{3-}$

ion (giving MX_4^-) from the organic molecule, leaving a carbonium ion. As well as the MX_4^- and MX_6^{3-} ions shown in the Table, the isolation of MCl_5^{2-} ions has been reported for indium and thallium. A crystal study shows that, in its tetraethylammonium salt, $InCl_5^{2-}$ is a square pyramid, similar to the structures described in section 12.6. This is

the first report of this structure for a main group element (not to be confused with species like IF_5 with a lone pair in addition). The form probably results from the way the ions pack with the large cation into the crystal. There is also one binuclear complex ion, $Tl_2Cl_9^{3-}$, in the III state. This has the structure shown in Figure 15.12.

The trihalides, and other trivalent MX_3 species, readily form tetrahedral complexes such as BH_4^-, $AlX_3.NR_3$, or $GaH_3.NMe_3$. A wide selection of 1:1 $BX_3.D$ complexes exist where D is a lone pair donor such as ammonia, amine, water or ether, phosphine, sulphide, etc., and X is halogen, hydrogen, or an organic group. The organic compounds $R_3B.D$ have been much studied to find the factors, such as electron attracting power and steric effects, which most influence Lewis acid-base behaviour. One anion of analytical importance is the tetraphenylboronate ion $B(C_6H_5)_4^-$, which forms insoluble salts with potassium and the heavier alkali metals and is used in their gravimetric determination.

Complexes of the elements other than boron include both tetrahedral types as above and also octahedral complexes, of which important examples are the β-diketone complexes shown in Figure 15.13 and the 8-hydroxyquinoline complex

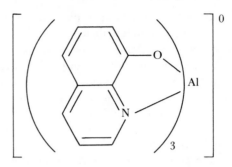

FIGURE 15.14 *The 8-hydroxyquinoline complex of aluminium*
This is commonly used to determine aluminium gravimetrically.

of Figure 15.14 which is used in the gravimetric determination of aluminium. These elements form hexahydrates, $M(H_2O)_6^{3+}$, which hydrolyse in solution. Also, hydrated salts with this cation and a variety of oxyanions are known. Aluminium also forms the well-known series of double salts, $MAl(SO_4)_2.12H_2O$, called alums. M is any univalent cation and the aluminium may be replaced by a variety of trivalent ions such as Cr^{3+}, Fe^{3+}, Co^{3+}, Ga^{3+} or Ti^{3+}. The crystals contain $M(H_2O)_6^+$, $Al(H_2O)_6^{3+}$—or other $M(H_2O)_6^{3+}$ ions—and sulphate ions.

All the boron Group elements form organic compounds, R_3M, and also mixed types, R_2MX and RMX_2, with halogens and related groups. The boron compounds are monomers with planar BC_3 skeletons, but the later members of the Group give dimeric or polymeric compounds. The halides dimerize through halogen bridges similar to those in Al_2X_6. The purely organic compounds polymerize through electron deficient carbon bridges, as in $Al_2(CH_3)_6$, shown in Figure 8.12, similar to the bonding in diborane. A recent refinement of the crystal structure indicates that, instead of the bridge being bonded Al—C—Al, the hydrogen is involved giving four-atom electron-deficient bridges Al—C—H—Al.

Organo-aluminium compounds are involved in a number of related, and commercially important, catalytic processes which arose from the work of Ziegler. The basic discovery was that Al—H bonds add across the alkene double bond coupled with the fact that aluminium metal reacts under fairly easy conditions with hydrogen in the presence of aluminium alkyl. Thus, Al—H bonds may be formed and converted into Al—C in the overall process shown in equation (1).

$$Al + \tfrac{3}{2}H_2 + 3RCH=CH_2 \xrightarrow{AlR_3} Al(CH_2CH_2R)_3 \ldots (1)$$

The addition is to terminal double bonds. The Al—C bond in turn will add across a terminal double bond, in a series of steps which leads to growth of the alkyl chain

$$R_2Al-R + R'CH=CH_2 \rightarrow R_2AlCH_2CH(R)R' \text{ etc.} \quad (2)$$

Finally, the process may be terminated by the reverse step to (1), yielding a long-chain α-alkene, or hydrolysis and oxidation is used to yield an alcohol

$$R_2Al-CH_2CHRR' \xrightarrow{O_2} R_2AlOCH_2CHRR' \xrightarrow{H_2O}$$
$$R_2AlOH + RR'CHCH_2OH \ldots (3)$$

All the Al—R links are eventually broken in step (3).

There are several applications of this process: firstly, using ethylene and high pressures at 160 °C, steps (1), (2) and (3) can be arranged to produce alcohols with chain length about C_{14} which are used in the production of bio-degradable detergents. Secondly, using ethylene at about 100 °C, steps (1), many repetitions of (2), terminated by the reverse of (1) produces polythenes with average chain-length about C_{200}. Thirdly, longer chain olefins may be dimerized, as in the formation of isoprene via the dimerization of propene

$$2CH_3CH=CH_2 \xrightarrow{(1) \text{ and } (2)} CH_3CH_2CH_2C(CH_3)=CH_2$$
$$\xrightarrow{heat} CH_4 + CH_2=CHC(CH_3)=CH_2$$

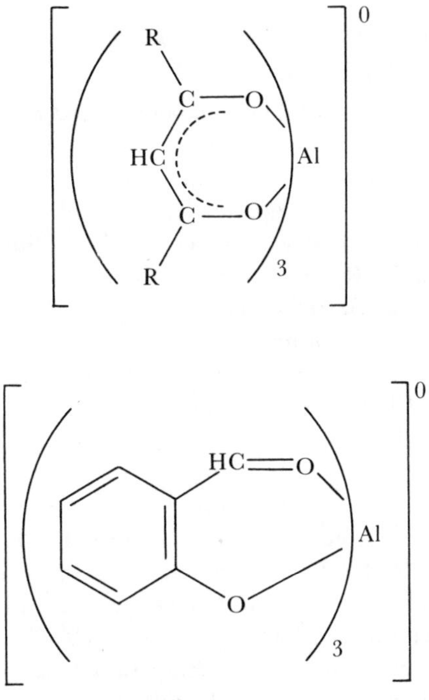

FIGURE 15.13 *β-diketone complexes of aluminium*

Isoprene polymerization itself takes place in the presence of catalysts including aluminium alkyls.

A further process, developed in part by Ziegler and partly by Natta, is an extension of the olefin polymerization. The process above, steps (1), (2) and the reverse of (1), gives a wide spectrum of chain lengths and disordered polymers which are soft and low-melting. In the Ziegler-Natta process, a transition metal halide, such as $TiCl_4$, is added to an aluminium trialkyl and the resulting reaction mixture is found to catalyse the polymerization of alkenes, giving a stereoregular product (one where, for example, all the sidechains lie the same way) and these regular polymers are much higher melting and more crystalline. For example, while ordinary polythene softens below 100 °C, polythene from the Ziegler-Natta process melts at 130–135 °C.

The exact nature of the catalytic process is still under study. A variety of transition metal halides may be used together with other active organometallic species in place of the AlR_3. For $TiCl_4$, it is established that the titanium is reduced to the III (or lower) state and one theory is that the catalysis takes place on the crystal surface of the reduced species. A chain growth process like (2) occurs but, as it is on a surface, the approach of the incoming olefin is oriented, giving a regular polymer. The aluminium alkyl acts as the reducing agent and also forms $Ti-R$ groups on the surface to provide growth sites. An alternative theory suggests some $Ti \ldots X \ldots Al$ bridge which provides the active site, where X may be an organic group or a halogen. Here again, the orientation of substituents is postulated to restrict the attacking alkene into a regular and repeatable orientation, giving a regular orientation of the product.

The I oxidation state and mixed oxidation state compounds

The I oxidation state is most important in thallium chemistry where it is the most stable state. The few common thallium(III) compounds, such as the oxide and halides, are strongly oxidizing, and the potential Tl^{3+}/Tl^+ of 1·3 V in acid solution makes thallium(III) in solution as oxidizing as chlorate or MnO_2. Thallium(I) compounds are stable and show some resemblances to both lead(II) compounds and to those of alkali metals. The oxide, Tl_2O, and the hydroxide, $TlOH$, are strongly basic, like the alkali metal compounds, and absorb carbon dioxide from the atmosphere. The halides resemble lead halides in being more soluble in hot water than in cold and behave in analysis like the lead or silver compounds. Tl^I forms a number of stable salts which are generally isomorphous with the alkali metal ones; examples include the cyanide, perchlorate, carbonate, sulphate, and phosphates. TlF has a deformed sodium chloride structure in the solid, while the other thallous halides crystallize with the cesium chloride structure. The chemistry of thallium is indicated in Figure 15.15 which shows the interrelation between the two oxidation states.

The I oxidation state becomes rarer and less stable as the Group is ascended. Indium forms reasonably stable monohalides but the evidence for other In(I) compounds is in doubt as not enough is known about indium chemistry.

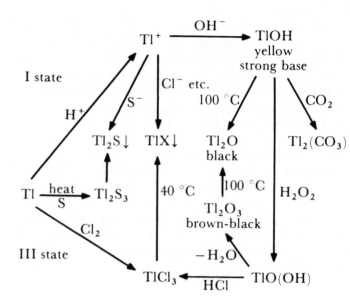

FIGURE 15.15 *Reactions of thallium compounds*

Gallium forms unstable monohalides and a sulphide, but no oxide. The apparent II state in the gallium dihalides, GaX_2, has been shown to be the result of a mixed-valency compound containing gallium(I) and gallium(III). These dihalides are tetrahalogallates(III) of the gallium(I) cation, $Ga^+(GaX_4)^-$. Related compounds with Ga^+ and other anions, such as $GaAlX_4$, are also found. $TlBr_2$ has a similar structure and it is probable that InF_2 will be similar. It is probable that In_2Cl_3 is to be formulated in a similar manner as $In_3^I(In^{III}Cl_6)$. The first report of gallium(II) has been made recently in the announcement of $Ga_2X_6^{2-}$ species for $X = Cl, Br, I$, prepared electrolytically from gallium in strong acid. The structure appears to be $X_3Ga-GaX_3^{2-}$, like the isoelectronic Ge_2Cl_6. The formal $+II$ state then arises because of the metal-metal bond.

Although Al_2O and AlO have been identified in the vapour phase above 1000 °C, no low-valent compound of aluminium exists at ordinary temperatures and aluminium chemistry is entirely of the III state.

Boron is found in low formal oxidation states in the hydrides and in a variety of halides. The latter include the dihalides, B_2X_4, and more complex molecules such as B_4Cl_4 and B_8Cl_8. All these species contain $B-B$ bonds and have the structures shown in Figure 15.16.

15.5. The carbon Group, ns^2np^2

References to the properties of the carbon Group elements which have already occurred include:

Ionization potentials	Table 2.8
Atomic properties and electron configuration	Table 2.5
Radii	Table 2.10
Electronegativities	Table 2.13
Redox potentials	Table 5.3
Hydrides	Chapter 8
Structures	Section 4.8

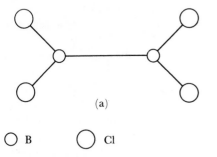

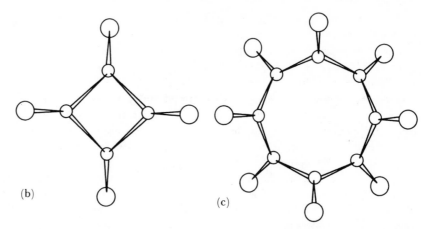

○ B   ◯ Cl

FIGURE 15.16 *Structures of polymeric boron halides :* (a) B_2Cl_4, (b) B_4Cl_4, (c) B_8Cl_8
The B atoms in B_8Cl_8 lie at the vertices of a distorted square antiprism.

(a)   (b)   (c)

Table 15.8 summarizes some properties of the elements, and the variation with Group position of ionization potentials, radii and oxidation state free energy is indicated in Figures 15.17 and 18. The use of carbon in metal extraction is discussed in section 7.7.
All the elements are common except germanium, which

occurs as a minor component in some ores, and also in trace amounts in some coals. It is becoming more freely available due to the large demand for it in semi-conducting devices. The structures of the elements are discussed in section 4.8 and illustrated in Figure 4.15.

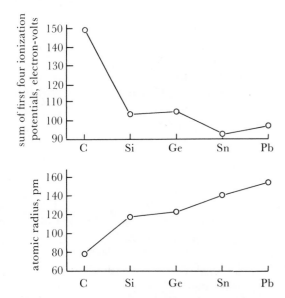

FIGURE 15.17 *Some properties of the carbon Group elements*
The Figure shows the variation, with Group position, of the atomic radii and the sum of the first four ionization potentials. The characteristic differences between first and second elements, and the similarity between second and third elements, are noticeable here.

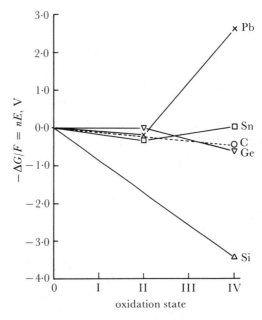

FIGURE 15.18 *The oxidation state free energy diagram for elements of the carbon Group*
It will be seen that Ge(II) is unstable, Sn(II) of nearly the same stability as Sn(IV), and Pb(IV) is very unstable relative to Pb(II).

TABLE 15.8 Properties of the elements of the carbon Group

Element	Symbol	Structures of elements	Oxidation states	Coordination numbers	Availability
Carbon	C	G, D	IV	4, 3, 2	common
Silicon	Si	D	IV	4, (6)	common
Germanium	Ge	D	(II), IV	4, 6	rare
Tin	Sn	D, M	II, IV	4, 6	common
Lead	Pb	M	II, (IV)	4, 6	common

G = graphite, D = diamond, and M = metallic forms

The elements may be produced by reduction of the oxides. Germanium and silicon, for use in semi-conductor devices, are produced in very high degrees of purity (better than $1:10^9$) by zone refining. In this process, the element is formed into a rod which is heated near one end to produce a narrow molten zone. The heater is then moved slowly along the rod so that the molten zone travels from one end to the other. Impurities are more soluble in the molten metal than in the solid and thus concentrate in the liquid zone which carries them to the end of the rod.

All these elements are fairly unreactive, with reactivity greatest for tin and lead. They are attacked by halogens, alkalies, and acids. Silicon is attacked only by hydrofluoric acid, germanium by sulphuric and nitric acids, and tin and lead by a number of both oxidizing and non-oxidizing acids.

Carbon reacts, when heated, with many elements to give binary carbides. Numerous silicides also exist and these are similar to the borides in forming chains, rings, sheets, and three-dimensional structures. Table 15.9 summarizes the various carbide types and Figures 15.19 and 4.13 give some of the structures.

The carbon Group shows the same trend down the Group towards metallic properties as in the boron Group. The II state becomes more stable and the IV state less stable from carbon to lead. Carbon is a non-metal and occurs in the tetravalent state. Silicon is metalloidal, but nearer non-metal than metal, and forms compounds only in the IV state, apart from the occurrence of catenation. Germanium is a metalloid with a definite, though readily oxidizable, II state. Tin is a metal and its II and IV states are both reasonably stable and interconverted by moderately active reagents. The Sn^{4+}/Sn^{2+} potential is -0.15 V and tin(II)

TABLE 15.9 Types of binary carbide

State of aggregation of the carbon atoms	Properties	Examples and structures
Single atoms		
(a) salt-like carbides	Yield mainly CH_4 on hydrolysis	Be_2C (antifluorite) Al_3C_4
(b) transition element carbides	(i) Conducting, hard, high melting, chemically inert	MC, M = Ti, Zr, Hf, Ta, W, Mo (sodium chloride) W_2C, Mo_2C
	(ii) Conducting, hard, high melting, but chemically active: give C, H_2, and mixed hydrocarbons on hydrolysis	Compounds of the elements of the later transition Groups, e.g. M_3C where M = Fe, Mn, Ni
Linked carbon atoms		
(a) C_2 units 'acetylides'	(i) CaC_2 type—ionic, give only acetylene on hydrolysis	MC_2, M = Ca, Sr, Ba: structure related to NaCl (Figure 4.13) Also Na_2C_2, K_2C_2, Cu_2C_2, Ag_2C_2
	(ii) ThC_2 type—apparently ionic, give a mixture of hydrocarbons on hydrolysis	ThC_2 and MC_2 for M = lanthanide element. Structure (Figure 15.19) also related to sodium chloride
Oxyanions		
(b) C_3 chain?	Gives allylene, $H_3C-C\equiv CH$, on hydrolysis	Mg_2C_3—may contain C_3^{4-} ions
(c) C_n chains	C–C spacing in chain is similar to that in hydrocarbons	Cr_3C_2. $-C-C-C-$ chains running through a metal lattice, compare FeB
Carbon sheets (lamellar structures derived from graphite)		
(a) Buckled sheets	Non-conducting, carbon atoms are four-coordinated	(i) 'graphite oxide' from the action of strong oxidizing agents on graphite. C:O ratio is 2:1 or larger and the compounds contain hydrogen. C=O, C–OH, and C–O–C groups have been identified (ii) 'graphite fluoride' from the reaction with F_2. White, idealized formula is $(CF)_n$
(b) Planar sheets	Conducting π system is preserved	(i) Large alkali metal compounds—of K, Rb, or Cs: e.g. C_8K. The metal is ionized and the electron enters the π system, while the metal ions are held between the sheets (ii) Halogen compounds. X^- ions are held between the sheets and positive holes are left in the π system which increase the conductivity

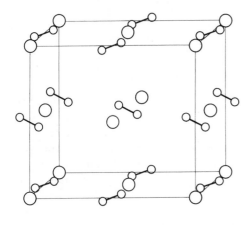

○ Th atoms

⊘ C_2 group

FIGURE 15.19 *Structure of thorium carbide*, ThC_2
As in CaC_2, the C_2 units occupy halide ion positions in a NaCl-type
structure but differ in being oppositely aligned in successive layers
(compare Figure 4.13).

in acid is well-known as a mild reducing agent. Lead is a
metal with a stable II state. Lead(IV) is unstable and
strongly oxidizing.

The elements of the carbon Group are particularly characterized by their tendency to *catenation*, i.e. to form chains
with links between like atoms. Carbon, of course, has this
property in an exceptional degree. In the hydrides, chains of
up to ten atoms are established for silicon and germanium,
as in $Si_{10}H_{22}$, and distannane, Sn_2H_6 is known. Silicon also
forms long chain chlorides but germanium is limited to
$GeCl_4$ and Ge_2Cl_6 as far as present studies go. However,
when the chain is fully substituted by organic groups, as in
$M_n(CH_3)_{2n+2}$, there is no apparent limit to n for M = Si,
Ge, and Sn. In these compounds, as with silanes and germanes, the restriction is the experimental difficulty of
handling high molecular-weight compounds. While the
hydrides readily oxidize, and the halides hydrolyse rapidly
in air, the organo-derivatives are moderately stable to attack
by air. Lead compounds are more restricted, but Pb_2R_6
species are well-known. In addition, these elements form
compounds $(MR_2)_n$ which are ring compounds. Rings with
$n = 4, 5$ and 6 are known for M = Si, Ge and Sn where R
is phenyl and an Sn_6 ring is also reported for R = Me.
There are indications of larger rings, especially for tin. Not
only are straight chains and rings observed, but there is also
definite evidence for the existence of branched chains. The
compound $(Ph_3Ge)_3GeH$ has been identified and branched
chain hydrides of M_4 and M_5 forms (M = Si, Ge) are indicated by chromatographic experiments. For tetrasilanes and
tetragermanes, the *n*- and *iso*- forms have been separated on
the macro-scale, as have two of the three Ge_5H_{12} isomers
and related silicon-germanium species (see section 8.5).
Recently, a number of compounds $(Ph_3M)_4M'$ (M = Ge,
Sn, Pb and M' = Sn, Pb) have been reported. The penta-

plumbane of this *neo*-form, $Pb(PbPh_3)_4$, is the only other
species with Pb — Pb bonds.

The IV state

The IV state is found for all the elements of the Group, and
is stable for all but lead. Its properties are well illustrated by
the oxygen and halogen compounds. Some properties of the
oxygen compounds are listed in Table 15.10. All the dioxides are prepared by direct reaction between the elements
and oxygen. They are also precipitated in hydrated form
(except CO_2 of course) by addition of base to their solutions
in acid. No true hydroxide, $M(OH)_4$, exists for any of the
elements.

The very marked effect of $p_\pi - p_\pi$ bonding on the structures
of the carbon, as compared with the silicon, compounds is
obvious, as is the tendency towards a higher coordination
number to oxygen for the heavier elements.

All the tetrahalides, MX_4, are found except for lead (IV)
which is too oxidizing to form the tetrabromide or iodide.
All may be made from the elements, from the action of
hydrogen halide on the oxide, or by halogen replacement.
All the carbon tetrahalides, all the chlorides, bromides, and
iodides, and also SiF_4 and GeF_4, are covalent, volatile molecules. The volatility and stability fall in a regular manner
with increasing molecular weight of the tetrahalide. By
contrast, SnF_4 and PbF_4, are involatile solids with melting or
sublimation points at 705 °C and 600 °C respectively. They
have polymeric structures based on MF_6 octahedra, with
partially ionic bonding, as has aluminium trifluoride. Thus
the tetrafluorides of the carbon Group parallel the trifluorides of the boron Group in changing from volatile to involatile and polymeric, but the change-over comes further
down the Group. Bond lengths are given in Table 2.10b.

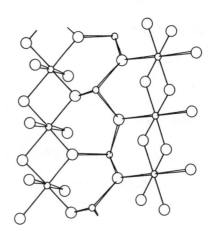

○ Pb

◯ O

FIGURE 15.20 *The structure of* Pb_3O_4
The structure contains $Pb^{IV}O_6$ octahedra linked together into
chains by sharing edges. These chains are, in turn, linked by
$Pb^{II}O_3$ pyramidal units which both link two of the Pb^{IV} chains and
form a chain of $Pb^{II}O_3$ units.

TABLE 15.10 Oxygen compounds of carbon Group elements

Compound	Properties	Structure	Notes
Dioxides			
CO_2	Monomer, weak acid	Linear, $O = C = O$	$p_\pi - p_\pi$ bonding between first row elements giving π-bonded monomer
SiO_2	Involatile, weak acid	4:2 coordination with SiO_4 tetrahedra (see Figure 4.4b)	There is some weak $p_\pi - d_\pi$ bonding in some $Si - O - Si$ systems, though not in the oxide.
GeO_2	Amphoteric	Two forms: one with a 4:2 silica structure, and one with the rutile (6:3) structure	The Ge/O radius ratio is on the borderline between six- and four-coordination
SnO_2	Amphoteric	Rutile structure	
PbO_2	Inert to acids and bases	Rutile structure	Strongly oxidizing
Other Oxides (excluding monoxides)			
C_3O_2	Carbon suboxide, prepared by dehydrating malonic acid $CH_2(COOH)_2 \xrightarrow[300\ °C]{P_2O_5} O = C = C = C = O$	Linear, $C - C$ and $C - O$ distances are intermediate between those expected for single and for double bonds	The molecule contains extended π bonding of the same type as in CO_2.
Pb_3O_4	Red lead: oxidizing, contains both Pb(IV) and Pb(II)	Figure 15.20. The structure consists of $Pb^{IV}O_6$ octahedra linked in chains; the chains are joined by pyramidal $Pb^{II}O_3$ groups	
Carbonate	CO_3^{2-}	Planar with π bonding	Again C and O give p_π bonds
Silicates	Wide variety (section 4.7)	Formed from SiO_4 units	
Germanates	Variety of species	Contain both GeO_4 and GeO_6 units	Compare the two forms of germanium dioxide
Stannates Plumbates } e.g. $M(OH)_6^{2-}$		Contain octahedral MO_6 units	
Oxyhalides			
carbonyl halides			Rapidly hydrolysed
COF_2	b.p. $-83\ °C$	All are planar $\overset{X}{\underset{X}{\diagdown}} C = O$	Very poisonous: has been used as nonaqueous solvent
$COCl_2$ (phosgene)	Stable		
$COBr_2$	Fumes in air		

Other, mixed, oxyhalides such as COClBr are known

silicon oxyhalides

These are all single-bonded species containing $-Si-O-Si-O-$ chains (for example, $Cl_3Si-(OSiCl_2-)_nOSiCl_3$ with $n = 4, 3, 2, 1$ or 0) or rings (for example $(SiOX_2)_4$ where X = Cl or Br)

Carbon tetrafluoride (and all the fluorinated hydrocarbons) and carbon tetrachloride are very stable and unreactive, though CCl_4 will act as an oxidizing and chlorinating agent at higher temperatures. Carbon tetrabromide and iodide are stable under mild conditions, but act as halogenating agents on warming, and are also decomposed by light.

The silicon tetrahalides, except the fluoride, are hydrolysed rapidly to 'silicic acid' which is hydrated silicon dioxide.

The heavier element tetrahalides also hydrolyse readily, but the hydrolysis is reversible and, for example, $GeCl_4$ can be distilled from a solution of germanium (IV) in strong hydrochloric acid.

Hydride-halides of the types MH_3X, MH_2X_2, and MHX_3 are also formed. Most representatives of these formulae (for X = F, Cl, Br, I) are found for silicon and germanium, but a few tin compounds, such as SnH_3Cl, are also known. Such compounds are key members of synthetic routes to organic and other derivatives, as in reactions such as:

$$SiH_3Br + RMgX = RSiH_3 \text{ (R = organic radical)}$$
$$\text{or} \quad GeH_3I + AgCN = GeH_3CN + AgI$$

It has recently been discovered that the higher hydrides of silicon and germanium behave similarly, and all the compounds M_2H_5X, for M = Si or Ge, and X = F, Cl, Br, and I, have been prepared.

The elements from silicon to lead have an extensive organometallic chemistry and a wide variety of MR_4 and M_2R_6 compounds exist, with sigma metal-carbon bonds of considerable stability. Organotin and organolead compounds have been studied for their pharmaceutical and biocidal properties and organogermanium compounds may have similar properties. Tetra-ethyl lead is manufactured on a large scale as an anti-knock agent for petrols. All the tetra-alkyl and tetra-aryl compounds are stable, although stability falls from silicon to lead and the aryls are more stable than the alkyls. For example, tetraphenyl-silicon, Ph_4Si, boils at 530 °C without decomposition, tetra-phenyllead, Ph_4Pb, decomposes at 270 °C, while tetra-ethyllead, Et_4Pb, decomposes at 110 °C. A wide variety of organocompounds, with halogen, hydrogen, oxygen, or nitrogen linked to the metal, is also known and this class includes the silicone polymers. These are prepared by the hydrolysis of organosilicon halides:

$$R_2SiCl_2 \rightarrow \begin{array}{ccccccc} & R & & R & & R & \\ & | & & | & & | & \\ -& Si & -O- & Si & -O- & Si & -O- \\ & | & & | & & | & \\ & R & & R & & R & \end{array}$$

This long-chain polymer is linked by the very stable silicon−oxygen skeleton and the organic groups are also linked by strong bonds so the polymer has high thermal stability. The organic groups also confer water-repellent properties. The chain length is controlled by adding a proportion of R_3SiCl to the hydrolysing mixture to give chain-stopping $-OSiR_3$ groups, while the properties of the polymer may also be varied by introducing cross-links with $RSiCl_3$:

$$R_2SiCl_2 + R_3SiCl + RSiCl_3 \longrightarrow$$

$$\begin{array}{c} | \\ O \\ | \\ Si \\ | \end{array}$$

$$\begin{array}{ccccccccccc} & R & & R & & R & & R & & O & \\ & | & & | & & | & & | & & | & \\ R_3Si & -O- & Si & -O- & Si & -O- & Si & -O- & Si & -O- \\ & | & & | & & | & & | & & | & \\ & R & & O & & R & & R & & R & \\ & & & | & & & & & & & \\ & & & R-Si-R & & & & & & & \\ & & & | & & & & & & & \\ & & & O & & & & & & & \\ & & & | & & & & & & & \end{array}$$

The elements from silicon to lead use their d orbitals to form six-coordinated complexes which are octahedral. All four elements give stable MF_6^{2-} complexes with a wide variety of cations. The MCl_6^{2-} ion is formed for M = Ge, Sn, and Pb, and tin also gives $SnBr_6^{2-}$ and SnI_6^{2-}. In addition, a variety of MX_4L_2 complexes are formed by the tetrahalides with lone pair donors such as amines, ethers, or phosphines. The chemistry of the tin compounds is particularly well-explored, and both *cis* and *trans* compounds are known. There are also a limited number of penta-substituted complex ions including MF_5^- (M = Si, Ge) and $GeCl_5^-$.

The d orbitals are also used in internal π bonding, especially in silicon compounds. The classic case is trisilylamine, $(SiH_3)_3N$ Figure 15.22. This has a quite different structure from the carbon analogue, trimethylamine, $(CH_3)_3N$, shown in Figure 15.21. The pyramidal structure of tri-methylamine is similar to that of NH_3 and reflects the steric effect of the unshared pair on the nitrogen. The NSi_3 skeleton, by contrast, is flat and the nitrogen in trisilylamine shows no donor properties. This is due to the formation of a π bond involving the nitrogen p orbital and d orbitals on the silicon, as shown in Figure 15.23. The lone pair electrons donate into the empty silicon d orbitals and become de-localized over the NSi_3 group, and hence there is no donor

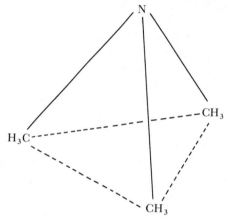

FIGURE 15.21   *The structure of trimethylamine, $(CH_3)_3N$*

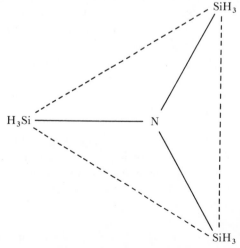

FIGURE 15.22   *The structure of trisilylamine*

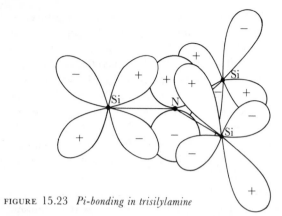

FIGURE 15.23 *Pi-bonding in trisilylamine*

property at the nitrogen. Structural evidence comes from the infra-red and Raman spectra and from the zero dipole moment. Tetrasilylhydrazine, $(SiH_3)_2NN(SiH_3)_2$, also shows differences in symmetry compared with the methyl analogue, arising from similar π bonding. Another clear case is the isothiocyanate, MH_3NCS. Where $M = C$, the $C-N-C-S$ skeleton is bent at the nitrogen atom due to the steric effect of the nitrogen lone pair while, when $M = Si$, the $Si-N-C-S$ skeleton is linear and there is $d_\pi-p_\pi$ bonding between the Si and N atoms. Similar effects are observed in $M-O$ bonds: dimethyl ether, CH_3OCH_3, has a $C-O-C$ angle of $110°$ which is close to the tetrahedral value, while the bond angle in disilyl ether, SiH_3OSiH_3, is much greater—$140°$ to $150°$—indicating delocalization of the non-bonding pairs on the oxygen.

This formation of π bonds using metal d orbitals and a p orbital on a first row element is most marked in the case of silicon, but there is some evidence that it occurs in germanium compounds as well. For example, the $Ge-F$ bond in GeH_3F is very short, which may indicate $d_\pi-p_\pi$ bonding but GeH_3NCO and GeH_3NCS have bent $Ge-N-C$ skeletons in contrast to their silicon analogues. Other evidence for π bonding by germanium is indirect and derived from acidities and reaction rates in substituted phenyl-germanes.

The II state

The II state is the stable oxidation state for lead, and the IV state of lead, like thallium(III) in the last Group, is strongly oxidizing. Pb^{2+} ions exist in a number of salts, though hydrolysis occurs readily in solution:

$$Pb^{2+} + 2H_2O = PbOH^+ + H_3O^+$$

and a further equilibrium is found:

$$4Pb^{2+} + 12H_2O = Pb_4(OH)_6^{2+} + 6H_3O^+$$

A considerable variety of lead(II) salts is known, and these generally resemble the coresponding alkaline earth compounds in solubility, e.g., the carbonate and sulphate are very insoluble. The halides are less similar and $PbCl_2$ is, like TlCl, insoluble in cold water though more soluble in hot water.

Addition of alkali to lead(II) solutions gives a precipitate of the hydrated oxide, which dissolves in excess alkali to give

plumbites. The hydrated oxide may be dehydrated to PbO, called litharge, which is yellow-brown in colour. The structure of PbO is an irregular one in which the lead is co-ordinated to four oxygen atoms at corners of a trigonal bipyramid, with the fifth position occupied by an unshared pair of electrons.

The II state of tin is mildly reducing but otherwise resembles lead(II). In solution, Sn^{2+} ions hydrolyse as in the first equation for Pb^{2+} above. Addition of alkali to stannous solutions precipitates $SnO.xH_2O$ (neither $Sn(OH)_2$ nor $Pb(OH)_2$ exist), and the hydrated oxide dissolves in excess alkali to give stannites. Dehydration gives SnO, which has a similar structure to PbO. Tin(II) gives all four dihalides and a number of oxysalts. In the vapour phase, $SnCl_2$ exists as monomeric molecules with the V-shape characteristic of species with two bonds and one lone pair. One molecule of water adds to this molecule to give the pyramidal hydrate, $SnCl_2.H_2O$.

Tin and lead form some complexes in the II state, including halogen complexes MX_3^- and MX^+. In addition, tin and lead give polyanions in liquid ammonia. The tin or lead dihalides react with sodium in ammonia to give, first, the insoluble Na_4M species, and these add extra metal ions to give intensely-coloured solutions of Na_4Sn_9 or Na_4Pb_9. The latter, and probably the tin compound, is formulated as an ionic species with Pb_9^{4-} ions, on the basis of conductance and electrolytic experiments. For example, electrolysis was found to deposit $2\frac{1}{4}$ lead equivalents for each faraday passed. The compounds are heavily ammoniated and decompose when the ammonia is removed. The structure of the Pb_9^{4-} ion may be the same as that of the isoelectronic Bi_9^{5+} ion shown in Figure 15.35.

Germanium gives a number of compounds in the II state; GeO, GeS, and all four dihalides are well established. These compounds all appear to have polymeric structures and are not too unstable, probably because attack on the polymeric molecules is relatively slow. The known structures are those of GeI_2, which has the CdI_2 layer structure, and of GeF_2 which has a long chain structure similar to that of SeO_2. The II state is readily oxidized to the IV state: thus the dihalides all react rapidly with halogen to give the tetrahalides, while the corresponding reaction of tin(II) compounds is slow. Yellow GeI_2 disproportionates to red GeI_4 and germanium on heating. The divalent compounds are involatile and insoluble, in keeping with a polymeric formulation. One complex ion of the II state is known, $GeCl_3^-$, in the well-known salt $CsGeCl_3$ and adducts $R_3P.GeI_2$ are also reported. Germanium(II) may also be obtained in acid solution, in absence of air, and addition of alkali precipitates the yellow hydrated oxide, $GeO.xH_2O$.

No compound of silicon(II) at ambient temperatures is known.

Reaction mechanisms at silicon

Work on inorganic reaction mechanisms is less developed than the corresponding area in organic chemistry, mainly because of the large variety of systems and of experimental difficulties. In particular, many non-organic reactions are extremely fast. Of all the p elements (for mechanisms at d elements, see section 12.9), silicon presents one of the most favourable cases for study, and we illustrate something of what is known about Main Group mechanisms by this outline of mechanisms at silicon.

Mechanisms are postulated (and remember that all reaction mechanisms are only hypotheses) on the basis of reaction kinetics, and study of silicon has the major advantage that kinetic work may be independently supported by evidence from optically-active compounds. The isolation and resolution of active silicon species has given a powerful tool which has been used, particularly by Sommer and his colleagues, to study mechanisms of substitution.

A number of optically active silicon species have been reported, one of the first being $Ph(\alpha\text{-}Nt)(Me)SiX$, where $\alpha\text{-}Nt = \alpha\text{-naphthyl}$. This was resolved using $X = (-)$ menthoxide (menthol being a naturally-occurring optically active species) by recrystallization from pentane at $-78\ ^\circ C$. We shall abbreviate the optically active species as R_3Si^*X. This isolation was greatly aided by the presence of bulky aromatic groups which reduce the rate of reaction. Even so, R_3Si^*X commonly reacts about a thousand times faster than similar carbon compounds.

It was first shown that stereospecific substitutions did occur by cycles of changes analogous to the Walden cycle, e.g.

$$(+)R_3Si^*H + Cl_2 = (-)R_3Si^*Cl \xrightarrow{LiAlH_4} (-)R_3Si^*H$$

$[\alpha]_D$
values: $+34^\circ$ $\qquad -6^\circ \qquad\qquad -34^\circ$

Thus one of these steps must occur with inversion, and one with retention, and both must be highly stereospecific. Later work showed that the same relative configuration occurred in the following species R_3Si^*X, shown with the rotations

X	H	Cl	OH	OMe	Br	F
$[\alpha]_D$	$+34^\circ$	-6°	$+20^\circ$	$+17^\circ$	-22°	$+47^\circ$.

Thus, the chlorination above is retention, while the reduction involves inversion of configuration.

These observations establish that stereospecific substitutions do take place. Extensive further work has led to the postulation of four main mechanisms at silicon. These are briefly outlined.

S_N2. This is similar to the mechanism at carbon, but is much faster. It is found for R_3Si^*X in polar, but poorly ionizing, solvents and particularly when X is a halogen. The reaction takes place with inversion of configuration, and is postulated to proceed through a trigonal bipyramidal intermediate conformation, in which the organic groups are in the central plane

Typical examples are hydrolyses, or other replacements of

Si—X by Si—OR, and the formation of the hydride above.

While such a process is assisted by using one of the silicon d orbitals to achieve five-coordination (which probably accounts for the speed of reaction) it does not necessarily follow that a stable intermediate forms. This could happen, or the effect of the d orbital may simply be to lower the activation energy compared with the carbon analogue.

S_Ni. When $X = OR$, and hydride or organometallic reagents are used in non-polar solvents, a slow reaction is found which proceeds with retention of configuration. This cannot be S_N2, as the intermediate would undergo fast loss of H or R (leading to racemization) rather than undergo cleavage of the very strong Si—O bond. It is therefore postulated that the reaction proceeds via a four-centre intermediate, and it is termed *internal nucleophilic substitution*. The intermediate may be represented

where E is the electrophilic and N the nucleophilic part of the reagent. Thus, for a Grignard reagent, $N = R$, and $E = MgX$; or for AlH_4^-, $N = H$ and $E = AlH_3$. The process may be understood as a nucleophilic attack assisted by the electrophilic coordination to oxygen which helps to overcome the strong Si—O binding energy. As an example,

$$R_3Si^*OMe + LiAlH_4 \xrightarrow{ether,\ 16\,h} R_3Si^*H$$

$[\alpha]_D = 16^\circ \qquad\qquad\qquad [\alpha]_D = 30^\circ\ (90\%\ \text{retention})$

A similar four-centred mechanism is postulated for the very wide range of reactions called *hydrometallations* in which an M—H bond adds across a double bond. These are found for many metals, M, of which the most important are for $M = B$, Al (see last section), Si or Sn.

S_N1. This is less common than in carbon chemistry, and is found typically for halides in polar solvents of high dielectric constant. Thus, while R_3Si^*Cl is recovered unchanged from solution in CCl_4 or an ether, when it is dissolved in acetonitrile or nitromethane (CH_3CN or CH_3NO_2, both with high dielectric constants) racemization takes place rapidly. This is postulated to proceed through a solvent-stabilized cation, $RR'R''Si^+(solv)$.

EO (expanded octet). One special mechanism is sometimes involved when $X = F$. For most reactions, fluorides behave as other halides and give the above mechanisms. However, the Si—F bond is much stronger than, for example, Si—Cl and this allows a further mechanism. An example is the reaction in which R_3Si^*F is racemized in dry pentane solution by the addition of MeOH. This reaction has the following characteristics which exclude any of the three mechanisms outlined above:

(a) the rate is retarded in formic acid, a solvent of high dielectric constant, hence the reaction is not S_N1. Further, addition of HF retards the reaction so that the reaction does not proceed by loss of F^- as this would be stabilized as HF_2^-.

(b) A mixture of R_3Si^*F and R_3Si^*OMe plus MeOH gives unchanged R_3Si^*OMe and racemic R_3SiF. Thus the racemization is not via R_3SiOMe or any species which could give rise to it, excluding S_N2 and S_Ni.

These features led to the postulate of an expanded-octet mechanism with a five- or six-coordinate intermediate formed by addition of OMe, and which subsequently loses OMe again:

$$
\begin{array}{ccc}
\overset{\displaystyle OMe}{\underset{\displaystyle F\quad R\quad R'\quad R''}{\overset{|}{Si^-}}} & \text{or} & \overset{\displaystyle F\quad OMe}{\underset{\displaystyle R'\quad OMe\quad R''}{Si^{2-}\!\!-\!R}}
\end{array}
$$

As the intermediates are labile or inactive, racemization occurs. There is no breaking of the very strong $Si-F$ bond. Formation of an expanded octet would be assisted by the presence of the fluorine substituent. Note that this mechanism must be rare or there could be no isolation of optically active silicon compounds at all.

These conclusions from optical studies may be supported by kinetic studies in favourable cases. Thus, the formation in a fast step of a relatively stable intermediate, followed by a slow dissociation to products

$$
A + X - Y \underset{\longleftarrow}{\overset{fast,\ k_1}{\longrightarrow}} A - X - Y \underset{\longleftarrow}{\overset{slow,\ k_2}{\longrightarrow}} A - X + Y
$$

would be characterized by a dependence on k_2' alone, and by the fact that the rate of consumption of A was not equal to the rate of appearance of Y. Thus, in S_Ni, EO, and some S_N2 reactions (if the intermediate was relatively long-lived) the above difference in rates would be detected.

Conversely, if the intermediate was unstable and immediately gave the product (i.e. if the second step above was very fast) the rate of appearance of Y would equal the rate of loss of A, and k_1 would be rate-determining. This reaction would thus be second order. Such kinetics would characterize the normal S_N2 reaction.

Finally, the S_N1 reaction is first order and the determining step is the dissociation into cation and anion.

While many mechanisms give rise to an intermediate kinetic picture (there may be a wide range of lifetimes for the $A-X-Y$ intermediate, for example) if kinetic and optical studies agree, the postulated mechanism is quite strongly supported. As far as the silicon mechanisms outlined above are concerned, such kinetic studies as are reported do validate the proposed mechanisms.

15.6 The nitrogen Group, ns^2np^3

References to the properties of the nitrogen Group elements which have occurred in the earlier part of the book include:

Ionization potentials	Table 2.8
Atomic properties and electron configurations	Table 2.5
Radii	Table 2.10, Table 2.11
Electronegativities	Table 2.13
Redox potentials	Table 5.3
Structures	Chapter 4

TABLE 15.11 Properties of the nitrogen Group elements

Element	Symbol	Oxidation states	Co-ordination number	Availability
Nitrogen	N	$-$III, III, V	3, 4	common
Phosphorus	P	$(-$III$)$, (I), III, V	3, 4, 5, 6	common
Arsenic	As	III, V	3, 4, (5), 6	common
Antimony	Sb	III, V	3, 4, (5), 6	common
Bismuth	Bi	III, (V)	3, 6	common

Table 15.11 lists some of the properties of the elements and the variation with Group position of important parameters is shown in Figure 15.24. The oxidation free energy diagram is shown in Figure 15.25.

Fixation of nitrogen is discussed under titanium (section 13.2) and nitrogen-complexes under ruthenium (section 14.6).

Of these elements, nitrogen and bismuth are found in only one form while the others occur in a number of allotropic forms. Nitrogen exists only as the triply-bonded N_2 molecule, and bismuth forms a metallic layer structure shown in Figure 15.26. A large number of allotropes of phosphorus have been reported, not all of which are well-characterized. In white phosphorus, and also in the liquid and vapour states, the unit is the P_4 molecule where the four phosphorus atoms form a tetrahedron. If the vapour is heated above 800 °C, dissociation to P_2 units starts and rapid cooling of the vapour from 1000 °C gives an unstable brown form of phosphorus which probably contains these P_2 units. When white phosphorus is heated for some time above 250 °C, the less reactive red form is produced. The structure of this form

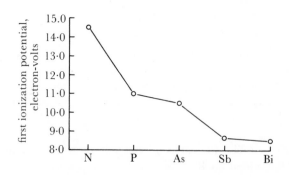

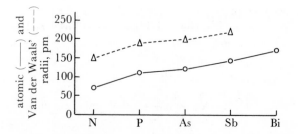

FIGURE 15.24 *Some properties of elements of the nitrogen Group*

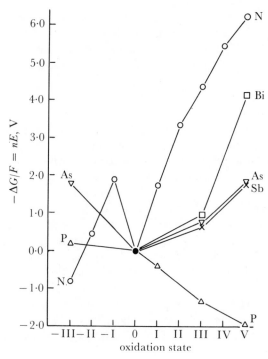

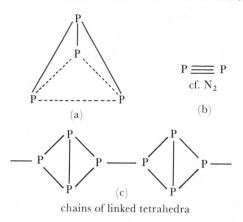

FIGURE 15.27 *Structures of phosphorus allotropes*: (a) *white phosphorus,* (b) *brown phosphorus,* (c) *red phosphorus (postulated)*

FIGURE 15.25 *Oxidation state free energy diagrams of elements of the nitrogen Group*
This is the most complex oxidation state diagram of all the Main Groups. The properties of nitrogen are the most individual, with the element and the − III states as the most stable. All the positive states between 0 and V tend to disproportionate in acid solution (though many form gaseous species in equilibrium with the species in solution) and the − I state is markedly unstable. The curves for P, As, Sb, and Bi form a family in which the − III state becomes increasingly unstable (values for Sb and Bi in this state are uncertain) and the V state becomes less stable with respect to the III state from P to Bi. All intermediate states of phosphorus tend to disproportionate to PH₃ plus phosphorus(V). The diagram also illustrates the very close similarity between As and Sb and the strongly oxidizing nature of bismuth(V).

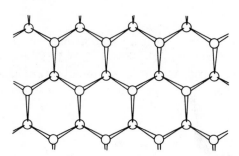

FIGURE 15.26 *The structure of bismuth*

is not yet established and it exists as a number of modifications with various colours—violet, crimson, etc. It is this form which provides most of the doubtful reports of phosphorus allotropes. These might be different structures or due to different crystal sizes, but they may also be due to the incorporation of part of the catalysts used in the transformation. When white phosphorus is heated under high pressure, or treated at a lower temperature with mercury as a catalyst, a dense black form results which has a layer struc-

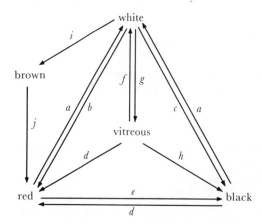

FIGURE 15.28 *The interconversion of the allotropes of phosphorus*
The reaction conditions are indicated by letters as follows:
(a) melting followed by quenching, or vacuum sublimation
(b) heating above 250 °C
(c) heating to 220 °C under pressure
(d) heating above 450 °C, or on prolonged standing at room temperature
(e) at 25 °C under high pressure
(f) vacuum sublimation
(g) heating above 250 °C under pressure
(h) heating to 400 °C under pressure
(i) rapid cooling of vapour with liquid nitrogen
(j) warming above liquid nitrogen temperature (− 196 °C)
Many of these interconversions may also be brought about by catalysis, especially by mercury.

ture like bismuth. A further vitreous form is also reported which results from heating and pressure. The structures of some of these allotropes are shown in Figure 15.27 and the interconversion of the allotropes in Figure 15.28. White phosphorus is the most reactive of the common allotropes, red phosphorus is much less reactive, and black phosphorus is inert.

Arsenic and antimony each occur in two forms. The most reactive is a yellow form which contains M₄ tetrahedral units and resembles white phosphorus. These yellow allotropes readily convert to the much less reactive metallic forms which have the same layer structure as bismuth.

Binary compounds of these elements are similar to those of previous Groups. Nitrides range from those of the active metals, which are definitely ionic with the N^{3-} ion, through the transition metal nitrides which resemble the carbides, to covalent nitrides like BN and S_4N_4. The phosphides are similar, and the heavier elements form compounds with metals which become more alloy-like as one passes from phosphorus to bismuth. One important feature is the appearance of ionic nitrides, as compared with monatomic carbides which are not ionic. Ionic nitrides include the Li, Mg, Ca, Sr, Ba, and Th compounds. They are prepared by direct combination or by deammonation of the amides:

$$3Ba(NH_2)_2 \rightarrow Ba_3N_2 + 4NH_3$$

By contrast, the corresponding carbides either contain polyanions, like the acetylides, or are intermediate between ionic and giant covalent molecules.

Nitrogen, because of the high strength of the triple bond (heat of dissociation = 962 kJ mol^{-1}) is inert at low temperatures and its only reaction is with lithium to form the nitride. At higher temperatures it undergoes a number of important reactions including the combination with hydrogen to form ammonia (Haber process), with oxygen to give NO, with magnesium and other elements to give nitrides, and with calcium carbide to give cyanamide:

$$N_2 + CaC_2 = CaNCN + C$$

The other elements react directly with halogens, oxygen, and oxidizing acids.

This Group shows a richer chemistry than the boron and carbon Groups, as there are more than two stable oxidation states and there is a wider variety of shapes and coordination numbers. Nitrogen shows a stable $-$III state in ammonia and its compounds, as well as in a wide variety of organonitrogen compounds. Phosphorus has an unstable $-$III state in the hydride and also forms acids and salts, which contain direct P$-$H bonds, in the III and I states. In the normal states of V and III a variety of coordination numbers is found. The MX_3 compounds, where M = any element in the Group and X = H, halogen or pseudohalogen, or organic group, are pyramidal with a lone pair on M. The MX_5 compounds are trigonal bipyramids in the gas phase and adopt a variety of structures in the solid. A number of MX_4^+ species are known, as well as MX_3.A (where A = any acceptor molecule such as BR_3) and these are tetrahedral.

A few $M^{III}X_5^{2-}$ complexes also exist and these are square pyramids with a lone pair in the sixth position. Finally, a variety of $M^VX_6^-$ and M^VX_5.A compounds are found which are octahedral. Many of these shapes are repeated in compounds which include π bonding. A full set of examples is gathered in Table 15.12.

This Group provides some examples of catenation, but the tendency to form chains is much less than in the carbon Group. Two nitrogen atoms are linked in hydrazine, H_2N-NH_2, and three in the azide ion, N_3^-. The fluorides, N_2F_2 and N_2F_4, also contain N$-$N links, as do a number of organic derivatives of hydrazine. In inorganic compounds, M$-$M links for the other elements of the Group are limited to the di-phosphorus halides, P_2X_4, and the very unstable hydrides, P_2H_4 and As_2H_4. Rather more stable compounds with M$-$M bonds are found in the organic analogues of these dihydrides, R_2M-MR_2, which are formed by all the elements, although the bismuth compounds are unstable.

Trends within the Group are similar to those observed in the boron and carbon Groups. The acidic character of the oxides, and the stability of the V state, decrease in going from nitrogen to bismuth, so that bismuth has only a handful of compounds in the V state and these are unstable and strongly oxidizing. Antimony(V) and arsenic(V) are moderately oxidizing. Phosphorus(V) is very stable, while nitrogen is again oxidizing in the V state, reflecting the differences in formulae and coordination compared with phosphorus. The III state increases in stability as the V state becomes unstable: P(III) and As(III) are reducing, Sb(III) is mildly reducing and Bi(III) is stable. The III state also becomes increasingly stable in cationic forms for antimony and bismuth. Bi^{3+} probably exists in the salts of strong acids, such as the fluoride, and Sb^{3+} may be present in the sulphate. Both elements exist in solution, and in many salts, as the oxycation, MO^+.

The V state

All the oxides of the V state are known and their structures, where known, are shown in Figure 15.29. N_2O_5 is made by dehydrating nitric acid with phosphorus pentoxide and has the symmetrical structure in the gas phase. The solid is ionized to $NO_2^+NO_3^-$, nitronium nitrate. Higher oxides of

TABLE 15.12 Coordination numbers and stereochemistry in the nitrogen Group

Number of electron pairs	Number of π bonds	Number of non-bonding pairs	Shape	Examples
4	0	0	Tetrahedron	NH_4^+, MR_4^+, PCl_4^+, PBr_4^+
4	0	1	Pyramid	MH_3, MR_3, MX_3 (X = all halogens)
4	1	1	V	NO_2^-
5	0	0	Bipyramid	MF_5, PCl_5, PBr_5, PPh_5
5	1	0	Tetrahedron	MOX_3, MO_4^{3-}, HPO_3^{2-}, $H_2PO_2^-$
5	2	0	Plane	NO_3^-
6	0	0	Octahedron	MF_6^-, PCl_6^-, $SbPh_6^-$?
6	0	1	Square pyramid	SbF_5^{2-}

(M = P, As, Sb and Bi, R = simple alkyl radical, Ph = phenyl)

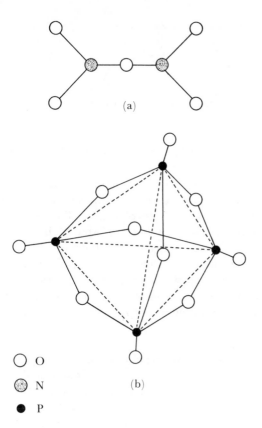

(a)

(b)

○ O
◉ N
● P

FIGURE 15.29 *Structures of pentoxides*, M_2O_5, *of the elements of the nitrogen Group:* (a) N_2O_5, (b) P_4O_{10}

nitrogen, NO_3 and N_2O_6, have been reported from the reaction of ozone on N_2O_5 but little is known of them.

Phosphorus burns in an excess of air to give the pentoxide which has the molecular formula, P_4O_{10}, and, in the vapour, the tetrahedral structure shown in Figure 15.29. This form is also found in the liquid and solid, but prolonged heating of either gives polymeric forms. The environments of each phosphorus atom in P_4O_{10} and in the polymeric forms are similar. In each structure, each phosphorus atom is linked tetrahedrally to four oxygen atoms, three of which are shared with three other phosphorus atoms.

The pentoxides, M_2O_5, of arsenic, antimony, and bismuth are made by oxidizing the element or the trioxide. Increasingly powerful oxidizing agents are needed from arsenic to bismuth, and Bi_2O_5 is not obtained in a pure stoichiometric form. All three pentoxides are solids of unknown structure, and all lose oxygen readily on heating to give the trioxides.

Oxyacids and oxyanions of the V state are a very important class of compounds in the chemistry of this Group. The nitrogen, phosphorus, and arsenic compounds are included in Table 15.13. Antimony and bismuth do not form acids in the V state. The oxyanions may be made by reaction of the pentoxides in alkali or by oxidation of the trioxides in an alkaline medium. The bismuthates are strongly oxidizing, the best-known being the sodium salt, $NaBiO_3$, which is used to identify manganese in qualitative analysis by oxidizing it to permanganate. The antimonates are oxidizing, but more stable than the bismuthates, and are octahedral ions,

$Sb(OH)_6^-$. This is in contrast to the tetrahedral coordination to oxygen shown in the phosphates and arsenates. Both phosphorus and arsenic form acids in the V state, both the mononuclear acid, H_3MO_4, and polymeric acids. The polyphosphoric acids include chains, rings, and more complex structures, all formed from PO_4 tetrahedra sharing oxygen atoms. A wide variety of polyphosphate ions also exists. These pyro-, meta- and other poly-phosphates are stable and hydrolyse only slowly to the ortho- acid, H_3PO_4. Polyarsenates also exist in similar forms, but these are much less stable and readily revert to H_3AsO_4. Nitrogen, of course, forms only one acid in the V state; nitric acid, HNO_3, and one ion, nitrate, NO_3^-. Here, the coordination number of only three to oxygen, and the planar structure, reflect the presence of $p_\pi - p_\pi$ bonding between the first row elements, nitrogen, and oxygen.

The V state halides are limited to the pentafluorides, PF_5, AsF_5, SbF_5, and BiF_5, together with PCl_5, $SbCl_5$ and PBr_5. This reflects the decreasing stability of the V state from phosphorus to bismuth: while the non-existence of $AsCl_5$ is an example of the 'middle element anomaly' already discussed. All the structures so far determined show that the pentahalides are trigonal bipyramidal in the gas phase, but these structures alter in the solid, reflecting the instability of five-coordination in crystal lattices. PCl_5 ionizes in the solid to $PCl_4^+ PCl_6^-$ while PBr_5 ionizes to $PBr_4^+ Br^-$. The cations are tetrahedral and PCl_6^- is octahedral. SbF_5 also attains a more stable configuration in the solid, this time by becoming six-coordinated through sharing fluorine atoms between two antimony atoms in a chain structure. Except for BiF_5, all the pentafluorides readily accept F^- to form the stable octahedral anion, MF_6^-. PF_5, especially, is a strong Lewis acid and forms $PF_5.D$ complexes with a wide variety of nitrogen and oxygen donors (D). The pentachlorides are similar but weaker acceptors. They do accept a further chloride ion, and $SbCl_6^-$ in particular is well established and stable: PCl_6^- is mentioned above. It has recently been shown that $AsCl_6^-$ may also be prepared if it is stabilized by a large cation, thus $Et_4N^+ AsCl_6^-$ has been prepared. All these pentahalides may be made by direct combination or by halogenation of the trihalides, MX_3. The pentafluorides, and PCl_5, are relatively stable, but PBr_5 and $SbCl_5$ readily lose halogen at room temperature (to give the trihalides) and are strong halogenating agents, as is BiF_5 which is by far the most reactive pentafluoride. A number of mixed pentahalides, such as PF_3Cl_2, are also known. They are formed by treating the trihalide with a different halogen:

$$PF_3 + Cl_2 = PF_3Cl_2$$

These are similar to the pentahalides in many ways. For example, PF_3Cl_2 is covalent on formation but passes over into the ionic form $PCl_4^+ PF_6^-$ on standing. Another example is AsF_3Cl_2 which also appears to be ionized to $AsCl_4^+ AsF_6^-$, showing that the second ionic component of the hypothetical $AsCl_5$ exists. It has also been shown that $AsCl_5.OPMe_3$ and $AsCl_5.2OPPh_3$ occur. The latter is conducting in acetonitrile and is formulated as an ionic species $(Ph_3PO)_2AsCl_4^+ Cl^-$

but the former may be an octahedral species containing the $AsCl_5$ unit. It is evident that, although free $AsCl_5$ definitely does not exist it, or its possible ionic components, can be stabilized by suitable donors or in the presence of suitable counter-ions.

Although nitrogen cannot form a penthalide, as only four valency orbitals are present in the second level, it is interesting that the cation NF_4^+ exists showing that nitrogen(V) can bond to fluorine, in a species where only sigma interactions are possible.

In their covalent forms, the mixed pentahalides PX_nY_{5-n}, such as PF_3Cl_2, have structures in which the most electronegative halogens occupy the two axial positions.

Nitrogen, in the V state, forms the oxyfluoride NOF_3 which is pyramidal with the double bond character in the NO bond. Phosphorus and arsenic form a range of oxyhalides in the V state. Three compounds of phosphorus are known, POX_3 where X = F, Cl, or Br, and one arsenic compound $AsOF_3$. $POCl_3$ may be made from PCl_3 or from the pentachloride and pentoxide:

$$PCl_3 + \tfrac{1}{2}O_2 = POCl_3$$
$$\text{or} \quad P_4O_{10} + 6PCl_5 = 10POCl_3$$

The other phosphorus compounds are made from the oxychloride. $AsOF_3$ is made by the action of fluorine on a mixture of $AsCl_3$ and As_2O_3. All these are tetrahedral, $X_3M{=}O$, with $p_\pi - d_\pi$ bonding between the O and M atoms (Figure 3.35). Phosphorus gives the corresponding sulphur and selenium compounds, PSX_3 and $PSeX_3$: again illustrating the marked stability of four-coordinated P(V).

Three pentasulphides are found in this Group. P_4S_{10} has the same structure as P_4O_{10}—and there is also a compound $P_4O_6S_4$ which again has the same structure, with the oxygen atoms bridging along the edges of the tetrahedron, and a sulphur atom attached directly to each phosphorus. The structures of As_2S_5 and Sb_2S_5 are unknown. All three sulphides may be formed by direct reaction between the elements, and arsenic and antimony pentasulphides are also formed by the action of H_2S on As(V) or Sb(V) in solution.

One final important class of phosphorus (V) compounds is that of the phosphonitrilic halides. If PCl_5 is heated with ammonium chloride, compounds of the formula $(PNCl_2)_x$ result:

$$PCl_5 + NH_4Cl = (PNCl_2)_3 + (PNCl_2)_4 + (PNCl_2)_x + HCl$$

The corresponding bromides may be made in a similar reaction, and the chlorines may also be replaced by groups such as F, NCS, or CH_3 and other alkyl groups, either by substitution reactions, or by using the appropriate starting materials. When x = 3 or 4, the six- or eight-membered rings shown in Figure 15.30, are formed. Similar rings have been identified for x values up to 17 in the case of chlorides and fluorides and for x = 6 for the bromides. In addition, for large values of x, linear polymers are formed, of accurate formula $Cl(PNCl_2)_xPCl_4$. In the ring compounds, the trimer and pentamer are planar, while the tetramer and

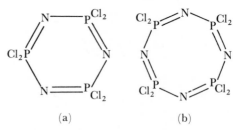

FIGURE 15.30 (a) *Trimeric and* (b) *tetrameric phosphonitrilic chlorides*

hexamer are puckered. The nature of the bonding in the rings, and also in the chain compounds, is not yet clearly determined but probably involves π bonding between nitrogen p orbitals and phosphorus d orbitals. In the trimeric chlorides, it has been suggested that this π bonding involves a strong interaction above and below the plane of the ring, as in benzene, and also a weaker interaction in the plane of the ring. In the non-planar tetramer, this second type of π bonding can make a stronger contribution.

The III state

The III state is reducing for nitrogen, phosphorus, and arsenic, and the stable state for antimony and bismuth. Among the oxygen compounds, all the oxides, M_2O_3, and all the oxyanions are known, but the free acids of the III state are found only for nitrogen, phosphorus, and possibly arsenic. This points to an increase in basicity down the Group, as expected.

The oxide of nitrogen, N_2O_3, is found as a deep blue solid or liquid. It is formed by mixing equimolar proportions of NO and NO_2, and it reverts to these two components in the gas phase at room temperature. It is thus the least stable of the nitrogen oxides. The solid is thought to exist in two forms; one with the symmetrical structure, $O-N-O-N-O$, and the other with a N−N bond. N_2O_3, or an equimolar mixture

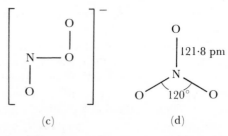

FIGURE 15.31 *The structures of the nitrogen oxyanions:* (a) *hyponitrite,* (b) *nitrite,* (c) *pernitrite,* (d) *nitrate*

of NO and NO_2, gives nitrous acid when dissolved in water, and nitrites when dissolved in alkali.

Phosphorus(III) oxide results when phosphorus is burned in a deficiency of air. In the vapour phase, it has the formula, P_4O_6, and a structure derived from that of P_4O_{10} by removing the terminal oxygen atoms, see Figure 15.29. The phosphorus trioxide is acidic and reducing, and dissolves in water to give phosphorous acid.

When arsenic is burned in air, the only product is the trioxide, which has the formula As_4O_6, and a similar structure to P_4O_6, in both the gas phase and in the solid. A second form also occurs in the solid, but this structure is not known. Arsenic trioxide is acidic.

Antimony and bismuth also burn in air to give only the trioxides. Antimony trioxide has the form, Sb_4O_6, both in the gas and in the solid, and there is a second solid form. This has a structure consisting of long double chains made up of $\dots -O-Sb-O-Sb-O- \dots$ single chains linked together

TABLE 15.13 Oxyacids and oxyanions of the nitrogen Group

Nitrogen

The nitrogen acids and anions all show nitrogen two- or three-coordinated to oxygen and all (except hyponitrous acid) have $p_\pi - p_\pi$ bonding between N and O.

$H_2N_2O_2$ hyponitrous	$N_2O_2^{2-}$ hyponitrite	Reduction of nitrite by sodium amalgam. Weak acid. Readily decomposes to N_2O.
HNO_2 nitrous	NO_2^- nitrite	Acidify nitrite solution. Free acid known only in gas phase. Weak acid. Aqueous solution decomposes reversibly, $3HNO_2 = HNO_3 + 2NO + H_2O$.
HOONO pernitrous	$(OONO)^-$ pernitrite	$H_2O_2 + HNO_2$. Free acid very unstable but pernitrites are found.
HNO_3 nitric	NO_3^- nitrate	Oxidation of NH_3 from Haber process. Strong acid. Powerful oxidizing agent in concentrated solution.

The structures of these species are shown in Figure 15.31.

Phosphorus

The phosphorus acids and anions all contain four-coordinate phosphorus. In the phosphorus (V) acids, all four bonds are to oxygen, while $P-H$ and $P-P$ bonds are present in the acids and ions of the I and III states.

H_3PO_2 hypophosphorous	$H_2PO_2^-$ hypophosphite	White P plus alkaline hydroxide. Monobasic acid, strongly reducing.
H_3PO_3 phosphorous	HPO_3^{2-} phosphite	Water plus P_2O_3 or PCl_3. Dibasic acid, reducing.
$H_4P_2O_5$ pyrophosphorous	$H_2P_2O_5^{2-}$ pyrophosphite	Heat phosphite: dibasic acid with $P-O-P$ link. Reducing.
$H_4P_2O_6$ hypophosphoric	$P_2O_6^{4-}$ hypophosphate	Oxidation of red P, or of P_2I_4, in alkali, gives sodium salt, which gives the acid on treatment with H^+. Tetrabasic acid with a $P-P$ link. Resistant to oxidation to phosphoric acid.
H_3PO_4 (ortho) phosphoric	PO_4^{3-} phosphate	P_2O_5 or PCl_5 plus water. Stable.

Also pyrophosphate $(O_3POPO_3)^{4-}$ and polyphosphates $(O_3P[OPO_2]_nOPO_3)^{(4+n)-}$ — Formed by heating orthophosphate. Cyclic polyphosphates with $n = 1, 2, \dots$ up to 6 and linear species with $n = 3, 4, 5$ have been identified (compare Figure 6.2). Higher polymers have probably three-dimensional cross-linked structures.

The structures of these phosphorus species are shown in Figure 15.32.

Arsenic

H_3AsO_3?, or $As_2O_3.xH_2O$ arsenious acids	$HAsO_3^{2-}$ and more complex forms arsenites	Formed from the trioxide or trihalides. The acid may simply be the hydrated oxide, but the arsenites are well-established, in mononuclear and polynuclear forms. The arsenic(III) species are reducing and thermally unstable.
H_3AsO_4 arsenic acid	AsO_4^{3-} arsenate	$As + HNO_3 \rightarrow H_3AsO_4.\frac{1}{2}H_2O$. Tribasic acid and moderately oxidizing. Arsenates are often isomorphous with the corresponding phosphates.
Condensed arsenates		A number of these exist in the solid state but are less stable than polyphosphates and rapidly hydrolyse to AsO_4^{3-}.

As far as they are known, arsenic anions and acids have the same structures as the corresponding phosphates.

Antimony and bismuth give no free acids, though salts of the III and V states are found, and are discussed in the text. Coordination to oxygen is always six, not four as with phosphorus and arsenic.

through an oxygen atom on each antimony. Antimony trioxide is amphoteric. Bi_2O_3 is yellow (all the other compounds are white) and exists in a number of solid forms. These are not known in detail, but some at least contain BiO_6 units in a distorted prism arrangement. Bismuth trioxide is basic only. Antimony and bismuth, in the III state, commonly exist in solution and in their salts as the MO^+ ion, as already noted.

The oxyanions and acids of phosphorus(III) and arsenic(III) are included in Table 15.13 and in Figure 15.32.

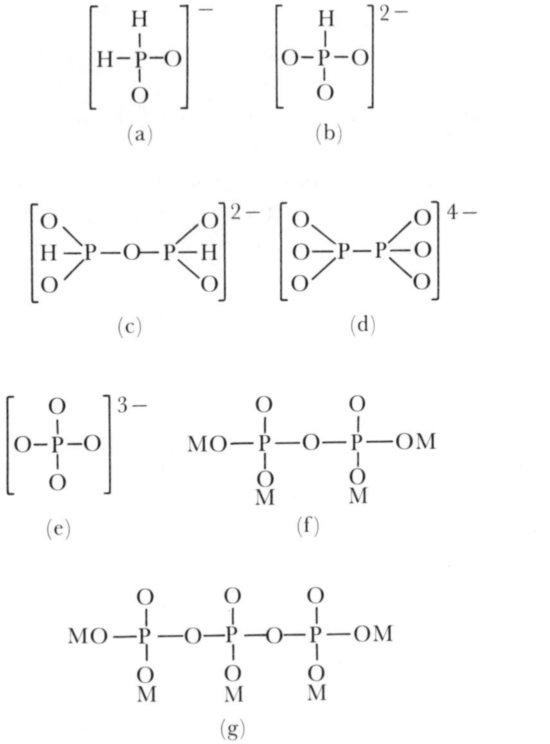

$$\left[\begin{array}{c} H \\ | \\ H-P-O \\ | \\ O \end{array}\right]^{-} \qquad \left[\begin{array}{c} H \\ | \\ O-P-O \\ | \\ O \end{array}\right]^{2-}$$

(a)   (b)

$$\left[\begin{array}{c} O \quad\quad O \\ H-P-O-P-H \\ O \quad\quad O \end{array}\right]^{2-} \qquad \left[\begin{array}{c} O \quad\quad O \\ O-P-P-O \\ O \quad\quad O \end{array}\right]^{4-}$$

(c)   (d)

$$\left[\begin{array}{c} O \\ | \\ O-P-O \\ | \\ O \end{array}\right]^{3-} \qquad \begin{array}{c} O \quad\quad O \\ \| \quad\quad \| \\ MO-P-O-P-OM \\ | \quad\quad | \\ O \quad\quad O \\ M \quad\quad M \end{array}$$

(e)   (f)

$$\begin{array}{c} O \quad\quad O \quad\quad O \\ \| \quad\quad \| \quad\quad \| \\ MO-P-O-P-O-P-OM \\ | \quad\quad | \quad\quad | \\ O \quad\quad O \quad\quad O \\ M \quad\quad M \quad\quad M \end{array}$$

(g)

FIGURE 15.32 *The structures of phosphorus oxyanions:* (a) *hypophosphite,* (b) *phosphite,* (c) *pyrophosphite,* (d) *hypophosphate,* (e) *orthophosphate,* (f) *pyrophosphate,* (g) *trimetaphosphate*

M = monovalent cation

All the structures known contain four-coordinated phosphorus or arsenic, and the III oxidation state results from the presence of a direct P−H or As−H bond (which, of course, does not ionize to give a proton). Although the stable form of phosphorous acid is the tetrahedral form shown in Figure 15.32, there is some evidence from exchange studies (compare section 2.3) for the transient existence of the pyramidal $P(OH)_3$ form. Organic derivatives of this form, $P(OR)_3$, are well-known. Phosphorous acid, and the phosphites, are reducing, and also disproportionate readily as the oxidation state free energy diagram, Figure 15.25, shows:

$$4H_3PO_3 = 3H_3PO_4 + PH_3$$

Arsenites are also mildly reducing with a potential, in acid, of 0·56 V with respect to arsenic(V) acid, so that arsenites are rather weaker reducing agents than iron(II) in acid solution.

Arsenic, antimony, and bismuth all form trisulphides,

M_2S_3. The arsenic compound exists as As_4S_6, while the antimony and bismuth sulphides have polymeric chain structures. All are formed by the action of H_2S on solutions of the element in the III state. Phosphorus does not form a simple III sulphide but, instead, gives three lower sulphides, P_4S_3, P_4S_5, and P_4S_7, whose structures are shown in Figure 15.33.

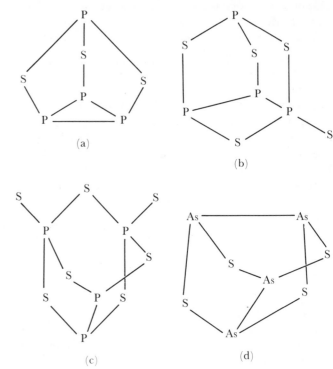

FIGURE 15.33 *The structures of* (a) P_4S_3, (b) P_4S_5, (c) P_4S_7, (d) As_4S_4

The recently-discovered fourth member of the series, P_4S_9, has a structure formed from P_4S_{10} by removing one of the terminal S atoms. The oxygen analogue, P_4O_9, is similar. All may be regarded as derived from the P_4 tetrahedron with varying numbers of sulphur atoms inserted in the edges or attached to the phosphorus atoms. Arsenic also forms a sulphide, As_4S_3, which may have the same structure as P_4S_3. A third arsenic sulphide, As_4S_4 (called *realgar*), is also known and its structure is shown in Figure 15.33.

The five elements of the nitrogen Group all give all the trihalides, MX_3. The least stable are the nitrogen compounds, where only NF_3 is stable. Nitrogen trichloride decomposes explosively, while NBr_3 and NI_3 can be prepared only as unstable ammoniates, such as $NI_3.6NH_3$, which detonate when freed from excess ammonia. Nitrogen trifluoride is a stable, pyramidal molecule, formed by the reaction of nitrogen with excess fluorine in the presence of copper. NF_3 has almost no donor power and has only a very low dipole moment, as the strong N−F bond polarizations practically cancel out the effect of the lone pair. The structure of one ammoniate, $NI_3.NH_3$, has been determined crystallographically. It consists of chains of NI_4 tetrahedra, formed by sharing corners, and the NH_3 molecules are bonded to the non-bridging iodines.

The other sixteen trihalides of the Group are all relatively stable molecules, with the expected trigonal pyramidal structure in the gas phase. The pyramidal structure is also found in many of the solids, but other solid state structures are also found, particularly among the tri-iodides which adopt layer lattices with the metal atoms octahedrally surrounded by six halogen atoms. In addition to the simple trihalides, MX_3, a wide variety of mixed halides, MX_2Y or $MXYZ$, are found.

All the trihalides are readily hydrolysed, giving the oxide, oxyanion, or—in the cases of antimony and bismuth—the oxycation, MO^+. They may act as donor molecules, by virtue of the lone pair, and PF_3 in particular has been widely studied. It is rather less reactive to water and more easily handled than the other trihalides. PF_3 complexes resemble the corresponding carbonyls, for example, $Ni(PF_3)_4$ is similar to $Ni(CO)_4$. The trihalides also show acceptor properties, especially the trifluorides and chlorides. Complex ions, such as SbF_5^{2-}, are formed, and SbF_3 also gives the interesting dimeric ion, $Sb_2F_7^-$, (Figure 15.34). The trihalides are common reaction intermediates, and, for example, react with silver salts to give products such as $P(NCO)_3$, and with organometallic reagents to give a wide variety of organic derivatives, MR_3.

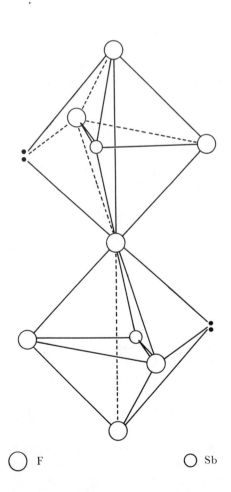

○ F   ○ Sb

FIGURE 15.34 *The structure of the ion,* $Sb_2F_7^-$

Other oxidation states

A number of oxidation states other than V and III are found, especially among the oxides and oxyacids. Nitrogen forms the I, II, and IV oxides, N_2O, NO, and NO_2 or N_2O_4. Nitrous oxide, N_2O, is formed by heating ammonium nitrate solution or hydroxylamine and is fairly unreactive. It has a π-bonded linear structure, NNO. Nitric oxide has already been discussed from the structural point of view. Although it has an unpaired electron, it shows little tendency to dimerize,

$$\begin{matrix} N & \cdots & O \\ | & & | \\ O & \cdots & N \end{matrix},$$

though some association to rather loose dimers,

occurs in the liquid and solid. It rapidly reacts with oxygen to form NO_2. This is also an odd electron molecule but does dimerize readily to dinitrogen tetroxide. The solid is entirely N_2O_4 and this dissociates slightly in the liquid and increasingly in the gas phase until, at 100 °C, the vapour contains 90 per cent NO_2. NO_2 is brown and paramagnetic and has an angular structure with ONO = 134°. N_2O_4 is colourless

and diamagnetic with the symmetrical structure

$$\begin{matrix} O & & O \\ & N-N & \\ O & & O \end{matrix}$$

and a very long N–N bond of 175 pm (compare 145 pm for the N–N single bond length in a molecule like hydrazine).

Phosphorus forms a third oxide, in addition to the pentoxide and trioxide, which is of formula PO_2 and is formed, along with red phosphorus, by heating the trioxide above 210 °C. This compound has a vapour density corresponding to P_8O_{18} and it behaves chemically as if it contains both P(V) and P(III). It may also contain P–P bonds as it reacts with iodine to give P_2I_4. Its structure is unknown.

Heating either Sb_4O_6 or Sb_2O_5 in air above 900 °C gives an oxide of formula SbO_2. This consists of a network of fused SbO_6 octahedra containing both Sb(III) and Sb(V). A corresponding AsO_2 may exist.

There are also two oxyacids of low oxidation states in the Group. These are hyponitrous acid, $H_2N_2O_2$, with nitrogen (I), and hypophosphorous acid, H_3PO_2, with phosphorus (I). These are included in Table 15.13. Nitrogen forms its low oxidation state in hyponitrous acid by $p_\pi - p_\pi$ and N–N bonding, while phosphorus in hypophosphorous acid is tetrahedral and the low oxidation state arises from two direct P–H bonds.

A number of lower halides are found in the Group. These are P_2Cl_4, P_2I_4, NF_2, N_2F_4, N_2F_2, and BiCl. The nitrogen fluorides are made by direct combination using less fluorine than required for NF_3. N_2F_2 has a planar structure

$$\begin{matrix} & & F \\ & & \nearrow \\ N & - & N \\ \nwarrow & & \\ F & & \end{matrix}$$ which is most stable in this *trans* form, but which

may also occur in the *cis* form $\begin{matrix} N-N \\ \diagup \quad \diagdown \\ F \qquad F \end{matrix}$. There is a π bond

between the two N atoms which have each a lone pair. N_2F_4

is a gas with a skew structure similar to that of hydrazine;

$$
\begin{array}{cc}
\text{F} & \text{F} \\
\diagdown & \diagup \\
& \text{N}-\text{N} \\
\diagup & \diagdown \\
\text{F} & \text{F}
\end{array}
$$

In the gas and liquid phases it undergoes reversible dissociation to NF_2:

$$N_2F_4 = 2NF_2$$

similar to that of N_2O_4. NF_2 is an angular molecule and contains an unpaired electron. For an odd-electron species, it has fairly high stability resembling NO, NO_2, and ClO_2 in this respect.

The two phosphorus halides have a direct $P-P$ bond·and have the X_2PPX_2 structure. As there is a lone pair on the phosphorus, these structures are probably skew as in hydrazine and N_2F_4.

The lower chloride of bismuth is much more complicated. Although a number of species are reported to be formed when bismuth is dissolved in molten bismuth trichloride, the only one whose structure is known is that resulting from a concentrated bismuth solution with the accurate formula $Bi_{12}Cl_{14}$. This is a complicated structure with 48 Bi atoms and 56 Cl atoms in the unit cell. These are arranged as $4Bi_9^{5+}$, $8BiCl_5^{2-}$, and $2Bi_2Cl_8^{2-}$ units. The $BiCl_5^{2-}$ ion is a square pyramid with Bi(III) and resembles the SbF_5^{2-} ion mentioned earlier. In the structure these units are weakly linked to each other to form a chain. The $Bi_2Cl_8^{2-}$ unit contains Bi(III) and consists of two square pyramids sharing an

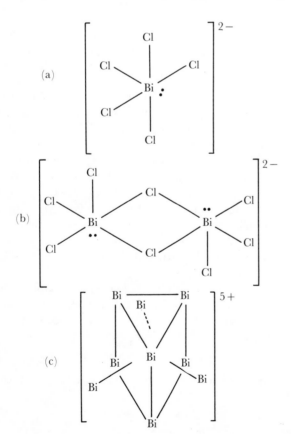

FIGURE 15.35 *The structures of* (a) $BiCl_5^{2-}$, (b) $Bi_2Cl_8^{2-}$, *and* (c) Bi_9^{5+}

edge of the base with their apices *trans* to each other. The Bi_9^{5+} unit has six Bi atoms at the corners of a somewhat distorted trigonal prism, and the other three Bi atoms above the rectangular faces. These units are shown in Figure 15.35. Thus the solid may be written $(Bi_9^{5+})_2(BiCl_5^{2-})_4Bi_2Cl_8^{2-}$.

Negative oxidation states appear in the hydrides, MH_3, and in the organic compounds, NR_3. The other elements, except possibly arsenic (see Table 2.13), are of lower electronegativity than carbon in alkyl groups, and their organic compounds correspond to positive oxidation states. This distinction is not a useful one and is best regarded—as in the case of carbon chemistry in general—as an accidental result of the definitions. The organic compounds, MR_3, and organohydrides such as R_2PH, behave in a similar way to the hydrides but with the $M-C$ bond stronger than $M-H$. An extensive organometallic chemistry of this Group exists which cannot be discussed here. As well as MR_3, analogous to MH_3, ions NR_4^+, PR_4^+, AsR_4^+, and SbR_4^+ exist, which are tetrahedral and analogous to NH_4^+. The phosphorus analogue of NH_4^+ appears to exist, for example in PH_4I, which is prepared from ·HI and PH_3. However, the phosphonium halides are relatively unstable and readily decompose to phosphine and hydrogen halide. They are much more covalent than the ammonium salts.

Although no pentavalent hydride, MH_5, exists, the pentaphenyls MPh_5 of P, As, Sb, and Bi exist, as does PMe_5. PPh_5 and $AsPh_5$ are trigonal bipyramids in shape, but $SbPh_5$ is probably a square pyramid. $SbPh_5$ reacts with PhLi to give the octahedral $SbPh_6^-$ ion.

Nitrogen is found in the $-II$ state in hydrazine and its organic derivatives, R_4N_2, and in the $-I$ state in hydroxylamine, NH_2OH. Nitrogen is found in a chain of three atoms in hydrazoic acid, HN_3, and in the azides. These are prepared from amide and nitrous oxide:

$$2NaNH_2 + N_2O = NaN_3 + NaOH + NH_3$$

The azide ion is linear, and isoelectronic with CO_2. The acid, and its organic derivatives, are bent at the nitrogen bonded to the substituent:

$$
\begin{array}{c}
\text{N}-\text{N}-\text{N} \\
\diagdown \\
\text{H}
\end{array}
$$

with the NNH angle equal to $110°$.

15.7 The oxygen Group, ns^2np^4

References to the properties of oxygen Group elements will be found in the following places:

Ionization potentials	Table 2.8
Atomic properties and electron configurations	Table 2.5
Radii	Table 2.10, 2.11, 2.12
Electronegativities	Table 2.13
Redox potentials	Table 5.3

Table 15.14 lists some of the properties of the elements and Figures 15.36 and 15.37 show the variations with Group position of a number of parameters and of the oxidation state free energies.

TABLE 15.14 *Properties of the elements of the oxygen Group*

Element	Symbol	Oxidation states	Coordination numbers	Availability
Oxygen	O	$-II, (-I)$	$1, 2, (3), (4)$	common
Sulphur	S	$-II, (II), IV, VI$	$2, 4, 6$	common
Selenium	Se	$(-II), (II), IV, VI$	$2, 4, 6$	common
Tellurium	Te	II, IV	6	common
Polonium	Po	II, IV		very rare

$$O_3 = 3/2 \, O_2 \quad \Delta H = -142 \text{ kJ mol}^{-1}$$

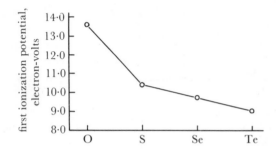

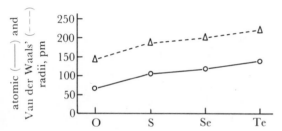

FIGURE 15.36 *Some properties of the oxygen Group elements*
The radii of the anions, X^{2-}, are almost identical with the corresponding Van der Waals' radii.

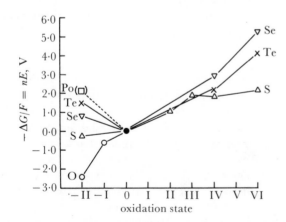

FIGURE 15.37 *Oxidation state free energies of the oxygen Group elements*
Oxygen shows only negative oxidation states. The $-II$ state becomes decreasingly stable from O to Po. The positive oxidation states show the drop in stability of the VI state after S and the tendency for intermediate states to be the more stable for Se and Te. Polonium values are not known.

Oxygen occurs both as the O_2 molecule and as ozone, O_3. O_2 is paramagnetic (section 3.4) and has a dissociation energy of 489 kJ mol^{-1}. It is pale blue in the liquid and solid states. Ozone is usually formed by the action of an electric discharge on O_2. Pure O_3 is deep blue as the liquid with m.p. $= -250$ °C and b.p. $= -112$ °C. It is diamagnetic and explodes readily as the decomposition to oxygen is exothermic and easily catalysed:

$$O_3 = 3/2 \, O_2, \; \Delta H = -142 \text{ kJ mol}^{-1}$$

Ozone has an angular structure with the OOO angle equal to 117°. Of the eighteen valency electrons, four are held in the sigma bonds and eight in lone pairs on the two terminal oxygens. Two are present as a lone pair on the centre oxygen, leaving four electrons and the three p orbitals perpendicular to the plane of the molecule to form a pi system. The three p orbitals combine to form a bonding, a nonbonding and an antibonding three-centre π orbital and the four electrons occupy the first two. There is thus one bonding π orbital over the three O atoms giving, together with the σ bonds, a bond order of about one and a third. The bond length is 128 pm which agrees with this; O—O for a single bond in H_2O_2 is 149 pm while O=O in O_2 is 121 pm. Ozone is a strong oxidizing agent, especially in acid solution where the potential is 2·07 V (Table 5.3). It is exceeded in oxidizing power only by fluorine, oxygen difluoride, and some radicals.

Sulphur shares with phosphorus the ability to form a wide variety of allotropic forms in all three phases. All the structures are not fully established. The main interrelations are shown in Figure 15.38. The commonest form is the puckered eight-membered ring, shown in Figure 15.39, which is found

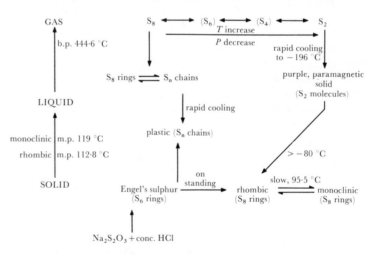

FIGURE 15.38 *The interconversion of the allotropes of sulphur*

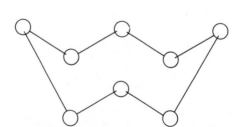

FIGURE 15.39 *The structure of the S_8 ring*

in the rhombic and the monoclinic forms of the solid which differ in the overall packing of the rings in the crystal. Other allotropes with S_8 rings are reported, but with little structural detail, and the eight-membered ring is found in the liquid and gas phases. Engel's sulphur contains S_6 rings, arranged in the chair form, while chains are found in plastic sulphur and in other forms. Recently, a number of larger rings have been made by reacting polysulphur hydrides and chlorides, as in the reactions

$$H_2S_6 + S_4Cl_2 = S_{10} + 2HCl$$
$$H_2S_8 + S_4Cl_2 = S_{12} + 2HCl$$

and the cyclodecasulphur and cyclododecasulphur exist as staggered rings. These ring forms differ in molecular parameters as follows:

	S−S (pm)	SSS angle	dihedral angle
S_8 orthorhombic	204·8	107° 54′	98° 42′
S_8 monoclinic	206		
S_6	205·9	102° 12′	74°
S_{12}	205·5	106° 30′	

Hexasulphur, S_6, is more reactive than octasulphur, and this reflects the strain in the smaller ring which is shown up in the small dihedral angle.

A further interesting modification of sulphur contains the S_2 unit which occurs in the gas at high temperatures. On rapid cooling, this may be condensed to a purple solid. Both the gaseous molecule and the solid are paramagnetic, like the isoelectronic O_2. Related to these forms of the element are the cations S_n^{2+}, $n = 4, 8, 18$, and possibly, S_8^+, which are rings and have aromatic character. Se_4^{2+} and Te_4^{2+} are well established and are planar, and it is likely that S_4^{2+} is similar.

Selenium also forms a number of allotropes but less is known about these. Two different red forms containing Se_8 rings are found but the stable modification is the grey form. This contains infinite spiral chains of selenium atoms with a weak, metallic interaction between atoms in adjoining chains. The chain form of sulphur also contains spiral chains but has no metallic character.

Tellurium has only one form in the solid. This is silvery-white, semi-metallic, and isomorphous with grey selenium, but with rather more metallic interaction as the self-conductivity is higher. In the vapour, selenium and tellurium have a greater tendency than sulphur to exist as diatomic and monatomic species.

Polonium is a true metal with two allotropic forms, in both of which the coordination number is six. Polonium is found in small amounts in thorium and uranium minerals (where it was discovered by Marie Curie) as one of the decay products of the parent elements, Th or U. It can now be more readily made by bombarding bismuth in a reactor:

$$^{209}_{83}Bi + ^1_0n = ^{210}_{83}Bi = ^{210}_{84}Po + ^0_{-1}e$$

All polonium isotopes have relatively short half-lives and this isotope, polonium 210, is the longest-lived with $t_{\frac{1}{2}} = 138$ days. The isotopes all have high activity and the handling

problems are severe so that polonium chemistry is not fully explored.

The trends observed in the nitrogen Group appear in a more marked degree in the chemistry of the oxygen Group. The −II state is well-established, not only for oxygen, but also for sulphur and even selenium, and it often occurs as the M^{2-} ion in the compounds of these elements. Apart from the Group state of VI, both the IV and the II states are observed in the Group, the IV state becoming the most stable one for tellurium, and the IV or II state the most stable one for polonium. Apart from the fluorine compounds, oxygen appears only in the −II state, except for the −I state in the peroxides. A variety of shapes is again observed. The VI state is usually octahedral or tetrahedral while the presence of the non-bonding pair in compounds of the IV state gives distorted tetrahedral shapes in the halides, MX_4, and pyramidal or V shapes in oxyhalides or oxides. Coordination numbers higher than six are uncommon but TeF_6 does add F^- to give TeF_7^- or TeF_8^{2-}.

Oxygen has little in common with the other elements in the Group apart from formal resemblances in the −II states. Polonium shows signs of a distinctive chemistry, similar to that of lead or bismuth, but the difficulties of studying the element mean that there are many gaps in its known chemistry. Tellurium resembles antimony in showing a strong tendency to be six-coordinate to oxygen, while sulphur and selenium show four-coordination. Sulphur is stable in the VI state, while selenium(VI) is mildly oxidizing.

The tendency to catenation shown by the elements is continued in the compounds, especially those of sulphur. Oxygen forms O−O links in the peroxides and in O_2F_2, and three- and four-membered chains may be present in the unstable fluorides, O_3F_2 and O_4F_2. Sulphur forms many chain compounds, particularly the dichlorosulphanes, S_xCl_2, where x may be as high as 100, and the polythionates, $(O_3SS_xSO_3)^{2-}$ where the compounds with $x = 1$ to 12 have been isolated. Chain-forming tendencies are slight for selenium and absent in tellurium chemistry.

It is convenient to discuss oxygen chemistry separately from that of the other elements.

Oxygen

Oxygen combines with all elements except the lighter rare gases, and most of the oxides are listed in Tables 12.3 and 15.2. Their properties have been discussed already. The change in acidity from the s element oxides to the oxides of the light p elements will already be familiar to the reader. Oxygen shares with the other first row elements the ability to form $p_\pi - p_\pi$ bonds to itself and to the other first row elements, and it is sufficiently reactive and electronegative to form $p_\pi - d_\pi$ bonds with the heavier elements.

As fluorine is more electronegative than oxygen, compounds of the two are oxygen fluorides and are discussed here. The halogen oxides of Cl, Br, and I are to be found in section 15.8. Four oxygen fluorides are well established— OF_2, O_2F_2, O_3F_2, and O_4F_2—and two more have been reported recently— O_5F_2 and O_6F_2. OF_2 is formed by the action of fluorine on dilute sodium hydroxide solution, while the other five result from the reaction of an electric discharge on an O_2/F_2 mixture. An increasing proportion of O_2 and decreasing temperature are required to make the highest members of the series. Thus O_5F_2 and O_6F_2 were prepared in a discharge at $-213\ °C$ using a 5/1 mixture of O_2 and F_2. The use of higher proportions of oxygen does not give O_7 or higher species at this temperature but it might be possible to make these by working at a still lower temperature. All six compounds are very volatile with boiling points well below room temperature.

OF_2 is the most stable of the four oxygen fluorides. It does not react with H_2, CH_4, or CO on mixing, although such mixtures react explosively on sparking. Mixtures of OF_2 and halogens explode at room temperature. OF_2 reacts slowly with water:

$$OF_2 + H_2O = O_2 + 2HF$$

and is readily hydrolysed by base:

$$OF_2 + 2OH^- = O_2 + 2F^- + H_2O$$

The structure of OF_2 is V-shaped, like H_2O, with the FOF angle $= 103.2°$ and the bond length, $O-F = 141.8\ pm$.

The other five oxygen fluorides are much less stable. O_2F_2 decomposes at its boiling point of $-57\ °C$, O_3F_2 at $-157\ °C$, O_4F_2 at $-170\ °C$, and O_5F_2 and O_6F_2 are stable only to $-200\ °C$. Indeed, O_6F_2 can explode if it is warmed rapidly up to $-185\ °C$. All are red or red-brown in colour. Electron spin resonance studies on O_2F_2 and O_3F_2 have shown that these contain the OF radical to the extent of 0·1 per cent for O_2F_2 and 5 per cent for O_3F_2. It is likely that the higher members also contain free radicals, and these may account for the colour. O_2F_2 has a skew structure,
$$\begin{matrix} & & F \\ O&-&O \\ F & & \end{matrix}$$
similar to that of hydrogen peroxide and the others may be chains, FO_nF.

When O_2F_2 is reacted at low temperatures with molecules which will accept F^-, such as BF_3 or PF_5, oxygenyl compounds result which probably contain the O_2^+ cation:

$$O_2F_2 + BF_3 = (O_2)BF_4 + \tfrac{1}{2}F_2 \ (\text{at} -126\ °C)$$

Oxygenyl tetrafluoroborate decomposes at room temperature to BF_3, O_2, and fluorine. The oxygenyl group may be replaced by the nitronium ion, NO_2^+:

$$O_2BF_4 + NO_2 = (NO_2)BF_4 + O_2$$

The oxygenyl ion may also be formed directly from gaseous oxygen by reaction with the strongly oxidizing platinum hexafluoride molecule:

$$O_2 + PtF_6 = (O_2)^+(PtF_6)^-$$

Oxygen forms two compounds with hydrogen, water and hydrogen peroxide, H_2O_2. Water, and its solvent properties, is discussed in Chapter 5 and hydrogen-bonding in $O-H$ systems is included in section 8.7. Here, we mention the recent work on an unusual form of water called 'anomalous water' or 'polywater'. It is found that water formed in quartz or glass capillaries has unusual properties, especially a high density of about $1.4\ g\ cm^{-3}$ and a freezing point of $-40\ °C$. Its structure is still debated but one suggestion is that it consists of symmetrical hydrogen-bonded $O-H-O$ units (isoelectronic with $F-H-F^-$; see section 8.7) which are formed into hexagons, perhaps on a template provided by the silica surface.

Hydrogen peroxide may be prepared by acidifying an ionic peroxide solution (section 9.2) or, on a large scale, by electrolytic oxidation of a sulphate system:

$$2HSO_4^- = S_2O_8^{2-} + 2H^+ + 2e^-$$

$$S_2O_8^{2-} + 2H_3O^+ = 2H_2SO_4 + H_2O_2.$$

(The intermediate is called persulphate.) The hydrogen peroxide is distilled off and may be concentrated by fractionation. Pure H_2O_2 is a pale blue liquid, m.p. $-0.89\ °C$, b.p. $150.2\ °C$, with a high dielectric constant and is similar to water in its properties as an ionizing solvent. It is, however, a strong oxidizing agent and readily decomposes in the presence of catalytic amounts of heavy metal ions:

$$H_2O_2 = H_2O + \tfrac{1}{2}O_2$$

Hydrogen peroxide has the skew structure shown in Figure 8.6b.

The ionic peroxides, and the peroxy compounds of the transition elements, have been discussed in previous Chapters 9 and 12). There are also a number of covalent peroxy species, acids or oxyanions, which may be regarded as derived from the normal oxygen compounds by replacing $-O-$ by $-O-O-$; just as $H-O-H$ is related to $H-O-O-H$. The best known examples are permonosulphuric acid (Caro's acid) H_2SO_5—which occurs as an intermediate in the persulphate oxidation above—, perdisulphuric acid (persulphuric) $H_2S_2O_8$, perphosphoric acid, H_3PO_5, and perdicarbonic acid, $H_4C_2O_6$. Only the structures of the sulphuric acids are definitely established and these are related to the oxygen compounds as shown in Figure 15.40. The others are probably $(HO)_2(HOO)PO$ and $(HO)_2C-O-O-C(OH)_2$.

There also exist a number of other compounds which are commonly termed per-acids, such as percarbonic acid and perboric acid, which may be only simple acids with hydrogen peroxide of crystallization. For example, the so-called percarbonic acid salts, such as $Li_2CO_4.H_2O$, are more probably carbonates, $Li_2CO_3.H_2O_2$. However, structural information is not available to finally determine these cases.

Finally, it must be noted that the terms, peracid and peroxide, are properly applied only to compounds which contain the $-O-O-$ group, which may be regarded as derived from hydrogen peroxide. In the older literature, and in much technical literature, some higher oxides such as MnO_2 are termed peroxides. This usage is incorrect on the modern convention of nomenclature.

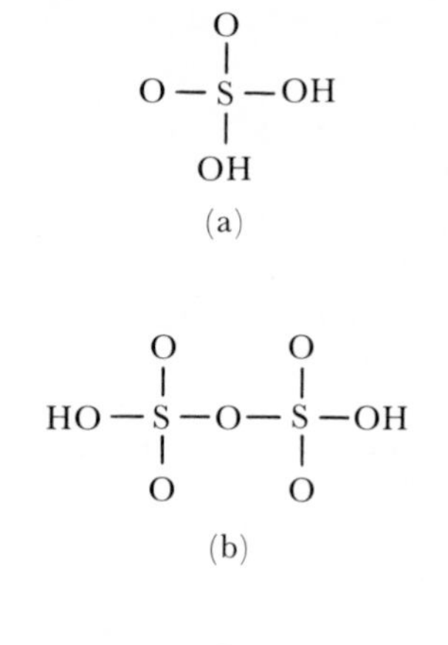

(a)

(b)

(c)

(d)

FIGURE 15.40 *The structures of the sulphuric and per-sulphuric acids* (a) Sulphuric acid, (b) pyrosulphuric acid, (c) peroxymono-sulphuric acid, (d) peroxydi-sulphuric acid.

The other elements: the VI state

The trioxides, MO_3, of the VI state are formed by sulphur, selenium, tellurium, and polonium, and all these elements (except possibly polonium) form oxyanions. The acids are included in Table 15.15. Sulphur and selenium are four-coordinated to oxygen in the acid (H_2MO_4) and in the anion (MO_4^{2-}), while tellurium forms the dibasic, six-coordinated acid, $Te(OH)_6$, and two series of six-coordinated tellurates, $TeO(OH)_5^-$ and $TeO_2(OH)_4^{2-}$. Sulphur, selenium, and tellurium all burn in air to form the dioxide, MO_2, and the trioxides are made by oxidizing these. Sulphuric acid, H_2SO_4, is formed by dissolving SO_3 in water, but selenic acid, H_2SeO_4, is more readily made by oxidizing the selenium(IV) acid, H_2SeO_3, made by dissolving SeO_2. Sulphuric acid and the sulphates are stable, while selenic acid, the selenates, telluric acid, and the tellurates are all oxidizing agents, although they are typically slow in reacting.

Only sulphur gives a condensed acid in the VI state, and it forms only the binuclear species, pyrosulphuric acid $H_2S_2O_7$. Condensed anions include $S_3O_{10}^{2-}$ and $S_5O_{16}^{2-}$ as well as $M_2O_7^{2-}$ (M = S, Se), but tellurium gives no polyanions at all. As condensation is also restricted in the IV state to binuclear compounds of sulphur and selenium, it will be seen that the tendency to condense is much slighter in this Group than it was in the nitrogen Group.

Oxidizing power of the VI state increases down the Group, and polonium(VI) is strongly oxidizing, so that the existence of the oxide and oxyanions is in some doubt.

The other main representatives of the VI state are the hexafluorides, MF_6. These compounds are all relatively stable but with reactivity increasing from sulphur to tellurium. SF_6 is extremely stable and inert, both thermally and chemically, SeF_6 is more reactive and TeF_6 is completely hydrolysed after 24 hours contact with water. The hexafluorides are more stable, and much less reactive as fluorinating agents, than the tetrafluorides of these elements. All the hexafluorides result from the direct reaction of the elements, and dimeric molecules, M_2F_{10}, are also found in the reaction mixtures. These compounds are rather more reactive than the hexafluorides, but still reasonably stable. The coordination is octahedral, F_5M-MF_5. They hydrolyse readily, with fission of the $M-M$ bond, and S_2F_{10} reacts with chlorine to give the mixed hexahalide, SF_5Cl. Tracer studies reveal the existence of a volatile polonium fluoride, probably PoF_6.

TeF_6 adds F^- to form a seven- or eight-coordinated anion and also reacts with other Lewis bases, such as amines, to give eight-coordinated adducts like $(Me_3N)_2TeF_6$.

The VI state is also found in sulphur and selenium oxyhalides, and in halosulphonic acids. Sulphur gives sulphuryl halides, SO_2X_2, with fluorine and chlorine, and also mixed compounds SO_2FCl and SO_2FBr. Only one selenium analogue, SeO_2F_2, is found. Sulphur also gives a number of more complex oxyhalides, some of which are shown in Figure 15.41. The oxyhalides are formed from halogen and dioxide, or from the halosulphonates. SO_2F_2 is inert chemically but the other compounds are much more reactive and hydrolyse readily.

Three halosulphonic acids are known, FSO_3H, $ClSO_3H$, and $BrSO_3H$. The structures are tetrahedral 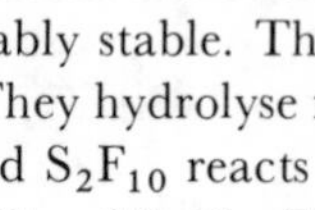 and are formed from sulphuric acid by replacing an OH group by a halogen atom. They are strong monobasic acids but

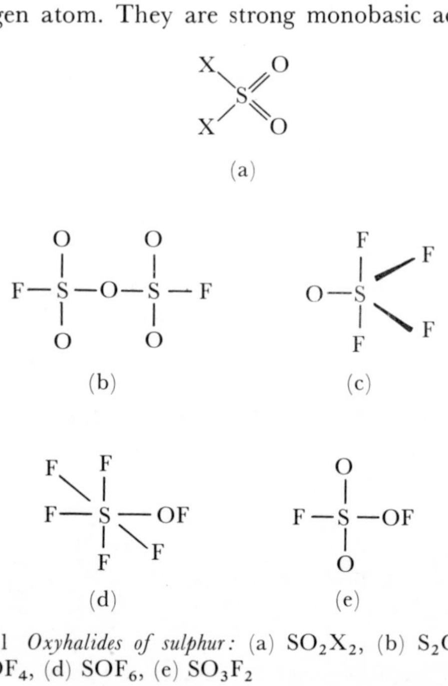

(a)

(b)

(c)

(d)

(e)

FIGURE 15.41 *Oxyhalides of sulphur:* (a) SO_2X_2, (b) $S_2O_5F_2$ (or Cl_2), (c) SOF_4, (d) SOF_6, (e) SO_3F_2

only fluosulphonic acid is stable and forms stable salts, fluo-sulphonates FSO_3^-, which are similar in structure and solubilities to the perchlorates.

The IV state

The IV state is represented by oxides, MO_2, halides, MX_4, oxyhalides, MOX_2, acids, and anions, for all four elements. Sulphur(IV) is reducing, selenium(IV) is mildly reducing (going to Se(VI)) and also weakly oxidizing (going to the element), tellurium(IV) is the most stable state of tellurium, while polonium(IV) is weakly oxidizing (going to Po(II)). In addition, a number of complexes are known.

The oxides are given in Table 15.2a and the acids and anions in Table 15.15. The dioxides result from direct reaction of the elements. SO_2 and SeO_2 dissolve in water to give the acids, but TeO_2 and PoO_2 are insoluble and the parent acids do not exist. Tellurium and selenium oxyanions result from the solution of the oxides in bases. Salts of one condensed acid exist: these are the pyrosulphites, $S_2O_5^{2-}$, which have the unsymmetrical structure with a $S-S$ bond, $^-O_3SSO_2^-$.

Figure 15.42 gives the structures of some of the oxides and oxions. The IV state oxides and oxyanions have an un-shared electron pair (in monomeric structures) and thus have unsymmetrical structures. The stable form of the di-oxide shows an interesting transition from monomeric covalent molecule, SO_2, through polymeric covalent, SeO_2, to ionic forms for TeO_2 and PoO_2. Sulphur, selenium, and tellurium dioxides are acidic and dissolve in bases. PoO_2 appears to be more basic than acidic and dissolves in acids as well as forming polonites with strong bases.

One compound intermediate between the IV and VI oxides is reported. This is Se_2O_5, formed by controlled heating of SeO_3. This compound is conducting in the fused state and it has been suggested that it is $SeO^{2+}SeO_4^{2-}$, a salt of Se(IV) and Se(VI).

The known tetrahalides are listed in Table 15.2b. The missing ones are SBr_4, SI_4 and SeI_4 while SCl_4 and $SeBr_4$ are also unstable, SCl_4 decomposing at -31 °C. The tellurium tetrahalides are markedly more stable, even TeI_4 being stable up to 100 °C. $TeCl_4$, $TeBr_4$ and $SeCl_4$ are all stable up to 200 °C and the tellurium compounds to 400 °C. All the tetrafluorides are known and these are rather more stable than the other tetrahalides to thermal decomposition. They are much more reactive than the hexafluorides and act as strong, though selective, fluorinating agents. All four polonium tetrahalides are known and resemble the tellurium analogues, though they seem to be rather less stable. The structures of four tetrahalides are known. SF_4, SeF_4 and $TeCl_4$, are all found as the distorted tetrahedron derived from the trigonal bipyramid with one equatorial position occupied by an unshared pair of electrons (see section 3.5). Tellurium tetrafluoride forms a polymeric structure in which square pyramids are linked into chains by sharing corners, $TeF_3(F_{2/2})$. Thus the lone pair is still sterically active. SCl_4 and $SeCl_4$, which exist only in the solid, give Raman spectra suggesting an ionic formulation, $MCl_3^+Cl^-$, but there is no other evidence supporting this. These two com-

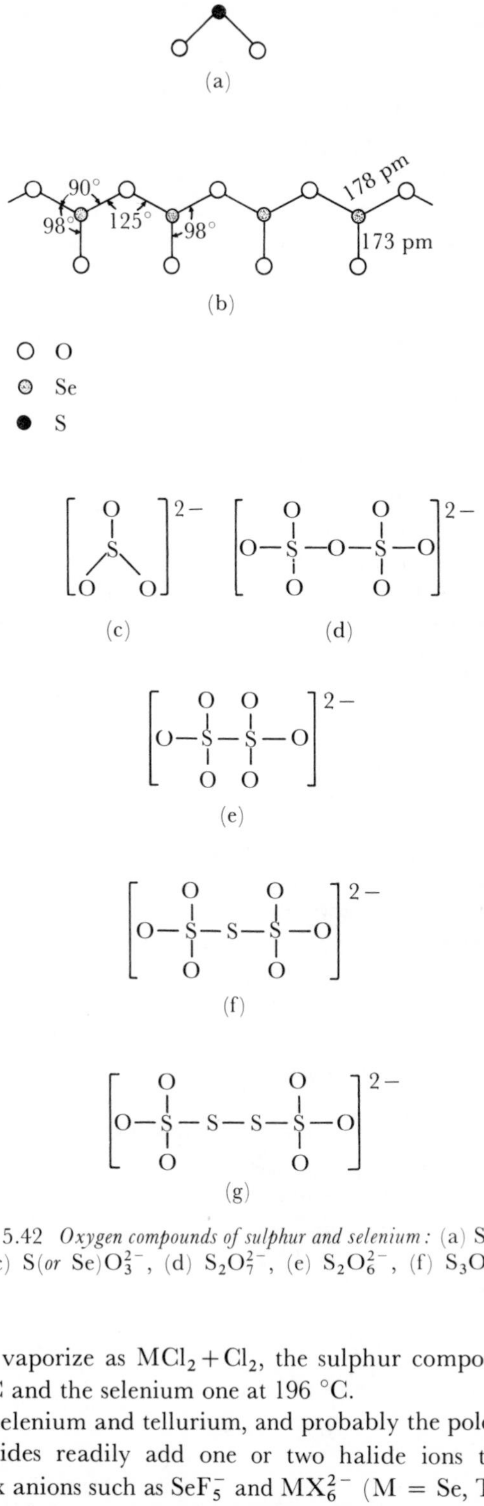

FIGURE 15.42 *Oxygen compounds of sulphur and selenium:* (a) SO_2, (b) SeO_2, (c) $S(or\ Se)O_3^{2-}$, (d) $S_2O_7^{2-}$, (e) $S_2O_6^{2-}$, (f) $S_3O_6^{2-}$, (g) $S_4O_6^{2-}$

pounds vaporize as $MCl_2 + Cl_2$, the sulphur compound at -30 °C and the selenium one at 196 °C.

The selenium and tellurium, and probably the polonium, tetrahalides readily add one or two halide ions to give complex anions such as SeF_5^- and MX_6^{2-} (M = Se, Te, Po: X = any halogen). These hexahalo (IV) complex ions are interesting from the structural point of view as they contain seven electron pairs. Structural determinations on a number of compounds show that the MX_6^{2-} group is octahedral, but the group may be regular or distorted. The electron pair repulsion theory, discussed in Chapter 3, would imply that the non-bonded pair occupied a spatial position and should lead to a distorted octahedron. It is possible, however, that with large, heavy, central atoms, the non-bonded pair might be accommodated in an inner orbital rather than in a particular spatial direction. Structural studies so far reported, indicate that two groups of structures are found. In the hexafluorides of the IV state, MF_6^{2-}, as in the isoelectronic IF_6^- and XeF_6, the structure is distorted octahedral, indicating a sterically active lone pair. However, in the heavier halides like MCl_6^{2-}, the structure is regular octahedral. It is proposed here that the dominant factor is repulsion between halogen atoms which is minimized in the regular structure. In this form, the lone pair would have to occupy an s orbital, presumably.

The structure of one pentafluoro complex has been determined. This is $KTeF_5$ which contains TeF_5^- ions which are square pyramids, with the lone pair to complete the octahedron. In this, as in TeF_4, the square pyramid has the Te below the base, that is, the base F atoms are bent up away from the lone pair position. It will be recalled from section 12.6, that the opposite distortion occurs when there are five ligands and no lone pair.

Oxyhalides of the IV state are formed only by sulphur and selenium and have the formula MOX_2. Thionyl halides, SOF_2, $SOCl_2$, $SOBr_2$, and $SOFCl$ are known while the selenyl fluoride, chloride, and bromide exist. Thionyl chloride is made from SO_2 and phosphorus pentachloride:

$$SO_2 + PCl_5 = SOCl_2 + POCl_3$$

The other thionyl halides are derived from the chloride. Selenyl chloride is obtained from SeO_2 and $SeCl_4$. These oxyhalides are stable near room temperature but decompose on heating to a mixture of dioxide, halogen, and lower halides. SOF_2 is relatively stable to water, but all the other compounds hydrolyse violently. These compounds have an unshared electron pair on the central element and are pyramidal in structure (Figure 15.43). In $SeOF_2$, FSeF = $92 \cdot 6°$ and OSeF = $105 \cdot 5°$. The oxyhalides act as weak donor molecules through the lone pairs on the oxygen atoms, and also as acceptors using d orbitals on the sulphur or selenium. The structure of one of the latter complexes, $SeOCl_2.2py$ (py = pyridine) has been determined. The lone pair occupies one octahedral position giving a square pyramid.

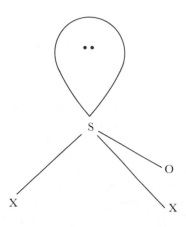

FIGURE 15.43 *The structure of thionyl halides,* SOX_2

The II state: the −II state

If poly-sulphur compounds are excluded, the II state is represented by a more limited range of compounds than the IV or VI states, but it is more fully developed than the I state of the nitrogen Group. Tellurium and polonium form readily oxidized monoxides, MO, and a number of dihalides, MX_2, exist. No difluorides occur (apart from a possible SF_2) but the general pattern of stability of the other dihalides resembles that of the tetrahalides, stability decreasing from tellurium to sulphur and from chloride to iodide. $TeCl_2$, $TeBr_2$, $PoCl_2$, and $PoBr_2$ are stable; $SeCl_2$ and $SeBr_2$ decompose in the vapour while SCl_2, the only sulphur dihalide, decomposes at 60 °C. No di-iodide is known although there is a lower iodide of tellurium, $(TeI)_n$. SCl_2 is unstable with respect to dissociation to $S_2Cl_2 + Cl_2$.

The −II state is found in the hydrides, H_2M, and in the anions, M^{2-}. With the more active metals, sulphur, selenium, and tellurium all form compounds which are largely ionic, although they have increasing metallic, and alloy-like, properties as the electronegativity of the metal increases. The tendency to form anions increases from the nitrogen Group (where the evidence for M^{3-} ions, apart from N^{3-}, is limited), through the oxygen Group, (where the M^{2-} ion is more widely found), to the halogens where the X^- ion is the most stable form for all the elements. This trend in anionic behaviour reflects the increasing electronegativity of the elements, and the greater ease of forming singly-charged species over doubly- or triply-charged ones.

Compounds involving catenation

A variety of sulphur compounds with $S-S$ bonds is found, including anions, hydrides, halides, and oxygen compounds. Polysulphides are formed by boiling sulphide solutions with sulphur and include cases like BaS_n ($n = 3$ or 4) and Cs_2S_6. Large cations appear to be necessary for stability, and decomposition to sulphide and sulphur occurs easily. The structures of these three polysulphides are known and all involve chains of sulphur atoms which have SSS angles ranging from $103°$ in S_3^{2-} to $109°$ in S_6^{2-}.

Polysulphur hydrides, called *sulphanes*, are formed either from the polysulphides or from the chlorosulphanes:

$$S_n^{2-} + 2HCl = H_2S_n + 2Cl^-$$
$$S_nCl_2 + 2H_2S = H_2S_{n+2} + 2HCl$$

Compounds up to $n = 8$ have been isolated and the existence of higher members has been demonstrated by chromatographic studies. All are yellow and range from gaseous H_2S to increasingly viscous liquids as the chain length increases. The sulphanes may be interconverted by heating—indeed any one member readily converts to an equilibrium mixture of the others—and all ultimately revert to sulphur and H_2S, although slowly. The structures are not known but are probably chains like the polysulphide ions.

Polysulphur halides, chlorosulphanes or bromosulphanes, include S_2Cl_2 and S_2Br_2—which are much more stable than the corresponding SX_2 compounds—and compounds S_nCl_2 and S_nBr_2. In the case of the chlorosulphanes, there is evidence that n may be as high as 100. However, individual members of the series have been isolated only up to $n = 8$ for both the chloro- and bromo-sulphanes. These compounds are all formed by dissolving sulphur in S_2Cl_2. The latter results from the reaction of Cl_2 on molten sulphur. It has a skew $Cl-S-S-Cl$ structure, similar to H_2O_2, with an SSCl angle of $103°$. The higher chlorosulphanes probably have chain structures similar to those of the anions. Reaction of a suitable chlorosulphane and a sulphane provides a specific route to some of the ring forms of sulphur by elimination of hydrogen chloride. For example,

$$H_2S_2 + S_4Cl_2 = S_6 + 2HCl.$$

Among other interesting ring formations are those involving metals. The ion PtS_{15}^{2-} contains three six-membered rings PtS_5 in which the Pt occupies one position in a chair conformation. The six S atoms, which are bonded to Pt from the three rings form an octahedron around platinum. A similar ring is found in $(C_5H_5)_2TiS_5$, and in the selenium analogue. Selenium also forms chain halides and the two compounds Se_2Cl_2 and Se_2Br_2 are well-established.

Among the oxides, only S_2O contains a $S-S$ bond. This compound was, for a long time, reported as SO but the most recent work has established the existence of S_2O and suggests that SO is a mixture of S_2O and SO_2. S_2O is formed by the action of an electric discharge on sulphur dioxide and it is unstable at room temperature. The structure SSO is proposed.

A variety of oxyacids and oxyanions with $S-S$ bonds exists, among which are the polythionic acids, $H_2(O_3S-S_n-SO_3)$ where n varies from 0 to 12, and a miscellaneous group of compounds including thiosulphate, dithionite and pyrosulphite. All these compounds are included in Table 15.15.

In the polythionic acids, there is a marked difference in stability between dithionic acid, $H_2S_2O_6$, and the higher members. As dithionic acid contains no sulphur atom which is bonded only to sulphur, it is much more stable than the other polythionic acids which do contain such sulphurs, see Table 15.15. Dithionates result from the oxidation of sulphites:

$$2SO_3^{2-} + MnO_2 + 4H_3O^+ = Mn^{2+} + S_2O_6^{2-} + 6H_2O$$

and the parent acid may be recovered on acidification. Dithionic acid and the dithionates are moderately stable, and the acid is a strong acid. The structure, O_3S-SO_3, has an approximately tetrahedral arrangement at each sulphur, with π bonding between sulphur and the oxygen atoms.

The reaction of H_2S and SO_2 gives a mixture of the polythionates from $S_3O_6^{2-}$ to $S_{14}O_6^{2-}$, while specific preparations for each member also exist, as in the preparation of tetrathionate in volumetric analysis by oxidation of thiosulphate by iodine:

$$2S_2O_3^{2-} + I_2 = S_4O_6^{2-} + 2I^-$$

These compounds have very unstable parent acids, which readily decompose to sulphur and sulphur dioxide, but the anions are somewhat more stable. The structures are all established as $O_3S-S_n-SO_3$ with sulphur chains which resemble those in sulphur polyanions. As with the other examples of sulphur chain compounds, each polythionic acid readily disproportionates to an equilibrium mixture of all the others.

Thiosulphate, $S_2O_3^{2-}$, is formed by the reaction of sulphite with sulphur. The free acid is unstable, but the alkali metal salts are well-known in photography and in volumetric analysis. The structure of the thiosulphate ion has not been determined, but is almost certainly tetrahedral, corresponding to sulphate with one oxygen replaced by sulphur. This conclusion is supported by exchange studies with radioactive sulphur, which have demonstrated the presence of two different types of sulphur atom.

Dithionite ions, $S_2O_4^{2-}$, result from the reduction of sulphite with zinc dust. The free acid is unknown, and the salts are used in alkaline solution as reducing agents. Dithionite (also called hyposulphite or hydrosulphite) decomposes readily:

$$2S_2O_4^{2-} + H_2O = S_2O_3^{2-} + 2HSO_3^-$$

and the solutions are also oxidized readily in air. The dithionite ion has the unusual structure shown in Figure 15.44. The oxygen atoms are in the eclipsed configuration and the $S-S$ bond is very long. The $S-O$ bond lengths show that π

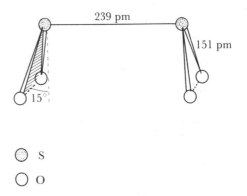

FIGURE 15.44 *The structure of the dithionite ion, $S_2O_4^{2-}$*

bonding exists between the sulphur and oxygen atoms, and there is also a lone pair on each sulphur atom. The S atoms must thus make use of d orbitals and it is proposed that the unusual configuration and the long $S-S$ distance arise from the use of a d orbital.

Sulphur-Nitrogen ring compounds

Sulphur and nitrogen form a number of ring compounds, of which the best-known is tetrasulphur tetranitride, S_4N_4. This is formed by reacting the chlorosulphane of approximate composition SCl_3 with ammonia. It is a yellow-orange solid which is not oxidized in air, although it can be detonated by shock. The structure is shown in Figure 15.45. The four nitrogen atoms lie in a plane, and the four atoms of sulphur form a flattened tetrahedron, interpenetrating the N_4 square. Alternatively, the structure may be regarded as an

TABLE 15.15 Oxyacids and oxyanions of sulphur, selenium, and tellurium

SULPHUR

Sulphur shows coordination numbers up to four, the most common being tetrahedral; S—S bonds are common among the lower acids.

H_2SO_4 sulphuric acid	HSO_4^- and SO_4^{2-} sulphate	Stable, strong, dibasic acid: formed from SO_3 and water. Structure, Figure 15.40.
$H_2S_2O_7$ pyrosulphuric acid	$HS_2O_7^-$ and $S_2O_7^{2-}$ pyrosulphate	Strong, dibasic acid: formed from SO_3 and H_2SO_4 (anions by heating HSO_4^-), loses SO_3 on heating. Sulphonating agent. Structure, Figure 15.40. Also anions $S_3O_{10}^{2-}$ and $S_5O_{16}^{2-}$.
H_2SO_3 sulphurous acid	HSO_3^- and SO_3^{2-} sulphites	Existence of free acid doubtful. $SO_2 + H_2O$ gives a solution containing the anions but this loses SO_2 on dehydration. Reducing and weak, dibasic acid. Structure, pyramidal SO_3^{2-} ion; lone pair on S.
	$S_2O_5^{2-}$ pyrosulphite	No free acid. Formed by heating HSO_3^- or by $SO_2 + SO_3^{2-}$.

polythionic acids

$H_2S_2O_6$ dithionic acid	$S_2O_6^{2-}$ dithionates	Acid stable only in dilute solution, anions stable. Formed by oxidation of sulphites and stable to further oxidation, or to reduction. Strong, dibasic acid. Structure, Figure 15.42.
	$S_nO_6^{2-}$ ($n = 3$ to 6) polythionates	Free acids cannot be isolated; anions formed by reaction of SO_2 and H_2S or arsenite. Unstable and readily lose sulphur, reducing. Structures contain chains of S atoms, Figure 15.42.
$H_2S_2O_4$ dithionous acid (or hyposulphurous)	$S_2O_4^{2-}$ dithionite	Acid prepared by zinc reduction of sulphurous acid solution, and salts (called also hyposulphites or hydrosulphites) prepared by zinc reduction of sulphites. Unstable in acid solution, but salts are stable in solid or alkaline media, powerful reducing agents. Decompose to sulphite and thiosulphate. Structure, Fig. 15.44, contains S—S link.
	$S_2O_3^{2-}$ thiosulphate	Prepared in alkaline media by action of S with sulphites. Perfectly stable in absence of acid, but gives sulphur in acid media. Mild reducing agent, as in action with I_2 which gives tetrathionate, $S_4O_6^{2-}$. Structure, Figure 15.42, derived from sulphate.
	SO_2^{2-} sulphoxylate	Best known as the cobalt salt from $CoS_2O_4 + NH_3 = CoSO_2 + (NH_4)_2SO_3$. The zinc salt may also exist. Unstable and reducing, structure probably V-shaped.

The peroxy acids H_2SO_5 and $H_2S_2O_8$, corresponding to sulphuric acid and pyrosulphuric acid, also exist. Structures are shown in Figure 15.40.

SELENIUM

Selenium commonly shows a coordination number of four: a smaller number of selenium acids than sulphur acids is found as the Se—Se bond is weaker.

H_2SeO_4 selenic acid	$HSeO_4^-$ and SeO_4^{2-} selenates	Formed by oxidation of selenites. Strong dibasic acid, oxidizing. Similar structures to sulphur compounds.
	$Se_2O_7^{2-}$ pyroselenate	No acid, formed by heating selenates.
H_2SeO_3 selenous acid	$HSeO_3^-$ and SeO_3^{2-} selenite	Selenium dioxide solution. Similar to S species, but less reducing and more oxidizing. Structure contains pyramidal SeO_3^{2-} ion.

Chain anions of selenium have not been found, but selenium (and tellurium) may form part of the polythionate chain; as in $SeS_4O_6^{2-}$ and $TeS_4O_6^{2-}$ where the Se or Te atoms occupy the central position in the chain.

TELLURIUM

Tellurium, like the preceding elements in this Period, is six-coordinate to oxygen.

H_6TeO_6 telluric acid	$TeO(OH)_5^-$ and $TeO_2(OH)_4^{2-}$ tellurates	Prepared by strong oxidation of Te or TeO_2. Structure is octahedral $Te(OH)_6$, and only two of the protons are sufficiently acidic to be ionized, and then, only weakly. The acid and salts are strong oxidizing agents.
	tellurites	TeO_2 is insoluble and no acid of the IV state is formed. Tellurites, and polytellurites, are formed by fusing TeO_2 with metal oxides.

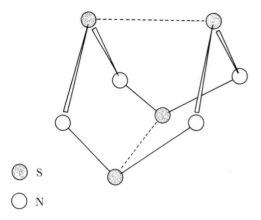

○ S
○ N

FIGURE 15.45   *The structure of tetrasulphur tetranitride*

S_8 ring with every second sulphur replaced by a nitrogen. $S-S$ distances are fairly short, 259 pm, showing that there is some weak bonding across the puckered ring between opposite sulphur atoms.

A number of other ring compounds exist, including $F_4S_4N_4$, where the substituent F atoms are on the sulphur atoms of the S_4N_4 ring, and $S_4N_4H_4$ with the H atoms on the nitrogens. A different eight-membered ring is found in S_7NH, again with H bonded to the nitrogen. This hydrogen is fairly acidic and can be replaced by a number of metals. Different ring sizes are also found, as in $S_3N_3Cl_3$ and S_4N_3Cl.

15.8 The fluorine Group, ns^2np^5 (the halogens)

Reference to properties of the halogens are included in the following Tables and Figures:

Ionization potentials	Table 2.8
Electron affinities	Table 2.9
Atomic properties and electron configurations	Table 2.5
Radii	Table 2.10, 2.11, 2.12
Electronegativities	Table 2.13
Redox potentials	Table 5.3
Uses as nonaqueous solvents	Chapter 5

Table 15.16 lists some of the properties of the elements, Figure 15.46 shows the variations with Group position of a number of important parameters, and Figure 15.47 gives the oxidation state free energies.

TABLE 15.16 Properties of the elements of the fluorine Group

Element	Symbol	Oxidation states	Coordination numbers	Availability
Fluorine	F	−I	1, (2)	common
Chlorine	Cl	−I, I, III, V, VII	1, 2, 3, 4	common
Bromine	Br	−I, I, III, V, VII	1, 2, 3, 5	common
Iodine	I	−I, I, III, V, VII	1, 2, 3, 4, 5, 6, 7	common
Astatine	At	−I, I, III?, V		very rare

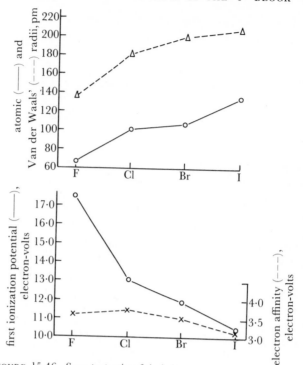

FIGURE 15.46   *Some properties of the halogens*
The radii of the anions, X^-, are almost identical with the corresponding van der Waals' radii. Notice also, that the electron affinities do not follow a regular trend from fluorine to iodine.

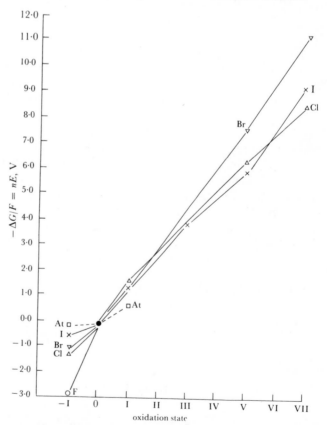

FIGURE 15.47   *Oxidation state free energies of the halogens*
This Group, like the earlier ones, shows a negative state which decreases in stability from the first to the last member of the Group. The positive states are all fairly similar: bromine has the least stable V and VII states while I^V is more stable relative to I^{VII} than is Cl^V relative to Cl^{VII}.

Astatine exists only as radioactive isotopes, all of which are very short-lived. Work has been done using either ^{211}At ($t_{\frac{1}{2}} = 7.2$ h) or ^{210}At ($t_{\frac{1}{2}} = 8.3$ h) and the very high activity necessitates working in 10^{-14}M solutions, and following reactions by coprecipitation with iodine compounds. The chemistry of astatine is therefore little known. Preparation is by the action of α-particles on bismuth, for example:

$$^{209}_{83}Bi + {}^{4}_{2}He = {}^{211}_{85}At + 2{}^{1}_{0}n$$

Coprecipitation with iodine compounds indicates the existence of the oxidation states shown in Table 15.16. The negative state is probably the At$^-$ ion and the positive states, the oxyions, AtO$^-$, AtO$_2^-$, and AtO$_3^-$. Astatine appears to differ from iodine in not giving a VII state, in which it parallels the behaviour of the other heavy elements in its Period, and in not forming a cation, At$^+$. Two interhalogens are known for astatine, AtI and AtBr, and there is evidence for polyhalide ions including astatine.

The other elements of the Group are all well-known and occur mainly as salts containing the halide anion. The free elements are too reactive to exist in nature. Free fluorine, F$_2$, is outstandingly reactive and can be handled only in extremely dry glass systems (otherwise it reacts with glass to give SiF$_4$), in Teflon, or in apparatus made of metals which form a protective layer of fluoride on the surface, such as copper, nickel, or steel. Provided these conditions are rigorously adhered to, fluorine can be handled quite readily in the laboratory.

Fluorine is the most electronegative element and can therefore exist only in the $-$I state. The $-$I state is also the most common and stable state for the other elements, although positive states up to the Group state of VII occur. The positive states are largely found in oxyions and in compounds with other halogens, but iodine does appear to exist in certain systems as the coordinated I$^+$ cation.

The reactivity of the elements decreases from fluorine to iodine. Fluorine is the most reactive of all the elements and forms compounds with all elements except helium, neon, and argon. In all cases, except with the other rare gases, oxygen, and nitrogen, fluorine compounds result from direct, uncatalysed reactions between the elements: the exceptions react in the presence of metal catalysts such as nickel or

TABLE 15.17 The oxides of the halogens

Oxidation state	Chlorine	Bromine	Iodine
I	Cl$_2$O (b. 2°)	Br$_2$O (d. -16°)	
IV	ClO$_2$ (b. 11°)	BrO$_2$ (d. -40°)	I$_4$O$_9$ (d. 130°)
			I$_4$O$_9$ (d. >100°)
V			I$_2$O$_5$ (d. >300°)
VI	Cl$_2$O$_6$ (b. 203°)	BrO$_3$ or Br$_3$O$_8$	
		(d. 20°)	
VII	Cl$_2$O$_7$ (b. 80°)	Br$_2$O$_7$ (?)	

(b = boils   d = decomposes, temperatures in °C

copper. Chlorine and bromine also combine directly, though less vigorously, with most elements while iodine is less reactive and does not react with some elements such as sulphur.

The positive oxidation states

The positive oxidation states occur in the compounds of chlorine, bromine, and iodine with oxygen and fluorine, and of the heavier halogens with the lighter ones. The oxides of the elements are shown in Table 15.17.

Stability of the oxides is greatest for iodine, then chlorine, with bromine oxides the least stable. The higher oxides are rather more stable than the lower ones. Typical preparations are:

$$2Cl_2 + 2HgO = Cl_2O + HgCl_2.HgO \text{ (also for Br}_2O)$$
$$2KClO_3 + 2H_2C_2O_4 = 2ClO_2 + 2CO_2 + K_2C_2O_4 + 2H_2O$$

(also with sulphuric acid: the method using oxalic acid gives the very explosive ClO$_2$ safely diluted with carbon dioxide).

ClO$_3$, BrO$_3$, and I$_4$O$_9$ are prepared by the action of ozone while Cl$_2$O$_7$ and I$_2$O$_5$ are the result of dehydrating the corresponding acids.

The most stable of all the oxides is iodine pentoxide, which is obtained as a stable white solid by heating HIO$_3$ at 200 °C. It dissolves in water to re-form iodic acid, and is a strong oxidizing agent which finds one important use in the estimation of carbon monoxide:

$$I_2O_5 + 5CO = I_2 + 5CO_2$$

The iodine is estimated in the usual way. The other iodine oxides are less stable and decompose to iodine and I$_2$O$_5$ on heating. I$_2$O$_4$ contains chains of IO units which are cross-linked by IO$_3$ units. The structure of I$_4$O$_9$ is less certain but it has been formulated as I(IO$_3$)$_3$. I$_2$O$_7$ has been reported as an orange polymeric solid formed by dehydrating HIO$_4$ with oleum.

The most stable chlorine oxide is the heptoxide which is the anhydride of perchloric acid, HClO$_4$. It is a strong oxidizing agent and dissolves readily in water to give perchloric acid. It has the structure O$_3$Cl$-$O$-$ClO$_3$ with tetrahedral chlorine. Cl$_2$O$_6$ is a red oil which melts at 4 °C. It is dimeric in carbon tetrachloride solution but the pure liquid may contain some of the monomer, ClO$_3$. It readily decomposes to ClO$_2$ and oxygen and reacts explosively with organic materials. Like the other chlorine oxides, it can be detonated by shock. In this, it is more sensitive than the heptoxide but more stable than Cl$_2$O and ClO$_2$. Chlorine dioxide is a yellow gas with an odd number of electrons. It has an angular structure with OClO angle = 118°. It dissolves freely in water to give initially the hydrate, ClO$_2$.8H$_2$O. On exposure to visible or ultraviolet radiation, this gives HCl and HClO$_4$ solution. Chlorine dioxide detonates readily but is gradually being used on a large scale in industrial processes. In these, it is made in situ and always kept well diluted. Dichlorine monoxide, Cl$_2$O is an orange gas which is soluble in water to give a solution containing hypochlorous acid, HOCl. It is a symmetrical, angular molecule with ClOCl angle of 110°. It is a powerful oxidizing agent and highly explosive — indeed, most manipulations of Cl$_2$O are carried out with a

TABLE 15.18 Oxyacids of the halogens

Type	Name		Stability
HOX	hypohalous	X = Cl, Br, I	Cl > Br > I. All are unstable and are known only in solution.
HOXO	halous	X = Cl	HBrO$_2$ possibly exists also.
HOXO$_2$	halic	X = Cl, Br, I	Cl < Br < I. Chloric and bromic in solution only but iodic acid can be isolated as a solid.
HOXO$_3$	perhalic	X = Cl, Br, I	Free perchloric, perbromic, and periodic acids occur.
Also (HO)$_5$IO and H$_4$I$_2$O$_9$ forms of periodic acid			

substantial wall between the experimenter and the compound.

The bromine oxides are the least stable of all the halogen oxides and all decompose below room temperature, though there is some indication that they may be less explosive than the chlorine analogues. Br$_2$O is dark brown, BrO$_2$ is yellow, and BrO$_3$ or Br$_3$O$_8$—it is not clear which is the correct formula—is white. The latter decomposes in vacuum with the evolution of Br$_2$O leaving a white solid which could possibly be Br$_2$O$_7$. No structural information is available on the bromine oxides. They dissolve in water or alkali to give mixtures of the oxyanions.

The oxyacids of the halogens are shown in Table 15.18. Most are obtainable only in solution, although salts of nearly all can be isolated.

Hypochlorous acid, HOCl, occurs to an appreciable extent, 30 per cent, in solutions of chlorine in water, but only traces of HOBr and no HOI are found in bromine or iodine solutions. All three halogens give hypohalite on solution in alkali:

$$X_2 + 2OH^- = XO^- + X^- + H_2O \ldots \ldots \ldots \ldots (1)$$

$$K = \frac{[X^-][XO^-]}{[X_2]} = 7 \times 10^{15} \text{ for Cl, } 2 \times 10^8 \text{ for Br, } 30 \text{ for I}$$

but the hypohalites readily disproportionate to halide and halate:

$$3XO^- = 2X^- + XO_3^- \ldots \ldots \ldots \ldots \ldots (2)$$

with equilibrium constants of 10^{27} for Cl, 10^{15} for Br, and 10^{20} for I. Thus the actual products depend on the rates of these two competing reactions. For the case of chlorine, the formation of hypochlorite by reaction (1) is rapid, while the disproportionation by reaction (2) is slow at room temperatures so that the main products of dissolution of chlorine in alkali are chloride and hypochlorite. For bromine, reactions (1) and (2) are both fast at room temperature so that the products are bromide, hypobromite, and bromate, the proportion of bromate being reduced if the reactions occur at 0 °C. In the case of iodine, reaction (2) is very fast and iodine dissolves in alkali at all temperatures to give iodide and iodate quantitatively:

$$3I_2 + 6OH^- = IO_3^- + 5I^- + 3H_2O$$

The only halous acid definitely established is chlorous acid,

HClO$_2$. This does not occur in any of the disproportionation reactions above and is formed by acidification of chlorites. The latter are themselves formed by reaction of ClO$_2$ with bases:

$$2ClO_2 + 2OH^- = ClO_2^- + ClO_3^- + H_2O$$

Chlorites are relatively stable in alkaline solution and are used as bleaches. In acid, chlorous acid rapidly disproportionates to chloride, chlorate, and chlorine dioxide.

For the halic acids, stability is greatest for iodine. Salts of all three acids are well-known and stable, with a pyramidal structure. The halic acids are stronger acids than the lower ones and are weaker oxidizing agents.

All perhalates exist and perchloric acid and perchlorates are well-known. HClO$_4$ is the only oxychlorine acid which can be prepared in the free state. It, and perchlorates, although strong oxidizing agents, are the least strongly oxidizing of all the oxychlorine compounds. The perchlorates of many elements exist. The ion is tetrahedral and has the important property of being extremely weakly coordinated by cations. It is thus very useful in the preparation of complexes, as metal ions may be introduced to a reaction as the perchlorates, with the assurance that the perchlorate group will remain uncoordinated—contrast this with the behaviour of ions such as halide, carbonate, or nitrite which are often found as ligands in the complexes, e.g.:

$$Co^{2+} + NH_3 + Cl^- \xrightarrow{\text{oxidize}} Co(NH_3)_6^{3+} + Co(NH_3)_5Cl^{2+} +, \text{ etc.}$$

$$\text{but } Co^{2+} + NH_3 + ClO_4^- \xrightarrow{\text{oxidize}} Co(NH_3)_6^{3+} \text{ only}$$

After eluding attempts to make it for many years, perbromic acid and its salts were synthesized in 1968. Perbromate, BrO$_4^-$ was prepared electrolytically, alternatively by oxidation with XeF$_2$, but the most convenient synthesis was found to be oxidation of bromate in alkali by molecular fluorine. Acidification yields perbromic acid which is a strong monobasic acid, stable in solutions up to about 6 M. KBrO$_4$ contains tetrahedral BrO$_4^-$ ions, and this species is predominant in solution with no evidence of a second form as found in periodates. The electrode potentials have been assessed for the reaction

$$XO_4^- + 2H^+ + 2e^- = XO_3^- + H_2O$$

TABLE 15.19 Interhalogens and polyhalide ions

AB	AB_3	AB_5	AB_7	A_3^-		AB_4^-	AB_6^-	A_n^-	A_n^{2-}
ClF	ClF$_3$	(ClF$_5$)	IF$_7$	Br$_3^-$	ClBr$_2^-$	ClF$_4^-$	IF$_6^-$	I$_5^-$	I$_8^{2-}$
BrF	BrF$_3$	BrF$_5$		I$_3^-$	IBr$_2^-$	BrF$_4^-$	(BrF$_6^-$)	I$_7^-$	
(IF)	(IF$_3$)	IF$_5$		BrCl$_2^-$	IBrF$^-$	IF$_4^-$		I$_9^-$	
BrCl	I$_2$Cl$_6$			ICl$_2^-$	IBrCl$^-$	ICl$_4^-$		I$_2$Cl$_3^-$	
ICl				BrI$_2^-$		ICl$_3$F$^-$		I$_2$Cl$_2$Br$^-$	
IBr									

as 1·23 V for X = Cl, 1·76 V for X = Br, and 1·64 V for X = I. Thus perbromate is a somewhat stronger oxidant than perchlorate or periodate, but its oxidizing reactions are sluggish. Thus, the oxidizing power is not the reason for the difficulties found in the synthesis of perbromate. It seems that the preparation from Br(V) requires the surmounting of an activation barrier, and any process proceeding by one-electron additions might have failed because of the instability of the intermediate species.

Thus, although the long-standing anomaly about perbromate no longer exists, it is clear that bromine(VII) is less stable in compounds with oxygen than either chlorine(VII) or iodine(VII). To this extent, bromine still reflects the middle element anomaly.

Periodic acid and periodates occur, like perrhenates, in both four- and six-coordinated forms. Oxidation of iodine in sodium hydroxide solution gives the periodate, Na$_2$H$_3$IO$_6$, and the three silver salts, AgIO$_4$, Ag$_5$IO$_6$, and Ag$_3$IO$_5$ may be precipitated from solutions of this sodium salt under various conditions. Deliquescent white crystals of the acid H$_5$IO$_6$ may be obtained from the silver salt and this loses water in two stages:

$$(HO)_5IO \xrightarrow{80\ °C} H_4I_2O_9 \xrightarrow{100\ °C} (HO)IO_3$$

Salts, such as K$_4$I$_2$O$_9$, of the binuclear acid may be obtained. The I$_2$O$_9^{2-}$ ion has the O$_3$IO$_3$IO$_3$ structure, of two IO$_6$ octahedra sharing one triangular face. In the periodates, the IO$_6$ group is octahedral and the IO$_4$ group is tetrahedral, as expected. The iodine oxyanions thus continue the trend, already observed in the earlier members of this Period, to become six-coordinated to oxygen.

Positive oxidation states for halogens are also found in the interhalogen compounds, where the lighter halogen is in the −I state and the heavier one in a positive state. A similar situation pertains for the mixed polyhalide ions. Table 15.19 lists these compounds.

All the interhalogens of type AB are known, although IF and BrF are very unstable. All the AF$_3$ and AF$_5$ types also occur, although, again, some are unstable. Only IF$_7$ is found in the VII oxidation state, and the only other higher interhalogen is iodine trichloride. All are made by direct combination of the elements under suitable conditions, apart from IF$_7$ which results from the fluorination of IF$_5$. All the compounds are liquids or volatile solids except ClF, which

A_2^+	A_3^+	AB_2^+	AB_4^+	AB_6^+
Cl$_2^+$ (?)	Cl$_3^+$	ClF$_2^+$	ClF$_4^+$ (?)	IF$_6^+$
Br$_2^+$	Br$_3^+$	BrF$_2^+$	IF$_4^+$	
I$_2^+$	I$_3^+$	FCl$_2^+$		
Also I$_4^{2+}$		BrCl$_2^+$		
ClF$^+$ (?)		ICl$_2^+$		

boils at −100 °C. Most boiling points fall between 0 °C and 100 °C.

The halogen fluorides are all very reactive and act as strong fluorinating agents. Reactivity is highest for chlorine trifluoride, which fluorinates as strongly as elemental fluorine. Reactivity falls from the chlorine to the bromine and iodine fluorides, and also falls off as the number of fluorine atoms in the molecule increases. BrF$_3$ and IF$_5$ are particularly useful as fluorinating agents for the production of fluorides of elements in intermediate oxidation states. These two interhalogens, along with the two iodine chlorides ICl and ICl$_3$, undergo self-ionization and are useful as solvent systems, as discussed in Chapter 5. The anions of these systems, BrF$_4^-$, IF$_6^-$, ICl$_2^-$, and ICl$_4^-$, respectively, are among the polyhalide ions in the Table. The cations, BrF$_2^+$, IF$_4^+$, I$^+$, and ICl$_2^+$, are less familiar but may be isolated by adding halide ion acceptors to the interhalogen, for example:

$$BrF_3 + SbF_5 = (BrF_2)^+(SbF_6)^-$$

Although these compounds are formulated as ions, structural studies show that there may be an interaction between the cation and anion. For example, ICl$_2$SbCl$_6$, (from ICl$_3$ and SbCl$_5$) consists of SbCl$_6$ octahedra and angular ICl$_2$ units, but with a weak coordination of two of the chlorines in an octahedron to two different iodines to give a chain structure, Figure 15.48. This structure seems to be quite general, BrF$_2^+$SbF$_6^-$ adopting a very similar form to that shown in Figure 15.48. In both cases, the AB$_2$ unit may be described as the cation, AB$_2^+$, forming two further weak bridges to the anion, SbX$_6^-$, or as covalent with a very distorted square planar AB$_4$ unit. The two A−B distances are sufficiently different to make the former description the more acceptable.

The other interhalogens of the AB type have properties and reactivities which are roughly the average of the properties of the constituent halogens.

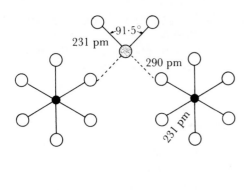

● Sb
◉ I
○ Cl

FIGURE 15.48 *The structure of* ICl_2SbCl_6

The structures of the interhalogens have been largely covered in Chapter 3. ClF_3 and BrF_3 are T-shaped, BrF_5, and probably the other pentafluorides, are square pyramids of fluorine atoms with the bromine atom just below the plane of the four fluorines of the base. IF_7 is approximately a pentagonal bipyramid. The remaining compound is iodine trichloride which has the dimeric structure shown in Figure 15.49.

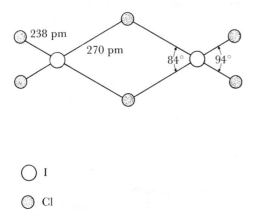

○ I
◉ Cl

FIGURE 15.49 *The structure of iodine trichloride,* I_2Cl_6

A variety of polyhalide ions is known. These are all prepared by the general method of adding halogen or interhalogen to a solution of the halide. The examples with three halogen atoms are linear with the heaviest atom in the middle in the mixed types (although it is not known whether $ClBr_2^-$ and BrI_2^- are symmetrical or obey this rule and have structures $I-I-Br$ or $Br-Br-Cl$). The three-halogen ion, $ClBrI^-$ forms a disordered crystal so its structure is not definitely known. The data are best fitted by the arrangement $Br-I-Cl^-$ with $I-Cl$ longer than $Br-I$ (291 pm and 251 pm respectively). These anions are AB_2L_3 structures, while the three-atom cations like ClF_2^+ have an electron pair less and are V-shaped. Established atom sequences include $F-Cl-F^+$ and $Cl-Cl-F^+$. The AB_4^- compounds are probably all planar ions, with two lone pairs in the

remaining octahedral positions. The structures of ICl_4^- and BrF_4^- have been determined and they are planar. The structure of one of the AB_4^+ cations, IF_4^+, with one pair of electrons fewer, has also been determined. This has the same form as SF_4, which is isoelectronic with it, and is a trigonal bipyramid with a lone pair in an equatorial position. IF_6^+ and IF_6^-, which again differ by one unshared electron pair, present an interesting structural pair. The cation, which has only the six-bond pairs around the iodine, is a regular octahedron as expected. The anion, with a lone pair in addition, is an example of AB_6L, and resembles other fluorides with this number of electrons in forming a distorted octahedron. That is, the lone pair is sterically active, as in XeF_6 or SeF_6^{2-}.

The other type of polyhalide is limited to iodine and the structures of I_5^-, I_7^-, I_8^{2-}, and I_9^- are all irregular chains. These compounds are only stable in presence of large cations such as cesium or ammonium and alkylammonium ions. The structures are shown in Figure 15.50. The shorter $I-I$ dis-

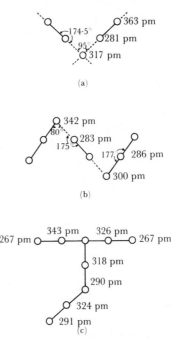

FIGURE 15.50 *Structures of polyiodide ions:* (a) I_5^-, (b) I_8^{2-}, (c) I_9^-

tances are similar to those in iodine but the longer ones correspond to only weak interactions, and all the polyiodides may be regarded as composed of I^-, I_2, and I_3^- groups weakly bonded together. None of the polyiodides survive in solution and they go to iodide, iodine, and triiodide. The recently discovered mixed species $I_2Cl_3^-$ and $I_2Cl_2Br^-$ have bent structures like I_5^-.

There are also a number of compounds which contain both halogen-oxygen and halogen-halogen bonds. These include the halyl fluorides, $XO_2F(X = Cl, Br, I)$, perchloryl hypofluorite, $O_3Cl(OF)$, halogen dioxytrifluorides, XO_2F_3, for $X = Br, I$, iodine oxypenta fluoride, IOF_5, and fluoro-iodate and fluoro-periodate ions, $IO_2F_2^-$ and $IO_2F_4^-$

The most stable of these compounds is perchloryl fluoride, formed by fluorinating perchlorate with fluorosulphonic acid. The structure is tetrahedral, O_3Cl-F, and the compound is an inert gas at ordinary temperatures. Chloryl fluoride, ClO_2F, from the fluorination of chlorate, is much more reactive. This, and the bromyl and iodyl fluorides, have probably the pyramidal structure derived from the halate ion by replacing one oxygen by fluorine, O_2X-F. The fluoro-iodate ion is formed by the reaction of hydrogen fluoride on iodate. The structure is derived from the trigonal bipyramid, with an equatorial lone pair on the iodine, and the two fluorine atoms axial, $F\!-\!\!-\!\!-\!I\!-\!\!-\!\!-\!F$. IOF_5 is formed

$$F\overset{\displaystyle\overset{..}{|}}{\underset{\displaystyle\underset{O\quad O}{}}{-\!\!\!-\!\!\!-I-\!\!\!-\!\!\!-}}F$$

during the preparation of IF_7, probably from traces of water. It has an octahedral structure.

Although all the compounds mentioned above contain halogens in positive oxidation states, all are essentially covalent. There is some evidence for the existence of positively-charged, coordinated ions among the halogens, especially for iodine. This feature may be regarded as a vestige of the tendency observed in the earlier Groups for the heavier elements to become metallic. Although no evidence is available, it is expected that astatine would show a stronger tendency to occur in cationic forms. A few compounds exist which contain iodine(III) and are ionic. These include the acetate, $I(OCOCH_3)_3$, and phosphate, IPO_4, and the fluosulphonate, $I(SO_3F)_3$. If a saturated solution of iodine triacetate in acetic anhydride is electrolysed, iodine is found at the cathode and, when a silver cathode is used, AgI is formed and electricity is used as required by the equation

$$I^{3+} + 3e^- + Ag = AgI.$$

Although no other structural evidence is available, these observations indicate that the compounds could be formulated as containing the I^{3+} cation, though such an ion would be expected to interact strongly with the anion and the solids are not to be regarded as simple salts. It is also possible to formulate the oxide I_4O_9 as $I^{3+}(IO_3^-)_3$.

A large number of well-characterized compounds are found which contain the I^+ ion stabilized by coordination. For example, the pyridine complexes $(Ipy_2)^+X^-$ are known for a wide variety of anions, and other lone pair donors coordinate similarly. In all these complexes the iodine appears at the cathode on electrolysis. Chlorine and bromine form analogous, coordinated X^+ species. In these compounds, the arrangement of the halogen and ligands is linear, $L-X-L^+$. One system which was thought for several years to contain the I^+ cation has recently been reformulated. This is the blue solution formed by iodine dissolved in oleum. A similar blue solution is found if iodine in IF_5 or HSO_3F is oxidized. It has now been shown that the blue species is not I^+ but I_2^+. Magnetic, conductivity and freezing point depression measurements all fit better for the I_2^+ species. Addition of further iodine gives rise to the I_3^+ and I_5^+ ions. Both the I_2^+ and the I_5^+ ions tend to disproportionate to the I_3^+ species. Later work established the existence of Br_2^+, and Cl_2^+ and ClF^+ have also been postulated though

not finally established. It is interesting that a structural study of Br_2^+ gives $Br-Br = 213$ pm, compared with 227 pm in neutral Br_2. This shortening is expected as the electron which is removed in forming the ion comes from an antibonding orbital.

The −I oxidation state

In this state, the chemistry of the halogens is well-known and many compounds have already been discussed in the earlier sections. Ionic compounds with X^- are formed with the s elements, except beryllium, and in the II and III oxidation states of the transition elements. The change from ionic to covalent character comes further to the left in the Periodic Table for the heavier halogens than for fluorine, as expected from the electronegativities. The region of change is marked by the occurrence of polymeric structures such as AlF_3, which is a giant molecule with $Al-F$ bonds which can be described as intermediate between ionic and covalent.

In the covalent halides, the main differences between the compounds of the different halogens may be ascribed to the differences in size and reactivity of fluorine compared to the heavier halides. This is often shown in the formation of hexavalent fluorides and tetravalent halides of the rest of the Group, or, similarly, by the formation of six-coordinated instead of four-coordinated complexes. Examples include the formation of SF_6 but only SCl_4, or of CoF_6^{3-} but only $CoCl_4^-$.

In this oxidation state, the halogens show their strongest resemblance and fluorine fits into place as the most reactive of all. This high reactivity derives in part from the relative weakness of the $F-F$ bond in fluorine (similar effects are found for the $O-O$ and $N-N$ single bonds in hydrogen peroxide and hydrazine). The heat of dissociation of F_2 is only $129{\cdot}3$ kJ mol^{-1}, compared with $237{\cdot}8$ kJ mol^{-1} for Cl_2, $188{\cdot}9$ kJ mol^{-1} for Br_2 and $147{\cdot}9$ kJ mol^{-1} for I_2. A value extrapolated from those of the heavier halogens would be about twice the observed F_2 value. The decrease is considerable when it is recalled that most element-halogen bond strengths are in the order $F > Cl > Br > I$.

Pseudohalogens or halogenoids

A number of univalent radicals are found which resemble the halogens in many of their properties, and the name pseudohalogen has been given to these. For example, consider the cyanide ion, CN^-. This resembles the halides in the following respects:

(i) it occurs as $(CN)_2$—cyanogen—and forms an HX acid, HCN
(ii) it forms insoluble salts with Ag^+, Hg^+ and Pb^{2+}
(iii) it also gives complex ions of similar formulae to the halogens, e.g. $Co(CN)_6^{3-}$ or $Hg(CN)_4^{2-}$
(iv) it forms covalent compounds and ionic compounds with similar ranges of elements as the halogens
(v) it gives 'interhalogen' compounds such as $ClCN$ or ICN

Other radicals with similar properties include cyanate, OCN, thiocyanate SCN, selenocyanate $SeCN$, and azidocarbondisulphide, $SCSN_3$: all these form R_2 molecules. In addition, the ions azide, N_3^-, and tellurocyanide, $TeCN^-$, act as pseudohalides although no molecule, R_2, is formed.

15.9 The helium Group

The elements of the helium Group are termed the rare gases, or the inert or noble (implying unreactive) gases. None of these terms is now particularly appropriate. The elements are rare only by comparison with the very abundant components of the atmosphere, oxygen and nitrogen. In terms of absolute composition of the crust and atmosphere, the lighter elements of this Group are common. Neither are the gases inert, as has recently been shown. Probably the term noble gases is least unsatisfactory but no general agreement has been reached as yet. The IUPAC recommended name is rare gas and this will be used here.

The rare gases occur as minor components of the atmosphere, ranging in abundance from argon (0·9 per cent by volume) to xenon (9 parts per million). Helium also occurs in natural hydrocarbon gases in some oilfields and is found occluded in some rocks. In both cases, this helium probably arises from α-particles emitted during radioactive decay. The heaviest member, radon, is radioactive and is found in uranium and thorium minerals, where it is produced in the course of the decay of the heavy elements. The other elements are usually produced by fractional distillation of liquid air. The main properties are given in Table 15.20. The low boiling points and heats of vaporization reflect the very low interatomic forces between these monatomic elements: the rise in these values with atomic weight shows the increasing polarizability of the larger electronic clouds.

The main isotope of helium is helium-4, and if this is cooled below 2·178 K surprising properties appear. In this form, called helium-II, the viscosity is too low to be detected, the liquid becomes superconducting, and it appears to flow in thin films without friction and is able to flow uphill from one vessel to another. No full theoretical explanation of these phenomena is yet available.

Until 1962, all attempts to form compounds of the rare gases had failed. Transient species, such as HHe, had been observed in electric discharges but these had very short lifetimes. The rare gases were also found in solids, such as $3C_6H_4(OH)_2.0·74Kr$, but these are not true compounds but *clathrates*. A clathrate is formed when a compound crystallizes in a rather open 'cage' lattice which can trap suitably sized atoms or molecules within them. An example is provided by para-quinol (p-$C_6H_4(OH)_2$; p-dihydroxybenzene) which, when crystallized under a high pressure of rare gas,

forms an open, hydrogen-bonded cage structure which holds the rare gas atoms in compounds like the krypton one above, or like $3C_6H_4(OH)_2.0·88Xe$. When the quinol is dissolved or melted, the rare gas escapes. That the clathrates are not true compounds is shown by the large variety of atoms and molecules which may enter the cages. Not only are quinol clathrates formed by krypton and xenon, but also by O_2, NO, methanol, and many others. The only requirement is that the clathrated species should be small enough to fit the cages and not so small that it can diffuse out: thus helium and neon are too small to form clathrates with p-quinol. Other compounds give clathrates with the rare gases. In particular, the reported hydrates of the rare gases are clathrates of these elements in ice, which crystallizes in an unusually open cage form. Although all clathrates do not involve hydrogen-bonded species, for example the benzene clathrate, $Ni(CN)_2.NH_3.C_6H_6$, clathrate formation by hydrogen-bonded molecules is common, as open structures are more readily formed.

All other attempts to form rare gas compounds, including many studies of possible donor action to yield compounds such as $Xe \rightarrow BF_3$, failed until 1962. Then Bartlett reported that xenon reacted with PtF_6 to form a compound which he formulated as $Xe^+(PtF_6)^-$. He was led to try this reaction after his discovery of $O_2^+(PtF_6)^-$—see section 14.8—by the consideration that the ionization potential of xenon was close to that (914 kJ mol^{-1}) of the O_2 molecule, so that if PtF_6 could oxidize O_2 to O_2^+, there was the chance of its oxidizing Xe to Xe^+. Further exploration of this field was extremely rapid. A fuller investigation of the reaction led to the discovery of XeF_4 in the second half of 1962. Interest was then concentrated on simple fluorides, oxides, oxyfluorides and species present in aqueous solution. The compounds of these classes so far discovered are listed in Table 15.21. The existence of $XeCl_2$ and $XeCl_4$ has been indicated in Mossbauer experiments using ICl_2^- or ICl_4^- as sources. These compounds decompose below room temperature, but $XeCl_2$ was found at 20 K in the products formed by passing Xe and Cl_2 through a microwave discharge. Spectroscopic studies indicate a linear structure.

This interest in the simpler compounds led to a relative neglect of the compounds with the transition metal fluorides, but it has more recently been shown that a number of other hexafluorides also react with xenon. In addition, the platinum hexafluoride reaction has been shown to give rise to

TABLE 15.20 Properties of the rare gases

Element	Symbol	B.p. (K)	Heat of vaporization (kJ mol^{-1})	Ionization potential (kJ mol^{-1})	Uses
Helium	He	4·18	0·092	2371	Refrigerant at low temperatures: airships.
Neon	Ne	27·1	1·84	2080	Lighting.
Argon	Ar	87·3	6·27	1520	Inert atmosphere for chemical and technical applications.
Krypton	Kr	120·3	9·66	1359	
Xenon	Xe	166·1	13·68	1170	
Radon	Rn	208·2	17·99	1037	Radiotherapy.

TABLE 15.21 Simple rare gas compounds

Oxidation State	Fluorides	Oxides	Oxyfluorides	Acids and Salts
II	XeF_2			
	KrF_2			
IV	XeF_4		$(XeOF_2)$	
	(KrF_4)			
VI	XeF_6	XeO_3	$XeOF_4$	$HXeO_4^-$
	XeF_7^-		$XeOF_5^-$	$(H_3XeO_6^{3-})$
	XeF_8^-		XeO_2F_2	(Ba_3XeO_6)
				$Xe(OH)_6$?
				$(BaKrO_4)$
VIII		XeO_4	(XeO_3F_2)	$HXeO_6^{3-}$
			(XeO_2F_4)	Ba_2XeO_6
				Na_4XeO_6
				(and similar salts)
				XeO_7^{6-}?
mixed	$XeF_2.XeF_4$	$XeO_3.K_4XeO_4$		

Notes: (1) A radon fluoride has been made on a small scale which was probably RnF_2.

(2) Compounds in brackets have not been isolated on a macro scale: compounds with a question mark have been reported but not yet confirmed.

TABLE 15.22 More complex compounds of xenon

Products from MF_6 reactions or their equivalent: $Xe(MF_6)_x$

for M = Rh, $x = 1 \cdot 1$

Ru, $x = 2$ to 3

Pt, $x = 1, 2$

Pu, $x = 1$?

The platinum system also yields $(XeF_5)^+(PtF_6)^-$ and Xe_2PtF_{17}. The XeF_5^+ ion is shaped as a square pyramid with a weak interaction to one of the fluorines of the octahedral PtF_6 group.

Adducts with covalent halides

(i) XeF_2 forms three series, $XeF_2.MF_5$, $XeF_2.2MF_5$, and $2XeF_2.MF_5$ with the pentafluorides of P, As, Sb and also I

Nb, Ta

Ru, Os, Rh, Ir, Pt

These are formulated respectively as

$XeF^+MF_6^-$, $XeF^+M_2F_{11}^-$ and $Xe_2F_3^+MF_6^-$.

Structures are indicated in the text.

(ii) XeF_4 forms a much more limited series but does interact with PF_5, AsF_5 and SbF_5 though not with other pentafluorides.

(iii) In the VI state, XeF_6 and $XeOF_4$ both form adducts. Examples include:

$XeF_6.BF_3$

$nXeF_6.MF_4$ for M = Ge, $n = 1, 2, 4$ and M = Sn, $n = 4$, but not with SiF_4

$nXeF_6.MF_5$ for M = As, Sb and $n = 1$, for M = V, P, As, Sb and $n = 2$,

$XeF_6.2SbF_5$, $XeOF_4.2SbF_5$, $2XeOF_4.VF_5$

more than one compound. A number of other complexes and adducts have also been formed and these more complex compounds of xenon are listed in Table 15.22.

The other xenon hexafluoro-metal compounds were made similarly to the hexafluoroplatinate:

$$Xe + MF_6 = XeMF_6$$

Only reactive hexafluorides oxidize xenon, more stable compounds, such as UF_6, do not react. Preparations of the fluorides are all by direct reaction under different conditions. Thus a 1:4 mixture of xenon and fluorine passed through a nickel tube at 400 °C gives XeF_2, a 1:5 ratio heated for an hour at 13 atmospheres in a nickel can, at 400 °C, gives XeF_4, while heating xenon in excess fluorine at 200 atmospheres pressure gives XeF_6. Most other compounds result on hydrolysis of the fluorides:

$$XeF_6 + H_2O = XeOF_4 + 2HF$$

with an excess of water,

$$XeF_6 + 3H_2O = XeO_3 + 6HF$$

or $\quad XeF_4 + H_2O = Xe + O_2 + XeO_3 + HF$,

here, about half the Xe(IV) disproportionates to Xe and Xe(VI) while the other half oxidizes water to oxygen, forming xenon. Xenon trioxide is hydrolysed in water, probably according to the equilibrium

$$XeO_3 \rightleftharpoons Xe(OH)_6$$

Interaction of XeO_3 and $XeOF_4$ gives XeO_2. The two xenon(VIII) oxyfluorides, XeO_2F_4 and XeO_3F_2, have been identified in a mass spectrometer but have not been prepared on the full scale. The report of XeF_8 has not been substantiated. The remaining oxyfluoride is formed from xenon and OF_2 in an electric discharge but its properties are not yet reported:

$$Xe + OF_2 = XeOF_2$$

The other oxide, XeO_4, is formed by acid on the perxenates, for example:

$$Ba_2XeO_6 + 2H_2SO_4 = XeO_4 + 2BaSO_4 + 2H_2O$$

Solution of xenon trioxide, or hydrolysis of the hexafluoride in acid, gives xenates such as Ba_3XeO_6. In neutral or alkaline solution, xenates rapidly disproportionate to perxenates, xenon(VIII). Like periodates, these show different stoichiometries, and salts of both the XeO_6^{4-} and XeO_7^{6-} ions are known.

Krypton compounds are much less stable than those of xenon—no reaction occurs between krypton and metal hexafluorides nor directly with fluorine. The tetrafluoride, KrF_4, has been made by the action of an electric discharge on a mixture of the elements, and the difluoride, KrF_2, has been identified by infra-red spectroscopy as a product of gamma radiation on a frozen mixture of the elements. Careful hydrolysis of KrF_4 with ice at $-60\ °C$ gives a small yield of an acid—called kryptic acid!—which gives a barium salt which may be $BaKrO_4$.

Tracer experiments with radon show the presence of a volatile fluoride from the reaction of radon and fluorine at 400 °C. This is a stable compound but its stoicheiometry is not yet established; RnF_4 is most likely.

The reactions of the fluorides are the most fully studied. These are all strong oxidizing and fluorinating agents, and most reactions give free xenon and oxidized products:

$$XeF_4 + 4KI = Xe + 2I_2 + 4KF$$
$$XeF_2 + H_2O = Xe + \tfrac{1}{2}O_2 + 2HF$$
$$XeF_4 + Pt = Xe + PtF_4$$
$$XeF_4 + 2SF_4 = Xe + 2SF_6$$

The three xenon fluorides are all formed in exothermic reactions (heats of formation of the gases are about 85, 230, and 335 kJ mol^{-1} for XeF_2, XeF_4, and XeF_6 respectively). The trioxide is endothermic by 402 kJ mol^{-1}—largely due to the high dissociation energy of O_2. The bond energies of the Xe—F bonds in the fluorides range from 120 to 134 kJ mol^{-1} from the difluoride to the hexafluoride, a difference which is not far outside the experimental errors. The Xe—O energy is about 85 kJ mol^{-1}.

The Xe—F bond lengths in the fluorides are also substantially the same, with a small shortening from 200 pm in XeF_2 to 191 pm in XeF_6. Kr—F in the difluoride is 187·5 pm. The fluorides are all white volatile solids with the volatility increasing from the difluoride to the hexafluoride. The trioxide is also white but non-volatile, while the tetroxide, XeO_4, is a yellow, volatile solid which is unstable at room temperature. The oxyfluorides are also white, volatile solids.

The structures of most of the xenon compounds are those predicted by the simple electron pair considerations outlined in Chapter 3, and correspond to the structures of isoelectronic iodine compounds. Some structures are shown in Figure 15.51. XeO_2F_2 has a structure like $IO_2F_2^-$ (Figure 3.25) with a nearly linear F—Xe—F arrangement. The most

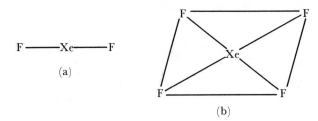

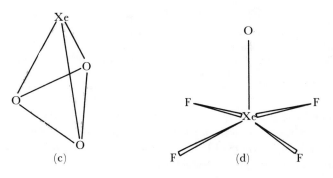

FIGURE 15.51 *The structures of xenon compounds:* (a) XeF_2, (b) XeF_4, (c) XeO_3, (d) $XeOF_4$
These structures are the same as those of the iodine analogues ICl_2^-, IF_4^- and IO_3^-.

interesting case is XeF_6. This contains eight valency electrons on the Xe, and six F electrons, giving seven pairs. This structure, as in a number of cases such as TeF_6^{2-} already noted, will be a distorted octahedron on the electron pair theory while alternative approaches suggest a regular octahedron. The concensus of a great deal of experimental work is that XeF_6 is a distorted octahedron with the unshared pair sterically active.

The structures of some of the adducts are known, and most attention has been paid to the XeF_2 species. These may be written as involving the XeF^+ ion, which is then weakly bonded to the fluoroanion through one fluorine, or they may be written with a fluorine bridge between Xe and the metal. The cation formulation is less justified here than was the case with the rather similar iodine and bromine species (see Figure 15.48) as the difference in bond length is less. An example is $XeF_2.2SbF_5$, shown in Figure 15.52. The shortest Xe—F distance of 184 pm is distinctly less than the other value of 235 pm, but this latter is much shorter than the van der Waals non-bonding contact distance of 350 pm. It is also interesting that the XeFSb angle is very close to the SbFSb angle at the fluorine which is bridging the two octahedra. The average non-bridging Sb—F distance is 180 pm.

XeF_2 also forms an interesting series of compounds when reacted with strong acids in a reaction which may be written

$$XeF_2 + HOY \xrightarrow{-75\ °C} HF + FXeOY + Xe(OY)_2$$

This has been carried out for Y = ClO_3, SO_2F and TeF_5. Compounds isolated include $FXeOSO_2F$, $Xe(OSO_2F)_2$ or $FXeOClO_3$. The structure of $FXeOSO_2F$ has been deter-

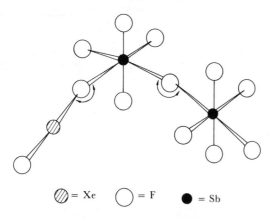

⬙ = Xe   ◯ = F   ● = Sb

FIGURE 15.52 *Structure of the adduct* $XeF_2.2SbF_5$
The structure is intermediate between that expected for a full covalent Xe—F—Sb bridge and that corresponding to $XeF^+Sb_2F_{11}^-$. Mean non-bridging Sb—F bond length = 183 pm.

mined and it involves a linear F—Xe—O unit bonded to a tetrahedral FSO_3 unit, through the oxygen, with an angle at this oxygen XeOS of 123°. The other species are probably similar with XeOCl or XeOS links.

Bonding in Main Group compounds: the use of *d* orbitals

The bonding in xenon compounds has been the subject of some controversy which raises the wider question of the degree to which the valence shell *d* orbitals participate in bonding in any compound where more than four electron pairs surround a Main Group atom. The case of XeF_2 is the simplest to discuss. The two ligands occupy axial positions and three lone pairs occupy equatorial positions in a trigonal bipyramid. To accommodate these five electron pairs requires five orbitals which are formed from the *s*, the three *p*, and a *d* orbital on the central atom. But it has been objected that, in the case of xenon, the energy gap between the *p* and *d* orbitals is very large, equal to about 960 kJ mol^{-1}, and it is unlikely that the bond energies are sufficient to compensate for the energy required to make use of a *d* orbital in such a scheme. Instead, three-centre sigma bonds between xenon and fluorine are proposed. The molecular axis is taken as the *z*-axis, as in Figure 15.53, and the relevant orbitals are the

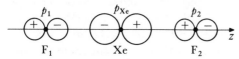

FIGURE 15.53 *The formation of three-centre bonds in XeF_2: the constituent atomic orbitals*

If the orbitals as drawn are taken as positive, the three-centre combinations which result are
bonding $\psi = p_1 + cp_{Xe} + p_2$
non-bonding $\psi = p_1 + p_2$
antibonding $\psi = p_1 - c'p_{Xe} + p_2$
where the constants *c* and *c'* are of similar size and the expressions are to be regarded as normalized. The bonding orbital and the non-bonding orbital both contain two electrons. As only one pair of bonding electrons exists between three atoms, the Xe—F bond order is only one-half.
XeF_4 may be described similarly using *p* orbitals in both the *x* and *y* directions.

xenon p_z orbital, which contains two electrons, and the p_z orbital on each fluorine which hold one electron each (the other six valency electrons on xenon and fluorine fill the *s*, p_x, and p_y orbitals on each atom, all non-bonding). The three p_z orbitals can be combined to give three sigma orbitals, centred on the three atoms, one of which is bonding, one non-bonding, and one anti-bonding (Figure 15.53). The four electrons fill the bonding and non-bonding orbitals to give an overall bonding effect. The position is similar to that in the three-centre B—H—B bond in the boranes, except that there are four electrons instead of two to be accommodated (compare section 8.5). The cases of $\overset{.}{X}eF_4$ and XeF_6 can be explained similarly using xenon p_x (or p_x and p_y for XeF_6) orbitals as well as p_z in the two cases.

Similar descriptions apply to the interhalogen compounds and ions such as ICl_2^-, ICl_4^-, BrF_5, and so on. All these species can be described either in terms of full electron pair bonds, plus non-bonding pairs, by using one or two of the *d* orbitals, or the use of *d* orbitals may be avoided by using polycentred molecular orbitals at the price of reducing the bond order. It has been reported by Wiebenga (see reference in Appendix A on p. 284) that calculations using *s* and *p* orbitals only are very successful in reproducing the bond angles and bond lengths found in interhalogen compounds.

The theory using three-centred orbitals formed by the *p* orbitals would predict that XeF_6, and the isoelectronic IF_6^-, should be regular octahedra while the electron pair repulsion theory suggests that these species with seven pairs around the central atoms should be distorted. It is now established that XeF_6, IF_6^- and TeF_6^{2-} are distorted octahedra, while the isoelectronic MX_6^{2-} species (where M = Se, Te, Po and X = Cl, Br, I but *not* F) are undistorted. This again shows the delicate balance which must exist between the various energies making up these two different structures.

Recent theoretical work on the general problem of *d* orbital contributions to bonding in Main Group compounds has made considerable progress. This is mainly because use of large computers has allowed work to be carried out with fewer initial assumptions and approximations, so that the answers derived by different workers are in better agreement. The present position may be summarized into two general classes.

(1) In compounds with second row and heavier elements bonded to highly electronegative elements like oxygen and fluorine, there is evidence of substantial *d* orbital participation in the bonding orbitals. For example, in PF_3, the population of the phosphorus orbitals is calculated to be:
sigma bonding
     $3s$ 1·51 electrons $3p$ 1·05 electrons $3d$ 0·13 electrons

pi bonding $3p$ 0·84 electrons $3d$ 0·62 electrons

Thus, the use of the *d* orbital gives a small stabilization to the sigma bonds, but makes a significant contribution to π bonds.

Notice that this contribution is made in a molecule with only four electron pairs on the phosphorus, that is where there are enough *s* and *p* orbitals to hold all the valency electrons. In higher oxidation state fluorides and oxides, such as PF_5, OPF_3, SF_6 or XeF_6, there is now support from calculations for d orbital involvement in the bonding in a substanial way with quite large populations of electrons in the d orbitals. This would also seem to be supported experimentally in the case of the distorted MF_6L species mentioned above.

In the above cases, the test of the calculation is its agreement with experimentally-determined ionization or promotion energies (showing the differences between orbital energies) and the agreement between calculated and experimental dipole moments (which reflect the distribution and spatial density of the electron clouds).

(2) This leads to the second class, of compounds where the Main Group element is bonded to elements of lower electronegativity like H or C. Calculations on cases such as these, for example PH_3 or H_2S, show a very low d orbital population of only 1–2%, and the introduction of d orbitals into the calculation gives no significant improvement to the agreement between calculated and experimental energy levels. However, this very low d population does make a substantial improvement to the calculation of the dipole moment. Thus, this class of molecule makes no real use of the central atom d orbitals in bonding and no extra stabilization results over that found simply using the s and p orbitals. However, introduction of the d orbitals into the calculation does give a much better description of the dispersion of the electrons in space, as reflected in the size of the dipole moment.

In summary, it seems that introduction of d orbitals (and other higher orbitals) into most systems gives a better picture of the electron distribution even though the actual electron population is only small. In addition, in high-valency compounds with substituents like F, O or Cl, there is a substantial contribution by the d orbitals, particularly to π bonding. In any particular individual case, the balance of contributions from s, p and d orbitals will be unique, and it is shown above that the bonding may be described in coordinations greater than four without invoking d orbitals if multicentre bonds are constructed. In the absence of clear experimental evidence or full calculation, the description of a molecule must be speculative, but these recent calculations give support to the simple models which made use of d orbitals in Main Group compounds when more than four orbitals were required to form the sigma and pi bonds required on a simple electron pair picture, together with the accommodation of any lone pairs.

It seems appropriate to end a modern textbook on inorganic chemistry with an account of rare gas compounds and use of Main Group d orbitals. The chemical inertness of the rare gases, based on many years of experimental attempts to form their compounds, has long been accepted as a fundamental property. The stability of the electron octet in the rare gas configuration has been the basis of most simple theories of chemical reactivity. Now, as in all advances, the simple concept has to be replaced by a more sophisticated one. Of course, the general ideas about 'forming a stable octet' were accurate, and the rare gas configuration is stable and unreactive, but no longer absolutely so.

The discovery and characterization of these compounds brought to a head a recurrent debate about d orbital participation in Main Group compounds, and this pressure, together with the advent of suitable computers, led to a concerted theoretical attack on the problem which has led —at the moment!—to a consensus of opinion.

It is unlikely, in chemistry, that these accepted pictures will not soon be subject to new criticism and further work which will, in turn, achieve a new, deeper, and more subtle level of understanding of the properties of materials.

Appendix A. Further Reading

(1) General Textbooks

For further information, or for a more detailed and sophisticated treatment, the student should next turn to the major inorganic textbooks written for the Honours courses. There is also a wide range of specialist texts, covering such topics as bonding, reaction mechanisms, individual groups of elements and so on. Such topics are also discussed in various review series and monographs. Various multi-volume works are available which aim to give a much more complete coverage of chemistry than is possible in a textbook, but these tend to be more-or-less out of date, although supplementary volumes may be produced at intervals.

Textbooks

Advanced Inorganic Chemistry, F. A. COTTON AND G. WILKINSON, Interscience, 2nd Edition, 1966
This text is an excellent modern inorganic textbook giving full coverage, at B.Sc. Honours level, of General Theory, Main Group, and Transition Element chemistry. References both to texts and to original papers give the reader a key to the recent literature on most of the important topics. The volume provides an excellent reference book for the beginner in inorganic chemistry, and is essential for the honours student, but the treatment in a number of areas—especially of transition element theory—is considerably more advanced than the general student requires.

Inorganic Chemistry, C. S. G. PHILLIPS AND R. J. P. WILLIAMS, Oxford, Clarendon Press, 1965–66, 2 volumes
This is the second major textbook of inorganic chemistry to appear in the last five years. It consists of three Parts—Principles, Non-metals (making up Volume I) and Metals (Volume II). Part I includes accounts of ionic, covalent and metallic bonding, atomic structure, solid structures, equilibria, kinetics, and electrode potentials. Parts II and III then deal with the chemistry of the elements including, in Part III, accounts of metals and alloys, complexes, spectra, magnetism and reaction kinetics and mechanism. This book is intended for Honours and Postgraduate readers and, to-

gether with Cotton and Wilkinson, provides the starting point for any search for further information.

New editions of both these texts are probably to be expected in the near future, incorporating developments in chemistry in the last few years. These two texts provide modern, and to a large extent, complementary treatments of Inorganic Chemistry and one or other of them will give information on almost any topic the general student wishes to follow up. Both strongly convey the excitement and intellectual appeal of modern inorganic chemistry.

Many other general texts will be available in a good chemistry library, most of which can be consulted with profit as long as due regard is paid to the date of publication. Among these are the following titles:

Inorganic Chemistry R. B. HESLOP AND P. L. ROBINSON, Elsevier, 3rd Edition, 1967
This book is written at a somewhat less advanced level than Cotton and Wilkinson, or Phillips and Williams, though still aimed at Honours students. It gives a good and very readable coverage of inorganic chemistry.

Inorganic Chemistry, T. MOELLER, Wiley, 1952
Although somewhat out-of-date, this text gives an excellent coverage of Main Group and Non-aqueous Solvent chemistry. It is also extremely useful for the extensive lists of data, such as ionization potentials, which it presents. Should a second edition appear, it would be well worth attention.

General and Inorganic Chemistry, J. R. PARTINGTON, Macmillan, 4th Edition, 1966
This was one of the best examples of the old-style inorganic text. It is not recommended for general use but will be found an adequate source for many of the minor pieces of information that do not appear in a modern text. Such items as the number of water molecules in a hydrated salt, or the formula of a mineral, or details of older industrial methods, will often be most readily found in texts of this

generation. Partington also includes a certain amount of historical detail, while his monumental *History of Chemistry* gives a complete treatment.

Among other standard inorganic texts are:

Modern Approach to Inorganic Chemistry, C. F. BELL AND K. A. K. LOTT, Butterworths, 2nd Edition, 1966. For general degree and higher national certificate students.

Inorganic Chemistry, E. B. BARNETT AND C. L. WILSON, Longmans, 2nd Edition, 1957

Among the longer textbooks and compendia may be mentioned:

Chemical Elements and their Compounds, N. V. SIDGWICK, Oxford University Press, 1950, 2 volumes

Treatise on Inorganic Chemistry, H. REMY (English translation by J. S. Anderson), Elsevier, 1956, 2 volumes

Comprehensive Treatise on Inorganic and Theoretical Chemistry, J. W. MELLOR, Longmans, 1922–37, 16 volumes and supplementary volumes which started in 1958

Nouveau Traité de Chimie Minèrale, P. PASCAL, Masson, 1956 onwards, about 25 volumes and further volumes appearing

Handbuch de anorganische Chemie, L. GMELIN, 1924 onwards, volumes and supplements appearing

The last three aim to be comprehensive accounts of the chemistries of the elements covered, and all three are being brought up-to-date although the immensity of the task means that coverage is very uneven.

Comprehensive Inorganic Chemistry, Pergamon, in preparation. This aims to cover inorganic chemistry in five large volumes, with probable publication date 1972. It will fall between Sidgwick and Pascal in depth of coverage, but will be much more up to date. When this appears, it will be the primary source for fairly detailed information, followed by Gmelin if a recent supplement has appeared in the area of interest.

We next come to a set of titles which cover a wide area of inorganic chemistry without attempting the complete coverage of the standard textbook.

Introduction to Physical Inorganic Chemistry, K. B. HARVEY AND G. B. PORTER, Addison-Wesley, 1963, Student Edition, 1965
This book provides an account of the physical aspects of inorganic chemistry at a level suitable for an Honours student. It covers bonding, structure, thermodynamics, electrolyte solutions, kinetics and mechanisms of reactions, crystal chemistry and complexes.

Modern Aspects of Inorganic Chemistry, H. J. EMELEUS AND J. S. ANDERSON, Routledge and Kegan Paul, 3rd Edition, 1962
This book forms a valuable supplement to the standard texts. It does not attempt complete coverage of Inorganic Chemistry but focuses on particular topics, which are usually given more detailed treatment than in the standard texts. For example, Chapters are devoted to Non-aqueous Solvents, Peroxides and Per-acids, Metals and Intermetallic Compounds, and Recent Chemistry of the Radioactive Elements, as well as to most of the commoner topics.

Inorganic Energetics, W. E. DASENT, Penguin, 1970
This useful paperback treats the energetics of atoms and ions, ionic crystals, covalent compounds, and of solutions. It gives a most valuable account of the thermodynamic aspects of the formation, stabilities, and binding energies of inorganic compounds, and provides full tabulations of data.

Some Thermodynamic Aspects of Inorganic Chemistry, D. A. JOHNSON, Cambridge University Press, 1968
For second and third year undergraduates.

New Pathways in Inorganic Chemistry, ed. E. A. V. EBSWORTH, A. G. MADDOCK AND A. G. SHARPE, Cambridge University Press, 1968
This volume was produced to mark the retirement of Professor Emeleus and consists of a series of reviews of many of the active areas of inorganic chemistry. While much of the treatment is at an advanced level, the student who dips into this volume will gain a view of current thinking. Articles include one on unusual oxidation numbers, several on transition and organometallic chemistry, on hydrides, on nonaqueous compounds, and a number on fluorine species.

(2) Reviews and Journals

More specific information may be obtained from reviews, and ultimately from papers in the journals. There are now many review series published which survey fairly specific areas of chemistry, such as organometallic chemistry, fluorine chemistry, transition metal compounds, spectroscopy, and so on. Two series of approximately annual volumes set out to cover the whole field of inorganic chemistry. These are:

Advances in Inorganic Chemistry and Radiochemistry, ed. H. J. EMELEUS AND A. G. SHARPE, Academic Press, 1959 onwards

Progress in Inorganic Chemistry, ed. F. A. COTTON, Interscience, 1959 onwards
Both these series give accumulated contents lists in the latest volume, and the tenth volume in each series has a full accumulated index.

Also important as initial sources of fuller information at a general level are *Quarterly Reviews of the Chemical Society,* 1949 onwards.

This sets out to publish articles on any part of chemistry, aimed at the general chemical reader. The student will find a number of useful articles, especially in the earlier volumes.

Chemistry in Britain, Royal Institute of Chemistry and Chemical Society

This is the monthly news-journal of these bodies, but it often carries useful articles about areas of current interest or controversy. Other sources of similar articles are *Education in Chemistry*, (Royal Institute of Chemistry) and *Journal of Chemical Education* (American Chemical Society)

(3) Bibliographies for Particular Sections of the Text

In addition to the appropriate sections of the advanced texts listed above, more information about particular themes may be gained from the following books and articles. This list is not intended to be exhaustive, but is designed to give a lead in to published material in each area. The more specialised and advanced texts are omitted, but will be referred to in many of the references given.

UNITS AND DATA COMPILATIONS (Chapter 1)

Physico-chemical Quantities and Units, M. L. McGLASHAN, Royal Institute of Chemistry, Monographs for Teachers No. 15, 2nd Edition, 1971
This discusses the basis for the SI system and defines all the basic and derived units in full detail.

Dictionary of Quantities and Units, J. V. DRAZIL, Leonard Hill, 1971

S.I. See *Chemistry in Britain,* especially 1969–70 for extensive correspondence and a number of articles: particularly D. J. G. Ives in November 1969, M. L. McGlashan in January 1970, and N. H. Davies in August 1970 issues.
SI Units and Conversion Tables. Machinery Publishing Co. of London, 1970
This is an inexpensive compilation in use at Queen Elizabeth College, London

Thermodynamic Tables in SI Units, R. W. HAYWOOD, Cambridge University Press, 1969

SI Chemical Data, G. H. AYLWARD AND T. J. V. FINDLAY, Wiley. 1971

Handbook of Chemistry and Physics. The Chemical Rubber Co., 52nd Edition, 1972
This is re-issued annually and is therefore quite up to date and gives the most extensive compilation of chemical, physical and mathematical data available in a single source.

An Introduction to Chemical Nomenclature, R. S. CAHN, Butterworths, 3rd Edition, 1968
A handy guide to inorganic and organic nomenclature.

THEORETICAL CHEMISTRY (Chapters 2 and 3)

The student should first refer to the appropriate section of the general textbooks listed in section (1) above. Introductory accounts, useful for revision, are given in

Introduction to Valence Theory, J. WORRALL AND I. J. WORRALL, Macdonald & Co., 1969

A Valency Primer, J. C. SPEAKMAN, Edward Arnold, 1968

The following titles are basic texts, for the less advanced student

Valence, C. A. COULSON, Oxford University Press, 2nd Edition, 1961, corrected revision, 1969
This is a very clear account of atomic and molecular structure, ranging over inorganic and organic molecules. The general student will find it useful if he is willing to 'read round' the more detailed sections.

Valency and Molecular Structure, E. CARTMELL AND G. W. A. FOWLES, Butterworths, 3rd Edition, 1966

Theoretical Inorganic Chemistry, M. C. DAY AND J. SELBIN, Van Nostrand Reinhold, 2nd Edition, 1969

Theoretical Basis of Inorganic Chemistry, A. K. BARNARD, McGraw-Hill, 1965

Principles of Atomic Orbitals, N. N. GREENWOOD, Royal Institute of Chemistry, Monographs for Teachers, 1964

The following titles are somewhat more specialised or more advanced:
Atomic Spectra and Atomic Structure, G. HERZBERG, Dover Paperback, 1944
Wave Mechanics and Valency, J. W. LINNETT, Methuen, 1960
This is a short monograph giving an excellent but mathematical treatment. See also the same author's
The Electronic Structure of Molecules, Methuen, 1964
Elementary Molecular Bonding Theory, P. G. PERKINS, Methuen, 1969
Electrons and Chemical Bonding, H. B. GRAY, Benjamin, 1964
Valence Theory, J. N. MURRELL, S. F. A. KETTLE, AND J. M. TEDDER, Wiley, 2nd Edition, 1970

The Nature of the Chemical Bond, L. PAULING, Cornell University Press (Oxford U.P. in England), 3rd Edition, 1961

This is Pauling's classical text on chemical bonding, which should be dipped into by every chemist. The approach to valency is, however, entirely via Pauling's own development of valence bond theory, which has been superseded largely by the more flexible molecular orbital approach. The text is also a useful source of data, and gives the key introduction to the oxidation state concept. See also *The Chemical Bond*, Oxford University Press, 1967, in which Pauling gives a shortened and updated survey of valency which does include the molecular orbital approach.

A further discussion of oxidation states is given in *Education in Chemistry*, **1**, 1964, 83, by D. B. SOWERBY AND M. F. A. DOVE.

The electron pair repulsion theory was introduced by N. V. Sidgwick and H. M. Powell, *Proc. Royal Soc.*, 176A (1940), 153, and developed extensively by R. J. Gillespie and R. S. Nyholm, as in *Quarterly Reviews*, 11 (1957), 261 and *Progress in Stereochemistry*, Ed. DE LA MARE AND KLYNE, Butterworths, **2**, 1958 261. See also the recent articles by GILLESPIE, *J. Chem. Education*, **47**, 1970, 18; **40**, 1963, 192, and *Advances Chem. Series*, **62**, 1967, 24.

Precise data on molecular structure, in the gaseous and condensed states, is to be found in The Chemical Society's Special Publications 11 and 18, *Interatomic Distances* and *Supplement*. Finally the reader's attention is drawn to the most attractive and imaginative impression of molecules to be found in *The Architecture of Molecules*, L. PAULING AND R. HAYWARD, W. H. Freeman, 1964

STRUCTURAL CHEMISTRY (Chapter 4)

The standard textbook on inorganic crystallography is *Structural Inorganic Chemistry*, A. F. WELLS, Oxford University Press, 3rd Edition, 1962

A much shorter, useful introductory text is the paperback *Structural Principles in Inorganic Chemistry*, W. E. ADDISON, Longmans, 1961

Ionic Crystals, Lattice Defects, and Nonstoichiometry, N. N. GREENWOOD, Butterworths, 1969
This paperback treats ideal crystals for about the first third of its length and then continues with the various types of defect structure.

Fundamentals of Inorganic Crystal Chemistry, H. KREBS, transl. P. H. L. WALTER, McGraw-Hill, 1968
This is a continental approach to crystal chemistry with a substantial introductory section on atomic structure and wave mechanics.
Stereochemistry of Ionic Solids, J. D. DUNITZ AND L. E. ORGEL, *Advances in Inorganic Chemistry and Radiochemistry*, **2**, 1960, 1— this also covers material in Chapter 12.

Lattice energies—a very detailed treatment with extensive data is given by T. C. WADDINGTON, *Advances in Inorganic Chemistry and Radiochemistry*, **1**, 1959, 158.

Models—see *Models in Structural Inorganic Chemistry*, A. F. WELLS, Oxford University Press, 1970; also the review of different commercially-available building sets by A. WALTON, *Progress in Stereochemistry* (Ed. B. J. AYLETT AND M. M. HARRIS) Butterworths, **4**, 1969, 335

SOLVENTS (Chapter 5)

The textbooks of MOELLER and EMELEUS AND ANDERSON, in the general bibliography, give fairly full accounts of non-aqueous solvents, and most of the general texts deal with aqueous solvents. Some more detailed texts on non-aqueous solvents are:

Non-aqueous Solvents, L. F. AUDRIETH AND J. KLEINBERG, Wiley, 1953
Covers nearly all the common solvent systems but is now a little out of date.

Non-aqueous Solvent Systems, Editor—T. C. WADDINGTON, Academic Press, 1965

This book presents a collection of reviews by acknowledged authorities on all the major solvent systems. The field covered is similar to that of Audrieth and Kleinberg, but the treatment is rather more advanced.

For Ladd radii see *Theoretica Chimica Acta*, **12,** 1968, 33, 3.

Non-aqueous Solvents, R. A. ZINGARO, Heath, 1969, and a book of the same title by T. C. WADDINGTON, Nelson, 1969, provide good and inexpensive surveys in paperback form

Chemistry in Non-Aqueous Ionizing Solvents, Editors—G. JANDER, H. SPANDAU, AND C. C. ADDISON, Vieweg, 1963 onwards
This is a series of volumes intended to deal exhaustively with each of the common classes of non-aqueous solvents. These are useful for reference mainly.

Among specific references may be mentioned:
An account of the physical nature of metal solutions in ammonia and related solvents, M. C. R. SYMONS, *Quart. Rev.*, **13**, 1959, 99

The Sulphuric Acid Solvent System, R. J. GILLESPIE AND E. A. ROBINSON, *Advances in Inorganic Chemistry and Radiochemistry*, **1**, 1959, 386

Acid-base theories:

The Use of the **Terms Acid and Base**, R. J. BELL, *Quart. Rev.*, **1**, 1947; 133. Also—*Acids and Bases* and the longer and more recent *The Proton in Chemistry*, both published by Methuen

These accounts by Bell give, in different degrees of detail, a very clear picture of acid behaviour in both aqueous and non-aqueous media, particularly when proton transfer is involved.

The Theory of Acids and Bases. F. M. HALL, *Ed. Chem.*, **1**, 1964, 91

The Coordination Model for Non-aqueous Solvent Behaviour, R. S. DRAGO AND R. F. PURCELL, *Prog. Inorg. Chem.*, **6**, 1964, 271
This article presents an alternative to the solvent-system model for non-protonic solvents.

EXPERIMENTAL METHODS (Chapter 6)

Two general texts are
The Synthesis and Characterisation of Inorganic Compounds, W. L. JOLLY, Prentice-Hall, 1970
This is an extensive text which ranges over thermodynamic and kinetic background, techniques of synthesis, characterisation including all the methods discussed in this chapter, and various excercises.

Practical Instrumental Analysis, J. KRUGERS AND A. I. M. KEULEMANS, Elsevier, 1965
A wide-ranging collection of reviews, aimed at organic as well as inorganic chemists.
A more advanced text, which will be useful for further study, is *Physical Methods in Inorganic Chemistry*, H. A. O. HILL AND P. DAY, Interscience, 1968

Ion exchange:
Quart. Rev., 1950, 2, 307, J. F. DUNCAN AND B. A. J. LISTER, or, the book by J. E. SALMON AND D. K. HALL. Butterworths, 1959; of this title

Solvent extraction:
Quart. Rev., 1953, **5**, 200, H. M. IRVING

Chromatography:
Quart. Rev., 1955, **7**, 307, R. A. WELLS
Gas Chromatography, J. M. KNOX, Methuen, 1962

Physical methods:
The Determination of Molecular Structure, P. J. WHEATLEY, Oxford University Press, 2nd Edition, 1968

X-Ray Crystallography, R. W. JAMES, Methuen, 5th Edition, 1961
A convenient small monograph which gives a useful introduction.

The Structure of Molecules, G. M. BARROW, Benjamin, 1962
This book gives a general account of spectroscopic methods, including rotational, vibrational, and electronic spectra.

Raman Spectra, L. A. WOODWARD, *Quart. Rev.*, 1959, **10**, 185

The Use of Nuclear Magnetic Resonance in Inorganic Chemistry, E. L. MUETTERTIES AND W. D. PHILLIPS, *Advances in Inorganic Chemistry and Radiochemistry*, 1962, **4**, 231

The Mossbauer Effect, E. FLUCK, *Advances in Inorganic Chemistry and Radiochemistry*, 1964, **6**, 433

Mossbauer Spectra of Chemical Compounds, N. N. GREENWOOD, *Chemistry in Britain*, **3**, 1967, 56

Magnetism and Inorganic Chemistry, R. S. NYHOLM, *Quart. Rev.*, 1955, **7**, 377

GENERAL PROPERTIES (Chapter 7)

Principles of Oxidation and Reduction, A. G. SHARPE, R.I.C. Monograph

Oxidation Potentials, W. M. LATIMER, Prentice-Hall, 2nd Edition, 1952
This is a complete survey of oxidation potential data, and the standard source (though more recent values are available especially for some of the rarer elements).

A Graphical Method of Representing the Free Energies of Oxidation-Reduction Systems, E. A. V. EBSWORTH, *Educ. Chem.*, 1964, **1**, 123

For properties and extraction of elements:

Principles of the Extraction of Metals, D. J. G. IVES, R.I.C. Monograph

Allotropy of the Elements, W. E. ADDISON, Oldbourne Press, 1963, and the general textbooks of the Bibliography.

Metals, the *Chemistry in Britain*, May 1969 number is devoted to metals and alloys.

THE CHEMISTRY OF THE ELEMENTS

In general, further information on the chemistry of the elements discussed in Chapters 8 to 15 is most readily found by reference to the general textbooks listed above. The following list of references has been confined to accounts of the chemistry of elements which are rare or demand special handling, or to particularly useful recent accounts which contain new information. The list is not exhaustive.

Topics of interest over the major part of the area of these chapters include

COORDINATION NUMBERS

High Co-ordination Numbers, E. L. MUETTERTIES AND C. M. WRIGHT, *Quart. Rev.*, 1967, **21**, 109

Eight-Coordination Chemistry, S. J. LIPPARD, *Prog. Inorg. Chem.*, 1967, **8**, 109

Mechanisms of Inorganic Reactions, F. BASOLO AND R. G. PEARSON, Wiley, 2nd Edition, 1967
This is the most far-reaching survey, but is directed largely to transition metal complexes.

Introduction to Inorganic Reaction Mechanisms, J. C. LOCKHART, Butterworth, 1965
A most useful introduction in paperback form.

Mechanisms of Inorganic Reactions in Solution, D. BENSON, McGraw-Hill, 1968 (Paperback)

ORGANOMETALLIC CHEMISTRY

Organometallic Compounds, G. E. COATES, 3rd Edition, Methuen 1968; Volume I with K. WADE, *The Main Group Elements*: Volume II, M. L. H. GREEN, *The Transition Elements*.
This is the basic wide-ranging account.

Principles of Organometallic Chemistry, G. E. COATES, M. L. H. GREEN, P. POWELL, AND K. WADE, Methuen, 1968
This is a shorter treatment directed at undergraduates.

Organometallic Chemistry, P. L. PAUSON, Arnold, 1967

HYDROGEN (Chapter 8)

Hydrogen Compounds of the Metallic Elements, K. M. MACKAY, Spon, 1966

Inorganic Hydrides, B. L. SHAW, Pergamon, 1967

f ELEMENTS (Chapters 10 and 11)

The Lanthanides, T. MOELLER, Chapman and Hall, 1964, paperback.

Unusual Oxidation States of some Lanthanide and Actinide Elements, L. B. ASPREY AND B. B. CUNNINGHAM, *Prog. Inorg. Chem.*, 1960, **2**, 267

Chapter 11:

Halides of the Actinide Elements, J. J. KATZ AND I. SHEFT, *Advances in Inorganic Chemistry and Radiochemistry*, 1960, **2**, 195
This article includes an account of experimental techniques for handling these highly radioactive elements.

Man-made Transuranium Elements, G. T. SEABORG, Prentice-Hall, 1963, Paperback in the 'Foundation of Modern General Chemistry' series.
This is a very interesting personal account of this work, for which the author won a Nobel Prize.

The Chemistry of Protoactinium, A. G. MADDOCK, *Quart. Rev.*, 1963, **17**, 289

Some Recent Chemistry of Protactinium, D. BROWN, *Advances in Inorganic Chemistry and Radiochemistry*, **12**, 1970

TRANSITION ELEMENTS (Chapters 12, 13 and 14)

An Introduction to Transition Metal Chemistry, L. E. ORGEL, Methuen, 2nd Edition, 1966
A short and very clear account of ligand field theory.

Coordination Compounds, S. F. A. KETTLE, Nelson, 1969
A slightly more advanced and general treatment than the above, clearly presented in paperback form.

Electronic Spectra of Transition Metal Complexes, D. SUTTON, McGraw-Hill, 1968

The Stabilisation of Oxidation States of the Transition Metals, R. S. NYHOLM AND M. L. TOBE, *Advances in Inorganic and Radiochemistry*, **5**, 1963, 1

The Stabilities of Transition Element Halides, P. G. NELSON AND A. G. SHARPE, *J. Chem. Soc.*, A, 1966, 501

The Chemistry of Transition Metal Carbonyls, E. W. ABEL AND F. G. STONE, *Quart. Rev.*, Part One, Structural Considerations, **23**, 1969, 325; Part Two, Synthesis and Reactivity, **24**, 1970, 498

Thermodynamics of Metal Complex Formation, G. BEECH, *Quart. Rev.*, **23**, 1969, 410

Stability Constants of Metal-Ion Complexes, Special Publications of the Chemical Society, Numbers 17 and 25

Enzyme Action: Views Derived from Metalloenzyme Studies, B. L. VALLEE AND R. J. P. WILLIAMS, *Chemistry in Britain*, **4**, 1968, 397: The Biochemistry of Sodium, Potassium, Magnesium and Calcium, *Quart. Rev.*, **24**, 1970, 331: These two articles, together with the appropriate sections from Phillips & Williams' textbook, give a useful introduction to the role of metals in biochemistry.

A considerable number of articles on the chemistry of individual transition elements will be found in the reviews listed in section (2) above.

MAIN GROUP ELEMENTS (Chapters 8 and 15)

Peroxides, Superoxides and Ozonides of the Metals of Groups I and II, N. VANNERBERG, *Prog. Inorg. Chem.*, 1962, **4**, 125

The Halides and Oxyhalides of Elements of Groups Vb and VIb, N and O Groups, J. W. GEORGE, *Prog. Inorg. Chem.*, 1960, **2**, 33
also, Phosphorus Group Halides, P, As, Sb, Bi, *Quart. Rev.*, 1961, **15**, 173

The Phosphonitrilic Halides, N. L. PADDOCK AND H. T. SEARLE, *Advances in Inorganic Chemistry and Radiochemistry*, 1959, **1**, 348. Also, *Quart. Rev.*, 1964, **18**, 168

Structures of Interhalogens and Polyhalides, E. H. WIE-BENGA, E. E. HAVINGA, AND K. H. BOSWIJK, *Advances in Inorganic Chemistry and Radiochemistry*, 1961, **3**, 133

Oxides and Oxyfluorides of the Halogens, M. SCHMEISSER AND K. BRANDLE, *Advances in Inorganic and Radiochemistry*, 1963, **5**, 42

The Chemistry of Gallium, N. N. GREENWOOD, *Advances in Inorganic Chemistry and Radiochemistry*, 1963, **5**, 91

The Inorganic Chemistry of Nitrogen, W. L. JOLLY, Benjamin, 1964
This includes an account of liquid ammonia as a solvent.

Halides of P, As, Sb, and Bi, L. KOLDITZ, *Advances in Inorganic Chemistry and Radiochemistry*, 1965, **7**, 1

Borides: Their Chemistry and Application, R. THOMPSON, R.I.C. Lecture Series, 1965. See also N. N. GREENWOOD, R. V. PARISH, AND P. THORNTON, *Quart. Rev.*, **20**, 441

The Chemistry of Fluorine and its Compounds, H. J. EMELEUS, Academic Press, 1969

Stereochemistry, Mechanism and Silicon, L. M. SOMME, McGraw Hill, 1965

Noble Gas Compounds, J. H. HOLLOWAY, Methuen, 1968

Finally, a fuller account of applied chemistry is to be found in *Industrial Chemistry-Inorganic*, D. M. SAMUEL, Royal Institute of Chemistry, Monograph for Teachers, 2nd Edition, 1970

GROUP THEORY (Appendix C)

Tables for Group Theory, P. W. ATKINS, M. S. CHILD, AND C. S. G. PHILLIPS, Oxford University Press, 1970
The best introduction is to be found in the following Honours and postgraduate text; the beginner can gain a feeling for the applications by reading the general sections.

Chemical Applications of Group Theory, F. A. COTTON, Interscience, 2nd Edition, 1970

Symmetry and Stereochemistry, J. D. DONALDSON AND S. D. ROSS, Intertext, 1972

PROBLEMS

Problems in Inorganic Chemistry, by B. J. AYLETT AND B. C. SMITH, English Universities Press, 1965.
This comprises a set of questions on qualitative and quantitative aspects of the inorganic and organometallic chemistry of the elements which are invaluable in directing the student's attention to important aspects of chemistry. Answers are provided which makes the book an important aid to students who have to study on their own.

Appendix B
Some Common Polydentate Ligands

Some common polydentate ligands

Ligand	Formula	Mode of Coordination
BIDENTATE LIGANDS		
ethylenediamine (en)	$H_2NCH_2CH_2NH_2$	Through both N atoms giving a five-membered ring (Figure 12.16a)
dicarboxylic acids (and S analogues)	$(CH_2)_n(COOH)_2$ ($n = 0, 1, 2$ etc)	A proton is lost from each COOH group and the two O^- atoms coordinate giving a $(5+n)$ ring (compare Figure 12·20 for $n = 0$)
acetylacetone (acac)	$CH_3C(OH) = CHCOCH_3$ (in enol form)	Enol proton lost and coordination is through both O atoms (Figures 10.2, 11.5, 13.10)
(and other β-diketones: also as a monodentate ligand in keto form)		
8-quinolinol (oxine or 8-hydroxyquinoline)	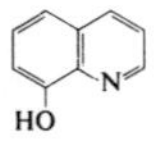	Proton lost and coordinates through O and N giving five-membered ring (Figure 12.16d)
biuret	$H_2NCONHCONH_2$	One NH_2 proton lost and coordinates through two outer N atoms to give a six-membered ring
dimethylglyoxime (DMG)	$CH_3C(=NOH)C(=NOH)CH_3$	One proton lost and coordinates through both N atoms giving five-membered ring. Further six-membered rings formed by hydrogen bonding between NOH and ON (Figure 13.25)
salicylaldehyde (salic)	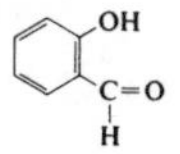	Hydroxyl proton lost and bonds through both O atoms giving six-membered ring (Figure 9.3)
(Similarly salicylaldimines: $C_6H_4(OH)(CH=NR)$ where R = H, alkyl or OH. Here bonding is through O and N giving six-membered rings. See Figure 12.16e)		
2:2'-dipyridyl (bipy) (Also numerous substituted dipyridyls and analogues)	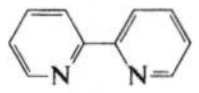	Through both N atoms giving five-membered ring (Figure 13.6)
1:10-phenanthroline (phenan)	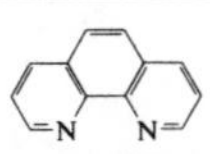	As above

| dithiols | $HS(CH_2)_nSH$; $HSCH_2(SH)CH_2R$;
 $o\text{-}(C_6H_4)(SH)_2$ 1,2 unsaturated thiols $\underset{\underset{HS}{\mid}\quad\underset{SH}{\mid}}{RC=CR}$, and dithioketones $\underset{\underset{S}{\parallel}\quad\underset{S}{\parallel}}{R-C-C-R}$ | Loss of one proton and coordination through two S atoms generally giving four-, five-, or six-membered rings |
| diphosphines (diphos) and diarsines (diars) | $R_2MCH_2CH_2MR_2$ or 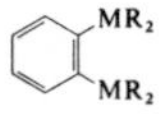 [structure with MR_2 groups]
 M = P or As | Through two P or two As (or P and As) giving five-membered rings |

TRIDENTATE LIGANDS

diethylenetriamine (dien)	$H_2NCH_2CH_2NHCH_2CH_2NH_2$	Through three N atoms to the same M ion giving two five-membered rings (Figure 12.16c)
2:6-bis(α pyridyl) pyridine (terpyridine or terpy)	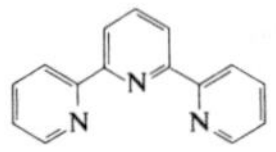 [structure]	To one metal through three N atoms giving two five-membered rings
triarsines (triars) Similar molecules with three As, three P or three (As + P) atoms	e.g. $Me_2As(CH_2)_3As(Me)(CH_2)_3AsMe_2$	To one metal atom through three As atoms giving two six-membered rings

QUADRIDENTATE LIGANDS

porphyrins and phthallocyanins	See Figure 13.19	Through four N atoms forming square-planar coordination to M
bis(acetylacetone)ethylenediamine (acacen) (A Schiff's base)	$\underset{\underset{HO}{\mid}\quad\underset{Me}{\mid}\qquad\underset{Me}{\mid}\quad\underset{OH}{\mid}}{MeC=CHC=NCH_2CH_2N=CCH=CMe}$	Two protons lost (from OH) and bonds through two O and two N atoms to an M ion forming two six-membered rings and one five-membered ring (compare Figure 12.17 for a related ligand)
triethylenetetramine	$H_2NCH_2CH_2NHCH_2CH_2NHCH_2CH_2NH_2$	Through four N atoms forming four five-membered rings around a metal atom
triaminotriethylamine (tren) Also similar tetrarsines and tetraphosphines where N is replaced by As or P	$N(CH_2CH_2NR_2)_3$ where R = H or alkyl	Through four N atoms as above
tris(o-diphenylarsinophenyl)arsine (QAS) Also the phosphine where the central As is replaced by P	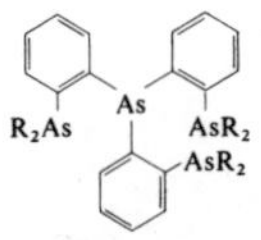 [structure with R_2As, As, AsR_2, AsR_2]	Through four As atoms giving four five-membered rings. The geometric requirements of the ligand give it a pyramidal coordination

QUINQUEDENTATE LIGANDS

| tetraethylenepentamine (tetren) | $H_2NCH_2CH_2NHCH_2CH_2NHCH_2CH_2NHCH_2CH_2NH_2$ | To one metal through all five N atoms |
| The Schiff's Base | $\underset{\underset{OH}{\mid}\qquad\qquad\qquad\underset{HO}{\mid}}{CH=NCH_2CH_2NHCH_2CH_2N=CH}$ | Loses two protons and bonds through three N and two O atoms. (Compare Figure 12.17 for the corresponding tetradentate ligand) |

SEXADENTATE LIGANDS

ethylenediaminetetraacetic acid (EDTA)	$(HO_2CCH_2)_2NCH_2CH_2N(CH_2CO_2H)_2$	Loses four H^+ and bonds through four O and two N atoms to give six five-membered rings. (Figure 9.4)
pentaethylenehexamine	$H_2NCH_2CH_2(NHCH_2CH_2)_4CH_2CH_2NH_2$	To one metal through all six N atoms
Schiff's Bases e.g.	$\underset{\displaystyle OH}{CH=NCH_2CH_2SCH_2CH_2SCH_2CH_2N=CH}\underset{\displaystyle HO}{}$	Loses two protons and bonds to one metal through two O, two S and two N atoms giving five rings

Also the corresponding compounds with NH in place of S, or with pyridine in place of the $-C_6H_4-$ OH ring. (Compare Figure 12.17 for a tetradentate analogue)·

An Octadentate Ligand ?

diethylenetriaminepentaacetic acid (Compare EDTA)	HOOCCH₂ ＞ NCH₂CH₂ HOOCCH₂ ＿ NCH₂COOH HOOCCH₂ ＞ NCH₂CH₂ HOOCCH₂	Bonding unknown but it is potentially eight coordinating through 5O and 3N atoms and it gives a 1:1 zirconium compound

Appendix C
Molecular Symmetry and Point Groups

The symmetry of a molecule or ion is a property of fundamental importance in more advanced study of its properties. For example, the choice of atomic orbitals which may be combined into molecular sigma or pi orbitals is restricted by symmetry considerations. Thus, the basic reason why the combinations indicated in Figure 3.12 are not allowed is that the two orbitals in each pair belong to different symmetry classes in a diatomic molecule. Similarly, the number and type of transitions expected in the electronic or vibrational spectra of a molecule depend fundamentally on the molecular symmetry. For example, in section 6.5, the predictions of the number of fundamentals in the infrared and Raman spectra which are indicated there are derived solely from the consideration of the molecular symmetry.

While it is not our purpose, in an introductory text, to discuss these topics it is useful for the student to be able to determine the symmetry of a molecule at an early stage. This may be done quite simply, as explained below, and can conveniently be tackled in conjunction with the determination of molecular shapes discussed in Chapter 3. Familiarity with the nomenclature of symmetry, and the ability to determine the formal symmetry of a molecule, are the necessary first steps to a fuller understanding of bonding and spectroscopy.

There are two steps in the process: first we define symmetry elements and symmetry operations and then the sum of the symmetry elements is used to determine the point group to which the molecule belongs.

Symmetry elements and symmetry operations

A symmetry *operation* is some transformation of the molecule, such as a rotation or a reflection, which leaves the molecule in a configuration in space which is indistinguishable from its initial configuration. A symmetry *element* is that point, line, or plane in the molecule about which the symmetry operation takes place. The number of symmetry elements and operations which apply to single, real molecules is quite small and these are detailed in Table C.1. (When dealing with a crystal, with extended repeated units, extra symmetry operations become possible which transform one molecule

into the neighbouring position which, in an infinitely extended lattice, also gives rise to an indistinguishable configuration. We shall not discuss such elements here.)

These operations will be more readily understood from a few examples. It is very much easier to follow the description using a molecular model and it would be well worth the reader's while to make models and carry out the symmetry operations described.

Consider first the ammonia molecule which is pyramidal in shape, Figure C.1. There is a three-fold axis passing

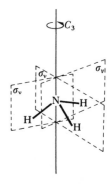

FIGURE C.1 *Symmetry elements of* NH_3
Ammonia belongs to the $\mathbf{C_{3v}}$ point group. The C_3 axis passes through N and the midpoint of the triangle defined by the three H atoms. There are three vertical planes, σ_v, each containing C_3 and one N—H bond and bisecting the opposite HNH angle.

through the nitrogen atom and perpendicular to the plane containing the three hydrogens. Rotation about this axis by $360°/3 = 120°$ leaves the N unchanged and moves each H into the position of the next. The resulting configuration is indistinguishable (though different) from the original one. The molecule may also be rotated twice about this axis by $120°$ to produce another indistinguishable configuration, but if the molecule is rotated three times successively by $120°$ it is returned to the original configuration. This symmetry element is labelled C_3, and the operations are distinguished

TABLE C.1. Symmetry elements and symmetry operations

Element	Symbol	Operation
identity	E	Leaves each particle in its original position
n-fold axis (proper axis)	C_n	Rotation about the axis by $360°/n$, or by some multiple of this. Only $n = 1, 2, 3, 4, 5, 6$ and ∞ need be considered for real molecules.
plane	σ	Reflection in the plane
centre	i	Inversion through the centre
n-fold alternating axis (improper axis)	S_n	Rotation by $360°/n$ (or by a multiple) followed by reflection in a plane perpendicular to the axis.

as C_3, C_3^2, and C_3^3. As the last produces the original configuration, we may write

$$C_3^3 = E$$

Similarly, $C_3^4 = C_3$, $C_3^5 = C_3^2$ and so on.

In the NH_3 molecule, there are also three planes of symmetry. Each one contains the N atom and one H, and bisects the angle between the other two H atoms. Reflection in this plane leaves the N and contained H unaltered, and exchanges the other two H atoms. Clearly, if this reflection operation is repeated, the original configuration is restored. That is, $\sigma^2 = E$.

The symmetry elements of the ammonia molecule are thus E, C_3, and the three planes. It is the convention to align the molecule so that the axis of symmetry is vertical. Then a symmetry plane which contains this axis is a *vertical plane*, symbol σ_v, while a plane perpendicular to the axis is a *horizontal plane* with the symbol σ_h. Thus the symmetry planes in ammonia are vertical planes.

If more than one symmetry axis is present, the one of highest order (highest value of n) is termed the *principal axis*,

and this is the one which is placed vertically. There are a few cases where a molecule has improper axes but no proper axis, and then the alternating axis of highest order is chosen as the principal axis.

As a further example, consider BF_3 which is planar, Figure C.2. This molecule has the C_3 axis through the B atom, and the three vertical planes, each containing one BF bond, which correspond to the symmetry elements found in NH_3. In addition, there are three C_2 axes, one along each B—F bond. A rotation of $360°/2$ about such an axis leaves the contained F and B unchanged, and interchanges the other two F atoms. Furthermore, the plane of the molecule is a symmetry plane as reflection in it leaves all the atoms unchanged. By the definitions above, the principal axis is the C_3 axis, as this is the axis of highest order. Thus the plane of the molecule is a horizontal plane, σ_h, while the planes through the BF bonds are vertical planes. Further, as there is a C_3 axis and a horizontal plane of symmetry, there is necessarily an S_3 axis coincident with the C_3 one. Thus, the symmetry elements of BF_3 are E, C_3, $3C_2$, σ_h, $3\sigma_v$, and S_3.

A centre of symmetry (or an inversion centre) is a point in a molecule such that if an atom is moved from its position, through the centre of inversion for an equal distance in the same direction, it lands in the position of an identical atom. For example, the centre of an octahedron or of a *trans*-MA_2B_4 species is a centre of symmetry. There is also a centre at the mid-point of the C—C bond in ethane in the staggered configuration. On the other hand, there is no inversion centre in a tetrahedron, or in ethane in the eclipsed configuration.

While an alternating axis is necessarily present when both a proper axis and a horizontal plane are present, as in the case of the S_3 axis in BF_3, the alternating axis is an independent symmetry element and it may be present when neither the proper axis nor the plane exist as symmetry elements. An example is provided by ethane in the staggered configuration. If the C—C bond is taken as an axis, rotation by 60° followed by reflection in a plane perpendicular to the C—C bond and containing its mid-point will exchange the two carbons and the hydrogens. Thus there is an S_6 axis along the C—C bond while the highest order proper axis is only C_3 along this bond. There is no horizontal plane of symmetry in staggered ethane. It may be noted that a centre of inversion, i, is equivalent to an S_2 axis.

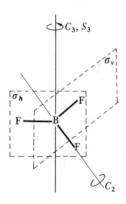

FIGURE C.2 *Symmetry elements of* BF_3
Boron trifluoride belongs to the point group $\mathbf{D_{3h}}$. This has the elements, C_3 (through B perpendicular to the plane of the molecule) and three σ_v (containing one B—F bond and bisecting the opposite angle) similar to those in $\mathbf{C_{3v}}$ (Figure C.1) together with a horizontal plane of symmetry, σ_h (which is the plane of the molecule), an S_3 axis coincident with C_3, and three C_2 axes, one along each B—F bond. Only one of the three vertical planes and one of the two-fold axes is shown.

Point Groups

The complete list of symmetry operations which may be performed on a molecule serves to define the *point group* to which the molecule belongs. Conversely, knowing the point group is equivalent to knowing all the symmetry operations. Thus the fact that ammonia may be subjected to the operations E, C_3, C_3^2 and three different σ_v reflections determines that NH_3 belongs to the C_{3v} point group.

The point groups are named by symbols, which are related to those for the symmetry elements, and are usually distinguished in print by the use of bold type. The point groups which span molecules are the following:

(a) C_n to which belong molecules with only a proper axis of symmetry C_n.

Note that C_1 contains only $E = C_1$, and is the point group in which molecules with no symmetry at all are placed.

(b) C_{nv} to which belong molecules with a C_n axis and n vertical planes only (note that if there is one plane, there must be n as the n-fold rotation about the axis transforms one plane into the next).

(c) C_{nh} to which belong molecules with only a C_n axis, a horizontal plane, σ_h, and the resulting S_n axis. Note that $C_{1h} \equiv C_{1v}$ is more commonly labelled C_s.

(d) D_n is the point group to which belong molecules containing only one C_n axis together with n C_2 axes (note that again, if there is one C_2 axis there must be n of them).

(e) D_{nd} covers molecules with one C_n and n C_2 axes, together with n vertical planes, and the S_{2n} axis which these imply. Note that the subscript is d (for dihedral) and not v.

(f) D_{nh} to which belong molecules with C_n and n C_2 axes, together with a horizontal plane. These elements imply the presence of an S_n axis and n vertical planes.

In chemical structures, axes of higher order than six are extremely rare, thus the point groups specified under (a) to (f) have $n = 1, 2, 3, 4, 5$ and 6, apart from linear molecules where n is infinity.

(g) S_n. A few cases exist where molecules have no planes of symmetry and the highest order axis is S_n, together with the $C_{n/2}$ which this implies. Only even values of n are found. S_2 is the group with one S_2 axis, which is equivalent to i, and this group is usually labelled C_i.

(h) T_d, O_h, I_h are respectively the labels for groups with full tetrahedral, octahedral, or icosahedral symmetry. These have

TABLE C.2. Some common point groups

Point group	Diagnostic elements	Other elements	Examples
C_1	E only		SiHClBrI
C_s	E, σ only		SiH$_2$ClBr
C_i	E, i only		*trans*-HClBrSiSiBrClH
C_2	E, C_2 only		H$_2$O$_2$ (non-planar)
C_{2v}	$E, C_2, 2\sigma_v$		H$_2$O, SiH$_2$Cl$_2$
C_{3v}	$E, C_3, 3\sigma_v$		NH$_3$, SiHCl$_3$
C_{4v}	$E, C_4, 4\sigma_v$	$C_2 = C_4^2$	BrF$_5$, SF$_5$Cl
C_{2h}	E, C_2, σ_h	i	*trans*-C$_6$H$_2$Cl$_2$Br$_2$
C_{3h}	E, C_3, σ_h		$B(OH)_3$ in form
D_2	$E, C_2, 2C_2$		
D_{2d}	$E, C_2, 2C_2, 2\sigma_v$	S_4	H$_2$C=C=CH$_2$
D_{3d}	$E, C_3, 3C_2, 3\sigma_v$	i, S_6	C$_2$H$_6$, Si$_2$Cl$_6$ (staggered)
D_{4d}	$E, C_4, 4C_2, 4\sigma_v$	$C_2 = C_4^2, S_8$	S$_8$ (puckered ring)
D_{5d}	$E, C_5, 5C_2, 5\sigma_v$	i, S_{10}	(C$_5$H$_5$)$_2$Fe (staggered)
D_{2h}	$E, C_2, 2C_2, \sigma_h$	$i, 2\sigma_v$	B$_2$Cl$_4$, *trans*-A$_2$B$_2$C$_2$M
D_{3h}	$E, C_3, 3C_2, \sigma_h$	$S_3, 3\sigma_v$	BF$_3$, PF$_5$
D_{4h}	$E, C_4, 4C_2, \sigma_h$	$i, S_4, C_2, 4\sigma_v$	PtCl$_4^{2-}$, *trans*-A$_2$B$_4$M
D_{5h}	$E, C_5, 5C_2, \sigma_h$	$S_5, 5\sigma_v$	C$_5$H$_5$, (C$_5$H$_5$)$_2$Ru (eclipsed)
D_{6h}	$E, C_6, 6C_2, \sigma_h$	$i, S_6, S_3, C_3, C_2, 6\sigma_v$	C$_6$H$_6$
T_d	$E, 4C_3, 3C_2, 3S_4$, and $6\sigma_v$		SiH$_4$, GeCl$_4$, TiCl$_4$
O_h	$E, 3C_4, 4C_3, 6C_2, i, 3S_4, 4S_6, 3\sigma_h, 6\sigma_v$		SF$_6$, ML$_6$, PF$_6^-$
I_h	$E, 6C_5, 10C_3, 15C_2, i, 10S_6, 6S_{10}, 15\sigma$		B$_{12}$H$_{12}^{2-}$, B$_{12}$ (Figure 15.7d)

large numbers of symmetry elements, listed in Table C.2, and it is impossible not to recognize them. There are a few related groups, such as **T** which has the axes but not the planes of a full tetrahedron, but these are very rarely encountered. In the discussion which follows, the groups S_4, S_6, S_8 from (g) and **T**, **T_h** and **O** which are related to those in (h) are omitted as they are seldom represented by real molecules in their most stable configurations.

Although the more symmetric groups are defined by quite large numbers of symmetry operations, the fact that the presence of the element automatically implies the presence of all the operations of that element (including those giving rise to related elements such as $C_4^2 = C_2$) means that the point group may be decided by considering the symmetry elements. Furthermore, as the presence of certain elements implies the presence of others (as the C_3 and σ_h in BF_3 implied the S_3), the point group to which a molecule belongs may be *diagnosed* by looking for a small number of elements *in a particular order*. If the diagnostic elements are present, then the other symmetry elements, and all the corresponding symmetry operations, are necessarily present.

It then becomes a simple matter to determine the point group to which a molecule belongs by looking for symmetry elements in a fixed order. This order is:

(1) Find the principal axis.
(2) Look for n C_2 axes perpendicular to the principal axis.
(3) Look for horizontal planes.
(4) Look for vertical planes.
(5) If none of the above are present, check for a centre of inversion.

It is important that the search is always carried out in this order or higher symmetry groups may be missed because too much weight is put on minor elements. Notice, for example, that **C_nh** groups have vertical as well as horizontal planes of symmetry so that, if vertical planes are looked for before the horizontal plane is excluded, **C_nh** might be mistakenly assigned as **C_nv**.

The process of diagnosing the point group may then be set out as follows:

(A) A molecule of high symmetry? — Assign as **T_d** (tetrahedron), **O_h** (octahedron) or **I_h** (icosahedron) after checking that all the requisite symmetry elements are present from Table C.2.

(B) Find the axis of highest order, C_n — If present, go to (C). If absent assign as **C_s** (plane only), **C_i** (centre of inversion, $i \equiv S_2$, only) or **C_1** (no element).

(C) Look for n C_2 axes perpendicular to the principal axis — If present, go to (F): if none, go to (D).

(D) Look for horizontal plane — If present, assign as **C_nh**, if absent, go to (E).

(E) If no horizontal plane, look for n vertical planes. — If present, assign as **C_nv**: if no planes then assign as **C_n**.

(F) If C_n and n C_2 axes are present, look for
   (i) a horizontal plane — Assign as **D_hh**.
   (ii) if no horizontal plane, look for n vertical planes
   (iii) If no planes at all — Assign as **D_n**.

A network form of this search is shown in Figure C.3, and

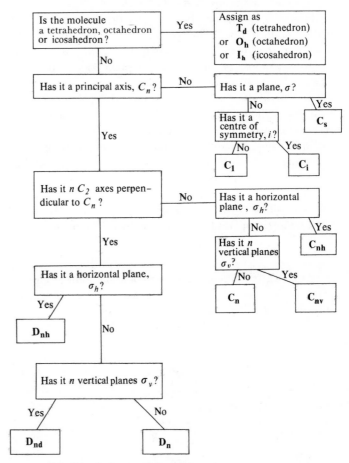

FIGURE C.3 *Diagnostic network for point groups*

Table C.2 lists all the symmetry elements of the chemically important point groups. It will be seen that while the simpler groups contain only the diagnostic elements, the more symmetric groups have a considerable number of consequent elements.

A linear molecule fits naturally into the above scheme once it is realized that the molecular axis is an infinity-fold one, because any rotation of the molecule about this axis—however small—produces an indistinguishable configuration. Thus the principal axis is C_∞ and a molecule like N_2 or CO_2 has an infinite number of C_2 axes, perpendicular to the principal axis, passing through the centre of the molecule (mid-point of $N-N$ or through the C in $O-C-O$). Thus the point group is of the D type and, as there is a symmetry plane, σ_h through the midpoint and perpendicular to the molecule axis, these molecules belong to the $\mathbf{D}_{\infty \mathbf{h}}$ group. Similar analysis shows that non-symmetric linear molecules like NO or N_2O (arranged $N-N-O$) are members of the $\mathbf{C}_{\infty \mathbf{v}}$ group.

For further use of the point group, it is necessary to refer to its *character table* which is effectively a summary of all the symmetry properties implied by the point group. For use of the character tables, the reader should refer to one of the more advanced treatments indicated in the references in Appendix A. However, for convenience of reference, we append the characters tables of those point groups which occur most commonly, Table C.3.

From the list, it may be seen that the character table is divided into sections. The symbol for the point group comes first, and then there is listed along the top row all the symmetry operations of the group. The main body of the table shows all the possible ways in which a function might transform under the operations of the group. Here, if the operation transforms the function into itself we find a 1, if the function is reversed we find -1, while if the function is mixed with others we find some other symbol. For example, think of the p orbitals on N in ammonia (that is, the unperturbed atomic orbitals before any combination or hybridization is carried out). Ammonia belongs to the point group $\mathbf{C}_{3v}$ and the p_z orbital will be taken to coincide with the C_3 axis. It can be seen that if the operation C_3, rotation by $120°$ about the z axis,

is carried out the p_z orbital is unaltered. Similarly the operations E or reflection in any of the σ_v planes leaves the p_z orbital unaltered. Thus we would write for p_z

$$
\begin{array}{ccc}
E & C_3 & 3\sigma_v \\
1 & 1 & 1
\end{array}
$$

corresponding to the row labelled A_1. On the other hand, rotation of the p_x orbital about the C_3 axis (i.e. the z axis) by $120°$ would bring it to a position which must be expressed as a combination of the p_x and p_y orbitals, $-p_x \cos 60 + p_y \sin 60$. Similarly, C_3 operating on p_y mixes it with p_x. Thus the p_x and p_y orbitals are degenerate in the $\mathbf{C}_{3v}$ point group and show the characters indicated in the table opposite E in the left-hand column. In general, the left-hand column in the character table gives symbols which summarize the symmetry properties of the different possible classes of functions in that point group. The exact nomenclature need not concern us, but note that non-degenerate classes are labelled A or B, doubly degenerate ones E and triply degenerate ones T (or F in the older nomenclature). The final two columns give functions which transform as the various classes. In the fourth column are vectors in the x, y or z directions (such as a p orbital or a translation along one axis) and rotations around each axis. In the last column are the second order products, such as the d orbitals.

TABLE C.3. Character tables for common point groups

$\mathbf{C_1}$	E		
A	1	all functions	

$\mathbf{C_s}$	E	σ_h		
A'	1	1	T_x, T_y, R_z	x^2, y^2, z^2, xy
A''	1	-1	T_z, R_x, R_y	yz, xz

$\mathbf{C_i}$	E	i		
A_g	1	1	R_x, R_y, R_z	$x^2, y^2, z^2, xy, xz, yz$
A_u	1	-1	T_x, T_y, T_z	

$\mathbf{C_2}$	E	C_2		
A	1	1	T_z, R_z	x^2, y^2, z^2, xy
B	1	-1	T_x, T_y, R_x, R_y	yz, xz

$\mathbf{C_{2v}}$	E	C_2	$\sigma_v(xz)$	$\sigma_v(yz)$		
A_1	1	1	1	1	T_z	x^2, y^2, z^2
A_2	1	1	-1	-1	R_z	xy
B_1	1	-1	1	-1	T_x, R_y	xz
B_2	1	-1	-1	1	T_y, R_x	yz

$\mathbf{C_{3v}}$	E	$2C_3$	$3\sigma_v$		
A_1	1	1	1	T_z	x^2+y^2, z^2
A_2	1	1	-1	R_z	
E	2	-1	0	$(T_x, T_y) (R_x, R_y)$	$(x^2-y^2, xy) (xz, yz)$

$\mathbf{C_{4v}}$	E	$2C_4$	C_2	$2\sigma_v$	$2\sigma_d$		
A_1	1	1	1	1	1	T_z	x^2+y^2, z^2
A_2	1	1	1	-1	-1	R_z	
B_1	1	-1	1	1	-1		x^2-y^2
B_2	1	-1	1	-1	1		xy
E	2	0	-2	0	0	$(T_x, T_y) (R_x, R_y)$	(xz, yz)

$\mathbf{D_{2h}}$	E	$C_2(z)$	$C_2(y)$	$C_2(x)$	i	$\sigma(xy)$	$\sigma(xz)$	$\sigma(yz)$		
A_g	1	1	1	1	1	1	1	1		x^2, y^2, z^2
B_{1g}	1	1	-1	-1	1	1	-1	-1	R_z	xy
B_{2g}	1	-1	1	-1	1	-1	1	-1	R_y	xz
B_{3g}	1	-1	-1	1	1	-1	-1	1	R_x	yz
A_u	1	1	1	1	-1	-1	-1	-1		
B_{1u}	1	1	-1	-1	-1	-1	1	1	T_z	
B_{2u}	1	-1	1	-1	-1	1	-1	1	T_y	
B_{3u}	1	-1	-1	1	-1	1	1	-1	T_x	

$\mathbf{D_{3h}}$	E	$2C_3$	$3C_2$	σ_h	$2S_3$	$3\sigma_v$		
A_1'	1	1	1	1	1	1		x^2+y^2, z^2
A_2'	1	1	-1	1	1	-1	R_z	
E'	2	-1	0	2	-1	0	(T_x, T_y)	(x^2-y^2, xy)
A_1''	1	1	1	-1	-1	-1		
A_2''	1	1	-1	-1	-1	1	T_z	
E''	2	-1	0	-2	1	0	(R_x, R_y)	(xz, yz)

D_{4h}	E	$2C_4$	C_2	$2C_2'$	$2C_2''$	i	$2S_4$	σ_h	$2\sigma_v$	$2\sigma_d$		
A_{1g}	1	1	1	1	1	1	1	1	1	1		$x^2+y^2,\ z^2$
A_{2g}	1	1	1	-1	-1	1	1	1	-1	-1	R_z	
B_{1g}	1	-1	1	1	-1	1	-1	1	1	-1		x^2-y^2
B_{2g}	1	-1	1	-1	1	1	-1	1	-1	1		xy
E_g	2	0	-2	0	0	2	0	-2	0	0	(R_x, R_y)	(xz, yz)
A_{1u}	1	1	1	1	1	-1	-1	-1	-1	-1		
A_{2u}	1	1	1	-1	-1	-1	-1	-1	1	1	T_z	
B_{1u}	1	-1	1	1	-1	-1	1	-1	-1	1		
B_{2u}	1	-1	1	-1	1	-1	1	-1	1	-1		
E_u	2	0	-2	0	0	-2	0	2	0	0	(T_x, T_y)	

D_{5h}	E	$2C_5$	$2C_5^2$	$5C_2$	σ_h	$2S_5$	$2S_5^3$	$5\sigma_v$		
A_1'	1	1	1	1	1	1	1	1		$x^2+y^2,\ z^2$
A_2'	1	1	1	-1	1	1	1	-1	R_z	
E_1'	2	$2\cos 72°$	$2\cos 144°$	0	2	$2\cos 72°$	$2\cos 144°$	0	(T_x, T_y)	
E_2'	2	$2\cos 144°$	$2\cos 72°$	0	2	$2\cos 144°$	$2\cos 72°$	0		(x^2-y^2, xy)
A_1''	1	1	1	1	-1	-1	-1	-1		
A_2''	1	1	1	-1	-1	-1	-1	1	T_z	
E_1''	2	$2\cos 72°$	$2\cos 144°$	0	-2	$-2\cos 72°$	$-2\cos 144°$	0	(R_x, R_y)	(xz, yz)
E_2''	2	$2\cos 144°$	$2\cos 72°$	0	-2	$-2\cos 144°$	$-2\cos 72°$	0		

D_{6h}	E	$2C_6$	$2C_3$	C_2	$3C_2'$	$3C_2''$	i	$2S_3$	$2S_6$	σ_h	$3\sigma_d$	$3\sigma_v$		
A_{1g}	1	1	1	1	1	1	1	1	1	1	1	1		$x^2+y^2,\ z^2$
A_{2g}	1	1	1	1	-1	-1	1	1	1	1	-1	-1	R_z	
B_{1g}	1	-1	1	-1	1	-1	1	-1	1	-1	1	-1		
B_{2g}	1	-1	1	-1	-1	1	1	-1	1	-1	-1	1		
E_{1g}	2	1	-1	-2	0	0	2	1	-1	-2	0	0	(R_x, R_y)	(xz, yz)
E_{2g}	2	-1	-1	2	0	0	2	-1	-1	2	0	0		(x^2-y^2, xy)
A_{1u}	1	1	1	1	1	1	-1	-1	-1	-1	-1	-1		
A_{2u}	1	1	1	1	-1	-1	-1	-1	-1	-1	1	1	T_z	
B_{1u}	1	-1	1	-1	1	-1	-1	1	-1	1	-1	1		
B_{2u}	1	-1	1	-1	-1	1	-1	1	-1	1	1	-1		
E_{1u}	2	1	-1	-2	0	0	-2	-1	1	2	0	0	(T_x, T_y)	
E_{2u}	2	-1	-1	2	0	0	-2	1	1	-2	0	0		

T_d	E	$8C_3$	$3C_2$	$6S_4$	$6\sigma_d$		
A_1	1	1	1	1	1		$x^2+y^2+z^2$
A_2	1	1	1	-1	-1		
E	2	-1	2	0	0		$(2z^2-x^2-y^2,$ $x^2-y^2)$
T_1	3	0	-1	1	-1	(R_x, R_y, R_z)	
T_2	3	0	-1	-1	1	(T_x, T_y, T_z)	(xy, xz, yz)

O_h	E	$8C_3$	$6C_2$	$6C_4$	$3C_2 (= C_4^2)$	i	$6S_4$	$8S_6$	$3\sigma_h$	$6\sigma_d$	
A_{1g}	1	1	1	1	1	1	1	1	1	1	$x^2 + y^2 + z^2$
A_{2g}	1	1	-1	-1	1	1	-1	1	1	-1	
E_g	2	-1	0	0	2	2	0	-1	2	0	$(2z^2 - x^2 - y^2,$ $x^2 - y^2)$
T_{1g}	3	0	-1	1	-1	3	1	0	-1	-1	$(R_x, R_y,$ $R_z)$
T_{2g}	3	0	1	-1	-1	3	-1	0	-1	1	(xz, yz, xy)
A_{1u}	1	1	1	1	1	-1	-1	-1	-1	-1	
A_{2u}	1	1	-1	-1	1	-1	1	-1	-1	1	
E_u	2	-1	0	0	2	-2	0	1	-2	0	
T_{1u}	3	0	-1	1	-1	-3	-1	0	1	1	$(T_x, T_y,$ $T_z)$
T_{2u}	3	0	1	-1	-1	-3	1	0	1	-1	

Index

Periodic Table of the Elements

	s^1	s^2	$d^n s^x$ (n = 1 to 10; x = 0, 1 or 2)										$s^2 p^1$	$s^2 p^2$	$s^2 p^3$	$s^2 p^4$	$s^2 p^5$	$s^2 p^6$
1s	1 H																	2 He
2s 2p	3 Li	4 Be											5 B	6 C	7 N	8 O	9 F	10 Ne
3s 3p	11 Na	12 Mg											13 Al	14 Si	15 P	16 S	17 Cl	18 Ar
4s 3d 4p	19 K	20 Ca	21 Sc	22 Ti	23 V	24 Cr	25 Mn	26 Fe	27 Co	28 Ni	29 Cu	30 Zn	31 Ga	32 Ge	33 As	34 Se	35 Br	36 Kr
5s 4d 5p	37 Rb	38 Sr	39 Y	40 Zr	41 Nb	42 Mo	43 Tc	44 Ru	45 Rh	46 Pd	47 Ag	48 Cd	49 In	50 Sn	51 Sb	52 Te	53 I	54 Xe
6s (4f) 5d 6p	55 Cs	56 Ba	57* La	72 Hf	73 Ta	74 W	75 Re	76 Os	77 Ir	78 Pt	79 Au	80 Hg	81 Tl	82 Pb	83 Bi	84 Po	85 At	86 Rn
7s (5f) 6d	87 Fr	88 Ra	89** Ac	104 †														

†Proposed name Khurchatovium

$f^p d^n s^2$ (p = 1 to 14, n = 0 or 1 (2 for Th))

*Lanthanide series 4f	58 Ce	59 Pr	60 Nd	61 Pm	62 Sm	63 Eu	64 Gd	65 Tb	66 Dy	67 Ho	68 Er	69 Tm	70 Yb	71 Lu
**Actinide series 5f	90 Th	91 Pa	92 U	93 Np	94 Pu	95 Am	96 Cm	97 Bk	98 Cf	99 Es	100 Fm	101 Md	102	103 Lw